Energy Conservation in Buildings –
Heating, Ventilation and Insulation

Energy R&D Programme of the
Commission of the European Communities

Energy Conservation in Buildings – Heating, Ventilation and Insulation

Proceedings of the Contractors' Meetings held in Brussels on 14–15 December 1981, 6–7 May, 24, 28 and 30 September, and 21 October 1982

Edited by

H. Ehringer, G. Hoyaux and P. Zegers

Directorate-General for Science, Research and Development, Commission of the European Communities

D. Reidel Publishing Company

A MEMBER OF THE KLUWER ACADEMIC PUBLISHERS GROUP

Dordrecht / Boston / Lancaster

Library of Congress Cataloging in Publication Data
Main entry under title:

CIP

Energy conservation in buildings.

At head of title: Energy R&D Programme of the Commission of the European Communities.
1. Buildings–Energy conservation–Congresses. I. Ehringer, H. II. Zegers, P. (Pieter), 1938- . III. Hoyaux, G.
TJ163.5B84E5264 1983 697 83-3231
ISBN 90-277-1578-5

Organization of the Contractors' meetings by
Commission of the European Communities
Directorate-General Science, Research and Development, Brussels

Publication arrangements by
Commission of the European Communities
Directorate-General Information Market and Innovation, Luxembourg

EUR 8463

Published by D. Reidel Publishing Company
P.O. Box 17, 3300 AA Dordrecht, Holland

Sold and distributed in the U.S.A. and Canada
by Kluwer Boston Inc.,
190 Old Derby Street, Hingham, MA 02043, U.S.A.

In all other countries, sold and distributed
by Kluwer Academic Publishers Group,
P.O. Box 322, 3300 AH Dordrecht, Holland

Printed in The Netherlands

P R E F A C E

The main objectives of the Community's energy policy are to secure an energy supply sufficient for the present and future demand of its Member States and to reduce the Community's dependence on imported energy through more rational use of energy and a broader diversification of supply. This requires a full set of common efforts at all levels, including energy research, development and demonstration.

In the framework of its strategy for scientific and technological research, the Commission of the European Communities has launched several R&D programmes in the field of energy. These programmes are implemented either directly in its Joint Research Centre or by concluding contracts with research institutions in the EC member countries.

In 1975, the Council of Ministers decided a first four-year Energy R&D Programme and in 1979, a second one, totalling 164 Mio ECU (corresponding to 200 million US $). This programme includes the five sub-programmes "Solar Energy", "Geothermal Energy", "Hydrogen as Energy Vector", Energy Modelling" and last but not least "Energy Conservation" in the three main energy consuming sectors: domestic, industry and transport.

The domestic and commercial sectors account for the largest part (40%) of final energy consumption in the European Community and this part is used mainly for heating purposes and hot water production.

In order to enhance rational use of energy and energy conservation in the EC Member States it is necessary, among other things, to promote the development of more efficient heating and ventilation systems and better heat insulating materials, and to assess the thermal characteristics of houses.

Therefore the Commission has concluded with industries, public organisations and universities of the EC Member States a number of R&D contracts dealing with improved heating and ventilation systems and heat insulating materials. The progress of work achieved in these contracts was reported in contractors' meetings, which took place on 14 and 15 December 1981, 6 and 7 May, 24, 28 and 30 September and 21 October 1982 in Brussels.

The papers presented at these meetings and the conclusions of the discussions are collected in the present book, which is intended to inform the Member States, research bodies and other interested parties about existing development work in this area.

I should like to address my sincere thanks to all the authors of papers and speakers ot these meetings. I should also thank the chairmen, Professor BOFFA, Professor KETELAAR and Mr. UYTTENBROECK, who animated the discussion with questions and comments. Last but not least, I extend may thanks to the editors of the proceedings, who summarized the content of the meeting and the discussions.

May I express the hope that the work reported will contribute to rapid progress in the development and practical application of improved heating and ventilation systems and better insulating materials, thereby contributing to the rational use of energy throughout the Community.

Dr. Albert STRUB

Director

Head of the Energy R&D Programme

ATTENDANCE LIST

1. CONTRACTORS

SESSION I : HEATING SYSTEMS AND VENTILATION

F.M. VAN BERGEN A.J. MEYNEN	TNO - Heat and Refrigeration Engineering Division
R. RAYMENT	BRE
K. MORGAN	BRE
H.B. AWBI	BSRIA
P.J. JACKMAN	BSRIA
P. DUBOIS	Laboratoires de MARCOUSSIS
Dr. J. SPITZER	BATTELLE-INSTITUT
Dr. R. LEOPARDO	OTTICO MECCANICA ITALIANA

SESSION II : HEATING CONTROL SYSTEMS

H. RADDE	PERIPHERE COMPUTER SYSTEME
E. NEUDORFER	PERIPHERE COMPUTER SYSTEME
P.L. SEGALINI	ADREATICA COMPONENTI ELETTRONICA
V. MAURO	CSEA
T. FULLER	NATIONAL INSTITUTE FOR PHYSICAL PLANNING AND CONSTRUCTION RESEARCH
H.J. NICOLAAS	TNO

SESSION III : HEAT STORAGE

P. DUBOIS	Laboratoires de MARCOUSSIS
J. DESPOIS	CEA
B. CONSEIL	ELF AQUITAINE

SESSION IV : INSULATING MATERIAL

P. DUBOIS	Laboratoires de MARCOUSSIS
M. MOREL	Laboratoires de MARCOUSSIS
Dr. V. CASTAGNETTA	INDUSTRIA ITALIANA PETROLI
G. ROSSI	LUIGI ROSSI - ARCO
G. ROUX	CSTB
M. ANDREANI	CEN Grenoble
Prof. D.M. FONTANA	ISTEDIL
Dr. R. ZARI	ISTEDIL

SESSION V : INSULATION

Dr. R.P. HOWSON	LOUGHBOROUGH UNIVERSITY
Y. LAGARDE	CEN Grenoble
M.R. BYBERG	TECHNICAL UNIVERSITY OF DENMARK
J. BEKKER	TNO
M. DUBBELD	TNO
Dr. J.G.F. LITTLER	POLYTECHNIC OF CENTRAL LONDON

SESSION VI : NATURAL VENTILATION

C. BRAYSHAW	BSRIA
J. DEWSBURY	BSRIA
Dr. M.D.A.E.S. PERERA	BRE
Dr. P.R. WARREN	BRE
P.F. COLLET	BYGGETECHNIK
J.L. BOY-MARCOTTE	BERTIN

SESSION VII : CHARACTERISTICS OF HOUSES

M. GUILLAUME	CSTC
M. GENGOUX	CSTC
J. UYTTENBROECK	CSTC
H. HEIKHAUS	CSTC
Prof. J. LEBRUN	Université de LIEGE
J.M. HAUGLUSTAINE	Université DE LIEGE
A. ZACCARIA	AGIP Nucleare
S. LOS	AGIP Nucleare
F. O'FARRELL	AN FORAS TALUNTAIS
G. LYONS	AN FORAS TALUNTAIS
J.A. CLARKE	UNIVERSITY OF STRATHCLYDE

SESSION VIII : DOMESTIC APPLIANCES

M. MORELLO	RIBER
J. NØRGARD	TECHNICAL UNIVERSITY OF DENMARK

2. COMMISSION

Dr. H. EHRINGER	DG XII
G. HOYAUX	DG XII
Dr. P. PILAVACHI	DG XII
P. ZEGERS	DG XII
M. PELLEGRINI	CCR ISPRA

3. CHAIRMEN

Prof. C. BOFFA	expert	Politecnico TORINO
J. UYTTENBROECK	expert	CSTC
Prof. Dr. J.A.A. KETELAAR	expert	UNIVERSITY OF AMSTERDAM

CONTENTS

SESSION II - HEATING CONTROL SYSTEMS

Contractors' meeting held on 15 December 1981

Chairman: C. BOFFA

SESSION III - HEAT STORAGE

Contractors' meeting held on 21 October 1982

Chairman: J.A.A. KETELAAR

SESSION IV - INSULATING MATERIAL

Contractors' meeting held on 24 September 1982

Chairman: C. BOFFA

SESSION V - INSULATION

Contractors' meeting held on 28 September 1982

Chairman: J. UYTTENBROECK

SESSION VI - NATURAL VENTILATION

Contractors' meeting held on 30 September 1982

Chairman: J. UYTTENBROECK

SESSION VII - CHARACTERISTICS OF HOUSES

Contractors' meeting held on 7 May 1982

Chairman: C. BOFFA

OPENING ADDRESS

by

Dr. Hermann EHRINGER
Commission of the European Communities
BRUSSELS, Belgium

Ladies and Gentlemen,

It is an honour and pleasure for me to open this contractors meeting on "Energy saving in buildings" and to welcome you on behalf of the Commission of the European Communities.

At present, about 55 - 65 % of the heat required in houses and buildings are lost through outer walls and the roof, and about 35 - 45 % through windows. It may be expected that about one third of these losses could be avoided by means of improved insulation of walls and windows. In practice, however, these energy-saving possibilities are offset to some extent because society tends to increase the volume of dwellings and the number of rooms which are to be heated.

Optimized utilization of energy in new and existing buildings by means of improved heating and ventilation systems and through better heat insulating materials are, therefore, major options for energy saving actions of the EC in this sector.

The purpose of the contractors meetings is to give the project leaders the opportunity to report on the progress of their work and on the results achieved. Moreover, these meetings serve to establish a link between contractors working in the same area and to strengthen cooperation on a European level.

The papers presented at this meeting and the discussion will be recorded in proceedings which will be published in order to meet the provisions laid down in the Council regulation EEC/2380/74 of 17 September 1974.

I hope that this meeting will be a successful one and that it will help to establish relations between contractors.

SUMMARY OF THE PROCEEDINGS OF EIGHT CONTRACTORS MEETINGS ON ENERGY SAVINGS IN BUILDINGS

P. Zegers

Commission of the European Communities

Introduction

In the framework of the European Community's second Energy R & D Programme, research work on energy saving in buildings is carried out in the sub-programme Energy Conservation R & D. The total cost of this research work is roughly 9 MioECU of which 50 % is paid by the Commission. The 40 projects have been subdivided in eight contractor groups dealing with : Heating and Ventilation, Heating Control, Heat Storage, Insulation materials, Insulation of windows, Air infiltration, Computer simulation of the thermal behaviour of buildings, and Domestic appliances. These contractor groups met in the course of 1982 to discuss the ongoing work and this book contains the papers presented during these contractors meetings together with the discussion.

In 1979 about 41 % of the total primary energy in the European Community was consumed in the domestic and commercial sector. Of this energy 26,5 % was used for heating and 14,5 % for the production of electricity for lighting, domestic appliances etc.

A reduction of the energy requirements for heating may be achieved in roughly three ways :

- More efficient heating may bring savings of 30-50 % (e.g. improved conventional heating, heat pumps, district heating). In these proceedings only improved conventional heating will be dealt with in the chapters I and II on heating and ventilation, and heating control. Daily and seasonal storage of heat is dealt with in chapter III. Work on heat pumps is also carried out in this programme but the progress of this work has been brought together in another book: EUR 8077 EN. It contains the proceedings of four contractors meetings on heat pumps. Work on district heating is not carried out in this programme as it has generally passed the R & D phase.
- Reduction of transmission losses through walls, roofs and windows may bring 20-30 %. This subject is discussed in the chapters IV and V on insulation material for walls and roofs (e.g. non-inflammability, aging, hygroscopic properties) and insulation of windows by shutters and coated windows (heat mirrors).
- Reduction of energy losses by decreased air infiltration may bring 10-20 %. In chapter VI problems connected with air infiltration will be discussed (e.g. measurement of air exchange rate in large buildings, problems connected with very tight houses, and the influence of human behaviour).

The work described above deals with specific energy saving techniques. Once these techniques are available a second step is their integration in buildings in such a way that energy use is minimized in an economically attractive manner. In such an integration computer models which simulate the thermal behaviour of buildings and their components play an increasingly important role, in particular in the energy saving design of new and energy auditing of existing buildings. Several of these simulation models and other work on thermal characteristics of houses will be discussed in chapter VII.

Finally as much as 30 % of the primary energy used for the building sector is required for the production of electricity. It is therefore essential to look into the energy saving possibilities for domestic appliances and lighting. Chapter VIII shows that the potential for energy saving is considerable.

Heating and ventilation systems and their control

This chapter deals only with conventional heating, ventilation systems and their control. Other more advanced systems, in particular heat pumps have been dealt with in four other contractors meetings mentioned previously. In this chapter a description is given of work carried out on the improved efficiency of boilers, the comparison of different heating systems and the development of two new heat meters, and a cheap plastic radiator.

TNO, Netherlands studied the possibilities of improving the efficiency of boilers for domestic heating (80° C, 100 kW). Calculations show that energy savingsup to 26 % are possible. Reduction of chimney losses by closing the chimney when the burner is not operating brings 7-10 %. A forced draught burner has a 7 % higher efficiency than normal draught boilers with regulation of the secondary air supply. In addition several other modifications bring minor contributions.

Four different heating systems (oil fired boiler, closed solid fuel stove, solid fuel fired boiler and high output boiler on an open fire) will be tested and compared by THE NATIONAL INSTITUTE FOR PHYSICAL PLANNING AND CONSTRUCTION RESEARCH, Ireland in a pair of houses. One of the houses is a reference house with electric heating and in the other house the heating systems will be tested. The pair of houses has been bought and is now being equipped. The computer system for data collection and treatment has been designed. In the coming two years the following measuring programme will be carried out :

- calibration and possible modification of the houses to ensure that their thermal behaviour is identical
- measuring the efficiency and ability of the four heating systems to meet a variety of heating regimes
- improve the heating systems.

In order to know how energy saving is influenced by comfort requirements it is important to know what the comfort criteria are. A study to that end was carried out by BUILDING SERVICES RESEARCH AND INFORMATION ASSOCIATION, United Kingdom. For reasonable comfort the dry resultant temperature should be about 21° C, the humidity between 40 and 70 % and the room air velocity smaller than 0,35 m/s. Subsequent air movement studies for air heating systems showed that conventional low level air inlet systems rarely meet these comfort requirements; high level air inlet normally complies with the comfort requirements.

A cheap plastic low temperature radiator (50° C) has been developed by MARCOUSSIS, France. Plastic material and production methods have been investigated on a laboratory scale. Radiator elements have been produced and tested with satisfactory results. The cost of these plastic radiators is 30 % lower than for conventional metallic ones. A drawback is the fact that the size is 60 % larger than metallic radiators with the same heat output at 50° C.

Cheap and reliable heat meters are of crucial importance. Up till now heat-meters are reliable but expensive or cheap but less reliable. In these two projects an attempt is made to develop both cheap and reliable heat meters. Generally the heat meter measures the heat consumed by determining the temperature between the incoming and outgoing water and the mass flow of the fluid. The difficult part of the heat meter is the determination of the mass flow. It is mostly measured with moving parts such as rotors which often cause difficulties due to fouling etc. Both methods developed in the EC programme are therefore based on the determination of the mass flow by static methods.

BATTELLE, Germany developed a heat meter where a small amount of heat is extracted from the incoming hot water tube to the outgoing cold water tube via a well defined copper slab. This heat flow can be determined from temperature difference of hot and cold water, the heat transfer coefficient (K) of copper and the dimensions of the copper slab. This heat flow equals the heat extracted from the incoming hot water which is given by $\dot{m}\ C_w\ (T_2 - T_3)$ where $\dot{m}$ is the mass flow which has to be determined. C_w is the specific heat of water and T_2 and T_3 are the temperatures of hot water before and after heat extraction by the copper slab. With the measured values of T_2 and T_3 and the given constants C_w and K the mass flow can be determined. This installation is now being tested.

OMI, Italy designed a heat meter based on the Lorentz force. In water normally a small quantity of ions is always present. In a flow of water which is subjected to a magnetic field perpendicular to the water flow direction,these ions experience a force which is perpendicular to both the magnetic field and the flow direction (Lorentz force). Electrodes would detect a small current,but with an electric field which compensates the Lorentz force the current can be made zero. The mass flow can now be determined if the magnetic and electric field are known. First measurements show a very satisfactory operation in particular at low flow values (0-0,5 m/sec) where most other flow meters are less accurate.

Heating Control

Heating control is a difficult and an expensive part of the heating system. In particular wiring may be very expensive when such a system is retro-fitted in existing houses. PERIPHERE COMPUTER SYSTEME, Germany has therefore investigated whether existing electric wiring could be used for signal transmission. A data transmission system which uses the electrical power wiring has been developed for a small building. It operated satisfactorily but due to the long pay back times the system is commercially not yet attractive. In large buildings with longer distances, with more equipment to be controlled and more devices connected for power use, the control system did not work satisfactorily.

Work by CSEA, a consortium of 23 small and medium sized electronic manufacturers in Italy showed that heating control in apartment buildings and schools can bring energy savings of 20 to 30 %. In an existing apartment building with 14 apartments a control system was installed consisting of a central computer and 14 local micro-processors. This system regulates the temperature in the apartments, taking into account the outside temperatures and the users' requirements. It learns the habits of the users and switches off the heating in an apartment when the users are not present even when they forget to put the heating off. This system of learning was very efficient in reducing the energy use. Also the operation of the boiler is optimized and the whole system is checked regularly to facilitate maintenance. This brought energy savings of about 20 %; the pay back time for new and existing buildings is 6 and 10 years respectively. For a school the pay back time was found to be even shorter (2.2 years). The installation of such a control system costs from 600 to 800 ECU per apartment. It is believed that this system is too expensive for single houses. Problems arose with the heating bills which were established with help of computer data. Often users did not agree. Finally the system turned out to be very reliable.

The reduction of energy use for cooling and lighting in an office building was studied by TNO, Netherlands. The building has a volume of 24 000 m3, double glazing, 40 % glass area and inside blinds. The influence was investigated of mechanical ventilation (not cooling) during office hours, shading and a reduction in lighting on the cooling requirements of the building.
It was found that no energy was saved with mechanical ventilation, as the energy consumed by the fan was about equal to the decrease in energy use for cooling. On the other hand a reduction of lighting from 20 W/m2 to 15 W/m2 brought considerable energy savings (± 25 %) due to the fact that cooling was no longer required and the energy for lighting was reduced. The reduction of lighting was brought about by a control system which switches off the light when the day light level is above 700 Lux. The project is now halfway and the remaining time will be spent on the development of a control system which integrates the air conditioning plant and the regulation of lighting.

Heat Storage

Large quantities of waste heat are presently discharged (e.g. in industry, power plant, urban incineration) which may be used in winter for heating if cheap long term heat storage would be available. One of the few options is storage of hot water in subsoil waterconducting layers of typically 20 m high which below and above are thermally insulated by natural clay layers. The hot water is brought in and extracted from the aquifer with a doublet of drilled holes. CEA/SNEA, France made an experimental study of such an aquifer heat storage system near Paris. The aquifer is 508 m deep and the feasibility of storing heat up to 180° C was investigated. The results of this study were promising and the construction of a heat storage system for about 5000 houses is now envisaged.

Another area where heat storage may be attractive is daily heat storage at 80° C in combination with an electrical heat pump, in countries where the electricity cost is low during night. In addition such a storage system would allow the heat pump to run continuously thus avoiding on/off losses in normal operation with a thermostat. These on/off losses can be as high as 10-15 %. MARCOUSSIS, France studied a storage system with paraffin. A system consisting of a 0,9 m3 container with paraffin in which a heat exchanger was immersed was found to operate satisfactorily. Tests for different weather conditions (climatic chamber) were carried out and the economy of such a system was calculated. The investment cost for such a storage system (assuming large scale production) is around 12,000 FF. With electricity prices of 0,38 FF/kWh (day) and 0,22 FF/kWh (night), heat storage was economically not attractive (pay back time : 20 years). In other countries with higher electricity prices the system may become interesting in particular as it seems to be possible to decrease the investment cost. The problem of deterioration of paraffin, which has been reported elsewhere, was not noticed. This was possibly due to the limited number of cycles.

Insulation Materials

Many insulating materials are already commercially available on the market but problems still exist. Most of these insulation materials are inflammable, prone to humidity and water vapour and their insulation properties deteriorate with time.

The problem of inflammability was tackled by MARCOUSSIS, France which developed a non-inflammable, glass fibre reinforced, insulating foam of the type NOVOLAQUE (phenolic-resins). Different properties (density, porosity, thermal conductivity, mechanical properties) of NOVOLAQUE have been tested as a function of different additives. The best material up till now had a density of 80 kg/m3 and a thermal conductivity of 0,04 W/m°K. The material was inflammable only under very extreme conditions. The cost per m3 of phenolic foams with a density of 80 Kg/m3 is 1,25 higher than polyurethane with a density of 40 Kg/m3 and 5 times higher than polystyrene which normally has a density of 20 Kg/m3. To improve economic feasibility phenolic resins will have to be developed with a lower density.

Another problem is the deterioration of the thermal insulating properties of polystyrene due to humidity and polluting agents in the air. INDUSTRIA ITALIANA PETROLI, Italy tries to solve this problem by coating the granular polystyrene with an impermeable and protective bituminous binder, which also binds the particles together into panels. The first results of this research look promising.

For polyurethane the insulating properties also deteriorate with time. CSTB/CENG, France found that after 14 days of accelerated aging at 70° and with a humidity of 95 %, the thermal conductivity increased by 20-30% (density $\pm$ 25 Kg/m3). The influence of different types and quantities of raw material for the production of polyurethane, in particular catalysts and surfactants, was also investigated. Unfavourable combinations resulted, after weathering, in a value for the thermal conductivity of 35 mW/m°K, whereas a suitable combination of catalysts and surfactants gave, after weathering, a much better thermal insulation (25 mW/m°K).

Work is also carried out on the influence of thermal stress and vibrations on different types of walls which contain granular or fibrous insulation material. Finally heat accumulating walls for application in glass houses are studied.

Insulation

In this programme window coatings and the use of thermally insulating shutters were investigated. The heat losses in buildings are made up of transmission losses through windows, walls, roofs etc. and ventilation losses due to air infiltration. About 30-40 % of the transmission heat losses are caused by windows. Also a large part of air infiltration occurs via the window frame. Work on the reduction of air infiltration and heat transmission heat losses through windows is therefore very important. Heat transmission can be reduced in three ways :

- decreasing thermal conductivity by using double or triple glazing
- coating windows with transparent heat reflecting layers
- using thermally insulating shutters during the night.

The CENG, France developed a process (PYROSOL) by which glass-panes are coated with layers which are transparent for visible light and which reflect heat radiation. In this way heat is kept in the house and the heat insulation of a single window pane with this type of coating equals roughly that of a double window. In this process tin oxide and indium oxide are used. A drawback of indium oxide is the high cost and for tin oxide the low performance. In order to reduce the cost and improve the performance, mixtures of indium and tin oxide (ITO) were investigated. Both tin oxide and ITO coatings have been tested on a double window where the external window was coated on the inside. The cost of this process is rather high but there is good hope that a considerable reduction is possible by using cheaper solvents (see table 1).

	Tin oxide (E = 0,2)	Indium Tin oxide mixture (E = 0,15)
Heat transmission coefficient	2,2 W/m2°C	1,9 W/m2°C
Transmission for visible light	69 %	75 %
Cost (now)	38 FF/m2	57 FF/m2
Estimated future cost	4 FF/m2	10 FF/m2

Table 1 Cost and properties of heat reflective layers.

Window coating can only take place in the factory. In existing houses therefore complete window panes have to be replaced. This is expensive and in order to avoid this problem it was decided to develop coated plastic sheets which can be easily fixed on windows. R and D by the UNIVERSITY OF LOUGHBOROUGH, United Kingdom resulted in a technique for coating plastic sheets. A problem was the low transparency for visible light which was only 60 % for $TiO_2/Ag/TiO_2$ coatings but which was improved to 75 % using mixtures of cadmium and indium oxides. It was also found that single windows, with a coated plastic sheet, reflect the heat radiation so well that they remain cold and are covered with condensation water which is opaque for heat radiation and spoils the effect of the heat reflective layer. Coated plastic sheets are therefore not very efficient heat mirrors when they are fixed on single windows. They may however be used as second windows. A combination of a glass pane with a coated plastic sheet at a distance of 20 mm has a heat transmission coefficient of 2 W/m2°K. Such a window has the advantage of being light and cheap. (The cost of coated plastic is 1 - 1,5 ECU/m2). As the coating is resistant against scratching it may also be used on blinds.

A second way to reduce heat transmission through windows is the use of shutters or blinds during the night. Here however human behaviour plays an important role : if the handling of the shutters is not very simple the shutters or blinds will not be used. The blinds or shutters should also be suitable for retrofitting in existing houses. PCL, United Kingdom developed a roller blind which can be drawn down at the inside of windows and which consists of a few millimeter of insulating material. Computer simulation showed that for a simple window with a rolling blind a heat transmission coefficient of 1 W/m2°K may be expected. This included the assumption that the blind is coated with heat reflective layers. The blind will be designed in such a way that the edges are sealed when the blind is completely drawn, so that air infiltration in the house is also considerably reduced (± 40 % for average U.K. houses). Different types of seals have been tested using a plastic film as a blind. Magnetic and EDDM synthetic rubber seals turned out to be most effective. A prototype blind is now being constructed.

In a project carried out by TNO/BRONSWERK, Netherlands different combinations of curtains, windows, windowsill and radiators have been studied. One of the conclusions is that an opening between the window and the windowsill above the radiator should be avoided. Measurements of the heat transmission coefficient for standard outside rolling shutters with single glass windows gave values of 3,5 W/m2°K.

DTH, Denmark studied the use of insulated shutters placed at the inside of the windows. A draw back of this system is that the handling of the shutters is not simple and that room must be available when the shutters are not in use.

Air infiltration

If transmission losses have been reduced by insulation of walls, roofs and windows, air infiltration which normally already accounts for 30-40 % of the heat losses in a house, will play an even more important role. Air infiltration losses can be reduced by making houses so tight that heat losses are considerably reduced. Much however depends on the behaviour of the persons living in a house; if they open windows and doors frequently the effect of a tight house is spoilt. Moreover for health reasons the exchange rate of air up till now is not allowed to be lower than 1/2 to 1 volume change per hour. Under these regulations, reduction of heat losses by air infiltration is only possible if the house is made as tight as possible and air infiltration occurs only by controlled ventilation where incoming air is heated by outgoing air via a heat exchanger. In smaller houses this again will not work due to human behaviour, and also the cost will be too high. A still different approach is being developed by BERTIN, France. It starts from the assumption that an air exchange rate of 1/2 to 1 h^{-1} for health reasons is a rather arbitrary measure. When nobody is in the room the air exchange rate may well be much lower. A better measure could be the CO_2 content in the different rooms. The control of the CO_2 level with a CO_2 detector which is coupled to simple ventilation holes, which may be opened and closed, would allow to keep the CO_2 level below an acceptable value ($<$ 1000 ppm). With such a control system the buildings can be very tight and a considerable amount of energy may be saved. Such a system may be suitable for large buildings such as conference buildings. In domestic buildings humidity may turn out to be a problem.

In a Danish study by the TEKNOLOGISK INSTITUT, the influence of exhaust ventilation and small ventilation holes was studied on very tight houses with a natural air exchange rate of 0,1 to 0,2 h^{-1}. This was done for different types of houses. It was shown that in these houses with a mechanical exhaust, an air exchange rate of 0,5 h^{-1} can be obtained with an under pressure in the house of 5 Pa. The distribution of air in the different rooms is no problem when the doors between the rooms are open. Several ventilation holes, which can be opened and closed, with a total surface of 300 to 400 cm3 for each apartment, are needed in order to have sufficient natural air exchange in the summer.

It is very important to develop good measuring techniques for the measurements of air infiltration in buildings. Normally this is done by mixing the air in a building with a tracer gas and measuring the tracer gas level as a function of time. Especially in large buildings this is not a simple problem. BRE, United Kingdom is developing a methodology for large and complex buildings. BSRIA, United Kingdom studied two new tracer gas detection methods. The first method is based on methane as a tracer gas measuring its concentration with an infrared laser technique. The second method is based on ethanol vapour measuring its concentration with a fuel cell detector.

Thermal characteristics of buildings and their simulation with computer models

In the previous chapters different energy saving techniques have been discussed. Once these techniques are available a second step is their integration in buildings in such a way that energy use is minimized in an economically attractive manner. This can be a complicated task in particular for large buildings such as schools, hospitals, etc. Computer models which simulate the thermal behaviour of buildings and their components are becoming increasingly important for this purpose; in particular for energy saving design of new houses and for energy auditing of existing buildings.

However these simulation models are still far from reliable as was shown by an IEA experiment where several computer models tried to simulate the thermal behaviour of a building. Also the experimental verification of these models is an arduous and time consuming task where reliable data are not easy to obtain and the interpretation of measuring results often is ambiguous. Finally simulation models, in particular for large buildings, are often complicated, require large computers and are expensive. That makes them less suitable for practical applications. For the daily work of architects, models must be developed which are reliable, simple to work with and which can be handled on small and cheap computers. These simple computer models are however likely to be even less precise than the bigger ones. It will therefore not be easy to achieve these objectives.

In the EC Programme several computer programmes are being developed. In these models the thermal characteristics of the building and the climate are taken into account. The UNIVERSITY OF LIEGE, Belgium is studying the basic characteristics of low cost housing with a view to reducing energy consumption. This project is carried out in collaboration with CSTC, Belgium. The temperature in houses without heating is a very important characteristic of a building and is needed to calculate the energy consumption in a house. Computer models have been developed with which the temperature without heating can be calculated. Factors such as external temperature, wind, direct and diffuse solar radiation can be taken into account. Models both for the static and dynamic thermal behaviour of the house have been developed. These models are being experimentally validated with one-m3 models and two full scale houses. In these houses also different types of heating systems are being tested. The final aim of this project is to develop guidelines for architects for energy saving design of buildings.

Another model describing the dynamic behaviour of buildings is being developed by ABACUS, UNIVERSITY OF STRATHCLYDE, United Kingdom. The ESP (Environmental System Performance) model simulates the energy transfer within, outwith and between a number of zones linked together to form a building. The model allows detailed modelling. It takes into account different climatic factors such as temperature, solar radiation and wind. These factors are brought together in one parameter which is called the Climatic Severity Index (CSI). The model is already used for energy saving design of buildings.

A simple computer simulation model was also developed by the TECHNOLOGICAL INSTITUTE, Denmark. The internal heat gain of buildings, which is a very important factor in all computer models for the energy use in buildings, is being measured in two dwellings by AN FORAS TALUNTAIS, Ireland. The resultant data will be processed to produce a theoretical simulation model for predicting internal ambient temperatures for different weather conditions.

Domestic appliances

The electricity production for the domestic and commercial sector requires around 14 % of the total primary energy. A large part of this power is used for domestic appliances and an effort should therefore be made to reduce their energy consumption. In three contracts work is carried out on :

- Energy saving in the catering sector
- Design and construction of energy saving domestic appliances
- A washing machine where the oxidiser is produced by electrolysis in the washing liquid at room temperature.

In the catering sector very little attention was paid to energy efficiency. A modest saving of 10 % in the UK would lead to an annual energy saving of 150 000 TOE. The objective of this project, carried out by BRITISH GAS, United Kingdom, is the investigation of appliance design, with a view to producing cost-effective reductions in energy consumption.

Activities are wide-ranging and include monitoring small commercial dish washers in operation; a survey of potential pan sensing techniques; efficiency studies of deep fat fryers, boiling tables, ovens, grills and griddles. These investigations are progressing at various speeds. At present the dishwasher monitoring exercise has been completed as also has that for the deep fat fryer design.

The main conclusions to date are that dish washing machines operate at only a fraction (25 %) of their full capacity and that they should be controlled automatically.

Conclusion for the deep fat fryers are that "cool zones" contribute neither to the life of cooking oils nor to the finished quality of food. Elimination of the "cool zone" could reduce the oil volume of the tank. It is expected that a redesigned tank could lead to energy savings of 10 %.

The TECHNICAL UNIVERSITY OF DENMARK evaluated the possibilities for energy saving design and construction of refrigerators, cooking pots, ovens and washing machines. The background of this project is an energy analysis for Denmark which included an evaluation of the energy saving potential for household appliances. It was believed that the energy consumption on the average could be reduced to 30 % of the energy use by the 1975 stock. The work was carried out in collaboration with manufacturers of these products. Work on the washing machine was not started because the target which was set by the DTH has already been reached by the manufacturer which was to be involved in the development work. Over the first months a cabinet for a 209 liter refrigerator with a 7 cm thick polyurethane foam insulation has been designed but not built. The refrigerator system is planned to have a COP of 1.5 or more, bringing the annual electricity consumption down to 100 kWh compared with 550 kWh for the 1975-stock and 250 kWh for the best models sold today. The economy in redesigning seems to be good; the calculated pay back time is two years. The refrigerator is now being constructed.

For cooking, first sketch designs have been made of a 2 liter cooking pot with double walls and a small 12 liter well insulated oven. An experimental set up has been designed to test a PID (proportional, integration, differentiation) control system for hot plates. This cooking equipment will be constructed in the next three months.

The objective of research carried out by RIBER, Italy is to develop a washing machine which operates at low temperatures. During the washing process active oxygen is normally produced from perborates (a detergent component). In order to obtain sufficient oxygen, temperatures of over 60° C are required for a certain period. In this project oxygen is produced from the washing water by electrolysis at room temperatures. A washing machine has been built where prewashing occurs with an electrolyser and is followed by washing between 40-90° C. As compared with traditional washing, energy savings of 25-35 % are possible with the new washing machine and the washing results are generally better.

SESSION I - HEATING SYSTEMS AND VENTILATION

Chairman: J. UYTTENBROECK

Preliminary investigation into the selection of boiler installations from the energy conservation point of view

Energy savings from secondary windows in experimental houses

Preliminary phases in the study of air distribution - Methods for domestic warm-air heating systems using low-grade heat sources

Water/air convector made with plastics of improved thermal conductivity

Measuring and control device for energy consumption of heating installations

Microprocessor-based heat meter

PRELIMINARY INVESTIGATION INTO THE SELECTION OF BOILER INSTALLATIONS FROM THE ENERGY CONSERVATION POINT OF VIEW

Author : Ir. F.M. van Bergen

Contract number : EEA-1-001-N

Duration : 30 Months 1 Febr. 1982 - 31 July 1984

Head of project : Ing. M Kiel

Contractor : Heat and Refrigeration Engineering Division TNO

Address : Heat and Refrigeration Engineering Division TNO
P.O. Box 342
7300 AH Apeldoorn
The Netherlands

Summary

Prior to the investigation "Development of a computer model for the selection of boiler plants from the view point of a low energy consumption", a desk study has been made dealing with the differences in energy consumptions of various boiler installations. For the purpose of this study use has been made of earlier developed formulea for determination of the efficiency of boilers under operating conditions. Some of the conclusions from the desk study is that it is worthwhile paying extra attention to the expected energy consumption of an installation under design. Calculations have shown that with the most economical design as much as 26% energy can be saved. A survey is given of the activities to be carried out in the framework of the current project for the development of a computer model.

1. Introduction

The activities in the framework of the project "development of a computer model for the selection of boiler plants from the viewpoint of a low energy consumption" started on 1st February 1982. This makes it impossible to give results of research for this project at this very moment.
Prior to this piece of research a preliminary research project "The selection of boiler installations from the energy conservation point of view" was carried out. This desk study involved calculations of the energy consumption of various boiler installations. The results of this preliminary investigation will be discussed, and subsequent activities to be carried out in the framework of the current project will be further gone into.

2. Preliminary research

2.1 Introduction

In this study the effect of different types of installation on energy consumption will be examined by means of examples of calculation. The installations considered are central heating installations in which, to simplify calculations, night reduction has not been applied.

2.2 Selection of installations

With regard to the design of boiler installations there are several possibilities :
- the number of boilers;
- the principle of regulation;
- the type of boiler;
- the type of burner.

Table I: Choice of possibilities

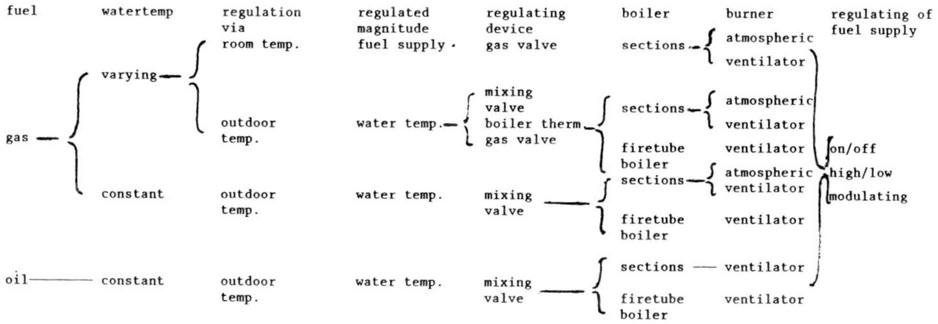

Table I shows that there are more possible variations in gas-fired installations than there are for oil. This is due mainly to the fact that oil-fired installations are equipped with a mixing water-regulation to reduce the chance of corrosion by acids. Therefore, the considerations following here apply mainly to gas-fired installations. Within the scope of this paper it is impossible to indicate and calculate all possible variations. For this reason several current types of installation have been selected from those used in actual practice.

For a building the energy consumption will be considered in case of:
1. an installation with one large natural-draught boiler with on/off regulation of the fuel supply;
2. an installation with a boiler with one forced-draught burner and on/off regulation of the fuel supply;
3. both above-mentioned types for two and for three boilers with on/off regulation of the fuel supply and connected in cascade;
4. all above-mentioned installation types for a watertemperature varying with the ambient temperature (weather-dependent control) and for a constant high boiler-water temperature;
5. a boiler with forced draught burner with high/low regulation compared with on/off regulation of the fuel consumption;
6. an installation with a butterfly valve in the return water circuit and a valve in the combustion gas discharge of the boilers.

2.3 Boiler characteristics

From results of research obtained by the TNO-Research Institute and from publications relating to the subject, some data have been summarized and these may serve as a basis for design.
In Table II the losses and efficiencies of three groups of gas boilers at full-load operation are given.
It should be noted that the figures apply to boilers heated to a water temperature of 80°C on an average.
Small natural-draught boilers are used in private installations in houses. Large natural-draught boilers and boilers with forced-draught gas burners are generally used in larger buildings and blocks of flats. This division is made because there is a distinct difference in the losses and efficiencies for the two groups of boilers. The data mentioned are averages.

Table III give the losses during a standstill period. The standstill loss, q, is divided into a transmission part and a ventilation part. These data also apply to a water temperature of 80°C. In small natural-draught boilers the spread in the standstill losses is so large that a range is indicated (within which the averages fall) rather then one average value.

Table II: Losses and e-ficiencies of gas boilers at full-load in percent of the load based on gross-calorific value

		Small natural-draught gas boilers	Large natural-draught gas boilers	Gas boilers with forced draught
Sensible heat loss	Q_v	7,0-16,0	11,3	8,3
Condensation heat loss	Q_c	10,0	10,0	10,0
Transmission loss	Q_t	4,0-8,0	2,5	0,8
Efficiency at water side	η_{vw}	65,0-79,0	76,2	80,9

Table III: Losses during standstill in percent of the full-load based on the gross-calorific value

		Small natural-draught gas boilers	Large natural-draught gas boilers	Gas boilers with forced draught
Standstill loss	q	2,5-5,5	2,3	0,7
Transmission part	q_t	1,3-3,0	1,15	0,35
Ventilation part	q_v	1,3-3,0	1,15	0,35

2.4 Method of calculation

The energy consumptions of the boilers have been calculated by using a method developed by Van der Heeden [1]. Although this method is mainly based on the results of measurements carried out for smaller CH-boilers it may be assumed that the method is also applicable to large CH-boilers. It will be discussed here point by point.

. Degree of utilization B

The degree of utilization is the percentage of the heating season during which the boiler is in operation for the production of useful heat. The heating season is to be understood here as the period in which the installation is available for the production of heat.

. Standstill losses q_c

Table III indicates the average losses occurring. They apply only for a constant, high water-temperature and no account is taken of the remainder of the installation of which the boiler forms part. The control response is determined by the installation as a whole.

The ventilation part of the standstill losses, q_v, depends on this control response. For correcting the standstill loss, an installation constant is calculated, namely

$$C = \frac{418{,}6 \times W \times \Delta\theta_t}{Q \times \frac{\eta_{vw}}{100}} \qquad \text{(see[1])}$$

W = water equivalent, in kg H_2O;
$\Delta\theta$ = regulating control span of thermostat, in K;
Q = boiler load based on gross caloric value, in kW;
η_{vw} = efficiency at the water side based on calorific value, in % of boiler load Q.

For the water equivalent of the installation without mixing regulation, the total water equivalent is taken into account. For installations with mixing regulation, only the primary water circuit, up to the mixing valves, is considered.
The standstill losses are corrected as follows :

$$q_c = q_t + 70 \sqrt{\frac{B}{C}}$$

where
q_c = corrected standstill loss, in %;
q_t = transmission standstill loss, in %;
B = degree of utilisation, in %;
C = installation constant.

If the installation is provided with means to stop water circulation through the boilers during the off-period, the boiler cools down and q_t is replaced by q_a. This is further explained in the example given in this paper.
If the ventilation flow through the boiler(s) is cut off by an air valve or combustion gas valve the standstill loss needs no correction. This is also treated in the relevant example.

. Correction of the degree of utulization B^1

The definition of the degree of utilisation refers to operation for the production of useful heat. To compensate for the standstill losses, however, the boiler must produce extra energy. The correction can be expressed by :

$$B^1 = B + (100-B)\frac{q_c}{100} \quad [\%]$$

. Utilisation efficiency η_g

When B and B^1 are known, the factor ψ (Dittrich) can be calculated :

$$\psi = \frac{B}{B^1}.$$

The utilization efficiency is then

$$\eta_g = \psi \times \eta_{vw}.$$

. Fuel consumption V

The fuel consumption per boiler can now be calculated :

$$V = \frac{P \times \tau_j \times B^1}{H_v \times \eta_g} \quad [m^3 \text{ gas}]$$

where
P = boiler capacity, in W;
τ_j = time the boiler is available for the production of heat, in s;
B^1 = corrected degree of utilization, in %;
H_v = gross calorific value of the fuel, in J/m^3;
η_g = utilization efficiency of the boiler, in %.

The total fuel consumption can be calculated approximately by summation of the fuel consumptions of the individual boilers.

2.5 Example

To illustrate what has been discussed above, the yearly fuel consumption of an arbitrarily chosen building has been calculated. The calculated heat loss of the building by transmission is 96 kW. To be able to heat up quickly after a break, and to allow for circuit losses, an installed power of 120 kW is chosen. To keep calculations simple we assume that no night and no weekend reduction takes place. We assume that the installation is equipped with large natural-draught boilers or boilers with forced-draught burners. The calculated fuel consumptions of the various designs, as indicated in Chapter 2, will be compared to examine the possibilities for an energy conscious design. It must be emphasized, however, that this example applies to one given situation only.
The results may not be considered as generally valid. The aim is to indicate a method of judging an installation design by calculation.

. Degrees of utilisation

If the heat demand is known the degree of utilisation can be calculated. It appears that in a normal heating season 100% of the demand must be met during 2544 hours. However, the 100% heat demand in this example is to be supplied by an installation operating at 80% of full-load capacity.
So the number of full-load hours, when one boiler is used, is $\frac{80}{100} * 2544 = 2035$ hours. The time the installation is available for operation during a heating season is 5900 hours, so

$$B = \frac{2035}{5900} * 100\% = 34,5\%.$$

For installations comprising 2 or 3 boilers the degrees of utilization are calculated in the same manner. The number of full-load hours will be dependent on the time during which the boilers are available for operation.
Using two boilers, each with half of the total capacity, one boiler is capable of meeting a demand up to $\frac{100}{80} * 50 = 62,5$ % of the nominal heat demand.

Using three boilers, each with one-third of the total capacity, one boiler can supply 41,6% of the heat, two boilers 83% and all three 100%.

In Table IV the calculated degrees of utilization are presented.

TABLE IV : degrees of utilization

installation	B_1	B_2	B_3	B_{high}	B_{low}
one boiler	34,5%	-	-		
two boilers	50,7%	18,3%	-		
three boilers	56,6%	39,9%	7,0%		
high/low boiler				44,2%	58,9%

In table IV also the degrees of utilisation are given for one boiler with high/low regulation. It is assumed that the output at low level is 40% of that at full load.

. The influence of water-temperature on standstill loss.

In the various designs, the following water-temperature conditions affect standstill losses:
- the water-temperature is kept at a constant, high value (80°C);
- the water-temperature varies because it is weather-dependently controlled;
- all boilers maintain circulation;
- the water circulation is stopped as soon as the fuel supply to a boiler stops.

The standstill losses at a constant, high temperature for boiler installations in which circulation is maintained are collected in Table III.

For calculation of the average stillstand losses at changing water temperatures the average water temperature during the heating season must be determined. In this example the average water temperature appears to be approx. 50°C. With the aid of measurements carried out by van der Heeden [1] the average stillstand losses at varying water temperatures have been determined. These average losses are summarised in Table V.

TABLE V : Average values of standstill losses with varying water temperatures

boiler type	standstill loss q (%)
large natural-draught boiler	1
forced-draught boiler	0,2

. Influence of stopping water circulation through the boiler.

The water circulation can be reduced or stopped by closing the butterfly valve in the return circuit of the boilers. An other possibility is to switch off the circulation pumps. In stead of transferring heat to the environment of the heating system, the water cools down in the boiler. The losses due to cooling during the off-periods, q_a, depend on:

- the water temperature with switched-off burners and discontinued water circulation;
- the water equivalents and time constants of the boilers;
- the degree of utilization, the number of switch-ons per unit of time and the average duration of the off-periods;
- the capacity of the installation;
- the number of boilers.

The most important factors will be discussed separately.

- Water temperature.
  In the case of installations operating at a constant, high water temperature of about 80°C this temperature is the same for installations with one or more boilers. If the temperature is weather-dependently controlled, however, the temperatures are not equal with boilers connected in cascade. The water temperature is determined by the capacity of the boiler and the heat demand to be met by the boiler or combination of boilers.
  In Table VI the supply-water temperatures at the end of the operating periods are given for boiler installations with on/off regulation.

TABLE VI Average supply-water temperature (°C) at the end of an on-period in installations operating at varying water temperatures.

One boiler	2 boilers		3 boilers		
	boiler 1	boiler 2	boiler 1	boiler 2	boiler 3
65	65	71	65	67	77

- Time constants of the boilers.
  The heat loss by cooling can be found by examining how many degrees the boiler temperature drops during an average off-period, τ_u.

  The time constant, $\tau = \dfrac{V_k x \rho x c_w}{k x A}$

  where
  V_k = volume of the boiler, in m^3;
  ρ = specific mass, in kg/m^3
  c_w = specific heat, in $J/(kgK)$;
  k = heat transfer coefficient, in $W/(m^2K)$;
  A = exterior surface area of the boiler, in m^2.

  These calculations include the water equivalent W. Substituting W for $V_k x \rho$ the formula becomes:

  $$\tau = \frac{W x c_w}{k x A}$$

  An approximate value of k can be calculated by using the transmission losses measured, q_t, and the surface area, A. $q_t = kxAx\Delta\theta$, where $\Delta\theta$ is the difference between the water temperature and the ambient temperature. For the large natural-draught gas boilers, $k = 4{,}6\ W/(m^2K)$, for the forced-draught gas boilers, $k = 2{,}6\ W/(m^2K)$.
  These values apply to the particular type and make of boiler used for this example. For every boiler the values of k will differ.

For every boiler or combination of boilers, the temperature drop during the average off-period τ_u can now be calculated.

$$\frac{\theta_a - \theta_\sim}{\theta_{am} - \theta_\sim} = e^{-\frac{\tau u}{\tau}}$$

where

θ_a = water temperature in the boiler at the end of the off-period, in °C;

$\theta_\sim$ = water temperature after $\tau_u = \sim$, in °C;

θ_{am} = maximum supply temperature;

τ_u = duration of the off-periods;

τ = time constant

We assume $\theta_\sim$ = 20°C.

After conversion we find:

$$\theta_a = 20+(\theta_{am}-\theta_\sim)e^{\frac{\tau u}{\tau}}$$

$$\Delta\theta = \theta_{am}-\theta_a \ [K]$$

The cooling loss during one cycle (τ_c) of one on-period and one off-period is:

$$q_a = \frac{W x c_w x \Delta\theta}{\tau_u} \times \frac{\tau_u}{\tau_c} \ [W]$$

$$\frac{\tau_u}{\tau_c} = \frac{100-B}{100}$$

so,

$$q_a = \frac{W x c_w x \Delta\theta}{\tau_u} \times \frac{100-B}{100} \ [W]$$

in per cent:

$$q_a = \frac{W x c_w x \Delta\theta}{\tau_u x Q} \times (100-B) \ [\%]$$

where Q = the load based on gross calorific value [W]

. Effect of a valve in the combustion gas discharge.

A valve in the combustion gas discharge is used with natural-draught gas boilers. It is assumed that the valve is placed before the draught interrupter. Until now the ventilation losses during standstill have not been measured for natural-draught boilers with a discharge valve. It is likely that these ventilation losses will be almost the same as those of a forced-draught boiler with its air valve closed during the standstill period. It is assumed that at the end of the operating periods the discharge valve and the air valve are both closed at the same moment the fuel supply is stopped. By closing the valves, the ventilation through the boiler is cut off. This means that it is not necessary to correct the standstill losses for the effect of a valve in the combustion gas discharge. Then $q_c = q$.

. Calculated fuel consumptions.

In Tables VII, VIII and IX the calculated fuel consumptions for the different installation types are listed. For the sake of convenience, the tables indicate whether or not the installation is equipped with valves.
A butterfly valve is a valve by which the return water flow can be closed partly or entirely. A discharge valve is a valve in the combustion gas discharge.
In natural-draught boilers such a valve is placed before the draught interrupter.
An air valve is a valve by which the air supply to the forced-draught burner is regulated. It is assumed that this valve is closed when the burner is out of operation.

TABLE VII : Yearly gas consumption when using large natural-draught gas boilers with on/off regulation ($*1000m^3$)

water temperature		High			Varying		
number of boilers		1	2	3	1	2	3
butterfly V.	discharge V.						
No	No	39,4	38,1	38,4	38,6	37,2	37,0
Yes	No	38,9	37,6	37,3	38,6	37,4	37,1
No	Yes	35,6	35,9	36,5	34,6	34,7	35,0
Yes	Yes	35,1	35,4	35,5	34,6	35,0	35,1

TABLE VIII : Yearly gas consumption when using forced-draught gas boilers with on/off regulation ($*1000m^3$)

water temperature		High			Varying		
number of boilers		1	2	3	1	2	3
butterfly V.	discharge V.						
No	Yes	32,7	32,7	33,0	32,1	32,1	32,3
Yes	Yes	32,6	32,8	33,2	32,3	32,4	32,7

TABLE IX : Yearly fuel consumption when using one forced-draught gas boiler with high/low regulation ($*1000m^3$)

water temperature		High	Varying
butterfly V.	discharge V.		
Yes	Yes	29,1	29,3

Note: Tables VIII and IX include the primary energy consumption of the forced-draught burner, expressed in m^3 gas.

2.6 Conclusions

The most important conclusion is that it is worthwhile paying extra attention to the energy consumption of an installation in the

design stage. From the example quoted, it appears that the fuel consumption of the most economical design can be about 26% lower than that of the design with the highest fuel consumption.
For building owners, it should also be a point of consideration to have their existing installations checked for economical energy consumption. From experiences in checking installations it appears that many installations can be improved on this point by relatively small alterations.
It is important to choose boilers with a high water-side efficiency, and to keep the capacity of the installation as small as possible. An oversized installation gives a lower degree of utilization and consequently a lower utilization efficiency.
An installation which does not have to produce energy for space heating purposes during the summer period should not be equipped with a hot water supply device. In the summer period the installation should be shut down.
The most interesting items of equipment discussed in this paper are:

- The valve in the combustion gas discharge, with which a 7 to 10% saving in fuel consumption may be reached during the heating season, provided the valve is regulated along with the fuel supply.
  This applies to natural-draught and possibly also to forced-draught boilers, the air valve of which is not closed during a break. In natural-draught boilers the valve must be placed before the draught interrupter.
  Under other conditions savings may be higher, for example, if heating continues for the whole year, or if night reduction is applied, in which case the degree of utilization of the installations is lower and the off-periods are longer.
- The boiler with forced-draught burner has a higher efficiency than the conventional natural-draught boilers without regulation of the secondary air supply. The difference in fuel consumption is about 7 to 8% on an average.
- A butterfly valve in the return water circuit has hardly any effect with varying water temperatures. With a high water temperature, 1 to 3% energy can be saved.
- High/low regulation with varying water temperatures does not give any saving. With a high water temperature, about 1,5% can be saved.
- The number of boilers has hardly any effect. Among installations with a valve in the combustion gas discharge or in the air supply and butterfly valves in the return water circuit, the single-boiler installation seems to be the most economical. If these valves are not present, a multiple-boiler installation is more economical.
- Using weather-dependent control, which allows operation with varying water temperatures, 0,4 to 2,6% energy can be saved.

3. Work to be carried out for the current project

In the preliminary study energy consumptions of various boiler installations have been calculated for one particular building with inner temperature regulation without night reduction.

It is useful, however, to have the facilities to calculate the energy consumption of boiler installations for any arbitrary design of building and mode of inner temperature regulation. This makes it necessary to develop a computer model which allows such calculations to be carried out guickly and with a high degree of flexibility.
Figure 1 gives the planning scheme for the current research project. Further to the preliminary study the principes of calculation used will have to be reviewed critically. As a matter of fact these principles include various empirically established formula which, for the greater part, have been derived from measurements carried out for the small CH-boilers of familyhouses. Measurements in practice will have to be made to find out if and to what extent these formula need modification for larger boilers. These measurements are indicated in figure 1 by "effect measurements".
In parallel herewith the computer model will be developed. The model will then be used to determine the hourly energy consumption of the boiler installation in relation to the heat demand of the building, using a "reference" year for the ambient conditions.
After the results of the measurements have been processed, the computer model can be adapted to the practical situation. The model will subsequently be used to make energy consumption calculations for an existing boiler installation applying different modes of regulation of the installation and inner temperatures. The results will be verified with those of measurements for the existing installation. These measurements are indicated in figure 1 by "verification measurements". If required, a final modification can be applied to the model, after which the computer program is available for the final application of calculating boiler installations.
The computer program will be used to make calculations for various boiler installations for a number of categories of buildings to find out what boiler installation is optimally suited for a particular category of buildings.

Figure 1 : TIME PLANNING SCHEME

"Computer model for the selection of boiler plant"

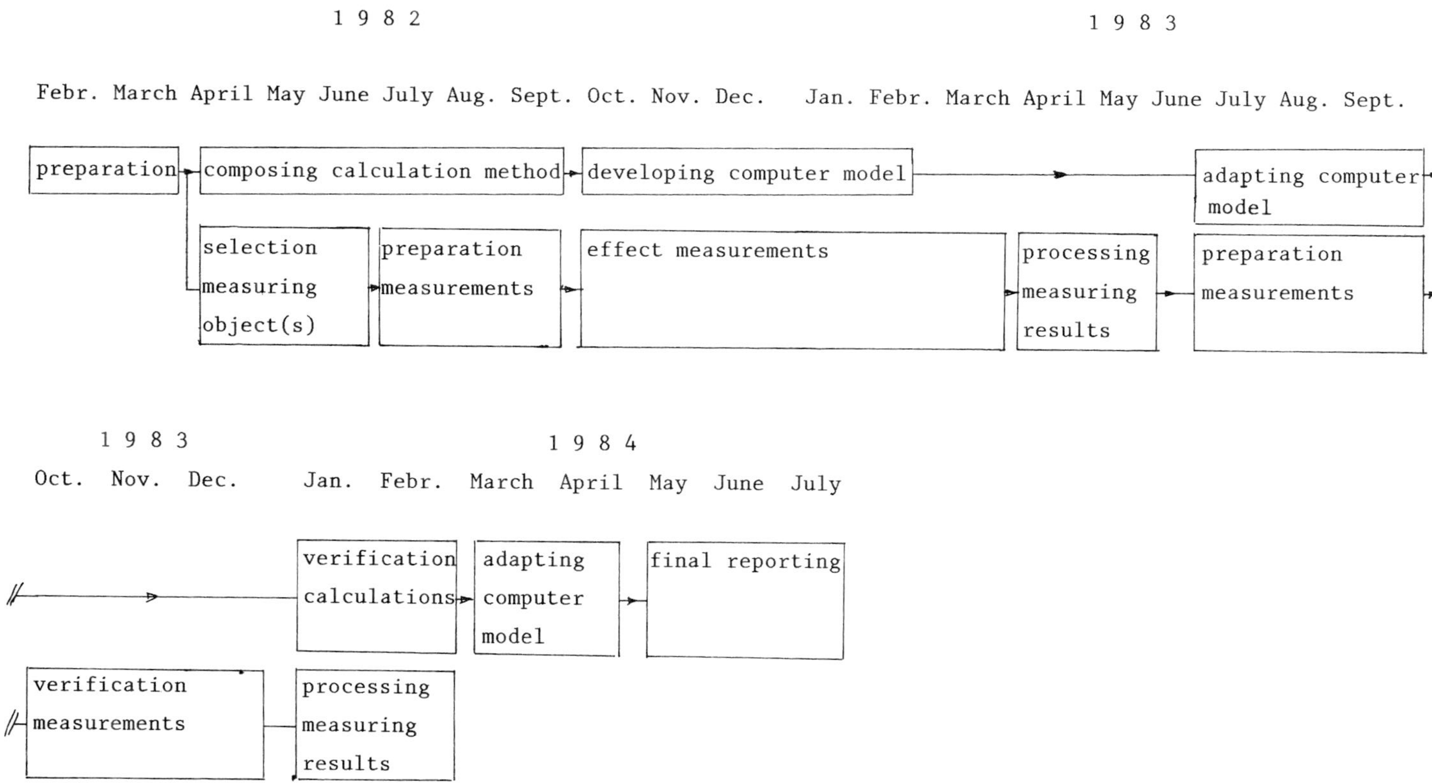

Reference

[1] The utilization efficiency of central heating boilers.
Het gebruiksrendement van cv-ketels.
Ing. D.J. van der Heeden
38e vacantieleergang voor de warmtetechniek 1976 pag. 5 t/m 17.

QUESTIONS AND DISCUSSION

Zegers (DG XII) : Will you include in your study the costs of the additional energy saving devices used to improve the boilers?

Answer : Cost has not been considered in the preliminary study, but will be included in the final computer model.

Rayment (BRE) : A study on seven commercial boilers has been made in the UK.

Spitzer (Battelle) : Do you think that your proposed formula and methodology will be applicable to small boilers in the domestic sector ?

Answer : The initially proposed formula (A. Dittrich) can be used also for small boilers ("on-off" type). One of the aims of the research is to generalise the method also for small boilers of other types.

Zegers (DG XII) : For important heating demands there is an advantage to install several boilers instead of one. This however is impossible for small heating demands.

Rayment (BRE) : If you take a good boiler and all the measures to avoid energy losses it is possible to obtain very good results even with one boiler.

Uyttenbroeck (Chairman) : In the TNO study, the starting point is the energy need of the building. But one should not forget that in the relation between the energy losses, of the building and its energy consumption other parameters than the boiler efficiency play a role (e.g. free heat-distribution losses, control efficiency, emission efficiency). It seems very dangerous to mix up all the effects of these parameters to come to an idea on the efficiency of the boiler. The efficiency of the boiler should be defined as the ratio between the amount of energy delivered in the boiler room to the system and the amount of energy consumed by the burner.
Mr Uyttenbroeck distributed a text (Revue CSTC no. 4 - 1981 - "Rendement saisonnier d'une chaudière").

ENERGY SAVINGS FROM SECONDARY WINDOWS IN EXPERIMENTAL HOUSES

Authors : R. RAYMENT, K. MORGAN

Contract number : EEA-1-002-GB

Duration : 36 months 1 July 1980 - 30 June 1983

Head of project : Dr. S.J. Leach, Building Research Establishment

Contractor : Building Research Establishment

Address : Building Research Establishment
Building Research Station
Bucknalls Lane
Garston,Watford, Herts. WD2 7JR

Summary

This paper is one of a series of reports for an EEC Contract on field, laboratory and theoretical investigations of energy saving methods in dwellings. Here, specially prepared matched-house pairs have been used to investigate the energy savings obtained from fitting secondary windows. This forms part of an ongoing programme designed to examine a range of energy conservation measures. The results demonstrate a mean reduction in energy consumption of 12% (±3%) for the secondary windows. The theoretical percentage reduction agrees well with that measured only if the air change rate is assumed to have been reduced by about one third. Energy consumption has been determined for a range of weather conditions which, from data gathered on the experimental houses in other heating seasons, should be representative of the average annual savings for this type of house and pattern of usage. Extension of the results to the housing stock of well heated semi-detached houses is not straightforward since differing air infiltration rates, window areas and insulation levels will affect the savings. In addition in this experiment it was not possible to arrange for air infiltration measurements to be made when the secondary windows were in place. Thus it is not certain how the infiltration influenced the results. Further comparative trials are being arranged in another pair of houses to provide additional information on which to base preditions of energy savings and thence calculations of cost-effectiveness, since double-glazing is recognised as an important area of activity.

INTRODUCTION

The BRE experimental semi-detached houses(1-3) were used during the 1980/81 heating season to measure the energy savings attributable to detachable secondary double windows. Since the internal heat gains from simulated occupancy devices were known along with the house construction and external climate a check could be made between the measured savings and those derived from the calculation procedure ennunciated by Uglow(4).

Though not all the relevant elements in occupants' behaviour were simulated, in particular windows were not opened, it is believed that the magnitude of the savings found give a very good indication of the attainable savings in well-heated semi-detached houses of this type, and fitted with this type of window.

BRE EXPERIMENTAL HOUSES

The pair of adjoining semi-detached houses were built in the late 1960s using conventional brick-cavity-block construction, each having a ground floor area of 56 m^2. Inside the space is divided up into a hallway, toilet, kitchen, dining room connected by large double doors to the lounge; a staircase leading to the first floor which contains a toilet, bathroom, 3 bedrooms and an airing/linen cupboard. End walls have no windows and a kitchen door gives access to a passageway screened by a garage and wall. 100 mm of roof insulation had been installed but no cavity fill has been used and the existing windows were single-glazed, glass and metal set in a wooden surround, with no extra weatherproofing or sealing. Orientation of the houses is such that the back of the houses face roughly south-east, so that the main living areas and bedrooms received most sunshine. The house nominated for the double window tests was chosen to be south-west of the reference house. Further details, required for carrying out thermal calculations, are given in reference (2) and included in current EEC Contract Reports.

SIMULATED INTERNAL HEAT GAINS

Heat gains from people and electrical appliances were simulated together with water run-offs at kitchen sink, bathroom basin and bath. All these elements including the movement of people around and in and out of the house plus curtain drawing were controlled by a programmable sequencer. Figure 1 shows a selection of occupancy gains and times used for the four member family simulated in these experiments. Table I shows a summary of the daily heat input from various sources, the total in these experiements amounting to 68.3 MJ/day.

HEATING SYSTEMS AND CONTROLS

Relevant elements of the heating system and controls are shown on Figure 2. The boiler is medium weight, wall hung and has a balanced, fan-assisted flue and uses electric spark ignition. In each house the boiler was placed on an outside kitchen wall.

Domestic hot water was supplied from an indirect cylinder, fully pumped from the primary heating circuit and located in the airing cupboard. Hot water temperatures were controlled by a cylinder-stat set to 60°C. A room-stat, located on an internal wall in the lounge and set to about 22°C, controlled the space heating. Control action was such that flow could be directed, through a 3-way valve, to supply hot water, space heating or both when called for by the thermostats. When both thermostats were satisfied the pump and boiler were automatically switched off. The room-stat was of the accelerator-heater type and both thermostats had vapour capsule sensors for responding to temperature.

The night switch-off of the heating, shown on Figure 1, was controlled by the sequencer.

INSTRUMENTATION AND LOGGING

The measured quantities of interest in this study were gas consumption, internal and external temperature, wind and solar radiation. All these were measured every 2½ minutes and 20 minute averages logged on 8" diskettes using a SOLARTRON SYSTEM 35 computer logger. Gas consumption was measured using a specially adapted Gas Board meter which had been calibrated against a certified 'wet' meter. Thermocouples housed in blackened table tennis balls were used for internal temperatures in every room in the houses (ie 8 in each house). These 'globe' thermometers being roughly equally responsive to air and radiant temperature. A platinum resistance thermometer and thermocouple, mounted on a weather-mast, were used to measure external temperature. All thermocouples had a common 'ice cell' reference. Wind speed and direction was found using a standard resolver set and solar radiation measured with two Kipp and Zonen solarimeters mounted vertically on the south-east and north-west walls of the houses level with the top of the ground floor windows. The manufacturers calibration as taken for the solarimeters.

CALIBRATION OF THE HOUSES

The principle of using matched house pairs relies on the existence of a high correlation between the energy consumption in the two houses when they are nominally identical with respect to their heating systems and controls, insulation and adventituous heat gains. Also the ventilation losses and solar heat gains, which are not necessarily symmetrical, must not be very different. Reference (3) and later discussions on solar gains show that the houses are satisfactory in this respect. A further requirement is that the set points on the various control thermostats be as closely identical as possible. For the cylinder-stats they were both set to 60°C and the boiler-stats were set to give flow temperatures, at switch off, of 85°C. The room-stat settings were adjusted for matching by measuring lounge temperatures for several weeks prior to the main experiment. Having completed this, six day calibration runs were made periodically through the heating season, with the two houses in their 'base' condition, ie with single glazing. The results of this calibration are shown by the black circles on Figure 3. A three-day overlapping average has been used to produce the points to reduce the effect of one house occasionally receiving a heat charge just prior to switch off at night. As shown on the Figure the regression equation for prediciting the consumption in the test house (G_T) is related to the consumption in the reference house (G_R) by:

$$G_T = -77.4 + 1.320\ G_R \qquad \text{MJ/day.}$$

Mean consumptions for the heating season together with other data are given in Table II.

SECONDARY DOUBLE WINDOWS

The construction of the secondary windows is shown on Figure 4. It can be noted that the original windows were metal framed but set into a wooden surrounding frame. This will give some reduction in the U-value compared with all metal framed windows. The windows were not sealed or weatherstripped. By contrast the secondary windows were wooden framed and made a good seal with the existing wooden frame. Since every window in the test house was fitted with these windows a noticeable reduction in air infiltration could be expected. The secondary windows were removed and refitted periodically through the heating season to allow calibration runs to be made.

RESULTS

The main results showing a reduction in energy consumption are plotted out using open circles on Figure 3. As indicated by the dashed line on Figure 3 they correspond to an approximate 12% saving with rather higher savings for high loads and hardly any savings at low loads. The latter result probably indicates that in warmer weather the reduction in solar gain counterbalances the reduced heat losses for the double windows.

Another way of looking at the results is to take the difference between the actual and predicted test house consumptions, for respectively with and without double windows, and add in the variance associated with the calibration regression. A statistical test can thereby be applied to the mean difference in consumption.

Let D be the actual consumption with the double windows in place minus the predicted single glazing consumption in the test house. From the points on Figure 3 the mean D is:

$$\bar{D} = -38.2 \text{ MJ/day.}$$

The variance about $\bar{D}$ plus that about the calibration regression leads to a standard deviation for D of $\sigma_D = 28.8$. So that for the 30 points on Figure 3 $\sigma_{\bar{D}} = \sigma_{\bar{D}}/\sqrt{30} = 5.26$ and the Student's t-value is $\bar{D}/\sigma_{\bar{D}} = 7.3$. This, as expected from an inspection of the regression lines, shows that the mean difference in consumption in going from single to double windows is highly significant in these experiments. The 95% confidence limits on $\bar{D}$ are -27.4 and -49.0 MJ/day.

The only noticeable effect of the double windows on house temperatures is to raise the night-time minimum internal temperature. This effect is in fact quite variable and 1°C is the maximum elevation observed. Apart from solar over-heating the room-stat controls the day-time internal temperature adequately in both situations.

COMPARISON OF THE RESULTS AGAINST STEADY STATE CALCULATIONS

The individual daily measurements for consumption, volume averaged internal temperatures, external air temperature, solar radiation and wind speed are shown in Table II. The whole season averages have been used for the comparison with the steady state calculations.

The consumption G_S, for space heating has to be derived from the total consumption G_{TOT}, so that

$$G_S = (G_{TOT} - 33.9/E_{hw})E_s \qquad \text{MJ/day}$$

where 33.9 MJ is the controlled daily energy content of the hot water run-off, E_{hw} is the hot water production system efficiency and E_s the space heating system efficiency. These efficiencies were not measured directly. As a starting point E_{hw} will be taken as equal to E_s. This will certainly not be the case if the hot water were considered separately but if space heating is being provided at the same time and the heat losses from the hot water circuits can be regarded as useful E_{hw} will approach E_s (see later comments). Thus

$$G_S \simeq G_{TOT} E_s - 33.9 \qquad \text{MJ/day}$$

If the consumption G_{SP}, is that predicted from a calculation based on steady state heat balances then

$$G_{SP} = 0.0864(R)\Delta T - I - S \qquad \text{MJ/day}$$

where the heat loss per °C (R) is given by:

$$R = \underbrace{\Sigma UA}_{\substack{\text{walls}\\ \text{roof}\\ \text{floor}}} + \underbrace{\Sigma UA}_{\substack{\text{uncurtained}\\ \text{windows}}} + \underbrace{\Sigma UA}_{\substack{\text{curtained}\\ \text{windows}}} + 0.34\ N.V \qquad W/°C$$

and U = thermal transmittances, $W/m^2 °C$
A = associated area, m^2
V = internal air volume, m^3
N = whole house air changes per hour
I = 68.3 = internal heat gains, MJ/day
S = solar heat gains, MJ/day
ΔT = internal-external temperature, °C

The value of R for a house is

$$R = 194.55 + U_{nc}x5.585 + U_c x12.65 + 91.8xN \qquad W/°C$$

where U_{nc} and U_c are the uncurtained and curtained U-values for the windows respectively.

For single glazing in the reference house, taking whole season averages, the solar gain S_S = 56.63 MJ/day, and from Table II ΔT = 19.89 - 6.93 = 12.95°C and G_{TOT} = 299.2 MJ/day. The space heating decimal efficiency for this case, E , is therefore

$$E_{s1} = 0.42 + 0.021xU_{nc} + 0.047xU_c + 0.34xN$$

The second case to be considered is for single glazing in the test house. For this G_{TOT} has to be predicted from the total consumption in the reference house and use made of the calibration regression. G_{TOT} then = 317.45 MJ/day. Letting the space heating efficiency here be E_{s2} gives

$$E_{s2} = 0.040 + 0.020xU_{nc} + 0.045xU_c + 0.32xN$$

not vastly different from the single glazing case in the reference house.

Lastly, the third case is for double windows in the test house. Here the solar gain S_D = 47.72 MJ/day, ΔT = 20.10-6.93 = 13.17°C and G_{TOT} = 280.9 MJ/day. Thus for E_{s3}

$$E_{s3} = 0.50 + 0.023xU_{nc} + 0.051xU_c + 0.37xN$$

where U_{nc} and U_c will now be for double windows and N will be an appropriately lesser value than for the other two cases.

A simple product-addition solution diagram for these equations is set out on Figure 5. From Reference (4) the lowest U-values that could be selected for windows are

U_{nc} single glazing = 4.3

U_c single glazing = 3.2

U_{nc} double windows = 2.5

U_c double windows = 2.1

Using these values and an N of 0.6 leads to E_{s1} = 0.86, E_{s2} = 0.82 and E_{s3} = 0.89. In general, winter time average heating system efficiencies of the order of 0.70 would be expected(5) so that these solutions are between 0.10 and 0.20 too high.

Another way of looking at the results is to take a more reasonable average system efficiency of say 0.70 and N of 0.6 and compare the predicted G_{TOT} against the actual. This produces

Case	Actual G_{TOT}	Predicted G_{TOT}	% over
1	299.2	372.1	24
2	317.45	372.1	17
3	280.9	354.4	26

noting that in Case 2 the actual G_{TOT} is that derived from the calibration regression equation.

If for the double window case N is taken as say 0.4 then E_{s3} becomes 0.81 and the percentage over-prediction reduces from 26 to 16% for a system efficiency of 0.70. Although a reduction in N will occur it is unlikely to be enough to redress the balance. It is interesting to note that for this latter calculation the predicted G_{TOT} is about 325 MJ/day which is nearly 13% less than the prediction for single glazing and, in percentage terms, nearly the same as that actually measured.

DISCUSSION

Since these particular experimental houses have had to be vacated it is not possible now to go back and discover the reasons for the over-estimates in system efficiency.

For the type of heating appliance used, laboratory tests(2) have shown that the steady state 'flue' efficiency can be of the order of 0.77. Thus for winter load patterns a mean efficiency of about 0.70 would not be unreasonable but values in excess of 0.80 are quite out of the question. Other reasons have to be looked for.

The solutions are quite sensitive to the internal, external temperature differences but errors are expected to be less than 0.1°C since, for the internal temperature an average of 8 readings is taken throughout the season. Also the external temperature is double checked and any random errors are virtually eliminated by the averaging. Alternative ways of averaging the internal temperature, such as weighting according to UA, produce negligible differences because the temperatures in the rooms are of similar magnitude. Errors in gas consumption should also be negligible since the meters are calibrated and checked against a specially certified 'wet' meter. All practical checks suggest that the causes for the over estimates in efficiency must lie elsewhere than with the measurements themselves. In the following discussion it is useful to remember that any assumption which increases the calculated energy consumption worsens the comparison with the actual consumption.

Two assumptions in the solution are that the hot water system efficiency E_{hw} is the same as for the space heating and that 50% of the energy content of the hot water run-off is useful. For the former, it can only be lower, and would worsen the situation. This is not to say it cannot be lower but other factors would have to more than compensate. For the internal gain from the hot water run-off an input to the house of about 17 MJ is taken. The 180 L used each day was fed in at about 55°C and allowed to drain away for between ½ and 1 hour, with an attendant drop in temperature of between 10 and 15°C for the bath and rather more for the other run-offs. If anything therefore the gains from the hot water were liable to be less than 17 MJ/day which would worsen the situation.

It has been observed that solarimeters can under-read by as much as 10% in some cases. Since the manufacturers calibration was taken in these experiments there is a chance that this could have occurred. However, increasing the solar gains by 10% only reduces the system efficiencies by about 2%.

An accurate air volume in the houses is not easily determined and the selection of 270 m^3 is a compromise. It might be argued that the internal floor's volume or some part of it should be subtracted so that the air volume could be reduced to about 260 m^3. Air infiltration losses would thereby be lowered from 91.8 N to 88.4 N but would only reduce the predicted system efficiencies by 0.01.

With regard to the U-values for windows and the magnitudes of N the lowest acceptable figures have been assumed.

Though this discussion is necessarily somewhat speculative the remaining candidates for scrutiny are the U-values of the roof, wall and floor. If overall such U-values were reduced by 10% this would lead to a reduction in the predicted system efficiencies of about 7% and very nearly bring the solutions into balance. What this suggests is that a programme of in-situ U-measurements will need to be included in future trials. If such a reduction in U-values could be shown to be necessary in general then this would have a pronounced effect on energy consumption calculation procedures.

CONCLUSIONS

These results have demonstrated a mean reduction in energy consumption of 12% (±3%, 95% confidence limits) for secondary double windows in these houses. The calculated percentage reduction agrees well with that measured provided a lowered air change rate is taken when the double windows are in place. However, the calculation procedure overestimates the consumption by of the order of 20% and it is postulated that the U-values for the walls in these test houses should be lower than the standards suggest.

The measured energy saving is likely to represent the annual average for the house since comparison with data gathered on the same houses in other heating seasons shows that the present mean consumptions for single glazing are not untypical. Before these results can be extended to the existing stock of well-heated semi-detached houses the following poiints must be considered.

(a) It was not possible to arrange for air infiltration measurements to be made when the secondary windows were in place, so that the influence of infiltration on the results is not known. However, from a full set of infiltration measurements made in a previous year which included calibration against wind speed, it is known that the average air changes per hour for single glazing has a value of about 0.6; a magnitude which is currently thought to be about the right minimum for an acceptable indoor air quality. A reduction below this figure could give rise to other problems.

(b) The heat loss through the single glazing was about 25% of the total design heat loss and the test houses had 100 mm of loft insulation but no cavity-fill. Other work has shown that the quality and extent of insulation measures can affect the energy savings from subsequent improvements.

(c) For security reasons it was not possible to include window and door opening in the occupancy simulation.

Comparative trials are being arranged in another pair of houses to provide further information on which to base predictions of energy savings and thence calculations of cost-effectiveness, since double glazing is recognised as an important area of activity.

ACKNOWLEDGMENT

The work described has been carried out as part of the research programme of the Building Research Establishment of the Department of the Environment and this paper is published by permission of the Director.

REFERENCES

1 Rayment R and Morgan K. The performance of heating control systems in experimental houses. Proc of Int EEC Seminar on New Ways to Save Energy, Brussels, 1979.

2 Rayment R, Cunliffe A and Morgan K. Basic characteristics of low cost houses in order to reduce the energy consumption for heating: Efficiency of heating plants and controls. EEC Contract (No 363-78-1), Final Report, 1981. To be published.

3 Rayment R and Morgan K. Comparing conventional and electronic room-stat performance in experimental houses. CIB S-17 Heating and Climitisation Conf, Delft, 1981.

4 Uglow C E. The calculation of energy use in dwellings. BSERT, Vol 2, No 1, 1981.

5 Bloomfield D P and Fisk D J. Seasonal domestic boiler efficiencies and intermittent heating. The Heating and Ventilating Engineer, September 1977.

QUESTIONS AND DISCUSSION

Jacqman (BSRIA) : Did you correlate the results with other parameters than outdoor temperatures (e.g. wind) ?

Answer : No.

Uyttenbroeck (Chairman) : In the BRE-programme the effect of double windows was measured by comparing the gas consumption in a house with single and with double windows. The risk of this method is that between these two extreme points there are a lot of not well known parameters. A simplier way could be the measurement of the temperature without heating (effect of free energy) in a house with single glazing and in another house with double glazing, and to compare the measured values.

Uyttenbroeck (Chairman) : Do you have information about the validity of ping-pong balls to measure the dry resultant temperature ?

Answer : Mr Humphreys has made a study on this problem and published a paper entitled : "The optimum diameter for a globe thermometer for use" (BRE - Current Paper November 1977).

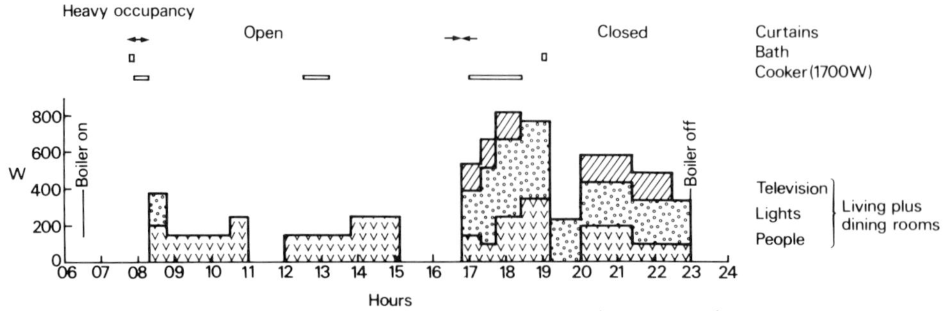

Figure 1 Selected occupancy gains and times

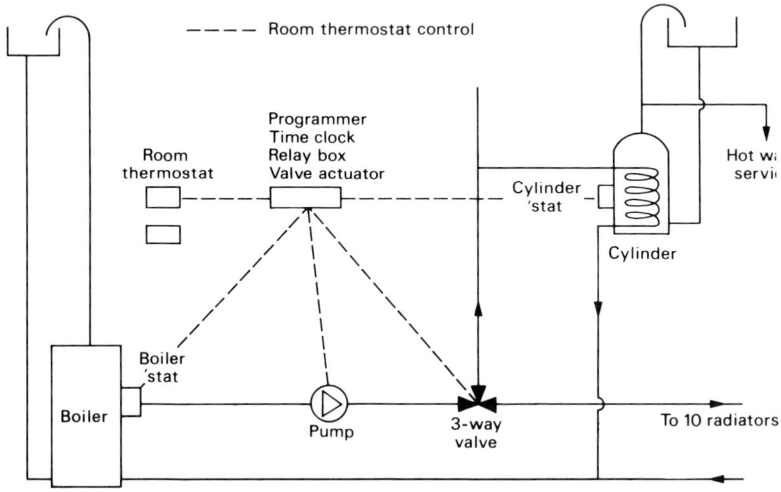

Figure 2 Heating system and controls

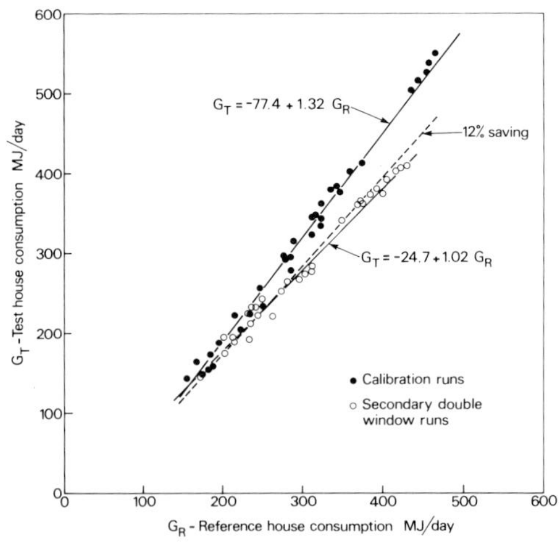

Figure 3 Daily consumptions during calibration and secondary double glazing experiments

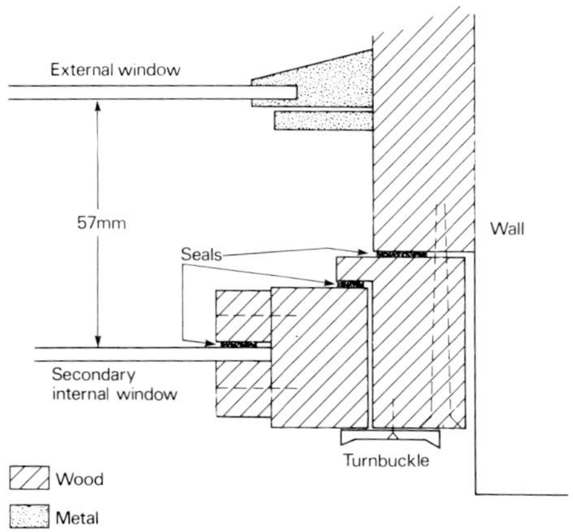

Figure 4 Schematic diagram of the construction and fixing of the secondary double window

Heat source	Heat gains/day MJ
People	19.7
Lights	9.0
Electric cooker)	
Kettle)	15.3
Washing machine)	
Refrigerator	5.2
TV (B/W)	2.2
Hot water*	16.9
Total	68.3

* Hot water gain taken as 50% of energy content of run-off

TABLE I Incidental heat gains simulated in the experimental houses

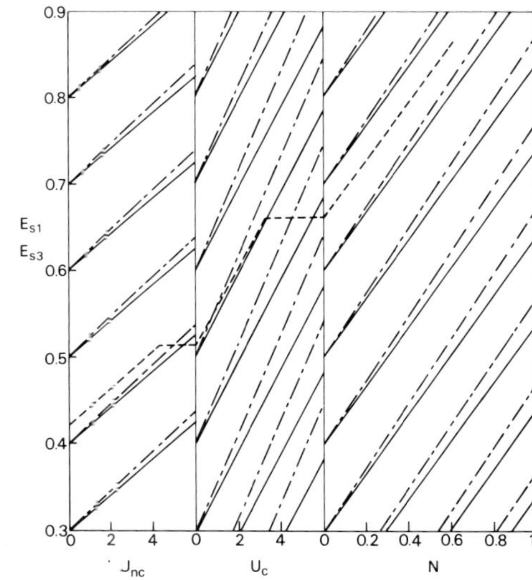

Figure 5 Solution diagram for the system efficiency equations E_{S1} [single glazing] and E_{S3} [double windows] given in the main text

——————— E_{S1}

- — - — E_{S3}

- - - - - Solution for E_{S1} using the U_{nc}, U_c and N values selected in the main text

TABLE II SECONDARY WINDOW TESTS

Date (Group)	G_T	G_R	T_T	T_R	T_X	W	SSE	SNW	S_D	S_S
17/10/80	289.7	325.7	19.62	19.32	7.0	3.3	0.38	0.54		
(I)	244.0	279.6	19.71	19.25	5.1	2.5	12.76	3.67		
	305.7	335.5	19.79	19.44	7.0	4.4	1.89	1.45		
	258.1	273.2	20.51	20.29	9.8	5.6	1.68	1.44		
	190.1	203.5	20.79	20.59	11.0	-	7.15	2.55		
22/10/80	188.3	220.7	21.05	20.96	13.4	-	0.84	0.82		
20/10/80	322.0	304.6	20.55	19.54	8.8	2.3	11.56	1.38		
(II)	250.2	291.4	20.42	19.92	8.4	2.5	1.67	1.22		
	250.8	318.0	20.64	20.49	11.0	3.9	0.72	0.72		
	163.3	176.9	21.66	21.64	15.1	5.6	1.33	1.42		
	167.9	201.5	21.64	21.61	12.8	3.9	0.16	0.24		
29/10/80	186.3	227.2	21.08	20.75	8.8	3.3	8.73	2.60		
10/01/81	411.0	437.6	18.95	18.49	1.4	4.6	3.80	0.98		
(III)	404.9	414.0	18.80	18.45	0.4	2.0	3.75	1.16		
	414.0	443.5	18.57	18.14	1.6	3.7	0.45	0.60		
	407.7	428.0	18.19	17.66	-0.0	3.3	7.42	1.29		
14/01/81	352.3	351.3	18.66	18.42	6.9	6.7	0.18	0.28		
16/01/81	431.0	446.5	18.34	18.00	2.3	3.3	0.27	0.31		
(IV)	379.8	396.6	18.55	18.30	4.6	5.2	2.85	0.83		
	400.8	416.1	18.98	18.63	3.0	2.5	1.04	0.71		
	337.9	350.0	19.09	18.97	5.4	3.7	3.54	1.22		
	350.4	355.5	19.31	19.21	3.3	1.6	4.12	1.60		
21/01/81	339.2	343.9	19.57	19.37	7.0	2.5	0.59	0.60		
13/02/81	342.1	384.8	19.10	18.66	1.0	3.2	11.88	2.72		
(V)	373.2	399.7	18.76	18.42	-2.0	0.6	3.56	2.47		
	410.0	418.2	18.78	18.40	-1.4	0.6	3.67	2.68		
	358.7	365.1	19.05	18.66	0.7	1.4	7.76	2.09		
	319.8	331.2	19.48	19.07	1.7	2.4	10.45	2.61		
18/02/81	413.8	417.9	18.76	18.76	1.1	3.1	1.09	1.99		
07/03/81	253.0	262.1	20.16	20.18	10.1	5.3	0.54	0.60		
(VI)	221.9	236.9	20.32	20.35	10.6	5.6	2.17	2.29		
	247.5	250.9	20.50	20.51	10.0	3.9	0.79	0.84		
	224.9	235.8	20.54	20.61	10.5	4.1	0.80	0.72		
	224.7	231.7	20.27	20.42	10.6	6.1	1.12	1.35		
12/03/81	224.7	230.6	20.17	20.33	8.6	2.4	3.59	2.27		
04/04/81	294.4	332.3	19.99	20.10	5.8	3.2	1.30	1.31		
(VII)	304.7	327.8	19.97	19.96	5.9	2.8	2.11	2.03		
	244.0	271.2	20.07	20.12	6.6	4.1	8.02	3.09		
	238.8	247.0	20.36	20.32	7.2	1.2	7.16	4.19		
	197.4	211.4	21.14	21.14	11.4	1.9	7.17	4.29		
09/04/81	147.0	149.0	22.59	22.59	13.6	1.3	11.33	7.21		
11/04/81	109.2	120.6	22.93	22.79	14.4	2.4	6.04	4.62		
(VIII)	139.6	161.9	22.07	22.05	11.2	2.2	4.87	7.45		
	196.3	228.8	21.53	21.33	9.1	4.6	3.62	3.60		
	234.9	262.0	21.02	20.86	8.7	4.5	3.46	8.72		
15/04/81	154.8	146.9	22.18	21.79	9.3	4.6	16.41	9.09		
Averages	280.9	299.2	20.10	19.89	6.93	3.4	4.26	2.30	47.72	56.63

Regression based on 3 day overlapping averages from each group:

$$G_T = -24.7 + 1.0186\ G_R \pm 9.6 \quad \text{(Correlation 0.9933)}$$

±9.6 is the standard deviation about the regression line

G = gas consumption MJ/day
T = temperature °C
W = wind speed m/s
SSE = solar radiation incident on south-east vertical plane MJ/day/m^2
SNW = solar radiation incident on north-west vertical plane MJ/day/m^2
S = solar gain MJ/day

Subscripts:

T = test house
R = reference house
X = external
D = double windows
S = single glazing

PRELIMINARY PHASES IN THE STUDY OF AIR DISTRIBUTION METHODS FOR DOMESTIC WARM-AIR HEATING SYSTEMS USING LOW-GRADE HEAT SOURCES

Author	:	H. B. Awbi
Contract number	.	EEA-1-010-GB
Duration	:	27 months 1 August 1981 - 31 October 1983
Head of project	:	P.J. JACKMAN , B.S.R.I.A.
Contractor	:	Building Services Research and Information Association
Address	:	Old Bracknell Lane West GB-BRACKNELL, Berks RG12 4AH

Summary

In this paper, a comfort criterion for sedentary persons in dwellings is specified. This will form the basis for assessing the significance of the parameters to be measured at later stages of a project concerning the study of air diffusion in domestic warm air heating systems that use low grade heat sources. The basic factors influencing the diffusion of air jets are described and the suitability of low level and high level air terminal devices for the present investigation are discussed. A brief description is also given of the test facility which was specifically constructed for the air movement studies that are to be conducted at later phases of the project.

NOTATION

A_e	effective area of jet outlet, m^2
Ar	Archimedes number $= \Delta\Theta_o \; \beta \; g \sqrt{A_e}/V_o^2$
b	effective slot width of linear diffuser, m
F	throw correction factor
g	gravitational acceleration, m/s^2
K	throw constant for a diffuser
l	distance of centre of jet from a surface, m
T	throw to 0.35 m/s, m
t_{res}	dry resultant air temperature, ^oC
Q	volume of air flow rate, m^3/s
U	thermal transmittance coefficient, W/m^2K
V_o	initial velocity of air jet, m/s
V_x	centreline velocity of air jet at a distance x, m/s
x	horizontal distance along jet axis measured from the air supply terminal, m
y	vertical distance of jet axis measured from the centre of air supply terminal, m
β	thermal expansion coefficient of air, K^{-1}
$\Delta\Theta$	temperature difference at a point in the jet $= \Theta-\Theta_a$, ^oC
$\Delta\Theta_o$	initial temperature difference of jet $= \Theta_o-\Theta_a$, ^oC
Θ	temperature at a point in the jet, ^oC
Θ_a	room air temperature, ^oC
Θ_o	initial jet temperature, ^oC

1. INTRODUCTION

Most existing domestic warm air heating systems use a high temperature heat source that supplies an intermittent or variable volume flow rate of air at relatively high temperature. Energy from alternative sources, such as solar, geothermal, heat pumps and heat recovery systems, is often available at more variable and lower temperatures. These alternative low grade heat sources could provide the basic heating energy requirement for low energy houses. A warm air heating system is probably best suited for distributing the heat from such low energy sources. However, to maximise their use it is necessary to know the limiting conditions of supply air flow rate and temperature beyond which it is no longer possible to maintain comfortable conditions within the living space. With this objective in mind, this project was commissioned by the Commission of the European Communities and the UK Department of Environment for BSRIA to undertake the investigation.

In this paper an account is given of the preliminary phases of this investigation. The parameters influencing the human thermal comfort are discussed and a comfort criterion suitable for domestic warm air heating application was derived. Some basic air jet diffusion data are presented graphically to show the significance of the various jet parameters. The test room which has been constructed specifically for this work is also described here.

2. TEST ROOM AND INSTRUMENTATION

A test room of 5.5m x 3.5m x 2.5m high has been constructed for conducting the air movement studies and is shown in Fig. 1 as a schematic diagram. The living room represents a critical part of the house with respect to thermal comfort because occupants may remain seated there for several hours at a time. Therefore, the test room was designed to simulate a typically sized living room subjected to a winter thermal environment (external air temperature of -1°C).

Insulation levels are expected to progressively increase above those stipulated in current national standards as the requirement for energy saving intensifies. In any case, a house using other than conventional heat sources is expected to have a high standard of thermal insulation. In view of this, the air distribution test room has been designed on the basis of higher than average thermal insulation in the exterior walls and ground floor for dwellings in the United Kingdom. The following thermal transmittance coefficients (U-values) have been chosen for the room:

exposed walls, U=0.43 W/m^{2}K
floor, U=0.57 W/m^{2}K
window (1.75m x 1m), U=2.9 W/m^{2}K

A comparison with the U-values specified by other European standards shows that these U-values represent well insulated dwellings in other European countries with temperate climate. Neglecting heat losses through the internal walls and ceiling and the heat generated by the occupants, the heating load for the room under design conditions is 624 W. Figure 2 shows the variation of supply air temperature for the room with volume flow rate for this heating load.

The thermally insulated test room has been constructed of timber with two external walls, floor and a window having the above U-values. The external conditions have been simulated by circulating cold air in a corridor outside the two exposed walls and around a void below the floor to achieve the same heat losses from the walls and floor as in an actual room.

Omnidirectional velocity transducers with spherical sensing heads will be used for measuring the room air velocity. The TSI model 1620 low velocity anemometer will be used to measure the time-mean air velocity and the DISA model 56C15, which has a much faster response, will be used where the measurement of turbulent intensity is required. Copper-constantan thermocouples will be used for measuring the room air temperature and also the dry resultant temperature by placing the thermocouple junction at the centre of a hollow black metallic sphere. The velocity and temperature sensors will be carried across the test room on an x-y traversing system that will be controlled by an Apple II microcomputer. The computer will also be used for the processing and collation of the experimental data. Further details concerning the test room and instrumentation are given in Reference (1).

3. PROGRESS OF WORK

3.1 Thermal Comfort

The first requirement for this investigation was to establish a thermal comfort criterion which will be suitable for domestic warm air heating application. For this purpose comfort data provided by the major building services guides and up to date published data were analysed in the light of the present requirements. As a result, a comfort criterion for sedentary persons in dwellings was specified and it is discussed in Section 4 of this paper.

3.2 Air Diffusion

The purpose of an air distribution system is to provide environmental conditions to a space that are acceptable to the occupants of that space. At present there are no simple means of evaluating the performance of an air diffusion device and for absolute evaluation of the diffusion quality of the device one has to resort to laboratory or site measurements of air velocity and temperature in the occupied space. An effort was made by Nevins and his co-workers (2) to provide a single number rating system for evaluating air distribution devices by proposing the Air Diffusion Performance Index (ADPI). This index was found to be a function of the type of air terminal device, the room heating/cooling load, the supply air flow rate and in some cases room geometry. To determine the value of ADPI, it is necessary to evaluate the distribution of air velocity and temperature in at least one plane in the room that includes the air supply device. One of the disadvantages of using ADPI to evaluate an air terminal device is its indiscrimination of the positions of discomfort in the occupied space.

Section 4 of this paper gives logarithmic plots of the basic jet diffusion equations which will be used as a guide for predicting the performance of the various diffusers to be employed in the test programme. The significance of these plots is explained in the same section.

3.3 Test Room Construction

The test room described in Section 2 has been constructed of timber for the basic structure and slotted steel sections for supporting the suspended floor. Wood-chip boards have been used for the inner and outer cladding of the walls, floor and ceiling. Isocyanurate foam

(Nilflam) has been generally used for insulating regions with large temperature differences and glass wool for other regions. Description of the air heating and cooling plants to be used is given in Reference (1).

4. ANALYSIS OF EXISTING DATA

4.1 Thermal Comfort Data

The quality of a controlled thermal environment of a space is assessed in terms of the degree of thermal comfort of its occupants. It has been shown in Reference (3) that in order to achieve human thermal comfort it is necessary to provide a condition of thermal neutrality as well as thermal uniformity in the controlled space. The condition of thermal neutrality is determined by the interaction of four physical variables: air temperature, mean radiant temperature, water vapour pressure and relative air velocity and two factors associated with the human occupants, namely the activity (metabolic rate) and the thermal resistance of the clothing. These six parameters are best represented by a single comfort equation due to Fanger (4) which is based on the thermal equilibrium between the body and the environment for a given activity level. Fanger obtained computer solutions for the thermal comfort equation and represented the results as general comfort charts. Figure 3 shows a selection of these charts, reproduced from Reference (4), for sedentary and medium activity. The shaded areas represent regions of practical significance.

Indoor design temperatures, as specified by building services codes and standards, provide conditions which satisfy the thermal neutrality of the occupants of a given space similar to those conditions predicted by the thermal comfort equation. However, it has been shown by several investigations (3) that thermal neutrality is not a sufficient condition and to provide thermal comfort it is also necessary to achieve thermal uniformity of the controlled space. To provide spatial uniformity of the thermal variables within the occupied space, it would be required to minimise the following:

(a) asymmetric radiant fields caused by hot or cold surfaces,
(b) local draughts created by natural convections or exessive air circulation,
(c) vertical air temperature gradients that may be caused by hot or cold surfaces or the mispositioning of air terminal devices.

Of particular interest to this investigation is the influence of air motion on thermal comfort. In order to meet the required heating load, it is necessary when using low supply air temperatures, to introduce and diffuse large quantitites of air into the occupied space. For this to be performed effectively without creating the conditions mentioned in (b) and (c) above, rigorous tests with various types and different locations of air terminal devices will be necessary to determine the limits of both room air flow rate and temperature differential in relation to selected values of the four thermal comfort variables. Building services codes normally recommend an elevation of the room air temperature to compensate for the cooling effect of high air velocities. Fig. 4 reproduced from the CIBS Guide (5) gives corrections to the dry resultant air temperature, t_{res}, as the air velocity increases above 0.1 m/s. From energy considerations, on the other hand, increasing the room air temperature means higher heat losses to the outside air. The ideal situation therefore would be to diffuse large quantities of air at a temperature only marginally above room air temperature in such a way that air velocities in the occupied zone would still be less than 0.1 m/s. This would obviously require reassessment of the current practice of diffusing the warm air in domestic heating systems in order to find new methods of diffusion without creating regions of large air velocities in the occupied zone.

After analysing the comfort data provided by the major building services guides (5-7) as well as the information available in the comfort literature (see Reference 3), the following comfort criterion is considered to be most suitable for the warm air heated domestic environment:

(i) The dry resultant air temperature (t_{res}) in the occupied zone of a living room should be 21°C with spatial variation not exceeding ± 1.5^{o}C for an outdoor temperature of -1°C.

(ii) Relative humidity should be within the range 40 to 70%.

(iii) The time-mean air velocity at any location within the occupied zone should not exceed 0.35 m/s but for V<0.1 m/s correction to the resultant air temperature should be made according to Figure 4.

(iv) The radiant temperature gradient within the room should not exceed 10°C.

4.2 Air Diffusion Data

4.2.1 Isothermal Jets

The velocity decay of an isothermal air jet can be estimated using one of the following equations (8):-

for a three-dimensional jet with $x/\sqrt{A_e} > 6.5$,

$$V_x/V_o = K \sqrt{A_e}/x \qquad \ldots \quad (1)$$

for a two-dimensional jet with $x/b > 6.5$, $V_x/V_o = \sqrt{K\, b/x}$ ... (2)

The value of the throw constant, K, depends on the type of opening and it varies from about 1.1 for a radial discharge ceiling diffuser to 7.9 for a circular side wall opening (6, 8, 9). Figure 5 is a logarithmic plot of V_x/V_o against $x/\sqrt{A_e}$ (or $\sqrt{x/b}$) for different values of K (or $\sqrt{K}$). It is evident from this figure that for a more rapid diffusion of the air jet into the room air, a diffuser with a low value of K would be required..

The effect of a surface on the propogation of a jet near it causes the jet to attach to the surface and entrainment of room air to occur on one side only, thus reducing the jet velocity on the entrainment side whilst on the surface side the jet velocity decreases less rapidly and this causes the maximum jet velocity to occur close to the surface. The effect of an adjacent surface on the diffusion of air jets was studied by Farquharson (8). He found that the proximity of a jet to a surface reduces the rate of diffusion, i.e. K increases. Figure 6 gives the correction factor, F, by which K should be multiplied to allow for the surface effect.

The volume flow rate through a diffuser is:

$$Q = V_o A_e = x\, V_x \sqrt{A_e} / K \qquad \ldots \quad (3)$$

By defining the throw, T, of a jet as the distance from the air supply terminal to a point along the jet axis where V_x = 0.35 m/s, the throw of a diffuser with known K value and effective area, may be calculated for different air flow rates. This value of V_x represents maximum air velocity in the occupied zone as stipulated by the thermal comfort criterion given in the previous section.

Low Level Air Terminal Devices

Figure 7 shows logarithmic plots of T vs Q for various A_e values of a typical low level side wall grille used in domestic warm air systems (K = 2.6). Assuming the boundary of the occupied zone to be at 0.15 m from the walls, it is evident from Figure 7 that a comfortable air velocity of less than 0.35 m/s throughout the occupied zone can only be achieved using very large grilles at very low air flow rates. A reference to Figure 2 shows that the air flow rates concerned are too low to meet the fresh air requirements for four occupants of the room under investigation. It may be concluded therefore that introducing air jets directly into the occupied zone will create regions of discomfort in the room and if low level diffusers are to be used these should be capable of diffusing the air along the walls but not directly into the occupied zone.

High Level Air Terminal Devices

For high level diffusers located outside the occupied zone and following the ASHRAE (6) recommendations by taking the throw as the distance between the air terminal and the nearest wall opposite, three throw values may be defined for the present test room:

for a centrally located ceiling diffuser, T = 1.75 m
for a wall diffuser, T = 3.5 m or 5.5 m

Figure 8 depicts logarithmic plots of equation (3) for these throw values. Examination of this figure shows that there is a range of diffuser sizes and throw constants that can be selected to meet the required flow rate and which are expected to produce comfortable velocities in the occupied zone.

4.2.2. Non-Isothermal Jets

In addition to momentum diffusion, non-isothermal jets also have thermal diffusion properties due to buoyancy forces and convective currents in the room. The decay of the difference between the jet air temperature and room air temperature can be expressed in a similar form to the velocity decay low (equation 1). Frean and Billington (9) found the following equation for a warm air jet:

$$\frac{\Delta\Theta}{\Delta\Theta_0} = 8.1 \left(\frac{\sqrt{A_e}}{x}\right)^{1.25} \qquad \ldots (4)$$

For the trajectory of a heated jet, they found the equation given below to correlate their data with other published results:

$$\frac{y}{\sqrt{A_e}} = 0.238\ Ar \left(\frac{x}{\sqrt{A_e}}\right)^{2.5} \qquad \ldots (5)$$

where $Ar = \dfrac{\Delta\Theta_0\ \beta\ g\ \sqrt{A_e}}{V_0^2}$ = Archimedes number

The role of the Archimedes number in the jet diffusion depends on the location of the air terminal device in the room. The general trend is for the room air velocity to increase with increasing Ar for low level diffusers and to decrease for high level diffusers.

The low level grilles that are commonly used in domestic warm air heating systems normally diffuse a small quantity of air at a relative high temperature (50 to 70°C) directly into the occupied zone.

This warm low velocity jet tends to rise rapidly after leaving the grille thus restricting the regions of thermal discomfort to the vicinity of the grille while still maintaining reasonably comfortable conditions in other parts of the room. This method of air diffusion would not be applicable for low supply air temperatures.

5. CONCLUSIONS

In this paper a brief outline has been given of the thermal comfort requirements for a domestic warm air heating system and then a suitable comfort criterion has been suggested. The primary factors controlling the human thermal sensation in this case were found to be the dry resultant temperature, the room air velocity and the air temperature gradient within the occupied zone. A range of values for these parameters, which are considered to be tolerable by most people, have been suggested and these will form the outer boundaries for evaluating the results from the air movement studies.

A consideration was also given to the effect of the various parameters of the isothermal jet flow equations which influence the diffusion of an air jet in a space. The effect of a surface parallel to the jet axis on the diffusion of the jet has also been considered and published data show that the proximity of the surface reduces the jet diffusion and effectively increases the throw constant. The isothermal jet results also showed that conventional low level grilles which project the air directly into the occupied zone would not meet the comfort criterion that has been stipulated. Consequently, other types of air terminal devices that diffuse the air along the walls may be required if low level air terminal devices are to be used effectively.

ACKNOWLEDGEMENT

The author wishes to thank Mr. P. J. Jackman, Deputy Head of Technical Division, BSRIA for editing this paper.

REFERENCES

1. AWBI, H. B., Air distribution methods for domestic room air heating systems using low grade heat sources - Test rig design. BSRIA Progress Report No. 2 for C.E.C. Contract No. EEA-1-10-81 UK (H), February 1982.
2. NEVINS, R. G., Air diffusion dynamics - theory, design and application, Business News Publishing Company, U.S.A., 1974.
3. AWBI, H. B., Air distribution methods for domestic room air heating systems using low grade heat sources - Comfort criterion. BSRIA Progress Report No. 1 for C.E.C. Contract No. EEA-1-10-81 UK (H), November 1981.
4. FANGER, P. O., Thermal comfort, Danish Technical Press, Copenhagen, 1970.
5. CIBS Guide, Section A1, 1978.
6. ASHRAE Handbook and Product Directory, 1977 Fundamentals volume.
7. DIN 1946 Part 2, Room ventilation technique, technical health principles (VDI Ventilation Rules), June 1979.
8. FARQUHARSON, I. M. C., The ventilating air jet, J.I.H.V.E., Vol. 19, pp 449 - 469, 1951/52.
9. FREAN, D. H. and BILLINGTON, N. A., The ventilating air jet, J.I.H.V.E., Vol. 23, pp 313 - 344, 1955.

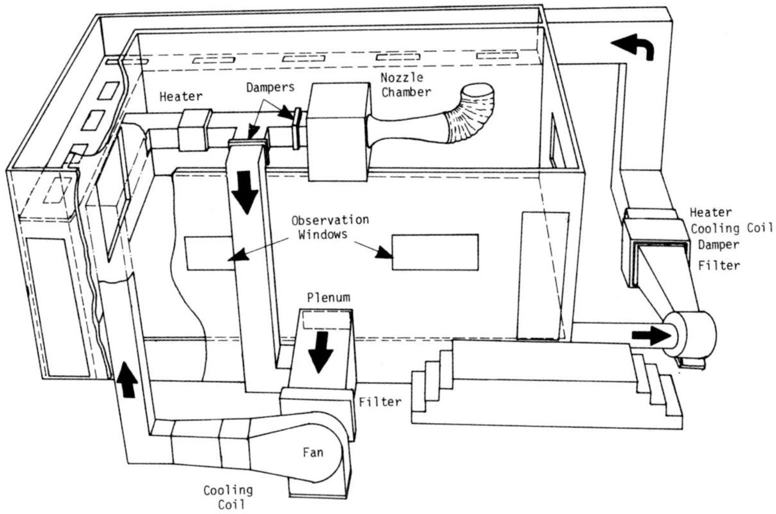

Fig. 1 Test room

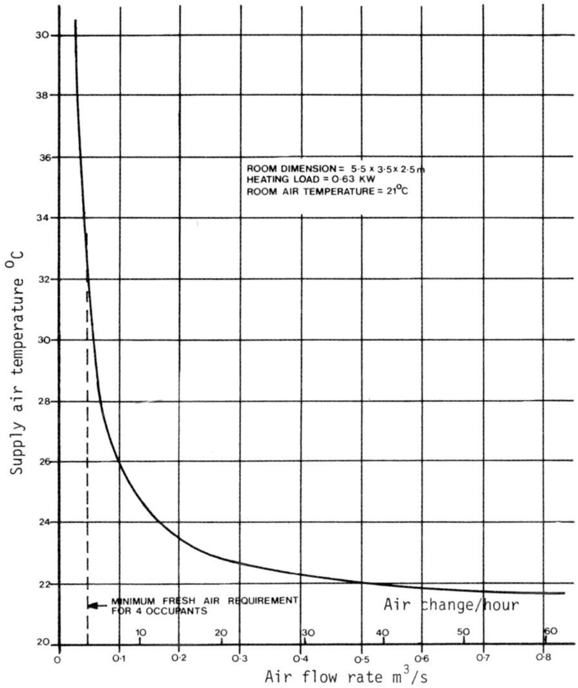

Fig. 2 Variation of supply air temperature with air flow rate for the test room

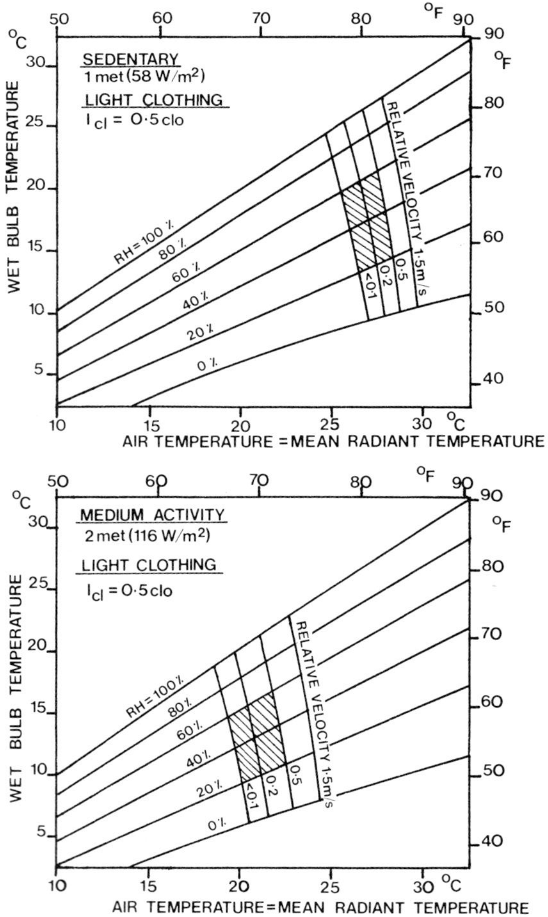

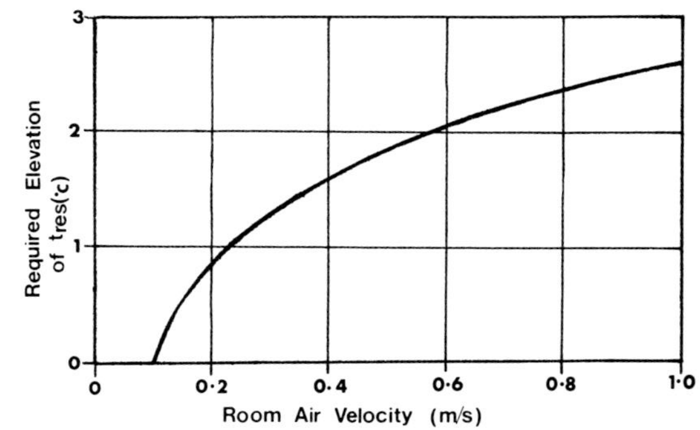

↑ Fig. 4 Corrections to the dry resultant temperature to take account of air movement (5)

← Fig. 3 Combined effect of air temperature, air velocity and humidity on thermal comfort (4)

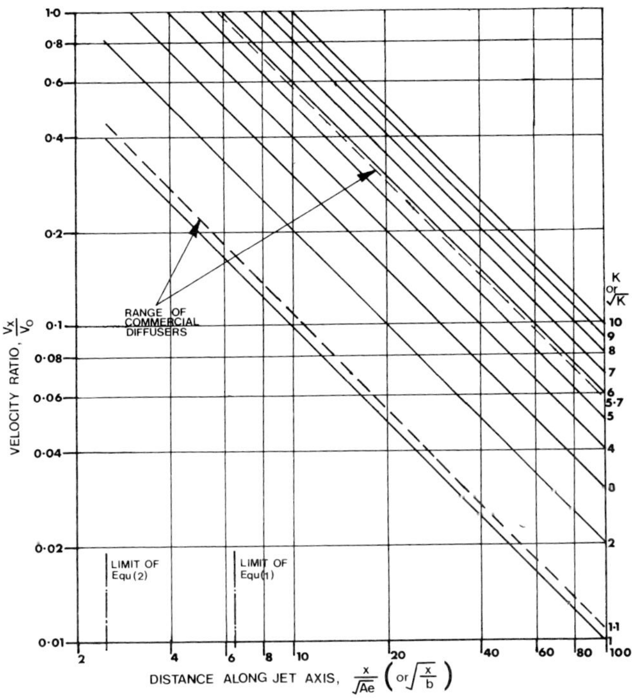

Fig. 5 Velocity decay for two- and three-dimentional jets

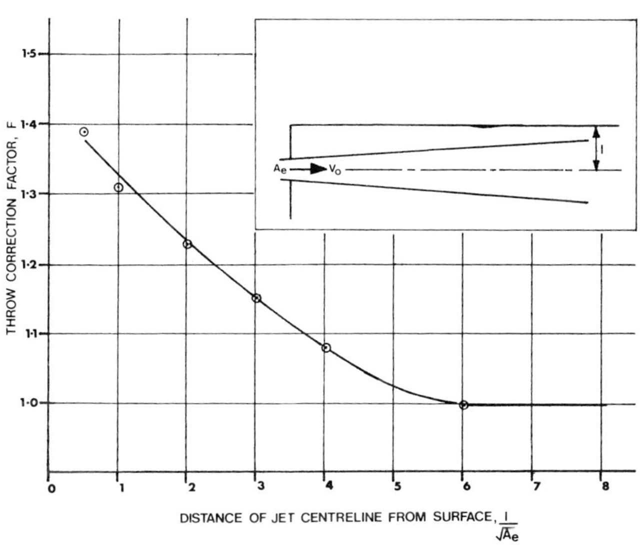

Fig. 6 Throw correction factor due to adjacent surface (8)

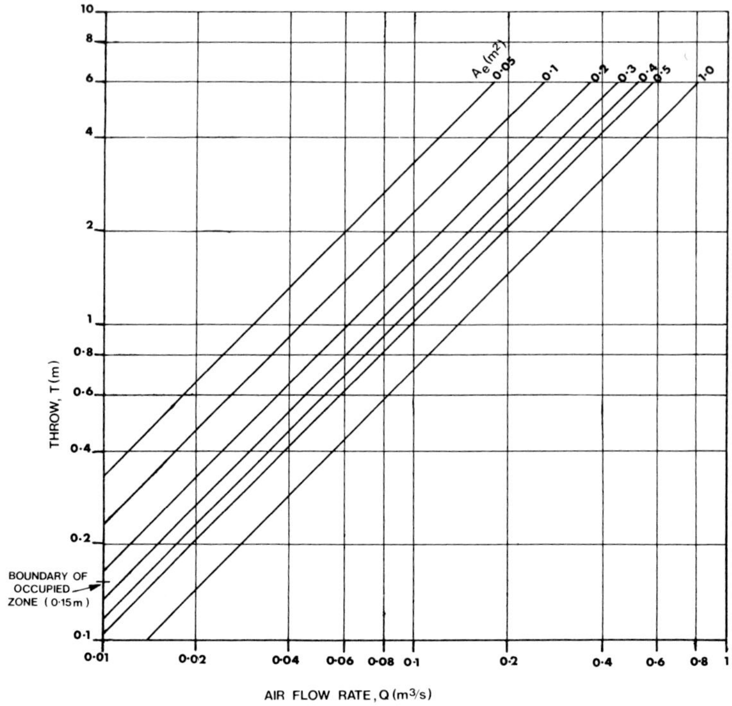

Fig. 7 Throw of a typical warm air grille (K = 2.6)

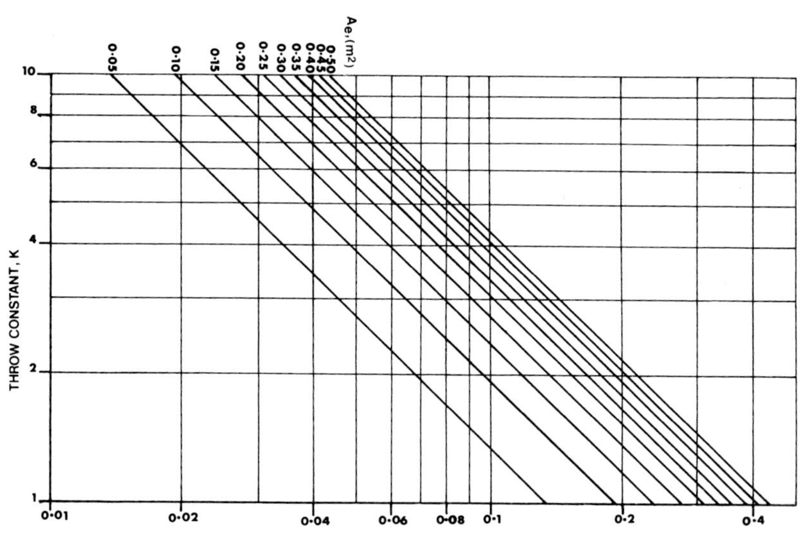

Fig. 8a Throw constant vs. air flow rate for T = 1.75 m

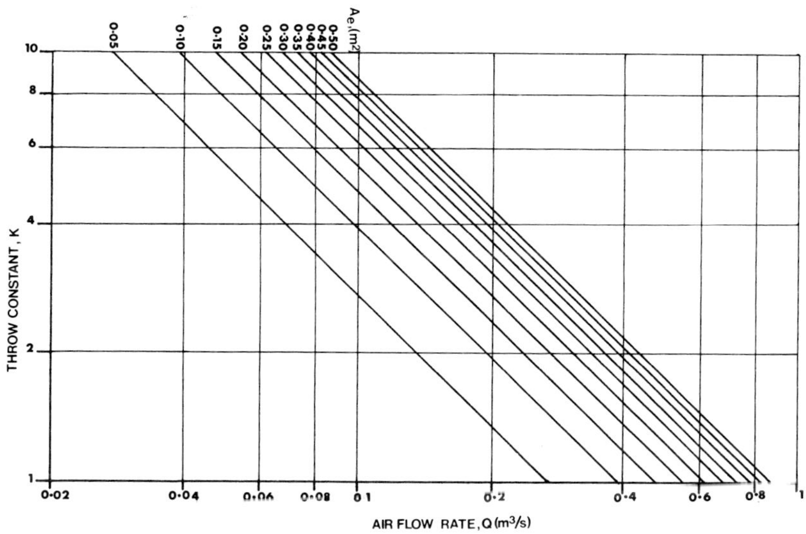

Fig. 8b Throw constant vs. air flow rate for T = 3.5 m

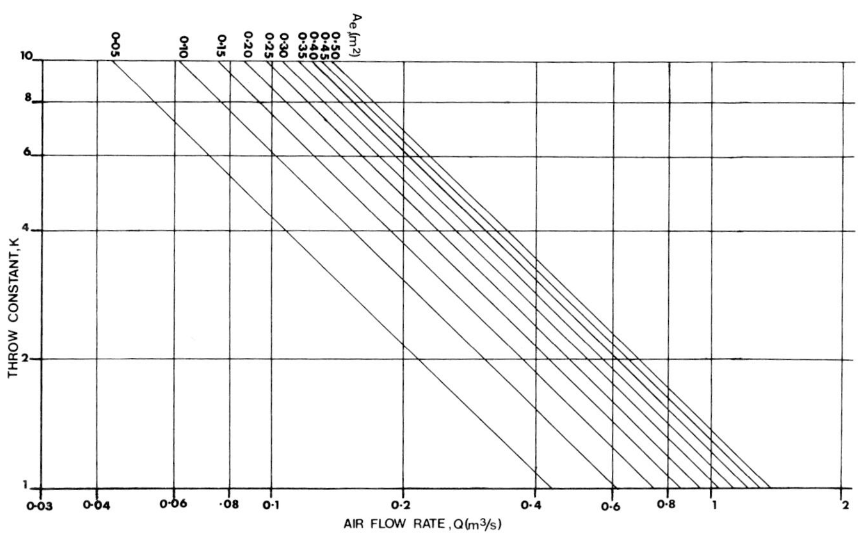

Fig. 8c Throw constant vs. air flow rate for T = 5.5 m

QUESTIONS AND DISCUSSION

Uyttenbroeck (Chairman) : You defined the confortzone as an area with vertical limits at a distance of 15 cm from radiator, wall or windows. These requirements are too severe.

Answer : The draft of ISO ITC 144/SCIN 89 concerning air distribution and movement in rooms indicates this distance.

Uyttenbroeck (Chairman) : In order to have comparable results we should have the same definitions for confort and confort zones.
Would it not better to have the limits of the confort zone at 50 cm from the wall, radiators and windows and 2 meter above the floor
The ISO/TC 144/SC is probably only concerned with air distribution and movement and not with heat sources such as radiators and so on.

Rayment (BRE) : Brit. Gas has taken confort measurements in two profiles far from the wall.

Spitzer (Battelle) : There is a DIN regulation which specifies at which height the temperatures in room should be measured. This DIN standard will be transmitted to Mr Uyttenbroeck.

The participants generally agreed that distances of about 50 cm are appropriate.

Zegers (DG XII) : Did you try to find the combination of temperature and airflow which requires minimum energy and still gives sufficient comfort ?

Answer : No.

Uyttenbroeck (Chairman) : High velocity air with the air flow injected above the occupation zone gives good comfort as the mixing is good; this however will give larger energy losses; in particular when the ceiling is an external wall and is badly insulated.

WATER/AIR CONVECTOR MADE WITH PLASTICS OF IMPROVED THERMAL CONDUCTIVITY

Authors : Pierre DUBOIS, Michel BERTHET and C. REINAUDO

Contract number : EE-A-A-003 - F (S.D)

Duration : 18 months

Head of project : Pierre DUBOIS

Contractor : Laboratoires de Marcoussis, Centre de Recherches de la Compagnie Générale d'Electricité

Address : Route de Nozay - 91460 Marcoussis (France)

Summary

The reported work has been conducted on water/air convectors for space heating by lower temperature water (< 50°C) made with plastics of improved thermal conductivity ; it includes :

- studies of technological problems which have to be solved for the fabrication of tubes or plates with water channels,
- computer model and design studies,
- test of a 300 watt convector,
- design of 1 kW convectors and economic assessement.

O. Introduction

The program relates to water/air convectors designed for space heating by water at a temperature not exceeding 50°C and made of plastic with improved thermal conductivity, incorporating a polyethylene matrix with an ethylene propylene elastomer and a carbon black load.

In this water/air convector the naturel convection of the air is actuated by a chimney effect induced in the casing which contains, in the lower part, the water/air exchanger and, higher up, the back up electrical heater (cf. diagram (a) of figure 1). The convector is designed to extract 1 kW thermal power from the water entering at 50°C.

The reported work includes :

- studies of the technological problems presented by the construction and assembly of tubes or of plates with water channels,
- the elaboration of a computer program which could be used as a design and dimensioning tool and the analysis of several preliminary 1 kW convector designs,
- the construction and testing of a 300 W convector model which demonstrated the validity of the computer program,
- the design of a 1 kW convector satisfactory from the technical and economical standpoints.

1. Studies of the technological problems presented by the construction and assembly of tubes and plates with water channels

1.1. Fabrication of tubes and plates by extrusion

We have studied the technological problems presented by the fabrication by extrusion of tubes or plates :

- by mixing the constituants of the compound in an industrial mixter,
- by extruding these mixtures to fabricate flat plates and tubes whose mechanical properties at 50°C are comparable to those obtained from mixtures made in a laboratory apparatus :

. strain at breaking point	110 kg/cm^2
. elongation	75 %
. modules of elasticity	3 300 kg/cm^2/

1.2. Tube and plate assembly

Assembly problems were studied :

- hot forming of plates with flowpaths from extruded flat strips,
- measure of the mechanical strength and aging of tube or plate bonding : standard test pieces were checked to ensure that the breaking load under traction of bonded parts stuck together with ECOBOND 45 LV glue was not modified by several hundreds or thermal cycles between 20 and 70°C,
- developing heat welding of hot-formed plates :
  - . the "heating strip" method consists of bringing the two parts to a temperature sufficient to soften the material (250 to 270°C) then removing the heating strips while applying a controlled pressure to the two opposite parts. Tensile and shear tests showed that unit assemblies of this kind could withstand without changes in their mechanical strengh 2000 hours of aging tests at a pressure of 10 bars in water at 50°C,

- in order to weld the hot-formed plates which, once assembled, form the plates with water channels for the 300 W model, two prints were made fitting the shapes of the hot-formed plates. They make it possible to keep the initial geometry of the plates during the softening part of process. Development of the welding process involved adjustment of parameters: duration of heating of the material, cooldown during welding and crushing of the material : since the material is highly loaded and contains a large quantity of elastomer, it is only slightly deformed during welding, which makes it difficult to adjust this parameter.

2. Computer program and premilinary designs of a 1 kW convector

A computer program was devised for the thermal and aeraulic calculation of a "chimney effect" convector. By means of this program, a first parametric analysis of the performances of the preliminary 1 kW convector design was carried out.

2.1. Setting up a computer program

The calculation allows for the following two related physical phenomena :

- air heating, which is a function of the air flow between the plates,
- the driving effect generated by air heating, which dissipates into pressure drops which are a function of air flow. These pressure drops were calculated by totalling :
  - the friction pressure drop between the flat parts of the plates (cf. diagram (b) of figure 1),
  - the extra pressure drop caused by the obstructions represented by the plates water channels,
  - the inlet and outlet pressure drops.

2.2. Preliminary 1 kW convector design

A preliminary design was made for a 1 kW convector of following dimensions :

Height : 0,75 m
Length : 1,50 m
Thickness : 0,20 m

The water/air exchanger consists of plates with water channels, vertically arranged by zigzagging the water flowpaths to obtain more regular air channels (cf. figure 2a). Each 15 cm high plate, obtained by heat welding of hot-formed plates, includes two water channels zone in a bend configuration (cf. figure 2b).

The exchanger incorporates fifteen 15 cm high plates with water channels of 10 mm outside diameter and 9mm spaces ; spacing between plates is 9,8 mm.

At a water flow of 100 litres/hour entering at 50°C and with air at a temperature of 18°C, the power extracted from the water is 965 watts without back up electrical heating and 1035 watts with 500 watts electrical heating.

2.3. Preliminary parametric study

The 1 kW convector design was subjected to an analysis of the influence of key parameters on the power extracted from the water :

- height of plates and distance between plates : the increase of extracted power with plate height is restricted by the reduction of air flow resulting from the increased pressure drop accompanying the increase of exchange surface (cf. figure 3),
- water flow : the increase in water flow increases extracted power but there is no point in exceeding about one hundred l/h (cf. figure 4),
- back up electrical heater : the use of back up electrical heating allows an increase in the power extracted from the water : +7 % for an electrical heating of 500 W and a plate height of 2 x 76 mm (cf. figure 5).

3. Construction and testing of a 300 W model

3.1. Model construction

A 300 W convector, which is a 1/3 model of the 1 kW convector previously designed (cf. 22) was constructed : the cross section of the convector is the same but its length is reduced to about 500 mm. Each plate consists in fact of two half-plates of 80 mm height, hydraulicaly in series, to ensure water circulation in a bend configuration. A photo of the model is shown in figure 6.

Nominal operating conditions are :

- water inlet temperature : 50°C
- air inlet temperature : 18°C
- water flowrate : 30 l/h
- distance of air inlet nozzle from ground : 90 mm.

3.2. Model tests

The results of the model tests demonstrate that :

- the power extracted from the water under nominal conditions is 325 watts, i.e slightly higher than the calculated value of 305 watts ; it goes up to 350 watts when the water flowrate increases from 30 l/hour to 45 l/hour (cf. figure 7),
- a back up electrical power of 170 W increases the power extracted from
- raising the distance from the ground from 90 to 150 mm does not improve performances but reduction to 50 mm causes a 7.5 % drop in the power extracted from water.

4. Design of a 1 kW convector and economic assessment

The acquired knowledge has been used to design a 1 kW convector satisfactory from the technical and economic standpoints. Two designs were made, differing from each other by the geometry of the heating unit, the mode of fabrication and the material thicknesses : one design has hot-formed and hot-welded plates and the other has finned plates or tubes produced by extrusion. Each design is for a convector whose outside dimensions are as follows :

- Height : 0,75 m
- Thickness : 0,20 m
- Length : 1,60 m

4.1. Convector incorporating water/air exchanger with ho-formed and hot welded plates

Given the technological constraints of heat-welding, an extruded plate thickness of 1 mm was chosen, leading to flat sections between water channels of 2 mm thickness. With a water channels internal diameter of 4 mm, the outer water channels diameter is 6 mm. With these data, the design leading to the minimum plastic weight for a 1 kW convector's a water/air exchanger consisting of 21 plates of 2 mm thickness, 50 mm height, each incorporating 4 channels with an outside diamter of 6 mm and an inside diameter of 4 mm separated by a flat section of 6,6 mm. There is a 7 mm spacing between the plates.

The plastic exchanger weighs 4,5 kg.

4.2. Convector incorporating water/air heat exchanger with exctruded plates or tubes

Fabrication by extrusion of the water/air exchanger components permits lower thicknesses than those used for the other design. In the case of extruded plates with water channels, the thickness of the flat sections between channels can be reduced to 1,6 mm ; the tubes have an inside diameter of 4 mm and an outside diameter of 5,6 mm. Study of the plate height and spacing leading to the minimum plastic weight for a 1 kW convector led to the following dimensions :
water/air exchanger incorporating 18 plates with a thickness of 1,6 mm and a height of 65 mm, each with three water channels with diameters of 4 mm-5,6 mm separated by a flat section of 16 mm. There is a 7 mm spacing between the plate.

Heating unit weight is slightly under 4 kg.

4.3. Economic assessment

Economic assessments were made for both the above-described convector designs :

- the factory fabrication price is 66 F (T.F) for the extruded water/air exchanger and 89 F (T.F) for the heat-welved water/air exchanger ; the weight of the materials is nearly the same in both cases ; the price difference is due to the difference in the estimated cost of fabrication.
- the sale price to the public, obtained by multiplying by 2,5 the fabrication price is 165 F (T.F) for one of the designs and 223 F (T.F) for the other. These prices should be compared to an estimated price of 310 F (T.F) for a stell heating unit.

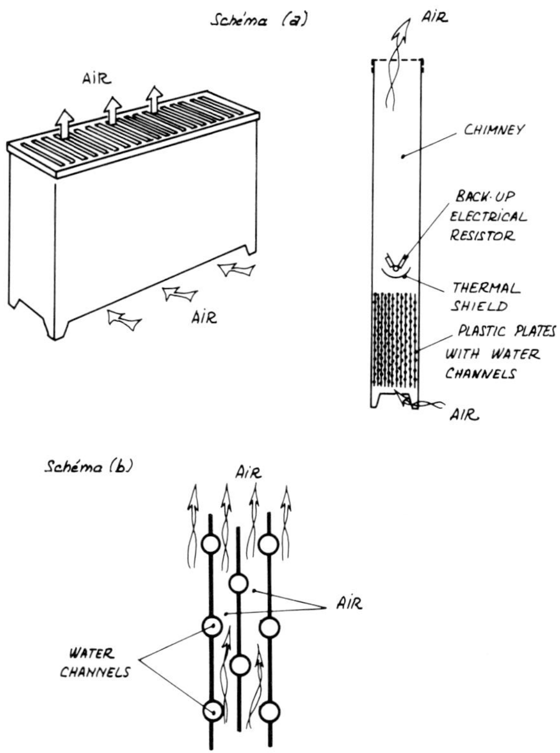

Fig. 1 Water/air convector for space heating

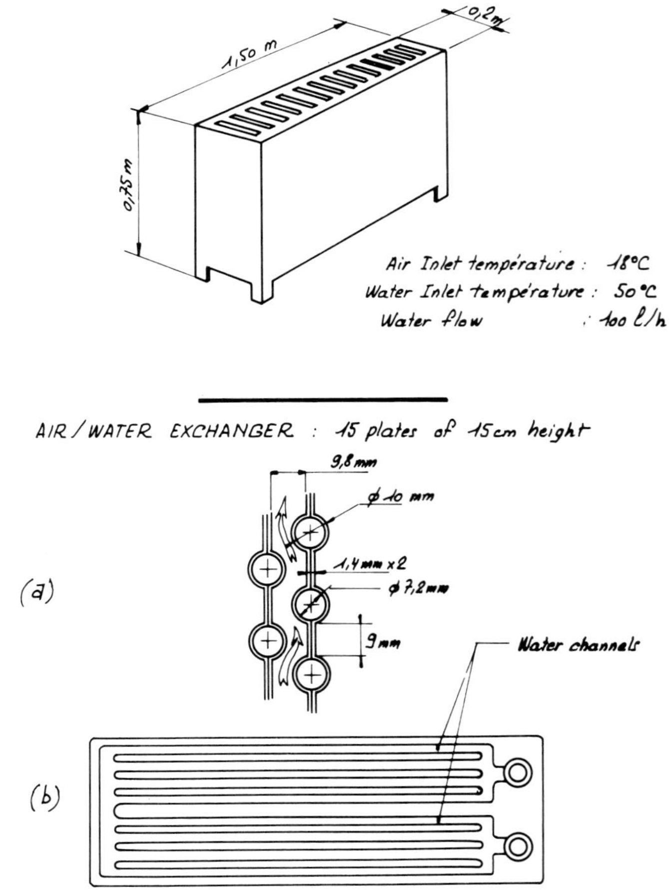

Fig. 2 1 kW convector - Preliminage design

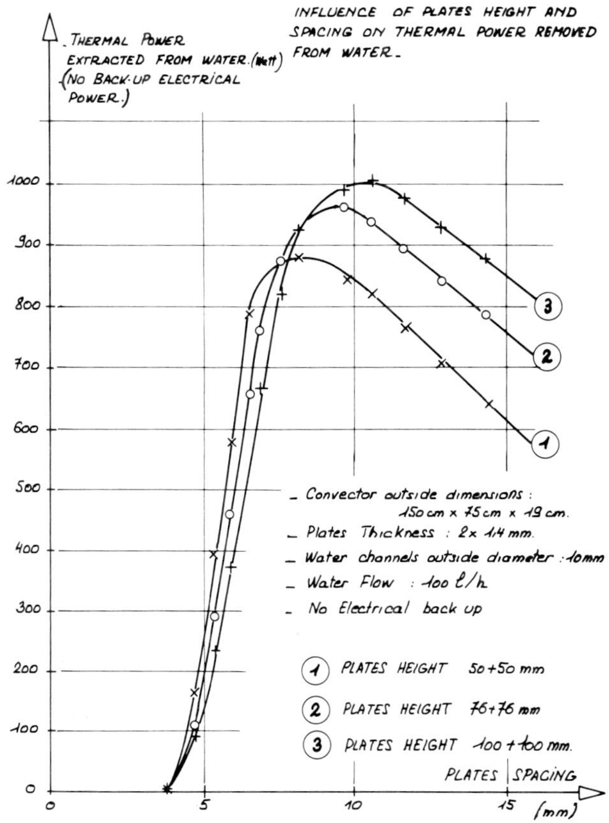

Fig. 3 Parametric studies of the water/air exchanger of a 1 kW convector

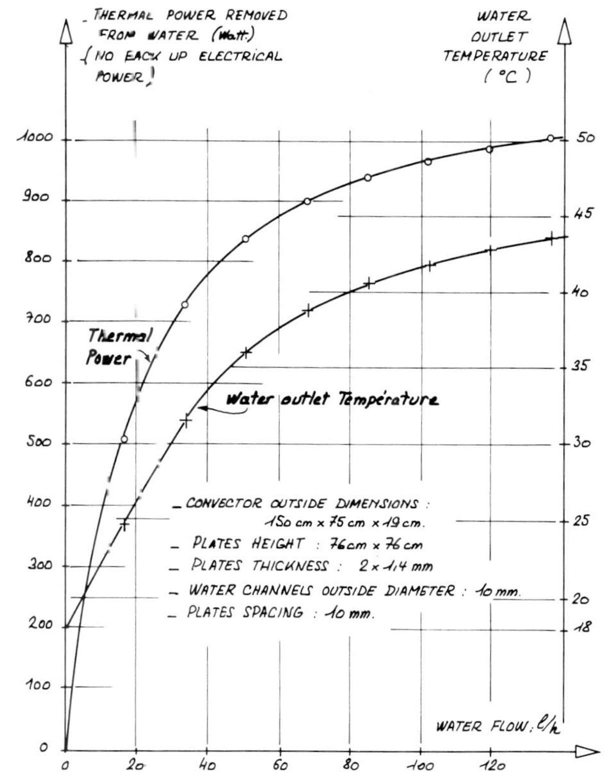

Fig. 4 Parametric study of the water/air exchanger of a 1 kW convector

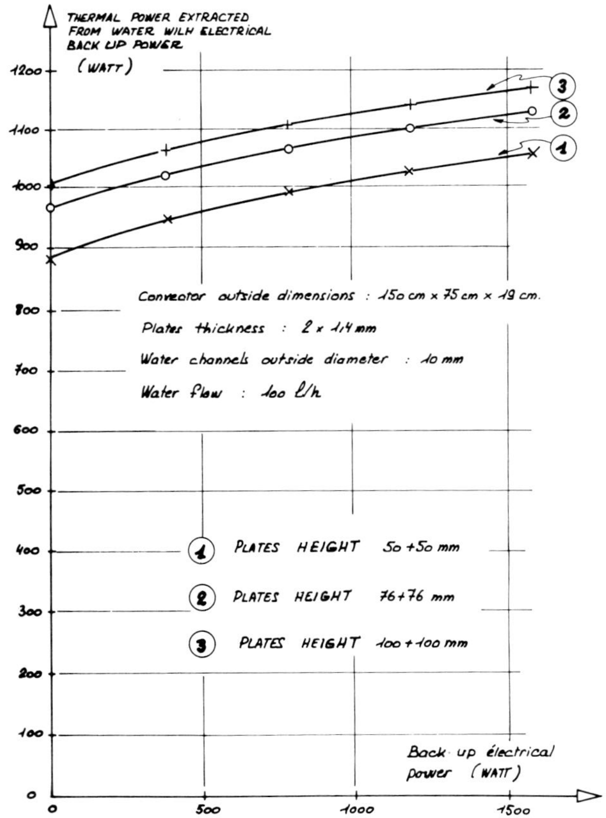

Fig. 5 Parametric studies of the water air exchanger of a 1 kW convector

Fig : a) PLATE WITH WATER CHANNELS

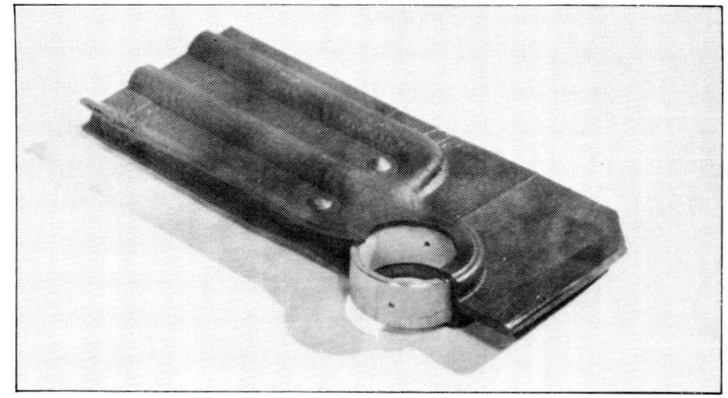

Fig : b) WATER / AIR HEAT EXCHANGER OF THE 300 WATT MODEL.

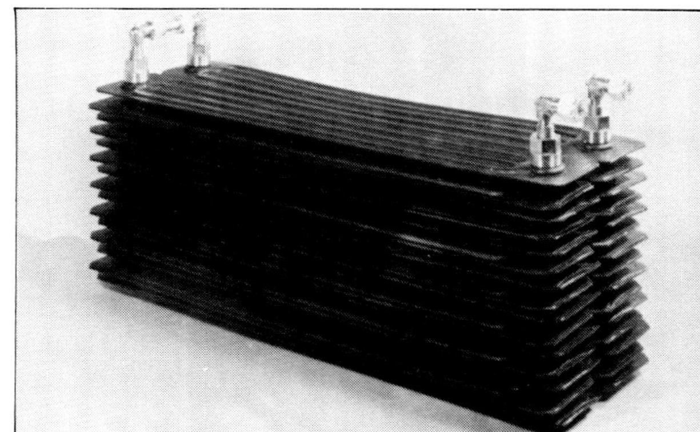

Fig. 6

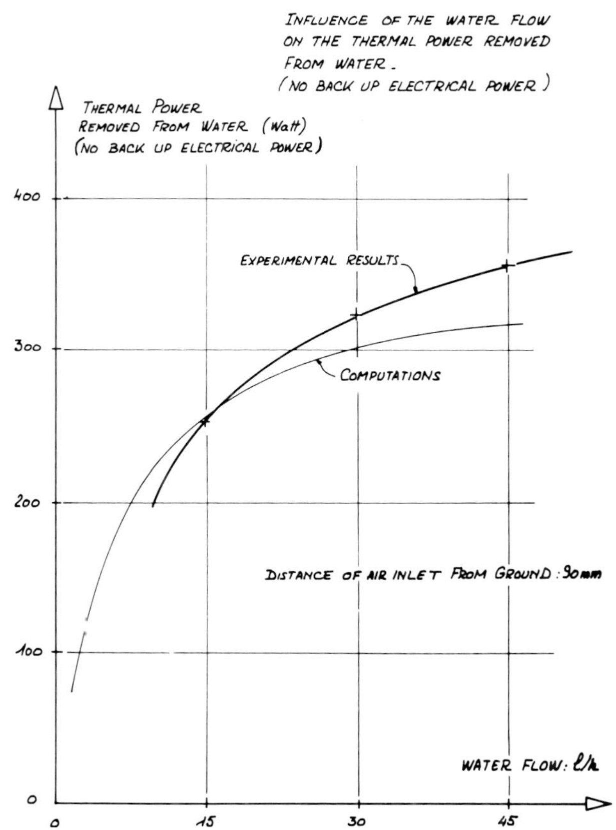

Fig. 7 "300 Watt" water/air heat exchanger

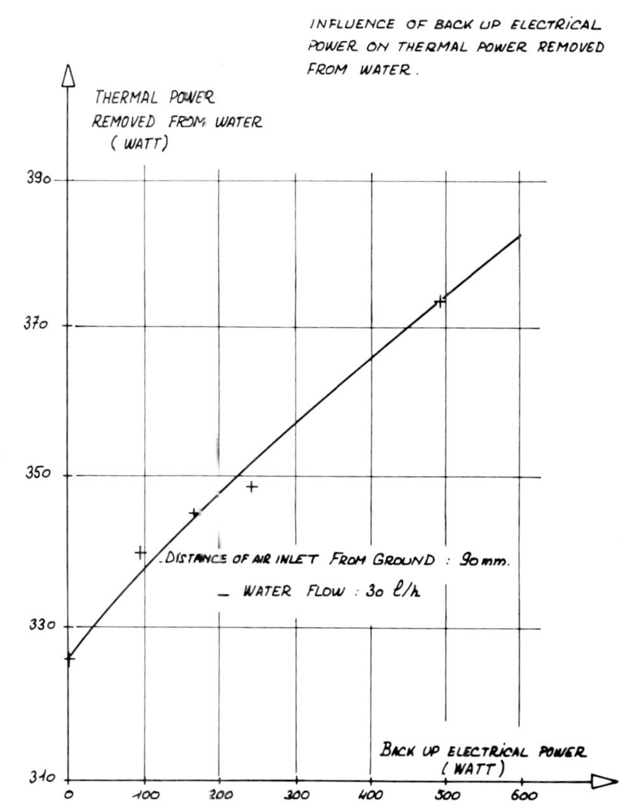

Fig. 8 "300 Watt" water/air heat exchanger

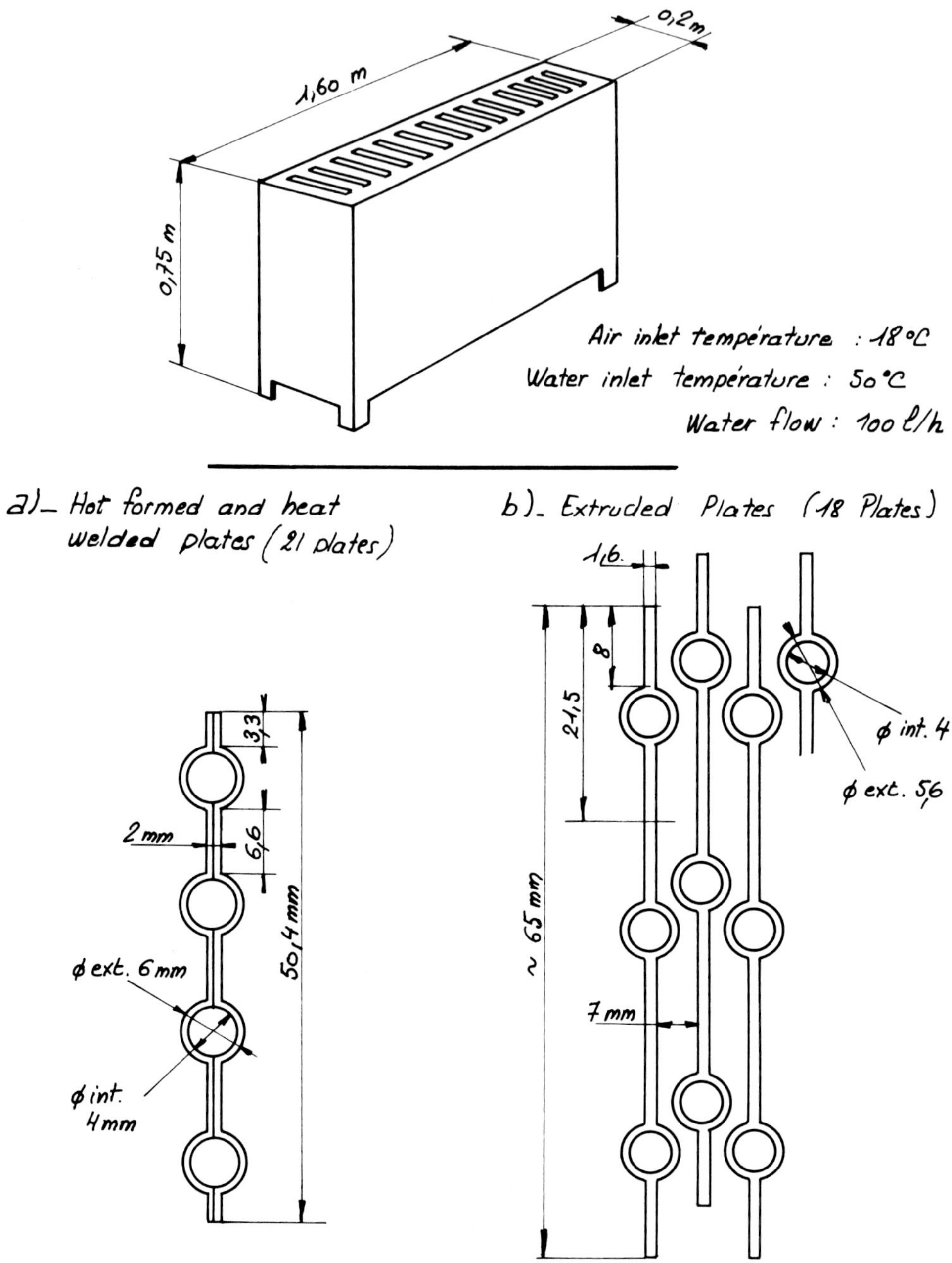

Fig. 9 1 kW convector

QUESTIONS AND DISCUSSION

Rayment (BRE) : Has the choice of 50° C, as a maximum temperature for the radiator, been made in view of the material properties of plastic ?

Answer : No, 50° C is the optimum temperature for use with heat pumps. The plastic radiators can be used up to 70° C without material problems.

Zegers (DG XII) : Is it possible in existing houses to use radiators designed for 70-80° C in combination with heat pumps which deliver heat of 50° C ?

Answer : Yes, for cases where the heat demand is considerably lower than the maximum heat output of the radiator at 70° C (e.g. installation of heat pumps in old houses using existing radiators in combination with improved insulation).

Pilavachi (DG XII) : Is the volume larger than for radiators with comparable heat output at 70° C.

Answer : Yes, 20 % larger, as compared to metal radiators. The cost of plastic radiators is considerably lower (30-50 %).

MEASURING AND CONTROL DEVICE FOR ENERGY CONSUMPTION OF HEATING INSTALLATIONS

Authors	:	J. Spitzer, W. Friedel, W. Gramatte
Contract number	:	EEA-1-o11-D
Duration	:	24 months, 1 Nov. 81 - 31. Oct. 83
Head of project	:	Dr. J. Spitzer
Contractor	:	Battelle-Institut e.V.
Address	:	Am Römerhof 35 6ooo Frankfurt am Main

Summary

The objective of the work is the design, construction and prototype operation of a heat meter based on a new principle based on temperature measurements only, i.e. with no moving parts. Compared with systems available today the heat meter being developed is expected to have a longer life time and a better accuracy. The initial development is aimed at a unit to be used to monitor the heat flow at district heating distribution points. It is expected that the design can be adapted later to systems allowing, for example, the measurement of the heat consumption in different flats within multi-family houses. Therefore it is intended to develop a modular system to cover the various application possibilities. The work completed so far concentrated on the design of a "heat bridge" between the hot and cold leg of a heating installation, in which three temperatures are measured allowing the online-calculation of the heat consumption. Tests of transducers, capable of performing the temperature measurements with the required accuracy ($\leq$ o.oo1 K) showed that resistance transducers of the PT 5oo/1ooo type will be suitable. The measurement positions and the algorithm for the on-line-calculations have been determined. Preparations for testing a prototype heat bridge in a flow test facility are under way.

2. Work program

The work program is divided into three steps

- design and optimization of the "heat bridge": geometry, materials, measurement positions and selection of temperature instrumentation
- design of the signal processing unit (micro-computer), construction and test operation of a prototype heat bridge in a test loop
- construction and test of the signal processor, construction and test of the complete unit, modification and optimization of the prototype unit towards a commercial unit.

3. Progress of work

The first step of the work program is completed. The following symbols will be used in the description.

C_w = specific heat of fluid
A = cross sectional area of bridge
$K(\dot{m})$ = heat transfer coefficient across bridge between hot and cold leg
$k_{1,2}(\dot{m})$ = coefficients
T = temperatures according to Fig. 1

Bridge Design

The design of the prototype heat bridge is shown in Fig. 1. The bridge is made of copper and the instrumentation consists of three PT 5oo transducers. The design was based on heat transfer calculations using a computer program that calculates the fluid temperatures and the temperature distribution within the bridge. The locations for the temperature measurements were selected such that the measurement range and the dependence on the parameters (flow rate and temperature difference hot/cold leg) was optimized.

Signal Evaluation

The evaluation of the heat flow to the user on the basis of the three temperature measurements T_{B_1}, T_{B_2}, T_{B_3} (see fig. 1) is as follows:

$$\text{heat flow } \dot{Q} = \dot{m}\, C_w (T_{F_2} - T_{F_3}) \tag{1}$$

$$\text{mass flow } \dot{m} = K(\dot{m}) \frac{A}{C_w} \frac{T_{F_2} - T_{F_3}}{T_{F_1} - T_{F_3}} \tag{2}$$

with the definitions

$$\Delta T_{F_L} = T_{F_2} - T_{F_3} = k_2 (T_{B_2} - T_{B_3}) = k_2 \Delta T_{B_L} \tag{3}$$

$$\Delta T_{F_S} = T_{F_1} - T_{F_3} = k_1 (T_{B_1} - T_{B_3}) = k_1 \Delta T_{B_S} \tag{4}$$

$$K_B(\dot{m}) = K(\dot{m}) \frac{k_2(\dot{m})}{k_1(\dot{m})} \tag{5}$$

one arrives at

$$\dot{Q} = \dot{m}\, C_w\, \Delta T_{B_L}\, k_2\,(\dot{m}) \tag{6}$$

$$\dot{m} = K_B(\dot{m})\, \frac{A}{C_w}\, \frac{\Delta T_{B_L}}{\Delta T_{B_S}} \tag{7}$$

The relationship for $K(\dot{m})$ and $k_2(\dot{m})$ are of an empirical nature and can be calculated based on heat transfer laws.

Thus the evaluation of the heat flow for a certain set of values for ΔT_{B_L} and ΔT_{B_S} is performed as follows.

- iterative determination of $\dot{m}$ using equ. (7) and the relationship for $K_B(\dot{m})$
- calculation of $\dot{Q}$ using equ. (6) and the relationship for $k_2(\dot{m})$ based on the value of $\dot{m}$ determined in the preceeding step.

During the operation of the heat meter this evaluation will be performed on-line by a built-in micro-computer.

Measurement Range

The selected district heating application requires that the heat meter covers the following parameter ranges

- volume flow (m^3/h) $\quad 0.05 \leqslant \dot{v} \leqslant 0.5$
- temperature difference (K) $\quad 5 \leqslant (T_{F_2} - T_{F_3}) \leqslant 80$

For these parameter variations the temperature measurements have to yield temperature differences as shown in Fig. 2.

Accuracy

Derived from equ. (6) the relative error for the heat flow $\dot{Q}$ with C_w and A constant is given as:

$$\frac{\Delta\dot{Q}}{\dot{Q}} = 2\,\frac{\Delta(\Delta T_{B_L})}{\Delta T_{B_L}} + \frac{\Delta(\Delta T_{B_S})}{\Delta T_{B_S}} + \frac{\Delta K_B}{K_B} + \frac{\Delta k_2}{k_2} \tag{8}$$

The heat transfer analysis of K_B and k_2 shows no temperature dependence of K_B and a linear dependence of k_2 so that the measurement error for the heat flow can be assumed as

$$\frac{\Delta\dot{Q}}{\dot{Q}} = \frac{4\Delta(\Delta T)}{\Delta T} \tag{9}$$

where $\Delta(\Delta T)$ is the error of the temperature difference measurement. Thus, based on the required accuracy for the heat flow measurement (according to OIML-draft, May 1980 SP 12-SR8) the required accuracy for the temperature difference measurements is shown in Table I.

Table I: Required accuracy of temperature measurement

Temperature difference hot/cold leg $T_{F_1} - T_{F_2} = \Delta T_F$	Req'd accuracy $\Delta\dot{Q}/\dot{Q}$ acc. to OIML-draft	Req'd accuracy $\Delta(\Delta T)/\Delta T = E$
$\Delta T_F < 10$ K	$\pm$ 6 %	$E < \pm$ 1.5 %
$10\ K \leq \Delta T_F < 20$ K	$\pm$ 5 %	$E < \pm$ 1.25 %
$20\ K \leq \Delta T_F$	$\pm$ 4 %	$E < \pm$ 1.0 %

Temperature transducers

Based on the boundary conditions for the measurement of the temperature difference the transducers have to meet the following requirements

- range 0.1 K to 5 K (ΔT_{BS})
- accuracy $\pm$ 1 % (0.001 K)

To select a suitable transducer, tests with two systems have been performed:

- semiconductor transducer (AD 590)
- resistance transducer (PT 500, PT 1000)

So far endurance tests under conditions specified in the OIML-draft of May 1980 SP 12-SR 8 have been completed: 10 cycles of heating the transducers up to 140 °C with subsequent determination of the resistance at 23 °C and the determination of the temperature coefficient over the range from 23 °C to 50 °C. The results of the endurance tests show that semiconductor transducers are not suitable, but that it can be expected that resistance transducers meeting the requirements will be available.

Test facility

The test operation of a prototype heat bridge during step two of the program will be carried out in a flow test facility consisting of two loops (Fig. 3). With the heater in one loop in operation the hot and cold leg of a heating installation may be simulated. The tests will be conducted with the heat bridge mounted between the two loops thus "connecting" the hot with the cold leg. During step one of the program the prototype heat bridge was designed and the preparation of the test section was completed.

4. Conclusions

Design calculations for a heat bridge led to requirements for the temperature measurement at three locations with respect to measurement range and accuracy. These requirements can be met with PT 500 transducers. Endurance tests (under OIML-draft-May 1980 SP 12-SR 8 conditions) with a number of these transducers have been performed with positive results so that the conclusion was reached that suitable transducers are available and thus work for step two of the programme will be continued.

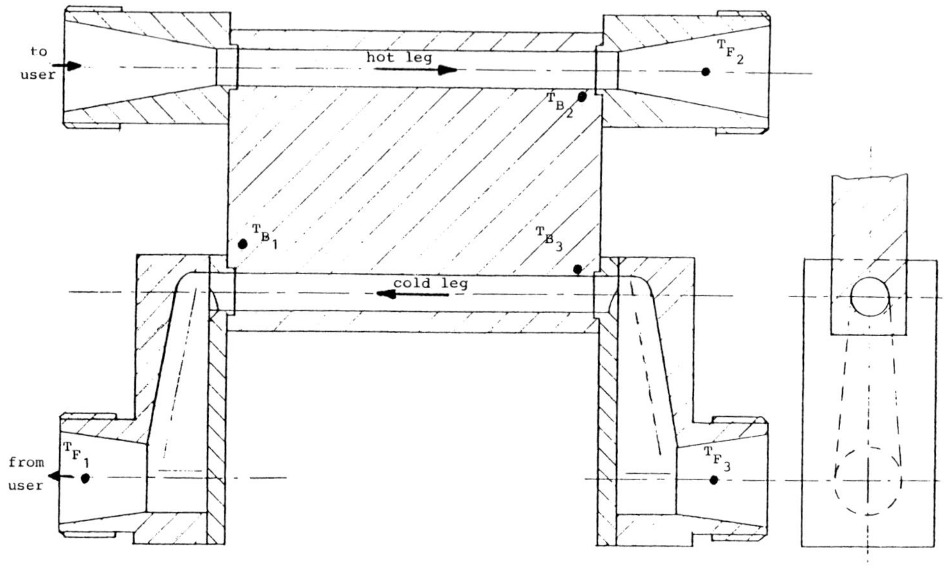

Fig. 1: Design of prototype heat bridge

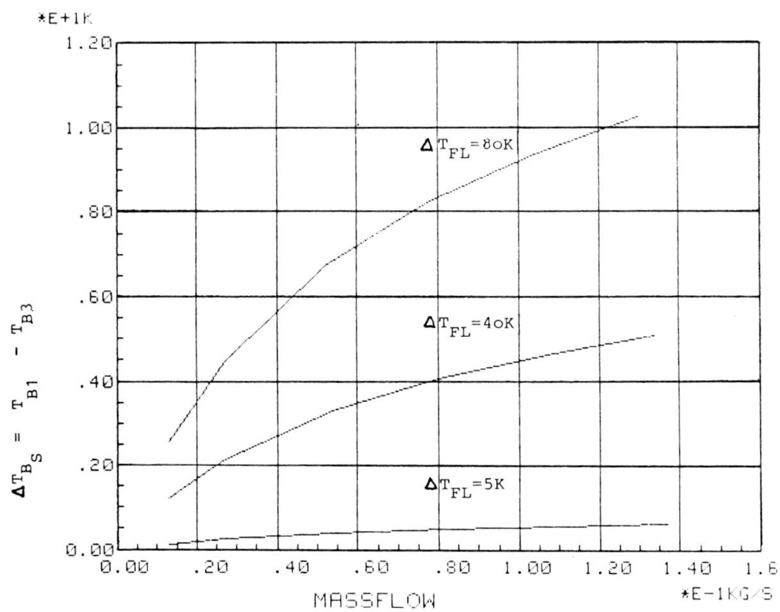

Fig. 2: Temperature difference ΔT_{B_S}

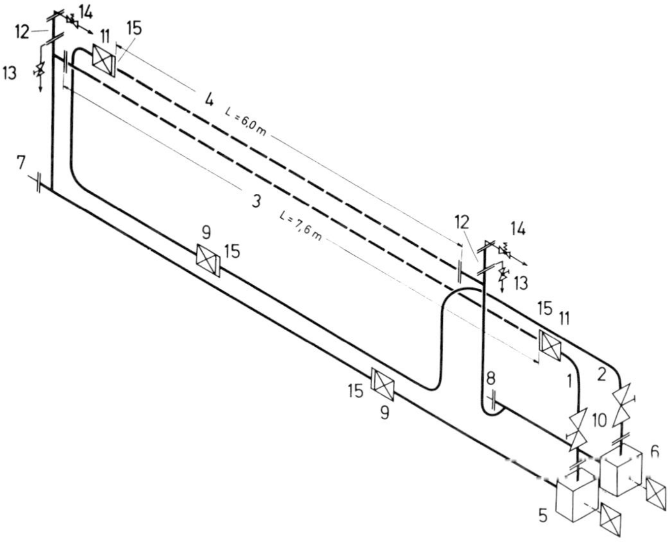

Fig. 3: Flow test facility

1 Loop 1: DN 150, PN 16
2 Loop 2: DN 150, Pn 16
3 Test section 1: DN variable
4 Test section 2: DN variable
5,6 Pumps 1,2, max 20 kW, max 570 m^3/h
7 Heater 1: max 49 kW
8 Heater 2: max 16 kW
9 Magn. inductive flow measurement
10 Throttle valve
11 Window
12 Header
13 Purge valve
14 Safety valve
15 Pressure and temperature transducer

QUESTIONS AND DISCUSSION

Zegers (DG XII) : I did not quite understand the principle of your heat meter.

Answer : In Fig. 1 heat is extracted from the hot incoming water and transferred by means of a copper slab to the colder outgoing water. The heat flow in the copper slab is given by $\dot{Q}_1 = K(\dot{m})\ A\ (T_{F_2} - T_{F_3})$.

$K(\dot{m})$ is heat transfer coefficient from the hot to cold water through the copper slab which has a cross sectional area A. This heat flow heats up the cold water in the cold leg from T_{F_3} to T_{F_1} and the absorbed heat in the cold water is $\dot{Q}_2 = \dot{m}\ c_W (T_{F_2} - T_{F_3})$.

$\dot{Q}_1$ must be equal to $\dot{Q}_2$ and with the constant K $(\dot{m})$, c_W and the temperature differences known the mass flow can be determined.

Rayment (BRE) : Is it possible to measure with an accuracy of 0,001 °C ?

Answer : Yes, that is possible. For temperature differences we use transducers from the same production batch.

Awbi (BSRIA) : Do you have a contact resistence between the transducer and the bridge ?

Answer : Yes, this resistance has to be measured carefully. The apparatus has to be calibrated carefully.

Jacqman (BSRIA) : Can this bridge be used in transient conditions; if the flow stops the temperature differences will remain for some time ?

Answer : If there is no flow, the temperature difference between T1 and T3 will be zero very soon and this will result in a zero indication for the energy flow.

Uyttenbroeck (Chairman) : Would it be possible to use this bridge in the laboratory ?

Answer : I think so. We are also investigating whether we can use this device for the measurement of mass flow.

Jacqman (BSRIA) : Is the energy consumption low ?

Answer : Yes, very low; the heat transferred from the hot leg to the cold leg via the bridge is very small.

Pilavachi (DG XII) : Is it possible to use the bridge with condensing steam ?

Answer : No the relation derived for hot water will become to complicated for condensing steam.

MICROPROCESSOR-BASED HEAT METER

Authors	: R. LEOPARDO, V. CENA
Contract number	: EEA-1-012-I(S)
Duration	: 42 months 1 January 1980 - 30 June 1983
Head of project	: Dr. R. Leopardo
Contractor	: Ottico Meccanica Italiana OMI S.p.A.
Address	: Ottico Meccanica Italiana OMI S.p.A. Via della Vasca Navale, 79/81 I - 00146 ROMA

Summary

The first half was carried out of a $3\frac{1}{2}$-year research whose scope is the development of a heat meter with the following characteristics: precise and reliable flowmeter, electronic integration of the product temperature difference x flowrate, microprocessor-based control unit capable of implementing complex control logics. The potential of such a device in the framework of the EEC energy saving programs was found interesting: therefore, a review was performed of the existing flowmeters. The electromagnetic principle was chosen as the most suitable, and several prototypes were built and tested. The development of the control unit is currently under progress. It can be concluded that there is the practical possibility of developing a sophisticated yet low-cost heat meter, but extensive tests will be needed to assess the reliability of the device.

1. Scope

Scope of the activities carried out in the framework of contract EEA-1-012-I(S) is to develop a heat metering system with the following features:

- precise and reliable flow meter
- electronic integration of the product temperature difference x flowrate
- microprocessor-based control unit, with complex control logics implemented via software.

2. Potential

In several countries of the European Economic Community codes and regulations have been issued with the aim of reducing the thermal energy consumption in residential space heating.

In most cases, the application of individual heat metering systems, one for each user, has been recommended.

Two main factors act against this recommendation:

- the very high cost of an industrial-quality flowrate/heat meter can not be mot by the single private user
- on the other hand, low cost meters are either unreliable (turbine flowrate meters) or imprecise (heat meters based on timers).

The development of an instrument capable of measuring the actual heat delivered to a single user with reasonable precision, reliability and cost would lead to a fast diffusion of the device in all Europe.

In the 9 countries of the EEC about 1.5 million new housing units are built each year.

Estimating that only 10% of these new units would be equipped with an individual heat meter, the potential market for such a device is of the order of 150,000 per year.

3. Review of existing flowmeters

A detailed review of the available flowmeters was carried out, from both the technical and the economic point of view.

The main advantages (A) and disadvantages (D) of each device are hereafter described.

Orifice

A : low cost of the sensor, precise, simple installation, easily available on the market, can be substituted if the flow rate range changes

D : the measurement is influenced by all the chemical and physical characteristics of the fluid, non-linear response, minimum threshold at 20% of full flow, high pressure drop at high velocity, requires installation on long straight pipe, subject to fouling and scaling, high cost of the signal transmitter.

Venturi

Similar to an orifice. Main differences:

A : more precise, lower pressure drop, can be used with fouling fluids

D : more costly, with larger overall dimensions, requires large flow rates.

Pitot

A : negligible pressure drop, suitable for large diameters and for very low static heads

D : requires clean fluids and the knowledge of the velocity profile, suitable only for steady state conditions, critical installation.

Rotameter

A : good precision, suitable for small flowrates, small and constant pressure drop, linear response

D : suitable only with clean fluids, not precise with oscillating flow rates, requires a complex indirect reading system and installation on a vertical pipe.

Oscillating disk

A : direct integration of the signal, precise even with small flow-rates

D : moving organ, not suitable for instantaneous flowrate, high pressure drop, subject to fouling and scaling, requires gas-free liquid.

Turbine

A : high precision, very fast response

D : very sensible to shocks and vibrations, subject to fouling, scaling and abrasion, high pressure drop.

Generation of vortices

A : good precision, no moving parts, sensor not in contact with the fluid, analog or digital signal

D : high pressure drop, suitable only for clean and non-viscous fluids, requires installation on a straight pipe.

Ultrasonic

A : not influenced by temperature and viscosity variations, can be mounted in any position, analog or digital signal, zero pressure drop, good precision, can indicate the direction of flow

D : high cost, suitable only for clean and non-viscous fluids, requires a symmetrical velocity profile and installation on a straight pipe.

Hot-film anemometer

A : small pressure drop, direct measurement of mass flowrate, fast response, good precision

D : suitable only for clean fluids, requires a specific calibration for each installation.

Magnetic

A : good precision, can measure the flow in both directions, linear response, zero pressure drop, not influenced by temperature and viscosity variations, simple installation

D : fluid must be electrically conductor, instrument unsuitable for temperatures above 200°C.

4. Principle of operation

The results of the preceding review led to conclusion that the best approach was to investigate the magnetic flowmeter principle, which is based on the Lorentz law.

When an electric charge q travels with velocity $\vec{v}$ in a magnetic field having induction $\vec{B}$, a force $\vec{F}$ is applied to the charge. This force is directed perpendicularly to the plane formed by the velocity and the magnetic induction, and its expression is

$$\vec{F} = q \, \vec{v} \times \vec{B}$$

Two electrodes are placed in the direction of the force by opposite sides with respect to the charge flow at distance D. One of them will collect the positive, the other the negative charges. The electrodes therefore behave as a capacitor, in which the electric field is given by

$$E = \frac{V}{D}$$

where V is the voltage between the two electrodes. The electric field acts on the charge q with a force

$$\vec{F} = q \, \vec{E}$$

In steady state conditions it will be $q \, E = q \, v \, B$, and therefore

$$V = v \, B \, D$$

This relationship allows the calculation of the velocity from the measurement of the voltage and the calibration of the constant (BD). From the velocity the flowrate can be easily derived:

$$Q = v \, \frac{\pi \, D^2}{4}$$

Measuring the temperature difference T between the hot and the cold pipe, the actual heat delivered between time t_1 and t_2 is given by

$$H = \int_{t_1}^{t_2} Q(t)T(t)\ dt$$

5. Prototypes of heat meters

Several prototypes were designed and built, with different geometrical and electronic solutions.

The main components of a heat meter are:

a) the flow-metering section, which is composed by a short pipe carrying the electrodes, the coils for the generation of the magnetic field and a magnetic circuit outside the section

b) two temperature sensors

c) an electronic equipment which acquires the flowrate and temperature signals, multiplies them by a constant and drives a mechanical counter which displays the integral

d) power supply for the magnetic coils and the electronic equipment.

Item c) was a temporary solution and was adopted only in the first prototypes. Presently, a more sophisticated, microprocessor-based unit is under testing. The main features of this unit are described in para. 7.

The problems which arose and were solved are hereafter described.

Magnetic field

The first prototypes were built with the magnetic field generated by permanent magnets, both external and internal to the pipe. It was found that the same process of collecting the electric charges causes the polarization of the electrodes.

For this reason the second generation of prototypes were equipped with coils, to which power was supplied in a square wave mode with frequency 3.125 Hz = 1/16 of the main frequency.

Material for the metering section

The section must be built with a non-ferromagnetic material, so that the lines of flux of the magnetic field can penetrate the inner part of the section. The first prototypes were built in polymethylmetacrylate (Plexiglas), which is transparent and therefore allows the inspection of the motion of the fluid and the detection of air bubbles. Eventually, polypropylene was adopted with satisfactory results.

Material for the electrodes

Several materials were tested in the typical conditions of the residential water-heating systems. The final choice fell on the AISI 316 stainless steel, which has the additional property of being non-ferromagnetic, and therefore does not disturb too much the magnetic field. In the present version, the electrodes do not protrude into the pipe and their surface is machined so that their curvature is the same of the pipe. This helps in keeping the instrument free from fouling.

Temperature sensors

The first prototypes used integrated sensors; eventually the more standard PT100 platinum thermoresistances were adopted.

Power supply to the coils

The final version of the power supply is composed by:
- a unit which accepts a 50 Hz-sine wave and transforms it in a 50 Hz-square wave
- a unit which divides the frequency of the square wave by a factor of 16 (final frequency 3.125 Hz)
- an instrumentation amplifieer with gain 40 dB, input impedance of the order of thousands megaohms, common mode rejection ratio 90 dB.

6. Preliminary tests

Figure I shows the performances of three prototypes (indicated by Mark 16, Saturn, Mark 17) with respect to the linearity of response as a function of the water velocity. The results are satisfactory, especially in the range 0-0.5 m/sec in which the majority of the flowmeters show marked deviations from linearity.
The tests were carried out in a loop having the following characteristics:

fluid : water
flowrate : 0-30 m^3/hr
velocity : 0-1.5 m/sec
total pressure drop : 600 mm H_2O

The complete heat meter was tested keeping the temperature sensor in thermostatic baths and integrating the "delivered" heat as seen by the electronic equipment.
Figure II shows the satisfactory linearity which was obtained between the flowrate and the release frequency of the counter, in the case of a temperature difference = 10°C.

7. Microprocessor-based control unit

The electronic block which acquires the signals and performs the multiplication and integration is being substituted by a microprocessor-based unit with more complex characteristics.
In particular, besides the cited mathematical functions, the unit will be responsible of a number of logic and even "policy-making" functions, and will command the actuators of the whole heating system.
The functions which are being implemented are the following:
- acquisition of the signal from two temperature sensors (water) and from the flowmeter
- acquisition of the signal from another temperature sensor (ambient

temperature) and from two pyrometers (monitoring the boiler)
- acquisition of the signal from a level transmitter
- autodial of a set of programmable telephone numbers if one or more of the above signals are outside the allowable range and transmission over the line of a code which identifies the alarm
- integration and display of the product flowrate x temperature difference
- switch on and off the burner and the pumps, according to a daily and/or weekly schedule
- adjust the internal set-points (maximum and minimum water temperature) as a preprogrammed function of the ambient temperature.

The most important function is the last one, because it leads to the adoption of energy-saving strategies. It is to be stressed that various strategies can be implemented without any modification of the hardware: actually they are "firmware" programs, resident on EPROM's (Erasable Programmable Read Only Memory). The private user will need only to specify the main parameters, pushing a few keys on the front panel. Even the programs can be modified just detaching the EPROM from the printed circuit and substituting it with an up-to-date one.

The main blocks of the control unit are the following:
- microprocessor based on the Fairchild F8 Central Processing Unit
- 256 bytes of RAM (Random Access Memory) which stores the working data
- 2048 (2 K) bytes of EPROM which stores the program which acquires the data, performs mathematical operations and logic functions, actuates external devices
- and A/D (Analog to Digital) converter for the acquisition of flowrate and temperatures
- three input ports (one for setting the set-points, one for the acquisition of the signals through the A/D converter and of the digital signals, one for the inclusion/exclusion of the various functions)
- two output ports (one for the display, one for the actions such as the autodialling, the switching of external devices, etc.).

8. Conclusions

The activities carried out in the first two years of the research project lead to the conclusion that there is the practical possibility of developing a low-cost heat meter with microprocessor-based logic.

In particular:
- the flowmeters perform satisfactorily and experimentation will proceed on different geometrical arrangements
- the control unit is in the first stages of development, but no major obstacle is foreseen
- extensive tests will be needed to assess the reliability of the whole system.

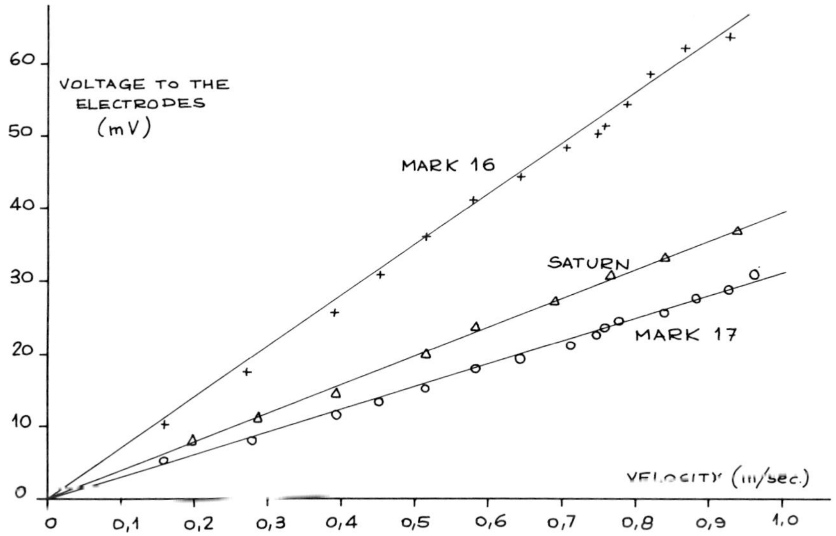

Figure 1

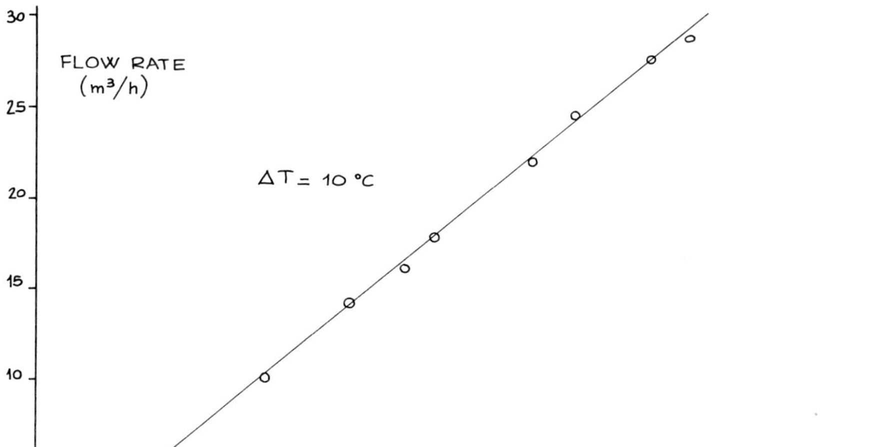

Figure 2

QUESTIONS AND DISCUSSION

Rayment (BRE) : Can oil and gas companies be forced to sell heat ?

Answer : Yes, by law.

Spitzer (Battelle) : Is the development going in the direction of district heating in Italy ?

Answer : I don't think so, but this heat meter is no only for district heating. It is meant to be used in apartments of apartment buildings where the energy use is managed by oil distributing companies.

Uyttenbroeck (Chairman) : Battelle and OMI are working in exactly the same field. I hope you have good contacts.

Answer : Yes.

Uyttenbroeck (Chairman) : The Battelle meter seems to be very simple and practical. What is the opinion of OMI on this Battelle heat meter ?

Answer : Below a certain energy flow the heat meter becomes very inaccurate. For example if a plant is working with a pump, and if the pump stops a water flow by gravitation can still persist. This small energy flow can not be measured accurately.

Spitzer (Battelle) : Our heat meter satisfies OMI standards.

SESSION II - HEATING CONTROL SYSTEMS

Chairman: C. BOFFA

Signal transmission via the electrical house wiring for energy saving in household and small business

Electronic heating control system

Computer controlled network for heating system

Performance of various heating systems

Reducing energy consumption of buildings by using microprocessors in climate control

SIGNAL TRANSMISSION VIA THE ELECTRICAL HOUSE WIRING FOR ENERGY SAVING IN HOUSEHOLD AND SMALL BUSINESS

Authors	:	E. NEUDORFER, H. RADDE, A. SEWALD
Contract number	:	EEA-I-04-D
Duration	:	12 months 1.12.80 - 30.11.81
Head of project	:	Dipl.-Ing. H. Radde, Development Division
Contractor	:	Periphere Computer Systeme GmbH
Address	:	Pfälzer-Wald-Straße 36 D-8000 München 90

Summary

PCS GmbH defined a module for an economic data transmission system with low mounting requirements, which supports the application of microelectronic control equipment for the purpose of optimization of energy consumption in domestic applications. The system will use the domestic AC power line network and is based on the carrier current technique. The "wireless" data transmission can help to reduce the effort of introducing microelektronic control equipment by avoidance of expensive additional installations, increase of users acceptance and variable connections. The project examined the state of the art, the requirements and the difficulties related to the carrier current technique in the heating control application. The module with transmission, logical and peripheral interface functions and an implemented transmission procedure is defined. A functional unit is designed. The transmission hardware is still too extensive for an economical conversion into a LSI circuit.

I. Introduction

The project's object is the definition of an economic data transmission system with low mounting requirements, which supports the application of control equipment for the purpose of optimization of energy consumption in domestic applications.

Of course, not the transmission system will save energy, but it can effect like an additional catalyzer to use control equipment. As a result of different scientific studies and in the common commercial opinion, by means of microelectronics, the reduction in heating energy consumption can be 15 to 20 % (of a potential of about 30 %).

It seems to be profitable to invest in development and manufacturing of tools promoting the use of microelectronics for that purpose. Such a tool will be a low cost data transmission system, realized also mainly by microelectronics.

The use of microelectronics for control purposes is attended by necessity of installations for controllers, sensors, actuators, displays, monitors and switches.

The "wireless" data transmission system can help to reduce the effort therefore by

- avoidance of expensive additional installations, especially in older buildings
- increase of user's readiness to accept control system (because the installation is more comfortable)
- optimization of adaption to local conditions by its characteristic of "variable installation".

To obtain the wireless transmission, the system will use the domestic electrical power line network, available nearly everywhere in buildings and outside. The transmission is based on the carrier current technique, known from simple alarm systems, interphone systems and remote-control equipments and on the high sophisticated microprocessor technique with the availability of low cost LSI-circuits.

The problem is the compromise to meet all the pretentious technical requirements with a low-cost solution, to select all the necessary features, and put them together to an optimum solution.

2. Research and experiments

2.1 Requirements

The intended application of the transmission and interface module is domestic energy management. It should be adapted especially to future multivalent heating system controllers. This means it has to support data acquisition and data output on distributed peripheral elements (sensors and actuators) and to transmit data from and to a central optimizer and monitor.

Therefore the module has to work in conversational mode. The thermal processes usually have time constants ranging from minutes to hours. Because of the sampling theorem transmission speed and telecontrol protocol have to perform response times less than 1 minute. Besides that response times are affected by measures to reduce the noise and bit error sensibility.

Transmission is disturbed especially by electromagnetical interference produced by switched appliances connected to the mains.

The transmission reliability must be improved by redundancy, an adapted protocol and a high sophisticated detection circuit.

Considering that domestic energy control systems could have up to 50 data sinks and sources, the data net rate has to be at least 16 bit per second with bit error probability of 10^{-3}.

Interfaces between data transmission modules and sensors, actuators and displays should be simple and easy to handle to find the acceptance of the manufacturer of those appliances. It would be desirable to connect peripheral devices with different not well regulated voltages.

The interface must be isolated from the power line. Concerning the use of HF power in most countries regulations have been given that have to be taken into consideration in order to get an official licence. For example in Germany this regulations have been laid down by the FTZ (Fernmeldetechnisches Zentralamt) in Darmstadt:

HF-output 5 mW	on asymetric artificial mains network
frequencies	30 to 146 kHz
interference	according to standards VDE 0875, FTZ 17R2014 and other FTZ regulations

2.2 State of the art

An extensive survey was made on the subjects concerning the task to perform. Fields of related remote control, data transmission and integrated circuits as well as existing devices were considered to get acquainted with the problems and to find suitable and approved solutions for some parts of the task.

Signal transmission via power line:

Electric supply companies use high voltage power lines for transmission of remote control and telephone signals. With HF power up to 40 VA and wavelengths carefully adapted to the distances they do obtain ranges up to 500 km. That is a total different order of physical magnitude and not transferable to the task.

These companies also run remote control systems in medium and low voltage power lines, the so-called ripple control installations.

They use them for tariff switching (space heaters, water boilers), switching of waterpumps, street lighting and sirens and peak-load power switching.

These systems cover medium areas and distances and use frequencies ranging from 100 to 1500 Hz. Beside other reasons, this technique can't be used because of the very simple transmission protocol and the very low speed.

In smaller scale there are systems and components for applications in household and office using the mains as transmission medium.

The best known may be the intercom for audio-signal transmission (e.g. for baby sitting).
In alarm systems informations are transmitted via mains from sensors to central station by means of HF signals. The sensors are detectors for smoke, movement, gas, sound. The signals are very simple (on/off) and its structures are insufficient for information interchange. In the meantime a few very interesting devices are available, based on the carrier current technique. One of the most powerful is a professional American modem from D.C.S. for bidirectional data transmission via mains with the following features:

- RS-232C interface,
- code and protocol transparent,
- 24 simplex or 12 duplex channels with 12 kHz bandwith within 100 to 500 kHz,
- transmission speed up to 9.600 Baud.

But it has no licence in Europe because of its transmitting power above the permissible level. And it's much too expensive for home use.

An other American system from Mountain Hardware is compatible to a personal computer and gets near to the present application, but the licence problems are the same.

A complete low cost system for simple light and appliance remote control functions from the U.S. Company B.S.R. has an European version, adapted to the voltage and to the requirements for an official licence. Furthermore the price and the information content comes very close to the heating control requirements. It has only one transmission direction, no peripheral interfacing and is a bit to slow for heating control purposes. But it's well worth to keep attention on its further development.

European companies offer different simple systems, not applicable to considered requirements. An own system comparable to the named American is not available today.

2.3. Analysis and experimental arrangement

A deep analysis of the devices pointed out before was done so far as the informations were available.

To prepare test, informations related to the power voltage network were collected and measurements were done. A main problem is to reproduce interference, impedance and attenuation concerning signal transmission via power lines, because the network characteristics are changing all the time. The electric supply companies refer to the standard VDE 0875. This is a regulation for the allowed maximum interference an electrical device may transmit but not a description of real electrical environment. Making measurements to obtain these characteristics needs a high expenditure both in time and money. Because of the varying characteristics probably the best approach is to use statistical data. Therefore a paper has been used with the results a research group obtained by evaluating thousands of measurements made in 1976 to 1978. (See figure 5) (By the way it's very hard to find bibliography concerning these problems.) These results were some sort of heuristic base for the additional work.
Impedance of the mains measured on a wall socket acts as a load for the transmitter. It is also varying in time. Because of the lack of real data the artificial mains network used in interference measurement has been taken as a standard.
The worst and most unknown characteristic is the attenuation of the HF signal on the mains. It depends on the loads and their distribution in network and varies heavy, too. For example: widly used radio interference filters have capacitors at their input and act like a short circuit to the HF signal.
Remedial measures (called mains conditioning) like additional inductors in series to the radio interference filter must be taken.

Two function units of the designed circuitry were built up (see figures 2 and 3).

Their behaviour with regard to filters for the modem and a digital demodulator which was found susceptible to interference were tested.
To the test arrangement a minicomputer was connected running transmission programs and monitoring the results.
The main purpose was to set up reproducable conditions for the test of the modem and the system. It was composed of

- a filter to separate the varying mains characteristics from the module under test,
- an artificial mains network to simulate reproducable mains characteristics,
- a line simulation,
- an interference simulator.

The experiences, which the performing development staff has made in connection with the design of data and energy transmission for so-called functional modules as an extension of an existing distributed building automation system, were used. (5)

3. Results

3.1. System concept (see figure 1)

A master-slave concept in partyline configuration has been chosen, because it is considered as the best choice for this application in respect to system complexity and costs.

Master may be for example the microprocessor controller of the heating system. Components close to it are connected to its I/O. Remote components e.g. outdoor temperature sensors are connected via transmission and interface module, mains (as a transmission medium) and modem. The controller polls the transmission and interface modules. The modules execute the commands and respond with its status or input values.

3.2. Transmission and interface module (see figure 2)

3.2.1. Modem

Frequency shift keying (FSK) modulation has been selected. It provides:

- save data carrier detection (DCD) in comparison to amplitude keying,
- easy modulation and demodulation in comparison to phase shift keying .

Modulation frequencies are: 45.45 kHz for logical 0
55.55 kHz for logical 1

They permit an operating range up to 500 m and the use of active filters, integratable on an LSI-chip. Besides of this they have a sufficient distance from mains frequency and harmonics.

The transmitter is composed of a digital type FSK modulator, a low pass filter in order to meet the interference regulations, a power amplifier (5 mW) and the coupling network to the mains.

The receiver: incoming signals pass through the coupling network. In order to get a wide dynamic range an AGC amplifier has been provided. After passing an active band pass filter the signal enters the demodulator. Two different demodulator circuits have been built up:

- PLL demodulator, which needs a minimum of external components but is sensitive to harmonics and supply voltage variations,
- envelope demodulator with 2 narrow band pass filters, which has a higher noise immunity (because of the distant carrier frequencies) but needs a higher expenditure in components.

A final decision will be made in coordination with LSI-experts.
The coupling network consists of a high voltage capacitor and a transformer for electrical insulation from the mains. A passive band pass filter cuts out the carrier frequencies. Provisions are made to protect the circuits from destruction (overvoltage protection).

3.2.2. Logical unit and protocol

Most of the modern 8 bit single-chip microcomputers (e.g. 6801 or 8051 made by Thomson-CSF or Siemens) may be applicable as base of the module's logical unit. A serial control unit based on the 3873 8 bit single-chip microcomputer from Mostek has been selected.
It contains:

- the processor,
- the ROM with the program for
    - interpretation of commands
    - execution of commands (simple input and output),
    - generation of responses,
    - control of local procedures like positioning of a heating valve (not yet implemented),
- the RAM for data storage,
- the UART for serial communication,
- parallel ports for controlling the module's inputs and outputs.

The general applicable <u>telecontrol protocol</u> with high transmission reliability for master-slave data transfer (polling mode) in party line configuration contains the following elements:

- start and stop information,
- address,
- command/status,
- data as specified by command,
- redundancy (LRC and parity) which results in Hamming distance = 4

3.2.3. Peripheral interfacing

Most individual part of the module is the interface to the connected devices and the function to be done, e.g.

- reading contacts like "window closed",
- reading analog values like temperatures,
- switching appliances like pumps or valves,
- displaying system states and values,
- reading keyboards,
- positioning e.g. heating valves.

To support easy special interfacing of the variety of devices to be connected a general applicable parallel interface up to 24 bits is provided which is electrically insulated from the mains and which is TTL and CMOS compatible.

4. Comments and conclusions

With regard of the physical requirements, the results are satisfactory:

- A data transmission system has been developed, which uses the electric power wiring in small houses. The system is applicable for heating control purposes.
- The systems response time is sufficient for thermal processes, occuring in energy generation and consumption.
- Its bit error probability is similar to data transmission via telephone line.

But from the technical point of view, a carrier current system is not an ideal signal transmission system. That is, because household wiring was not designed to carry rf signals. Every device, like lamps or appliances, connected to the line, will short out some of the rf energy. The main problem is, that we don't know at any time, which devices actually are connected. It means the network state changes all the time.

Therefore remarkable efforts for line conditioning could be necessary in certain installations. In this case it is usually better to use other transmission mediums.

But particularly in small buildings a few things can be done. That is the use of available components like phase coupling elements and noise suppression plugs.

Beside that low transmission frequencies were selected in order to minimize the effect of line length.
A good answer therefore could be to use temporary variable frequencies. It's worth to analyse the necessary expense. Because of this project's closed time limit it wasn't done.
So the rather low transmission frequency of about 50 kHz has been chosen in order to handle the wiring as a some undefined network consisting of inductances, capacitors and resistors. It isn't as bad as to handle the wiring as a long transmission line. (Due to the standing waves on the line, not terminated in its characteristics impedance, voltage and current will vary drastically.)

Unsatisfactory to this moment is the expense for the modem circuitry. That could make the realization of a LSI-chip too expensive for that application. More work must be done to reduce this expense. To decrease this problem and to minimize the effort for line conditioning, the described transmission protocol was selected right for parallel periphery interface with so many bits. An other reason, probably being more and more important in the future, is the easy interfacing to home computers.

The results of the project together with components, available in the meantime and applicable for carrier current control, confirm the feasibility of application to energy management in small buildings.

A good compromise will be to use carrier current systems in special cases where separate control wires are not practical. It seems to be realistic to get a compatible transmission system consisting of carrier current technique components, of supply voltage and data transmitting party-line connections like provided for the mentioned functional modules, beside classic wiring of course.

There is absolutely no doubt, that techniques like that seeming to be somewhat phantastical today, will turn over in consumer products tomorrow.

That's indicated by applications of carrier current technique in low cost products of other fields and the promotion of the present economical microcomputer power.

Momentary the development divisions in the automobile industry make an effort to reduce the car cabling by"intelligent"microelectronic signal transmission.
The data transmission via power wiring could help to avoid cable trees in future sophisticated energy control systems.

Literature

1) Benndorf, H. et.al.: Fernwirktechnik
Düsseldorf 1975

2) VDE 0875 / 0876

3) SCU2o Serial Control Unit, Operations Manual,
Mostek Corporation

4) Rehder, H.: Störspannungen in Niederspannungs netzen
etz. Bd. 100 (1979) Heft 5

5) Spezifikation der Funktionsmodulen PM-G (1)
für GEAZENT 8000 DI (PCS GmbH 1981)

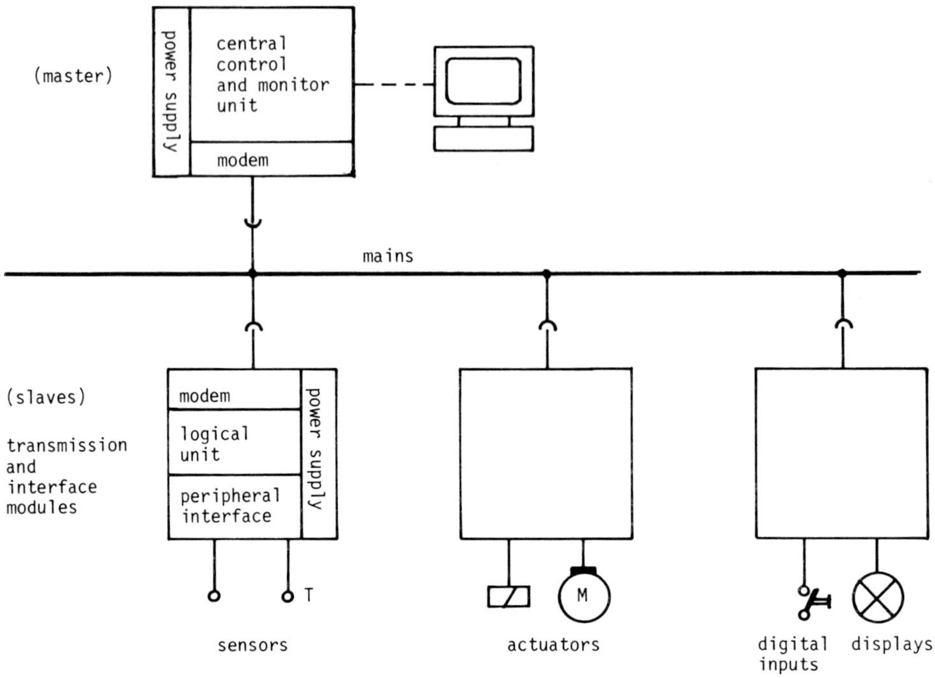

Fig. 1 System Concept

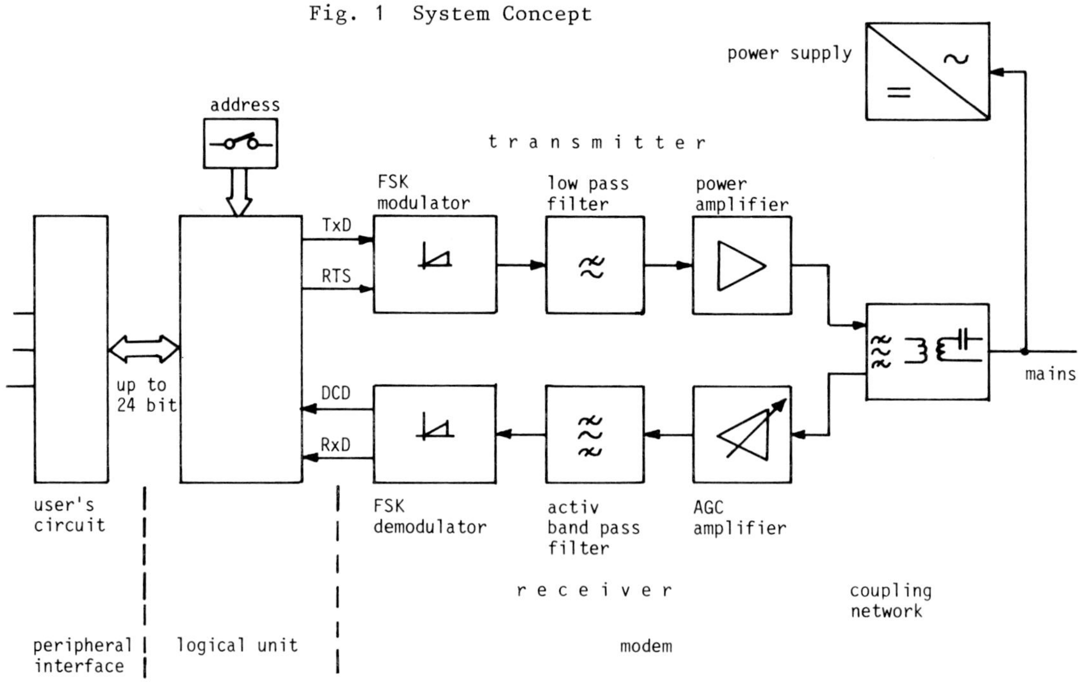

Fig. 2 Diagram of the Transmission and Interface Module

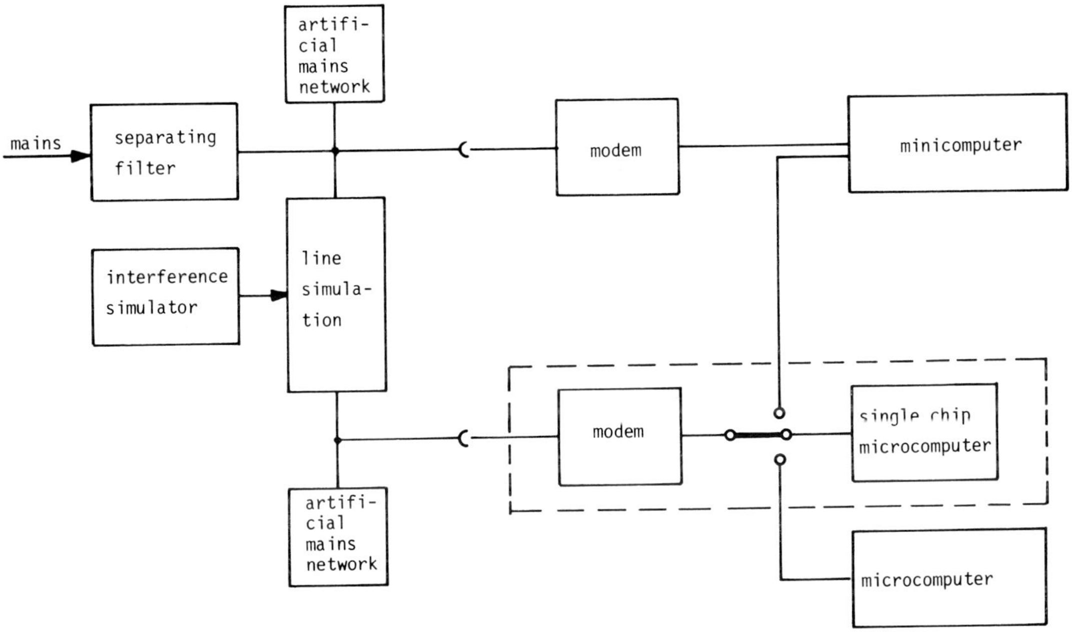

Fig. 3 Experimental Arrangement for Signal Transmission via the Electrical House Wiring

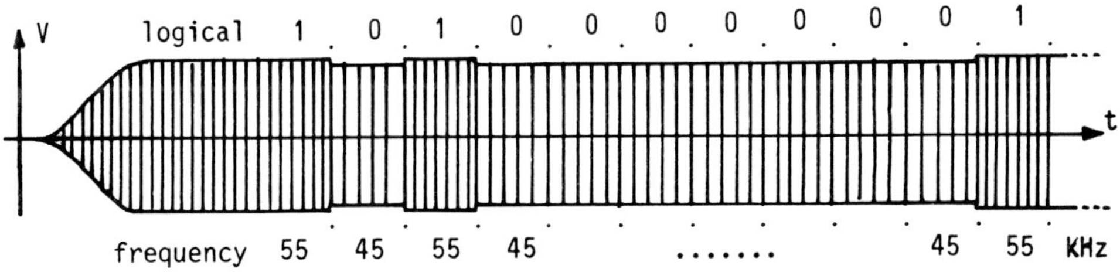

Fig. 4 Transmitted signal (1ste byte) measured at the input of AGC amplifier

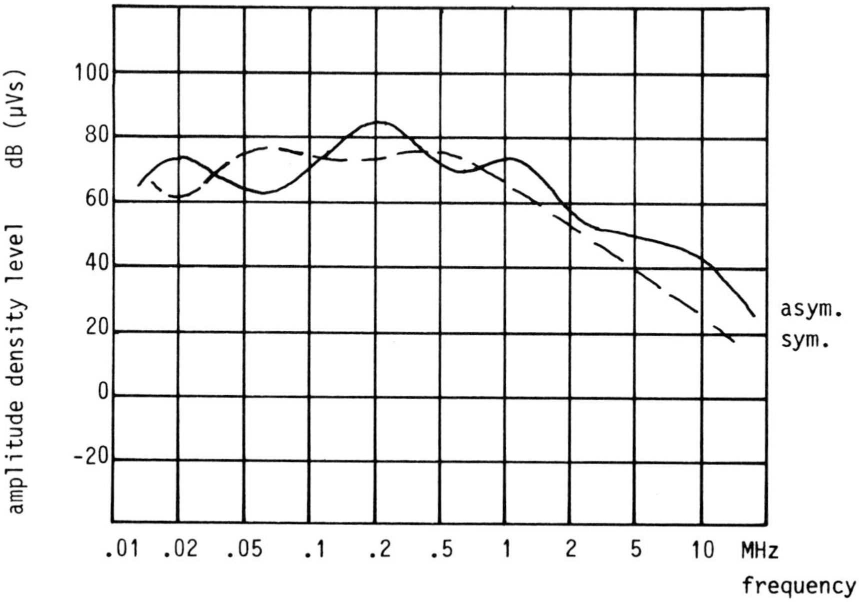

Fig. 5 Interference level on the mains
Envelopes of amplituted density of thousands of measuring points as a function of frequency
Source: Rehder, H.: Störspannungen in Niederspannungsnetzen

ELECTRONIC HEATING CONTROL SYSTEM

Authors	:	FULVIO FONZI, PIERLUIGI SEGALINI
Contract number	:	EEA-1-006-I
Duration	:	36 months 1 January 1980 - 31 December 1982
Head of project	:	Dr. Fulvio Fonzi, Adriatica Componenti Elettronici
Contractor	:	Adriatica Componenti Elettronici S.p.A
Address	:	Adriatica Componenti Elettronici S.p.A Viale della Repubblica, 12 I - 67039 SULMONA (AQ) Tx (0043) 600049, ph (0039864) 33441

Summary

The research studies the possible energy savings using devices wich at different level of complexity execute the control and maintain the heating system in condition of maximum efficiency.
Furthermore the devices must allow a rational use of heating system having thermal energy generation only when there is a real request.

1. INTRODUCTION

In reply to the unfavourable evolution of the cost of the heater nearby the traditional control systems of the burner numerous devices have been developed such as thermostatic valves, thermic energy meter, stations which accomplish the thermoregulation.

Most of these systems are electromechanic and therefore they effect a minor flexible control, they do not allow expansions, they are not easily interface with the user, with the growth of the controlled variants they become rapidly inadequate both functionally and economically.

Furthermore both the traditional systems and the more recent ones have the common characteristic of intervening in a sectorial way over one or more of the variant of the heating system.

L'ACE has in mind with this research to obtain a system which, accomplishing both the control of the burner and the thermoregulation, represents an overcoming of the actual systems.

Furthermore already in this phase, the system is planned in such a way so that new functions can be added without redoing the preceding parts.

2. OBJECTIVE

The purpose of the device in development is the reduction of the energetic waste having in mind the growing costs and the disposability of the fuel, which is always more precarious.

We intend reaching such a goal through two actions:

° control of the combustion so that it happens in stechiometric conditions;

° control of the heating system including the management of situations in which nearby the traditional burner we have other generators such as heating pumps and/or solar collectors.
Every time the optimal configuration of the heater is decided, according to the following basic considerations:

- inserted load
- external climatic conditions
- cost of the used energy.

The achievement of this objective has another remarkable advantage the reduction of the heating pollution.

The project has in mind therefore the following objectives:

° major security of functionality through the computerized control;

° major flexibility and easier use even if the realized functions are more complex;

° division of the heating cost according to the actual waste to be obtained with the accounting of the heat.

3. WORK DESCRIPTION

The work is based on the development of three kinds of prototypes:

° HECOS1 (HEating COntrol System 1)
  Low cost device that control only the activation and desactivation of the burner.

° HECOS2
  This device is based on one more powerful microcomputer and execute all the functions achieved by HECOS1 plus the thermoregulation. Furthermore the control unit, is designed with additional I/O lines for future expansion.

° HECOS3
  Structured identical at HECOS2 employs same I/O lines for operations requested by combustion control. Software is formed by module contained in HECOS2 plus module for combustion control.

4. STATE OF THE WORK

At the present time we have developed two prototypes respectively for device HECOS1 and HECOS2.

The third device will be prepared only after the definitive refinement of two preceding.

4.1 HECOS1 - Block diagram description

This device execute exclusively the following functions:

° Timing of the ignition control according to the type of furnace.

° Control of the circulating of the burned gas through the position of the rolling shutter on the smoke conduit.

° Control of the flame.

° Control of the pressure detectors.

° Self diagnosis of the electronic part.

In fig.1 we represent the device block diagram.

The operation indicated are accomplished by the software, now resident in EPROM, that present the following structure:

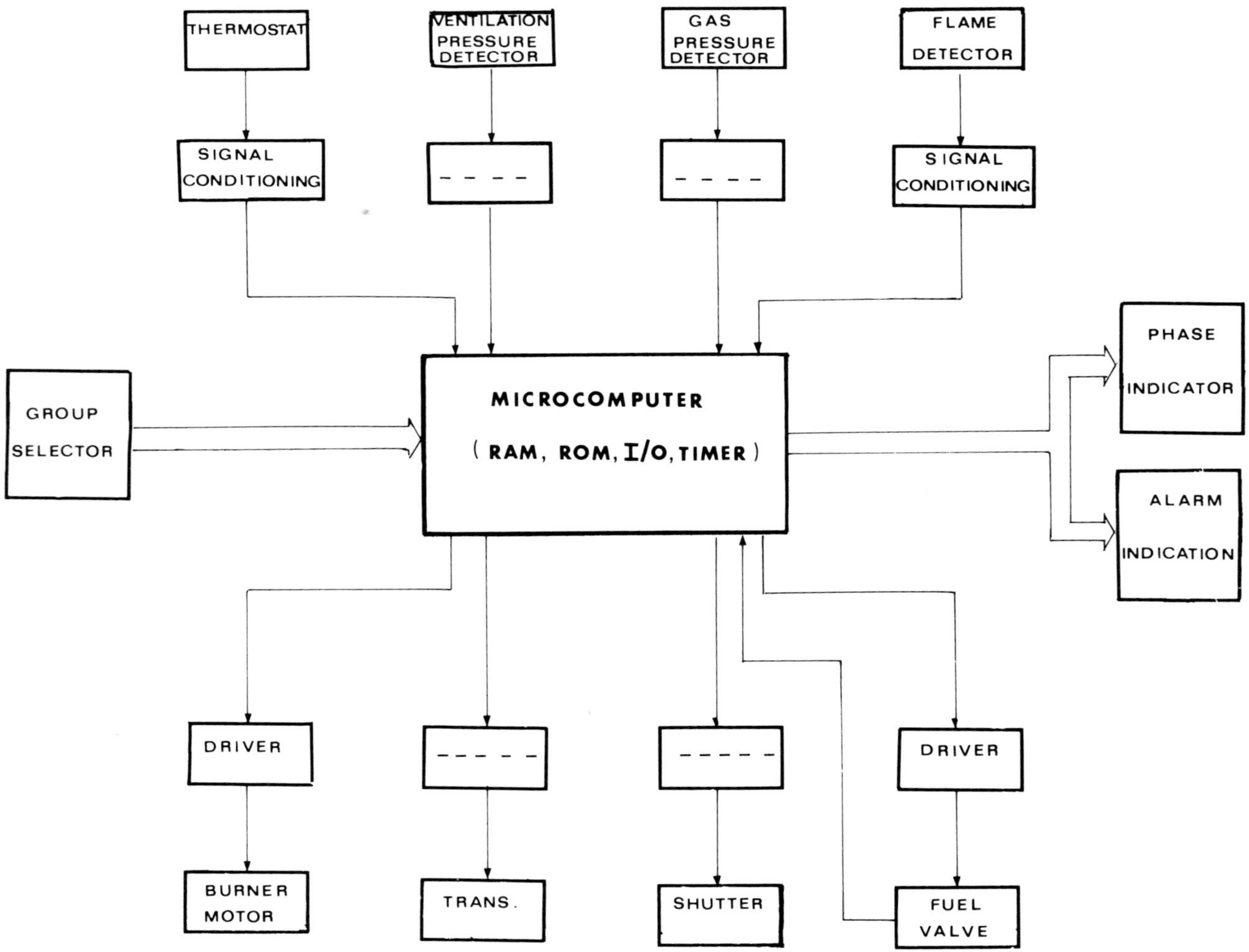
THERMOSTAT
VENTILATION PRESSURE DETECTOR
GAS PRESSURE DETECTOR
FLAME DETECTOR
SIGNAL CONDITIONING
SIGNAL CONDITIONING
GROUP SELECTOR
MICROCOMPUTER
(RAM, ROM, I/O, TIMER)
PHASE INDICATOR
ALARM INDICATION
DRIVER
DRIVER
BURNER MOTOR
TRANS.
SHUTTER
FUEL VALVE

FIG. 1

```
Init
Loop
   Loop
      Read  group selector
      Read  burner status
      If alarm display alarm type
   Repeat until Thermostat  is  off

   Active motor burner and open shutter
   Read status
   If alarm (Air pressure) pass to alarm manager

   Loop until preventilation time is elapsed

   Active transformer and close shutter
   Loop until T2 is elapsed
   Active EVG, open shutter

   Loop until transformer time activation is elapsed
   Desactive transformer

   Loop
      Read status
      If alarm pass to alarm manager
   Repeat until Thermostat is on

Repeat
```

4.2 HECOS2 Block diagram description

This device besides executing all the functions of HECOS1, described in the paragraph 4.1, has the possibility to be programmed by the operator.

Indeed, as it visible in block diagram of fig.3, the device is equipped with a console for interface with the user.

The panel of the device is completely represented in fig.4.

The user has the possibility to plan different types of operations trough six switches located in upper left-hand side of panel.

- WATCH
  With this switch active the operator can update internal watch, visible in fig.3.

- CURVE
  This switch give the possibility to establish the relation between external temperature and temperature of water envoyed in the plant to obtain a nominal internal temperature of 20°C.
  This curve can be different for each kind of plant. An example is given in fig.2. The operator with "curve" switch can store in the device the six value corresponding to three points P1, P2, P3, of

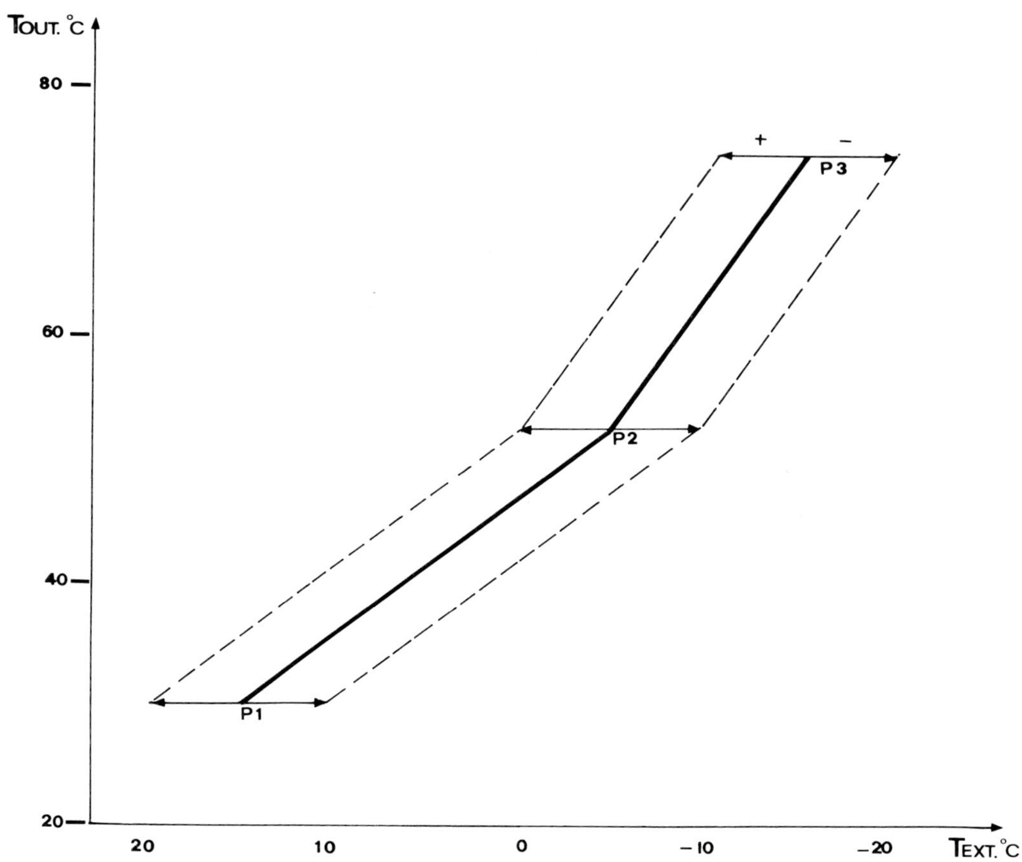

FIG. 2

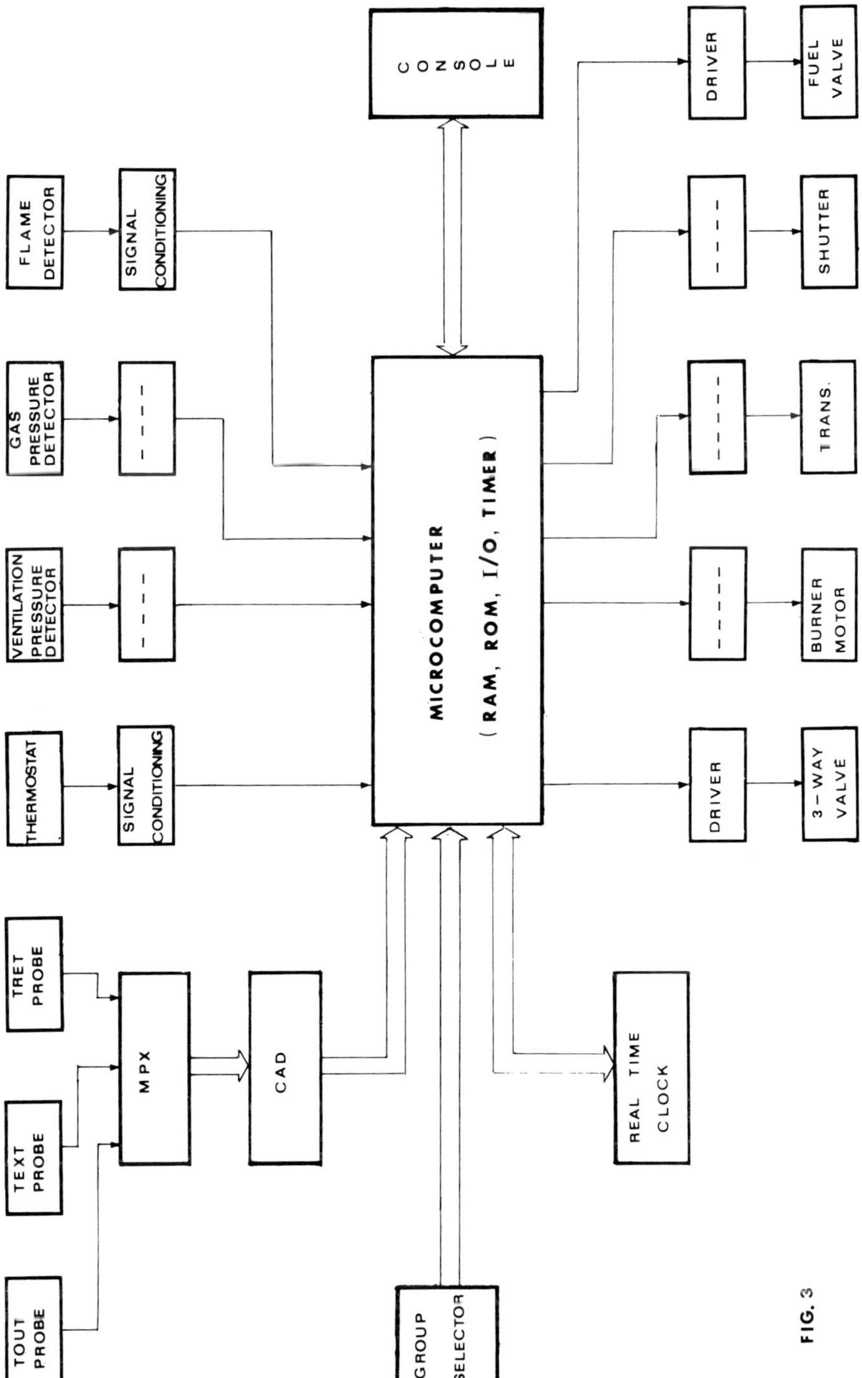
CONSOLE
FLAME DETECTOR
SIGNAL CONDITIONING
GAS PRESSURE DETECTOR
VENTILATION PRESSURE DETECTOR
THERMOSTAT
SIGNAL CONDITIONING
MICROCOMPUTER
(RAM, ROM, I/O, TIMER)
DRIVER
FUEL VALVE
SHUTTER
TRANS.
BURNER MOTOR
DRIVER
3-WAY VALVE
TRET PROBE
TEXT PROBE
TOUT PROBE
MPX
CAD
REAL TIME CLOCK
GROUP SELECTOR

FIG. 3

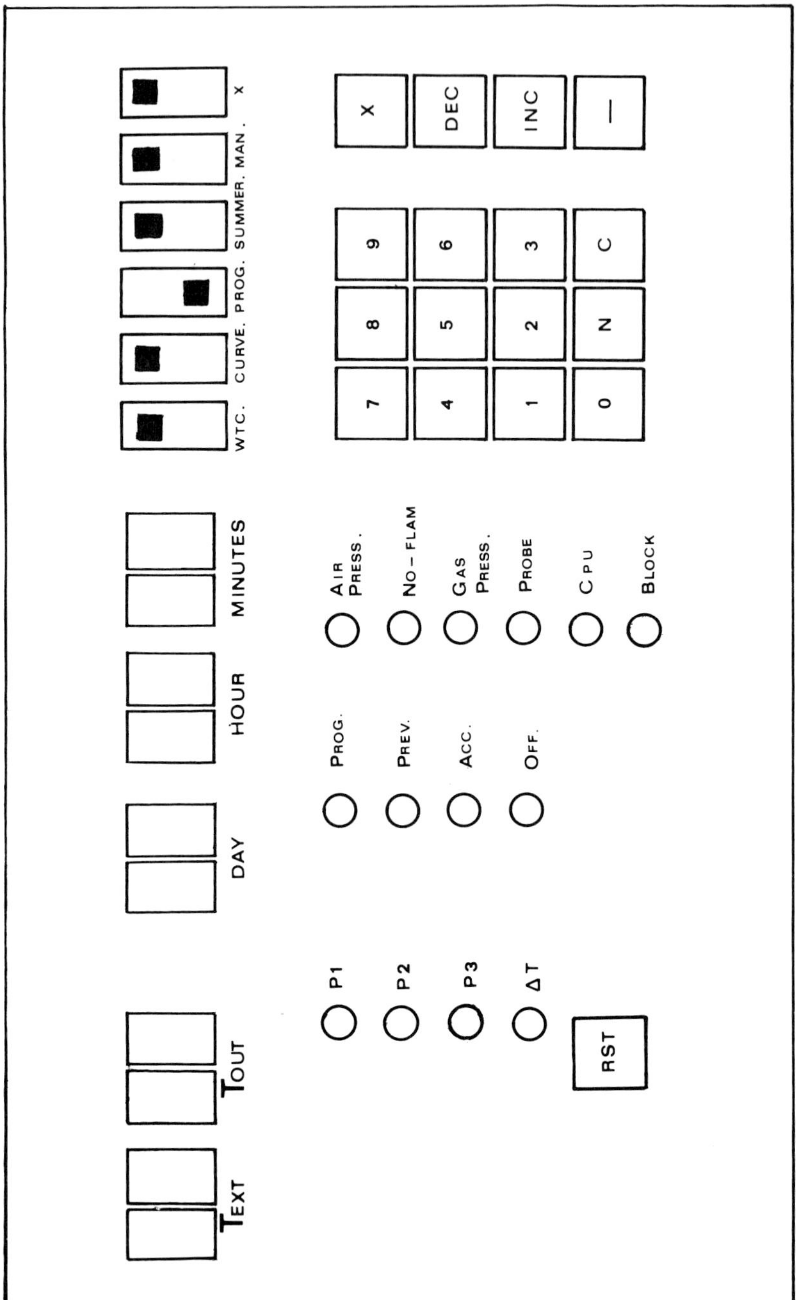
TEXT
TOUT
DAY
HOUR
MINUTES
WTC.
CURVE.
PROG.
SUMMER.
MAN.
X
P1
P2
P3
ΔT
RST
PROG.
PREV.
ACC.
OFF.
AIR PRESS.
NO - FLAM
GAS PRESS.
PROBE
CPU
BLOCK
7
8
9
4
5
6
1
2
3
0
N
C
X
DEC
INC
—

FIG.4

fig.2. Furthermore it is possible to cause a shift upper or under of the curve for successive refinement of the system.

° PROGRAM
With this switch active the user has the possibility to set up the desired heating for a maximum of 15 days with a maximum number of intervals equals to 200 memory locations,fed by standby batteries, reserved for such a purpose.
The desired temperature is obtained regulating the 3-way valve according to the user's choice and the outside temperature.

° SUMMER
The summer switch being set the 3-way valve closes the conveyance toward the plant and the thermic generator attends only the production of sanitary water.

° MANUAL
If this switch is active the 3-way valve is not controlled anymore by the main unit and its position is regulated manually by the user.

The software contained in HECOS2 has a structure constituted by two module.

The first, having the module related with burner control identical to HECOS1, is the following:

```
Init
Loop
   Read panel switches
      If "WATCH" SET OFF BURNER, UPDATE HOUR AND MINUTES

      If "CURVE" SET OFF BURNER, STORE  P1, P2, P3

      If "PROGRAM" SET OFF BURNER,TAKE AND STORE DESIRED TEMPERATURE

   Loop
      EXECUTE BURNER CONTROL
   Repeat until "WATCH", "CURVE", "PROGRAM" is OFF


Repeat
```

The second module is constituted by interrupt routine activated every 30 sec.

Schematically we have the following structure:

Save previous status

Read panel switches
  If "SUMMER" close 3-way valve and exit
  If "MANUAL" exit

Find slopes, S1 e S2, determined by P1, P2, P3

Measure T (e,s)

Let T (e,c) = T (e,s) + (20 -IT)

If T (e,c) < T (e,p2) then S = S2
  else S = S1

Let T (o,d) = T (e,c) * S

Measure T (o,s)

Calculate T (o,s) - T (o,d)

Apply appropriate algorithm to finded value and derive command for 3-way valve.

Return from interrupt

Legenda:

T (e,s)	= Temperature measured with external probe.
T (e,c)	= Temperature external, correct.
IT	= Internal desired temperature.
T (e,p2)	= Temperature external at P2.
T (o,d)	= Temperature out, desired
T (o,s)	= Temperature out from probe.

5. ANALISYS OF THE RESULTS AND COMMENTS

The two devices presented in the previous paragraph, have the same structure hardware and software.

In particular from a point of view of the hardware we have HECOS2 with all the elements of HECOS1 plus the circuit for the treatment of temperature (T_{OUT}, T_{IN}, T_{EXT}) and the circuits for link keyboard and displays.

Furthermore we have a watch that permit to have the desired temperature each moment for every day.

The watch generate an interrupt every 30 sec.

The routine associated with the interrupt accomplish the 3-way valve setting in according with the desired temperature programmed by the user.

In this manner the routine can be nested in HECOS1 software without substantial change.

6. CONCLUSION

At the present time we are in the phase of engineering of two prototypes and we plan to reach two goals:

- ° high grade of reliability;
- ° cost reduction.

Therefore we will deal the phase of inserting in the project the combustion control with wich we are expecting a conspicuous increase of efficiency.

Now the prototypes developed have the following advantages:

- ° use facility
- ° maintenance facility
- ° consumption reduction rising from a rational use of heating system and from maintain the furnace in condition of maximum efficiency.

QUESTIONS AND DISCUSSION

Pilavachi (DG XII) : What is the reliability of the system ? Which computer and which computer language do you use ?

Answer : - At present time we have no data on system reliability because we are testing these aspects of the devices now.
- The system is based on microcomputer 8050.
- For program development we make use of assembly language.

Mauro (CSEA) : You said that, for the regulation of the water temperature, by the three way valve, you make use of a open-loop control. What is the reason for not using a closed-loop method ?

Answer : The 3-way valve control implemented in our device can be evaluated as open loop because we do not measure directly the actual position of valve, but we have an indirect measure of valve position by reading temperature of water delivered at the plant.
From this point of view the control can be considered as closed loop, as is shown in figure.

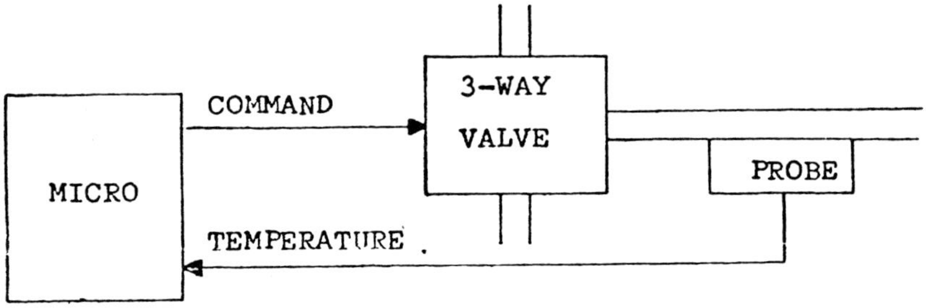

COMPUTER CONTROLLED NETWORK FOR HEATING SYSTEM

Author : V. MAURO

Contract number : EEA/1/009/I (S)

Duration : 14 months - 1.3.80/1.5.81

Contractor : C.S.E.A. - Consorzio per lo sviluppo dell'elettronica e l'automazione

Address : C.S.E.A.
Via Ventimiglia n. 201
TORINO

Summary

The paper presents the results of a research on a distributed system for energy saving in buildings heating.
The experiments have been carried out by installing and testing the system for one season in two different buildings.
The research has shown the system effectiveness in energy saving and has pointed out some of its peculiar characteristics.
The results of the research have been very useful for stimulating some improvements of the system.
After the experimental work, CSEA has decided to proceed toward the industrial production: in winter 81/82 an engineered pre - series will be installed in 10 more buildings.

1. Introduction

The system which has been under investigation during this research, is intended to save heating energy in buildings with a centralised heating system (hot water) and many individual users.

Typical applications are represented by block of flats for residential use, schools, office buildings, etc.

It is well known (1) that a good plant regulation, together with a distributed control of ambient temperature may give a substantial saving.

Moreover, the houses are never fully occupied (both in time and in space): temperature reduction during empty hours is another way of saving energy.

Finally, direct involvement of users in temperature regulation via a direct connection between consumption and individual expense may substantially reduce the total consumption.

"Heat Robot" has been designed to accomplish these needs. Before this reseach was started, a feasibility study (2), and a simulation anasysis made on the basis of average italian data (1), gave a forecasting of the energy saving induced by a system of this type as high as 30%.

2. Description of the system

2.1 System philosophy

Energy saving in "Heat Robot" is obtained by means of the following actions:

a) Diversification

the system permits a programming of differing temperatures for each individual room and for each hour of the day. In this way unused (or little used) rooms are not heated unnecessarily, while occupied rooms (chosen by the user) are assured a comfortable temperature for the period (and only that) of use.

The user may then graduate the comfort in the various rooms, by simply asking for higher or lower temperatures.

Technical data:

* Minimal interval of programming: 1 hour
* Three types of days: weeks days, half holidays (Saturday), holidays (Sunday)
* Temperature regulation: variations of 0.5 degrees

b) Accounting

The system measures (or estimates) the quantity of heat actually consumed bu the user and therefore permits a more accurate division of expenses. The consumption data are always accessible to the user, besides condensed information which indicates whether, in the most recent period, he is consuming proportionally more or less than the average tenant. This is done in order to stimulate the user to adopt his own politics of energy saving which will include the attempt to programme requests for heat and other ways of saving energy which may occur to him (double glazing, insula

tion, etc.). In calculating the amount of energy consumed, the so called "heat thefts" (heat from a heated apartment which is lost because of an adjoining unheated one) are kept in mind.
Technical data:

* The individual consumptions are estimated every minute, by taking into account various temperatures, the state of single valves the value of building parameters (a priori known and identified)
* individual consumptions are centralised totalised and, when needed, corrected by comparison with total energy produced
* "heat thefts are computed centrally
* individual consumptions are centrally stored
* expenses are subdivided with the help of rules which can vary from house to house

c) Signalling of absence

The user may, with simple commands, signal to the system that the apartment (or even a single room) will not be used for a certain period. This function is particularly useful to effectuate further saving in cases in which the house will be empty for periods of several hours or days. On the other hand, the possibility of signalling unforeseen re-entry, that is, outside the usual programme, has been allowed for in a way in which it is possible to obtain a comfortable temperature in a short time.
Technical data:

* Minimum absence or presence which can be signalled: 1 hour
* Maximum absence or presence which can be signalled: 99 days

d) Learning

The system is able to define in an autonomous way the heating program which is adapted to each individual user: this feature is useful because, in a long term, it is easily predictable that very few users will continuously change and adatpt their heating programs. To accomplish this, Heat Robot analyses all the user communications to make a time map for the various rooms. This map is continuously upgraded in case of lack of communications, it defines "minimum-risk-minimun-energy" policies.
Technical data:

* Time map for the various rooms (time interval = 1 hour), for weeks days; it is updated locally with bayesian algorithms based on user intervention and user programs
* The set-point for local temperatures is dynamically decided, on the basis of the time map, and of the nominal value, which also is locally learned
* Individual maps are centrally stored, for back - up purposes

e) Adaptability

The Caldo Robot is an "intelligent" system: which means it is not limited to executing an imposed heating programme but that it can adapt itself and learn information autonomously

For example, the system "learns" how the heating plant is made, the orientation of the rooms and other data in order to correct possible errors in the design of the plant itself.
Technical data

* Self - tuning of control parameters
* On - line identification of average nominal power for various rooms

f) Optimisation

The functioning of the central boiler is controlled on the basis of requests for heat from all the apartments served, in such a way as to maintain a high level of efficiency of the boiler, and to decrease the stress placed on it and so increasing its life
Technical data

* Plant temperature is continuously regulated
* Boiler temperature is set to the lower limiting value, compatible with safety requirements and with one hour forecasting

2.2 Description of the hardware

The "Heat Robot" (see fig. 1) is made up of a network of microprocessors(one for each apartment) connected to a central microcomputer. The apartment microprocessor measures the internal temperature and command, directly, the special electrothermic valves placed on the radiators.

The central microcomputer measures the external temperature and the temperature of autgoing and incoming water of the plant and in consequence acts both on the lighting of the burner and on the mixer valves.

All the functions of the processing and memorisation are distributed in a suitable way between the various units of the apartments and the central unit.

It is important to note that the system is protected from temporary power failure of the network (blackouts, mor or less programmed) it contains a non-volatile memory bank which maintains the data until a return to normality.

Technical characteristics

Local unit	Microprocessor 8031 16 K EPROM 1 K RAM Serial data link (1200 baud) Max 7 temp. probes 8 DAC channels 20 character alpha-numeric display VI Numeric and function keys (25)
Control capability	8 rooms, 3 radiators per room
Valves	Type electrothermical ON-OFF Hydraulic compatibility with manual valves

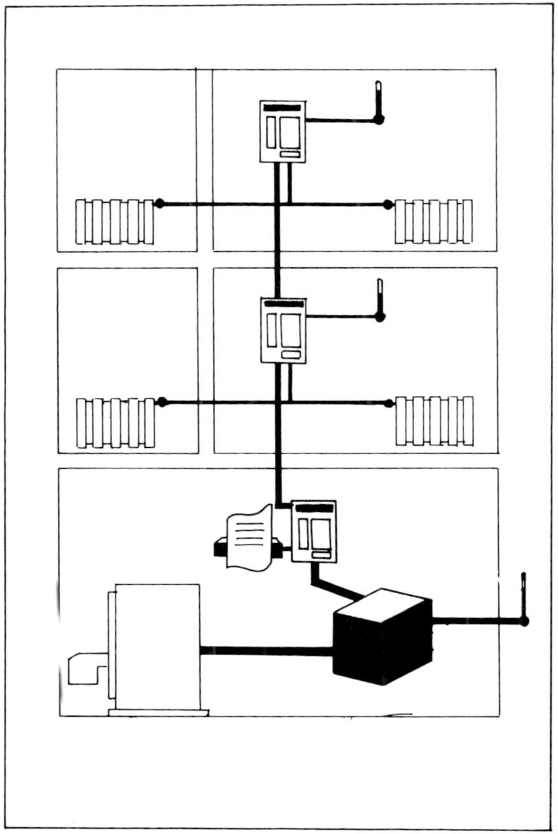

Fig. 1 The system

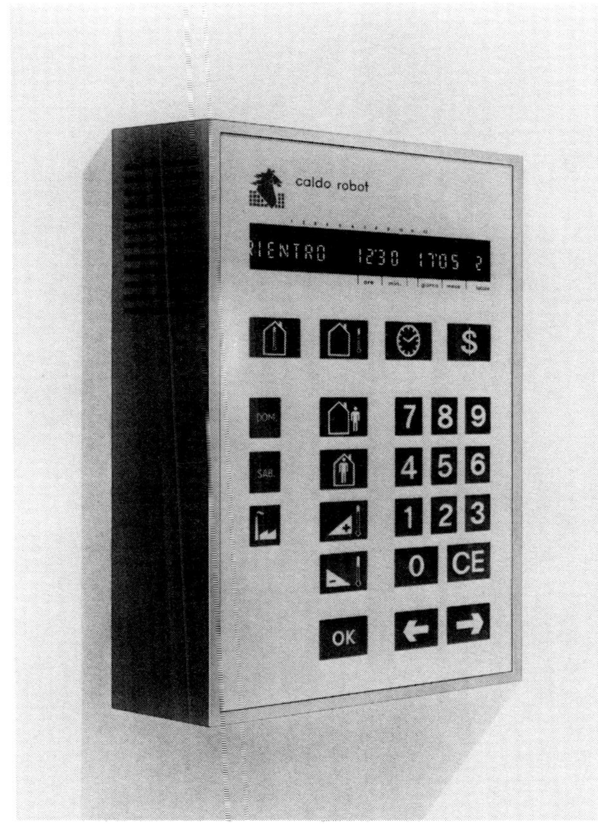

Fig. 2

Central unit	Microprocessor 8085 A2 128 K bubble memory 48 K EPROM 16 K RAM Serial data link Absolute clock
Operator terminal	Same as local unit, with impact printer

Installation

The system can be installed in old or new buildings. In the case of old buildings it involves:

- installation of the apartment unit (of reduced dimensions and an attractive appearance) generally in a suitable position (in the entrance hall); the predicted dimensions are about 15 x 20 x 4 cm. (see photograph in fig. 2)
- substitution of the manual valves with electrothermic ones on each radiator
- electrical connection between the valves and the command unit through a cable of reduced dimensions (to this end it is possible to utilise, where possible, existing pathways). The system of communication between the local units and valves permits both daisy chain and star connections
- electrical connection between all the apartment units and the central unit
- installation of the central unit in the boiler room (or in another suitable common room)
- installation of the external probe, of the thermal flow measure and connections to the central unit
- electrical connection between the central unit, the mixer valve and boiler thermostat

In the case of new buildings the operation of installation is simplified, especially in respect of the valves (in as much as substitution if not necessary) and electrical connections (foreseen in a way which is analogous to that of the mains tension).

3. The experimental work

In the winter 1980/81, the system has been installed in three buildings and fully tested in two of them, having the following characteristics

a - residential building in Melegnano (Milano)
14 apartments (av. 100 m^2.)
5 radiators per apartment

b - School building in Torino
326 radiators
26 zones (26 local units)

In this section two main results of the research will be presented, with reference to aspects of energy saving, and effectiveness of the learning algorithms

3.1 Energy saving in case a

The research period was subdivided in two parts: in the first the system was not used and in the second it was: by comparison of the daily consumption in the two periods and by taking into account the external temperatures it is possible to deduce the amount of energy saved via a linear regression model (see (3) for details).

Experimental results give for the model of total building consumption the following expressions:

$$C_1 = 137 - 6\ T_e \qquad (3\text{-}1\text{-}1)$$

where

C_1 (lt) is the daily consumption without Heat Robot

T_e (°C) is the day-average external temperature

$$C_2 = 115 - 6\ T_e \qquad (3\text{-}1\text{-}2)$$

C_2 (lt) is the daily consumption with the Heat Robot

The difference is 22 lt/day or, in percentage, 20%. This estimate has to be considered as conservative, because

- no economic incentive for saving was given to the users
- presence was above the average italian value, due to the high percentage of retired people in the house

In fig. 3, indeed, the average declared presence (derived from user programs) is reported, versus time-of-day.

The continuous line indicates the percentage of flats with at least one room occupied; the dotted line indicates the percentage of house volume occupied.

Both lines are definitely near to the 100% level

3.2 Energy saving in case b

In the latter case, the comparison has been made on the basis of total consumption and period averaged external temperature: again, the season has been subdivided in two period, as for the former case.

Experimental results give a percentage saving of 30%(see 3 for details)

3.3 Effectiveness of the learning algorithm

As in has been pointed out is section 2, one of the major novelties in Heat Robot design is represented by its ability to learn: during the research the effectiveness of this feature has been investigated.

Since the learning algorithm operates in a long-term, the

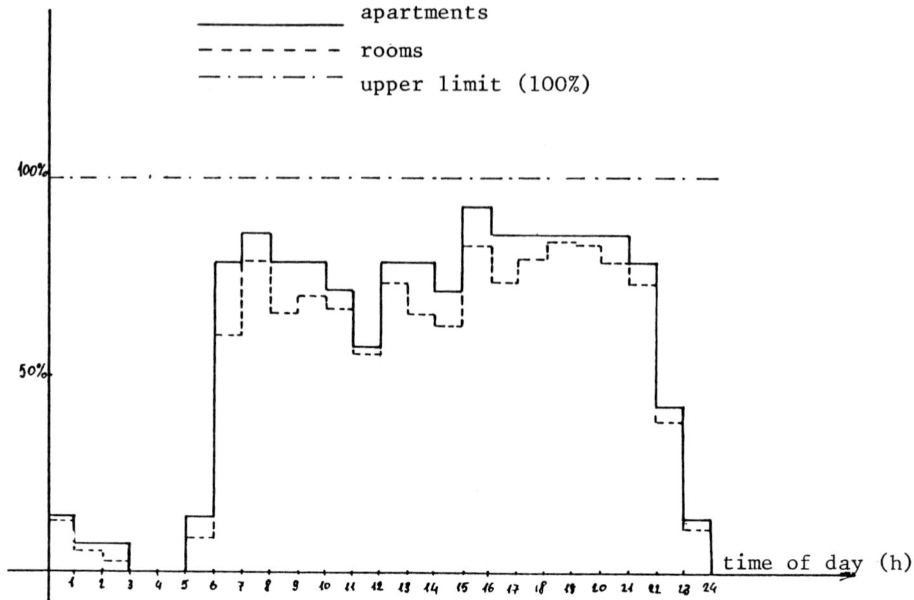

Fig. 3 Percentage of occupation declared

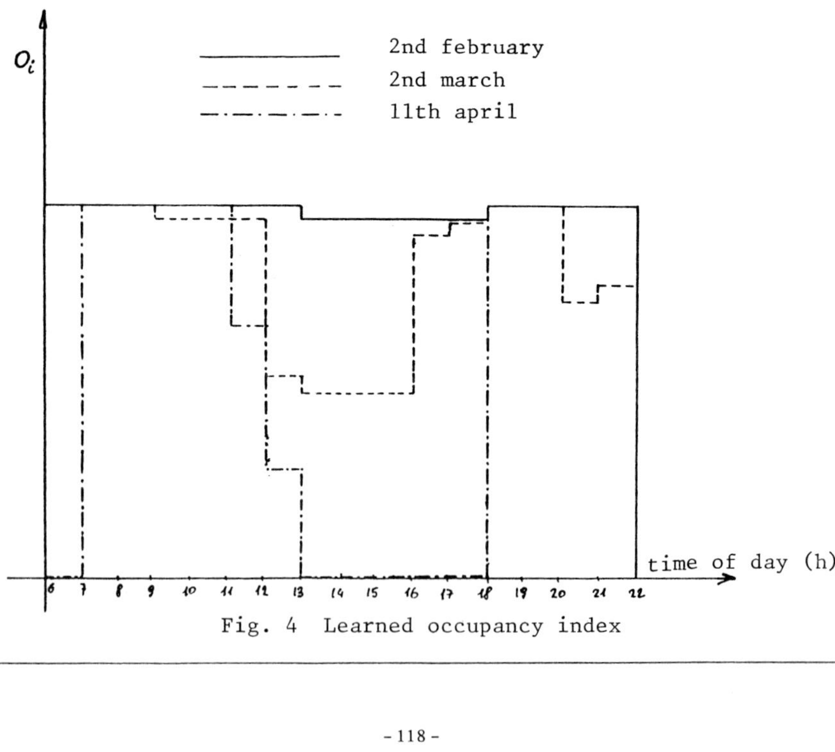

Fig. 4 Learned occupancy index

results can be seen by looking at the time-map after same months of operation.

To this end, in this section the dynamic of a "presence index" will be analysed, for different apartments of the Melegnano house.

Let us define

$$O_i = \frac{1}{S} \sum_{j=1}^{N} S_j \cdot \alpha_j \qquad (3-3-1)$$

where α_j is 1 if the jth room is considered as "occupied" at i^{th} hour and 0 elsewhere

S_j is the surface of radiators of ith room

S is the total surface of radiators in the apartment

N is the number of rooms

O_i can be considered as a "occupation index" at i^{th} hour

Let then

$$\overline{O} = \frac{1}{T} \sum_{i=1}^{T} O_i \qquad (3-3-2)$$

where T are the boiler's working hours in a day

$\overline{O}$ is then the day-averaged occupation index

In fig. 4 the index O_i is reported in three significant periods for apartment n. 1;(user of apartment n. 1, as shown by questionnaires, in one of the best users of the local unit).

Table I reports the index $\overline{O}$, at the same times, for five apartments.

Apartment	febr.12	march 2	april 4
1	0.99	0.83	0.56
3	0.89	0.94	0.05
9	0.87	0.53	0.06
10	0.75	0.66	0.69
13	0.87	0.50	0.35

Table I - Values of $\overline{O}$ (occupation index)

The table I and the figure 4 confirm two facts:

- the users tend to "overprogramme", by declaring presence when it is not needed
- the learning algorithm estimates the room's "energy requirement per hour", rather than presence/absence: in hot season

and in the warmer hours it will simply declare "heating not needed"

3.4 Other experimental data

The large amount of data collected by continuous inspection at the experimental sites, by contacting the users, by questionnaires, fault analysis, etc., have given also other interesting results:

- all the original algorithms and methods designed for Heat-Robot (including methods for estimating individual consumption and expenses), are completely satisfactory: some minor modifications which have been shown useful were done on field during experimental interval and have improved the system efficiency
- the friendliness of methods for user interaction with the system is of great importance for "Heat Robot". The small Keyboard of the local units, together with the colloquium itself has proven to be unsatisfactory. CSEA has substantially modified the system, in view of industrial development
- the reliability of a system such "Heat Robot" has proven to be an important factor of success: the user could never become confient in a system which is not highly reliable and "fail proof". Again, in view of industrial development, this is a crucial point
- the experimental system have shown a major drawback in their difficult installation, testing and in plant initialisation. These drawbacks increased the time and cost of system start-up, as well as the level of user unsatisfaction during the first weeks of experiments. Same substantial improvements of the design have been derived from these observations
- during the last weeks of experiments, questionaires were distributed to the individual users. The analysis of answers has shownthat, after a "rude" first approach due to the above mentioned difficulties, the "Heat Robot" has been friendly accepted by the users

4. Conclusion

In the winter 1980/81 the system named "Heat Robot" has been installed in three buildings and fully tested in two of them.

Testing has been intended to verify:

a - system capabilities
b - energy saving
c - user reaction

Conclusion that can be briefly drawn from the experimental period 80/81 are the folloquing:

- the "Heat Robot" system is in condition to work as foreseen

with simulation and to obtain saving estimated.
In fact even in the limiting conditions of the experiment (prototype system, no economic incentive for user) saving has been obtained higher than

20% (Melegnano)
30% (Torino school)

- algorythms realized, with improvements already effected or at least planned and simulated, are amply sufficient to ensure the systems performance and its diffusion
- hardware used, with the improvements found necessary during experimentation,allows functioning of all the algorythms foreseen and system's utilisation by the user
- accessories installed (valves, thermal probes, etc.), have suitable characteristics for integration in the system

Experimentation has therefore proved satisfactory for principal requirements and has shown that "Heat Robot" can offer combined saving, comfort, simple use.

It is therefore no longer strictly necessary to prolong experimentation. At the time of the pre-series installation (season 81/82) parallel research work can be effected on the following points:

- condition for and testing of installations
- methods for subdividing expense and relative additional saving

Experimentation carried out has however, even regarding these points, given the CSEA work group important items of information, which can be considered almost definite project solutions for pre-series 81/82.

The system, after the research work, is now ready for industrial production: indeed, a small number of systems (10 buildings, about 300 microprocessors) are in production to enlarge the base of installations and to reinforce the cost-benefit analysis.

This new series has been redesigned, by taking into account all the research results: this series is intended to work also as a small scale example of industrial production, in view of the market introduction of "Heat Robot" which is now foreseen, as for as Italy is concerned, in the first month of 1982.

List of references

(1) CNR - PFE - Risparmio di energia nel riscaldamento degli edifici - Vol. 2 (1979) - Vol. 3 (1980) PEG editrice

(2) CSEA - Studio di fattibilità per il CALDO-ROBOT - Internal report - 1980

(3) CSEA - Final report for CEE contract n. EEA/1/009/I

QUESTIONS AND DISCUSSION

Zegers (DG XII) : What is the cost of the control system ?

Answer : The installation of such a control system in apartment buildings costs 600 to 800 ECU per apartment. It is believed that this system is too expensive for single houses.

Zegers (DG XII) : How is the heat consumption for each individual apartment established and did you encounter any problems ?

Answer : The bill is established with help of the computer data. However often users do not agree and complain.

PERFORMANCE OF VARIOUS HEATING SYSTEMS

Authors	T. FULLER, P.J. MINOGUE, M. FINN
Contract Number	EE-A-3-007-EIR(N)
Duration	33 months 1 October 1980 - 30 June 1983
Head of Project	P. T. Pigott, An Foras Forbartha
Contractor	An Foras Forbartha (National Institute for Physical Planning and Construction Research)
Address	An Foras Forbartha, St. Martin's House, Waterloo Road, Dublin 4, Ireland.

Summary

The purpose of this project is to study the capability of conventional heating systems to meet various temperature regimes and their efficiencies and to introduce innovations in control and heat storage to these systems to improve their performance.
This paper discusses the experimental arrangement for the study and the theoretical framework within which the results will be interpreted.

1. INTRODUCTION

The work to be carried out in this project will involve setting up two test houses with instrumentation for detailed monitoring of internal and external climate, and the performance of heating systems and simulated occupancy. In general, the aim of the work is to study a number of conventional heating systems from the points of view of efficiency and ability to meet various temperature regimes. It is proposed that modifications will be introduced to some of the heating systems with the aim of improving their efficiency and more fully meeting user requirements. The data on system performance will be used to propose suitable use patterns and methods of control for the various heating systems and these will be evaluated in the test houses.

2. EXPERIMENTAL ARRANGEMENT

2.1 Houses

A pair of new, three-bedroomed, adjacent, semi-detached houses have been acquired in a large private housing development about ten miles from Dublin city centre. These houses were chosen because they were deemed to be typical of large number of houses currently being built in the private market. Figure 1 shows a schematic diagram of the test houses.

Modifications were carried out to the houses to facilitate monitoring and to allow various heating systems to be installed in one of the houses. Also the wall insulation was increased to meet new insulation standards of $0.6 W^2/m^2 {}^oC$, expected to come into force in Ireland on 1 January 1982.

2.2 Data Logging and Control Equipment

Transducers have been installed in the houses to monitor the following parameters:

- air and globe temperature
- water temperature
- wall surface temperature
- relative humidity
- solar radiation
- windspeed and direction
- electrical energy
- energy flow in pumped hot water circuit
- energy content of hot water storage tanks

Also occupancy will be simulated in the house in the following respects:

- heat gains from occupants
- lighting
- appliance usage
- hot water usage
- curtain drawing

2.3 Computer System

The data logging and control functions will be controlled by a computerised data logging and control system - the COMPULOG SYSTEM 3. A block diagram of the system is shown in Figure 2.

A real time operating system, RS X 11S, has been implemented on the system and a suite of programs developed to perform the various data logging control functions.

3. CALIBRATION OF HOUSES

The first phase of the experimental work will be to calibrate the houses. Direct electric radiator heating will be used in both houses during calibration and both houses will be maintained at the same temperatures. Because the project is studying the performance of heating systems for different heating patterns, the houses will be calibrated for different temperature regimes. The regimes will be discussed below.

For each temperature regime, the energy consumption in the test house, E_1, will be related to the energy consumption in the control house, E_0 by a regression relationship of the form:

$$E_1 = aE_0 + b \qquad 1.$$

A previous study using a twin house arrangement (1) has shown this type of model to give a good fit for prediction purposes.

It is also intended to fit a basic degree day model of the form:

$$E_i = a_i\,(T_{ini} - T_{out}) + b_i \qquad 2.$$

where $i = 0$ for the test house,
and $= 1$ for the control house,
and T_{ini}= internal temperature,
T_{out}= external temperature.

This model will be used as a measure of the accuracy of Model 1. From the results of (1) it is expected that Model 1 will provide a more accurate prediction of the energy consumption in the test house.

Model 2 will be refined by the inclusion of terms involving solar gain and wind velocity with a view to reducing the variability in models of this type. This is, in a sense, a by product of this experimental arrangement in that data on energy consumption and internal and external climate from the control house will be available for testing models relating energy consumption and other relevant parameters. This work will be carried out in cooperation with the University of Liege using their LPB suite of simulation programs.

Model 1 will be used to assess the performance of the heating systems under study when they are installed in the test house. If for a given period E_0' is the energy consumption in the control house and E_1 is the energy consumption of the heating system being tested in the test house, then Model 1 will give:

$$E_1' = aE_0' + b$$

the energy consumption which the test house would have had using direct electric heating and this can be compared with E_1.

However, the test heating system may not achieve the same temperatures as the control heating system and this must be taken into account when comparing heating systems.

This will be done by defining measures of temperature achievement. The temperature regimes will be defined in terms of desired temperatures, T_D, in specified rooms for specified time periods. The systems will be turned on to allow a fixed preheating time before the desired temperature is required.

If T_D is desired temperature for period of length t, and T_{inij} are achieved temperatures measured over the period in house i, then the index of underheating is defined by

$$U_i = {}^1\!/_t \sum_j d\,(T_D - T_{inij}) \quad T_D > T_{inij} \qquad 3.$$

and the index of overheating is defined by:

$$O_i = {}^1\!/_t \sum_j d\,(T_{inij} - T_D) \quad T_{inij} > T_D \qquad 4.$$

During the calibration these indices will be calculated for both houses and relationships will be fitted to the data which express U_1 in terms of U_0 and O_1 in terms of O_0.

It might be expected if a model of the form 1 relates the energy consumption of house 1 to that in house 0, then similar linear models will relate the temperature achievement indices.

When a heating system is being studied in the test house these indices can be compared to those predicted by the control house. These indices will also be used to evaluate innovations in control to the heating systems being studied, where it is desired to decrease energy consumption while achieving desired temperatures.

4. TEMPERATURE REGIMES

On the basis of previous studies carried out on domestic heating patterns in Ireland,(2) and (3), three temperature regimes have been identified as being typical of those found in this country:

(a) Whole house heating, 16 hours per day.

(b) Whole house heating part of the day and partial heating at other times.

(c) Continuous partial house heating, 16 hours per day.

The desired temperatures for rooms when heated will be those given in the CIBS Guide A1 1978 (4) as shown in Table 1.

Table 1 : Desired Temperatures in Heated Rooms

Room	Desired Temperature
Living room	21°C
Bedrooms	18°C
Kitchen *	16°C
Passage and stairs	16°C
Bathrooms	22°C
Toilets and cloakrooms	18°C

*In these houses the kitchen forms part of the living room and hence will be maintained at living room temperature.

For temperature regime (a), the temperatures in Table 1 will be desired in all rooms for the period 07.00 to 23.00 while for temperature regime (b) these temperatures will be desired in all rooms for the periods 07.00 to 09.00 and 18.00 to 23.00 while for the period 09.00 to 18.00 only the living rooms have the desired temperatures. For temperature regime (c) the living areas only will have the desired temperatures for the period 07.00 to 23.00

REFERENCES

1. R. RAYMENT, K. MORGAN. The performance of heating control systems in experimental houses. International Seminar "New Ways to Save Energy" Commission of the European Communities, Brussels, 1979.
2. P. MINOGUE. Occupancy Patterns and Energy Use in Housing. Proceedings of a Symposium - Thermal Performance and Energy Use in Housing. An Foras Forbartha, Dublin 1977.
3. T. FULLER, P. MINOGUE. Efficiency and Energy Conservation Potential of Domestic Heating Systems. An Foras Forbartha, Dublin 1981.
4. CIBS GUIDE A1. Environmental Criteria for Design. The Chartered Institute of Building Services, 1978.

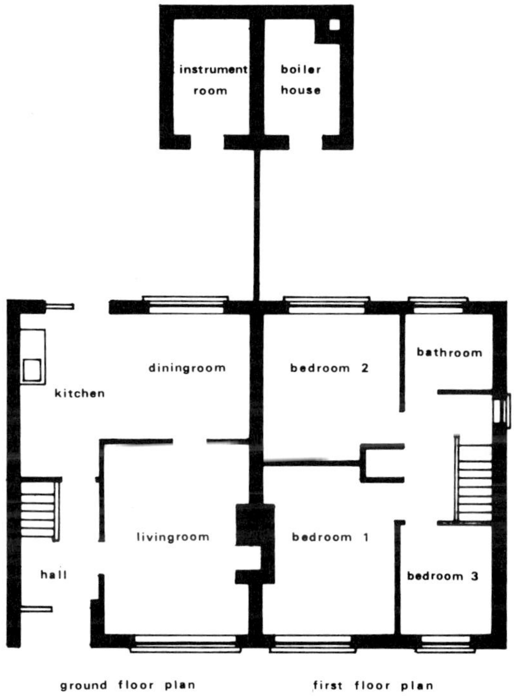

Fig. 1 Test houses and instrument room/boiler house

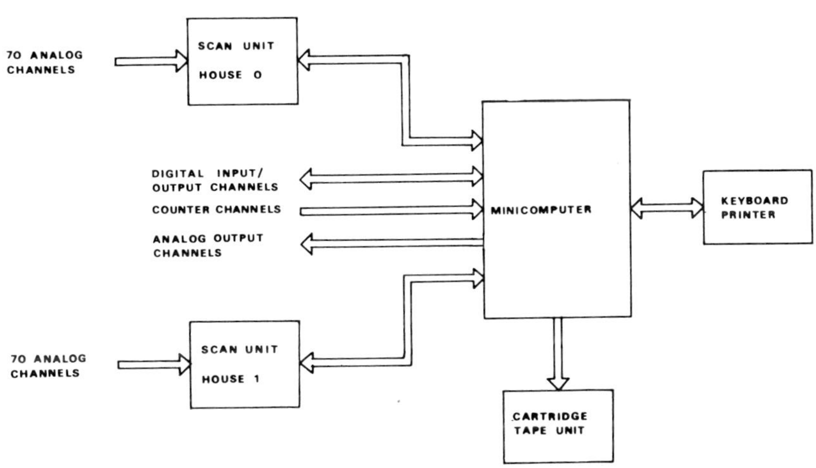

Fig. 2 Datalogging and control system

REDUCING ENERGY CONSUMPTION OF BUILDINGS BY USING MICROPROCESSORS IN CLIMATE CONTROL

AUTHORS : H.J. Nicolaas

CONTRACT NUMBER : EEA-1-006-NL

DURATION : 42 months, January 1, 1980 - June 30, 1983

HEAD OF PROJECT : Ir. P. Euser, Institute of Applied Physics TNO-TH

CONTRACTOR : Technisch Physische Dienst TNO-TH
(Institute of Applied Physics TNO-TH)

ADDRESS : P.O. Box 155
2600 AD DELFT, The Netherlands

Summary

In this paper a short state of the art will be given of applications of microprocessors in the climate control of buildings for a number of relevant industries.
Results of computer calculations will be given, explaining the influence of better control systems by means of microprocessors.
The calculations are carried out with computer programmes for the calculation of yearly heating and cooling loads in a building and the yearly energy consumption for various zones in a building with different airconditioning systems. Some applications will be treated in this paper such as:

- The possibility of mechanical ventilation during the night to cool down the building mass and to reduce the maximum cooling load and the maximum inside air temperature during daytime.
- Control of a lighting system by means of on/off switching depending on daylight level. This lighting system is tested during a year in an office building. With computer calculations maximum energy saving possibilities with an evaluated control lighting system are carried out.

1. Introduction

An important application of microcomputers in the climate control of buildings is based on the possibilities of reducing energy consumption. To see what these possibilities are, the state of the art in the relevant industries is considered. Information of ten firms, which are working in the field of heating, cooling and airconditioning in the Netherlands, was collected and a comparison will be made in the next chapter [1]. After that, attention will be paid to some specific aspects of reducing energy consumption by means of microprocessors.

2. State of the art

The applications which are given by the relevant industries can be distributed under special topics, such as: heating, cooling, airhandling, lighting, sunshading devices, maintenance and alarm.
The various subjects in each topic are:

Heating:
- regulation of gas-air mixtures in burners
- regulation of fans
- regulation of supply water depending on solar radiation, outside air temperature, windspeed and orientation
- duty cycling of heaters

Cooling:
- regulation of fans for cooling water
- regulation on temperature of supply cooling water

Airhandling:
- enthalpy regulations. By ventilating with outside air when the enthalpy of the outside air is lower than the enthalpy of the return air.
- using mechanical ventilation during night and weekend to cool down the building mass
- speed regulation of fans depending on heating and cooling demands
- duty cycling of fans
- day, night and weekend regulation (night set-back) depending on acceptable inside air temperatures
- start/stop regulations for heating and cooling depending on beginning and end of working time.

Lighting:
- switching and regulation of artificial light depending on daylight level
- switching on/off at the beginning and at the end of normal working hours

Sunshading:
- regulation of sunshading devices depending on solar radiation and lighting during working hours
- regulation of sunshading depending on heating and cooling demand during weekends and outside working hours.

Maintenance:
- praeventive maintenance lists depending on working hours of the installation

Alarm:
- mentioning of alarms
- counting of frequency of alarms
- restarting programmes after alarms

The here mentioned applications can be classified in three general items namely: monitoring, control and optimizing possibilities.

3. Monitoring

In the descriptions of the microprocessor applications in climate control of buildings, monitoring always plays a very important part. The measuring and registration of temperatures, water and air quantities, electrical consumption and so on gives the possibility to calculate the energy consumption of the various parts of the total system.

These monitoring functions, mostly embodied in the so-called 'Building Management Systems' become more and more important for energy conservation strategies. It is important to know how the total energy consumption of a building is distributed among the various energy users before conservation applications can be started.

4. Control

The second application for a microprocessor in climate control in buildings is for control and regulation functions of processes. All on/off functions such as mechanical switches and mechanical clock functions can be replaced by microprocessor control. The conventional pneumatic control systems can be replaced by so-called Direct Digital Control with a microprocessor. This contains the control of a process by means of a a microprocessor. The computer compares the signal of the sensor with information in the programme, performs calculations based on the control algorithm used and sends the action signal to the end-device. This control algorithm can for instance be a proportional, integral or derivative action or a combination of these, [2, 3].

5. Optimizing

The third important way in which the microprocessor can be used in relation to the control function are the possibilities of optimizing the system. With a good mathematical description of the building and the air-conditioning system it is possible to make predictions of the influence of certain factors on the energy use of the building. These factors are: outside air temperature, solar radiation. This mathematical model, built in a microprocessor can give information concerning the handling of the air-conditioning system, the artificial light and the control of the sun-shading. A few of these potential applications are:

- optimum airhandling
- optimum starting time of the airconditioning plant
- optimum use of solar radiation in relation to the control of the sun-shading during and outside working hours
- use of mechanical ventilation during the night and weekend to cool down the building mass
- lighting control depending on daylight level and use of sunshading

Some of these applications are realised on a small scale such as optimizing the stop/start time of the airconditioning plant. Others are carried out on laboratory scale such as, optimum air handling [4]. We will now go into the last two subjects. To get an idea how big the influence of these optimizing possibilities is on the energy consumption of a building computer calculations have been made with a set of computer programmes [5, 6]. With these computer programmes heating and cooling demands, energy consumptions and inside air temperatures can be calculated depending on the different control and optimizing possibilities.

6. Control of mechanical ventilation outside working hours to cool down the building mass

The study will be carried out for two different types of installations:

a. A four pipe induction system,
b. A heating system with mechanical ventilation (no cooling)

These systems will be placed in an oblong building (figure 1).

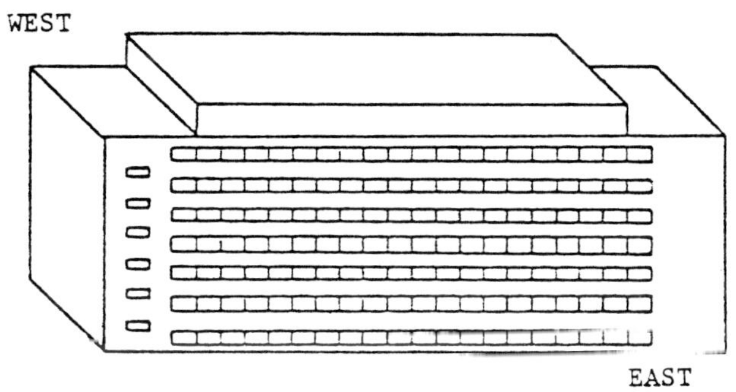

Figure 1: The 7 storey building.

The building has a volume of 24 000 m^3, double glazing, 40% glass area and inside blinds. Each floor has 30 rooms (15 on each side) with dimensions of 3.5 x 5.8 m. We will consider now a system with night ventilation in summer to cool down the building mass with outside air.

a. Four pipe induction system

Figure 2 gives the yearly cooling demand of the building depending on the number of hours the fans operate during the night period.
With a computer it is possible to minimize the cooling demand and the number of hours the fans are on during the night period by varying:

- the starttime of night ventilation in relation to the outside air temperature,
- the number of night ventilation hours depending on the temperature difference between inside and outside air temperature.

With such an optimized control system the energy consumption is calculated for a four pipe induction system.

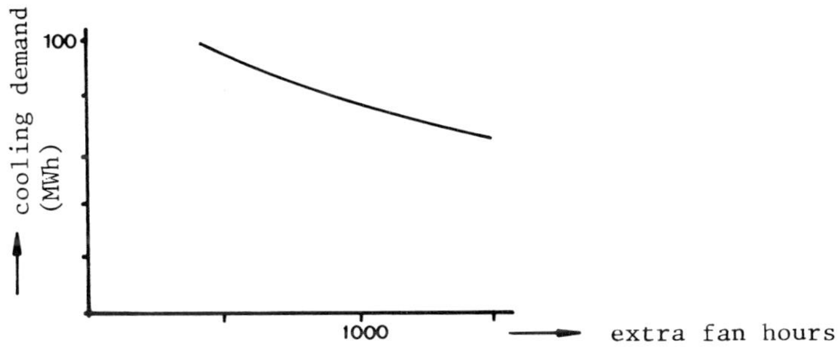

Figure 2: The cooling demand of the building depending on the extra number of fan hours.

In figure 3 the total primary energy consumption is shown with and without night ventilation.

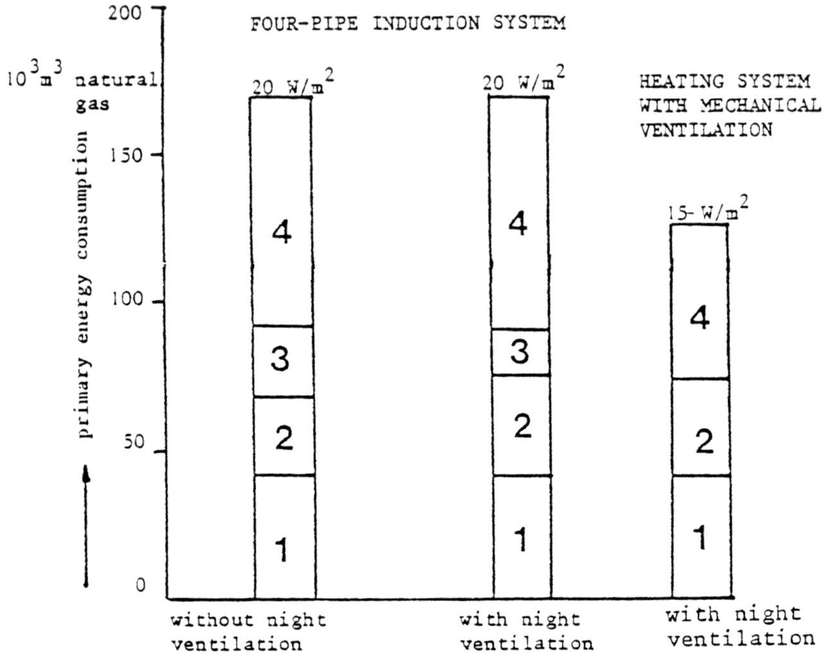

Figure 3: Energy consumption with and without night ventilation.

1 - Heating
2 - Transport
3 - Cooling
4 - Lighting

The total energy consumption is the same with the four pipe induction system: the increase of the fan energy consumption is as big as the decrease of the cooling energy consumption.

b. Heating system with mechanical ventilation (no cooling)

When in the same building a heating system is placed with mechanical ventilation during office hours (same air quantity as for the four pipe induction system) it is possible to calculate the frequency of the appearing inside air temperatures. In figure 4 the frequency of the inside air temperatures depending on the control system is shown.
When there is no night ventilation the inside air temperatures are intolerably high during a number of hours (400 hours higher than 26 ^oC). With night ventilation the number of hours during which the temperature is higher than 26 ^oC decreases considerably (80 hours). However, when it is possible to decrease the lighting level from 20 to 15 W/m^2, it will only be 10 to 20 hours a year that the inside air temperature is higher than 26 ^oC. For this case the energy consumption has been calculated (primary energy, see figure 3). The lower energy consumption is partly due to a lower lighting level and partly to the absence of the cooling system. With such a system it is possible to have an acceptable climate with a lower energy consumption and investment costs which will be about 25% lower than with a four pipe induction system.

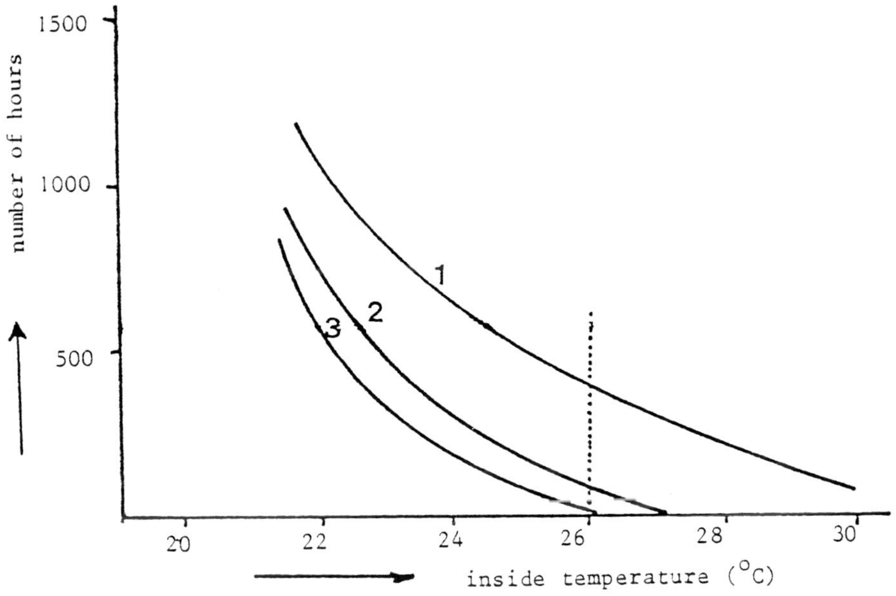

Figure 4: Frequency of the level of inside temperatures depending on the control system.

	Lighting
1 - without night ventilation	20 W/m^2
2 - with night ventilation	20 W/m^2
3 - with night ventilation	15 W/m^2

After finishing the study there are plans for starting a demonstration project to control these computer calculations in a building under actual weather conditions.

7. Lighting control depending on daylight level

In one of the office building of N.V. Nederlandse Gasunie in Groningen (The Netherlands) a demonstration project 'Energy Saving from Optimum Design and Regulation of Lighting in Office Buildings', was executed. This project is financially supported by the Commission of the European Communities, Directorate General for Energy in Brussels, and by Dutch institutions and firms.

In this project five different lighting systems, each installed on one floor of the office building, are tested under comparable conditions (see figure 1). All the different lighting systems have the same kind of control:

- The light is controlled in two zones, an outer and an inner zone (fig. 5) in each room.
  Switching off: central, above a certain daylight level (750 lux)
  Switching on : individual, manual, provided clearance has been given below a certain daylight level (700 lux)
- After working hours possibly central switching off by the porter.
- The control of the local lighting was effected individually by manual on/off switching.

The control of the lighting system was based on a constant daylight ratio. This is the ratio between the daylight level at a place of the central sensor (which gives the control signal to the lighting control system) and the daylight level at two fixed points in the room (see figure 5, nr. 1 and nr. 2) on the same orientation. In spite of the variations in this daylight ratio and a not optimal functioning of the control system, the decrease in energy consumption for lighting was more than 30%. When considering the maximum number of lighting hours of 2600 (50 weeks of 5 working days of 10 hours a day = 52 x 5 x 10 = 2600) the reference floor with no daylight control has 1430 hours of lights on, whereas the floors with daylight level control have less than 650 hours with lights on [7].

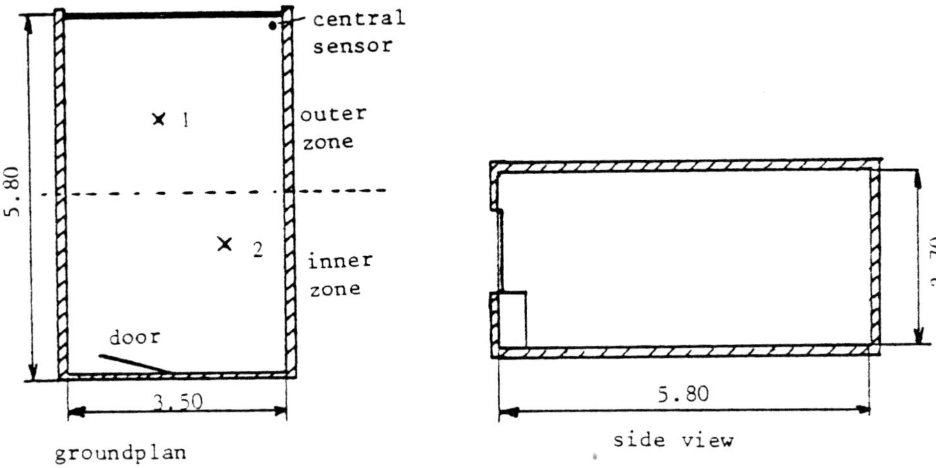

Figure 5: Dimensions of a room

The control apparatus which has been developed for this project was of a conventional analogue type. Special for these lighting control possibilities is direct digital control by using a microprocessor having a lot of profits such as:

- Programming of time delays for on/off switching is much better than the use of drop out relays with variable adjusting time.
- The control of daylight levels and changing setpoints is much easier in a programmable way.
- Only with memory functions it is possible to know how much time the daylight level is under or above a certain minimum level and to involve these functions in the control system.

Further investigations and evaluation of measurements gives possibilities for better control and predictions of daylight level at a certain place in a room by means of a central sensor [8] placed on the same orientation as the room. This method, based on measuring the solar radiation on the vertical plane gives the possibility to calculate the maximum energy saving by lighting control for this specific building.

This method is controlled with measurements in the building both with the sunshading up and with the sunshading down (45°). Figure 6 shows the influence of the solar radiation on the lighting level for the two working places in outer and inner zone of the room. The lines give the minimum values of the lighting level for a certain solar radiation.

Starting with a minimum lighting level of e.g. 750 lux on the two spots in the room during working hours (2600) (figure 5) and a central lighting control system, which switched the artificial light off, above a certain daylight level (> 750 lux) and switched the light on, under a certain daylight level (< 750 lux) it is possible to calculate the number of hours that the lights are off during an average year. This is carried out with weather tapes (10 years of hourly measured weather data) of the Dutch Meteorological Institute. Results of these calculations will be given in figure 7.

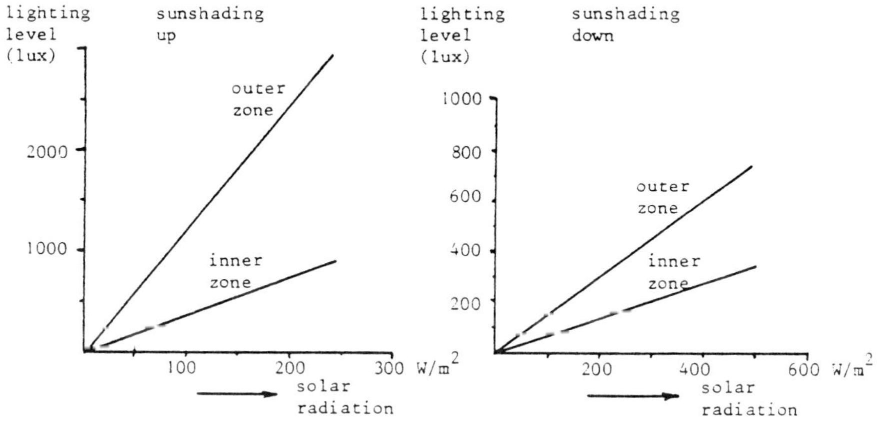

Figure 6: Lighting levels in the room depending on solar radiation

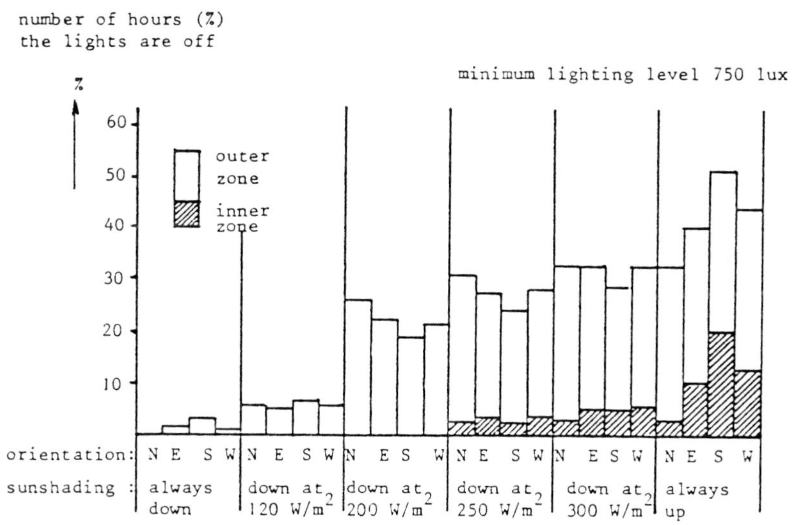

Figure 7: Number of hours (in %) that the lighting level is higher than 700 lux on the two fixed spots in the room, depending on the use of the sunshadings.

In the last column, the number of hours the lights are off can be read for the different orientations and with the sunshading always up. When the sun-

shading is switched depending on a certain level of solar radiation (120, 200, 250 and 300 W/m^2) the number of hours the lights are off, rapidly decreased.
The number of hours the lights are off in the inner zone, is relatively small. The influence of the lighting level on the number of hours the lights are off is shown in figure 8. In this figure this is done for three minimum required lighting levels, namely 500, 750 and 1000 lux on the working spots. In figure 8 there is a strong relationship to be seen between the minimum required lighting level and the number of hours the lights are off.
Investigations will be carried out to look at the influence of these lighting control systems on the yearly heating and cooling loads of the building and the yearly energy consumption.

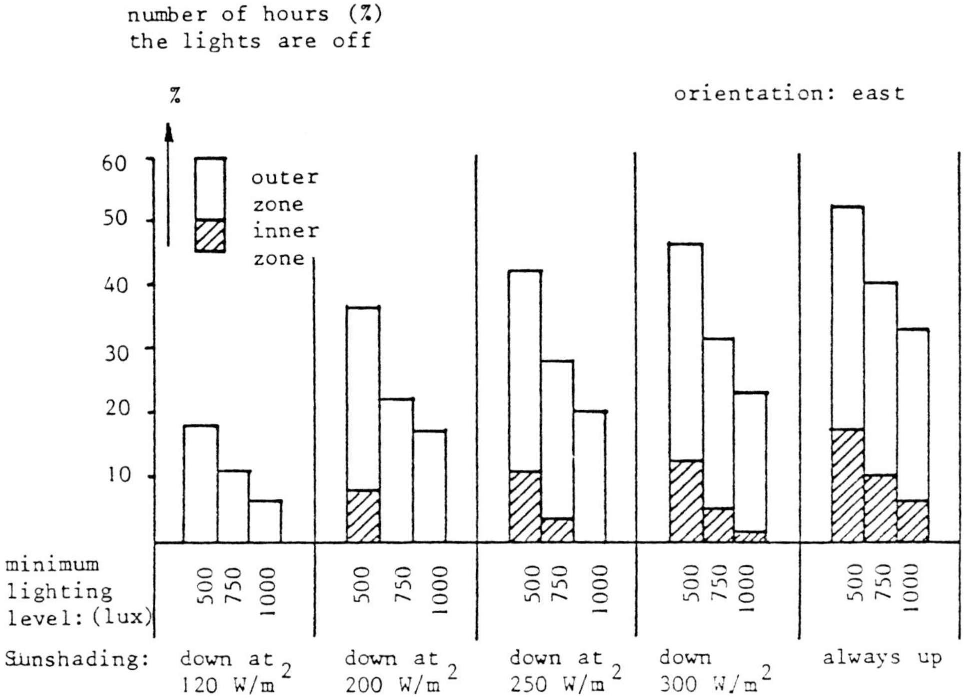

Figure 8: The number of hours the daylight level is higher than the lighting level required on the working spots (in % of the maximum number of hours (2600)).

REFERENCES

[1] Nicolaas, H.J., J. de Boer and J.B. de Wit
'Vooronderzoek naar de toepassing van microprocessoren bij de klimaatregeling in gebouwen en woningen ter vermindering van het energieverbruik', TPD-TNO report no. 900.211, Delft.

[2] Sams, Geoffrey, A., Ph. Eng.
'The design of direct digital control building management systems', Ashrae Transactions 1980, Volume 86, Part 1.

[3] Farris, Donald, R.
'Adaptive optimal control - an algorithm for direct digital control', Ashrae Transactions 1980, Volume 86, Part 1.

[4] Paassen, A.H.C. van, J.A. van der Kooij and J.C.H. Huiberts
'Een energetisch optimale luchtbehandelingsinstallatie gestuurd door een microprocessor', Klimaatbeheersing 8 (1979), nr. 5.

[5] Nicolaas, H.J., K.Th. Knorr and P. Euser
'A digital computer programme for the calculation of the yearly room energy demands and temperature exceeding rates, using hourly weather data'.
Energy Conservation in Heating, Cooling and Ventilating Buildings, 1 (1978), Hemisphere Publishing Corporation.

[6] Final Report, 'Comparision of Load Determination Methodologies for Building Energy Analysis Programs', Preparted for the International Energy Agency US.DOE/CE/20184-1), Washington D.C. 20585.

[7] Final Report, Demonstration Project: 'Energy saving from optimum design and control of lighting in office buildings', prepared for the Commission of the European Communities, contractnumber: EE/215/79.

[8] Nicolaas, H.J.
'Ontwikkeling centrale lichtsensor voor voorspelling van de verlichtingssterkte op een werkplek in een vertrek', TPD-TNO report 021-1, Delft, The Netherlands.

QUESTIONS AND DISCUSSION

Radde (PCS) : Were your measurements made in order to find something like standard values ?

Answer : The measurements and the calculations we are making at the moment are not aiming at the development of standards. We try to find out in what cases microprocessor applications in the field of heating, cooling and airconditioning are useful and to get an idea of the profits (finance and energy) it will give.

Mauro (CSEA) : Which mathematical model did you use in your simulation for the cooling control problem ?

Answer : The mathematical model is based on a thermal response factor method. The response factors are determined by means of a program in which the equations for the heat transfer processes in the rooms are solved by an implicit differential method. A detailed description of the program is published in "Energy Conservation in Heating, Cooling and Ventilation in Buildings", edited by C.J. Hoogendoorn and N.H. Afgan. Hemisphere Publishing Corporation.

SESSION III - HEAT STORAGE

Chairman: J.A.A. KETELAAR

Heat storage by paraffin wax fusion for space heating by a heat pump

Evaluation du comportement d'un stockage en aquifère captif utilisé comme réservoir géothermal

HEAT STORAGE BY PARAFFIN WAX FUSION FOR SPACE HEATING BY A PUMP

Authors	:	Pierre DUBOIS and Michel BERTHET
Contract number	:	EE-E-402 F
Duration	:	18 months 1 july 1980-31 december 1981
Head of Project	:	Pierre DUBOIS, directeur de la Division Energie
Contractor	:	Laboratoires de Marcoussis Centre de Recherches de la C.G.E.
Address	:	Route de Nozay 91460 Marcoussis (France)

Summary

The reported work relates to heat storage by paraffin wax fusion for water space heating, the heat beeing supplied by a heat pump. Such a heat store should improve the average performance coefficient of the heat pump due to a smaller number of stops and goes and it should reduce the cost of space heating by shifting electricity consumption from higher to lower tariff hours. The following tasks have been completed:

- research on cheap and thermaly efficient ways of paraffin wax packaging,
- theoretical and experimental study of the kinetics of heat transfer between paraffin and water during heat storage and heat discharge,
- design of a heat store for a lodging of 120 m2,
- test on a water loop of a heat pump with a 1 m3 heat store containing .9 m3 of paraffin ; the conclusion is that a paraffin heat store operates properly but is unfortunatly too expensive for an individual house compared to the electricity savings it provides with today's french electricity tariffs. Heat storage by paraffin fusion would be more interesting, from an economie point of view, in the case of air space heating systems including solar collectors.

O. Introduction

The object of the research was to study a heat store for space heating, in which the storage material is paraffin wax and the heat-transfer fluid is water. A store of this type, combined with a heat pump, should permit :

- an improvement in the APC of the heat pump through a smaller number of stops and goes,
- operation of the heat pump exclusively or mainly during the night, enabling the user to take advantage of more favourable electricity prices.

In the course of the 18 month programme, from the 1st of july 1980 to the 31st of december 1981, conducted with the Laboratoire de Thermocinetique of the CNRS at Nantes, we have :

- studied and developed high-performance and inexpensive technologies for the encapsulation of paraffin wax,
- modelled the behaviour of stores of different designs and studied the storage and discharge kinetics experimentally,
- designed a store of the appropriate size for a space of 300 m3 based on the results of simulations over one heating season, and we have built and tested, on a water test-bench, a 1 m3 store containing 0.9 M3 of paraffin associated to a heat pump,
- assessed the economic value of paraffin stores combined with heat pumps for space heating of individual houses.

1. Study of paraffin wax encapsulation technologies

We have fully investigated the encapsulation of paraffin by plastics. We studied encapsulation by rigid plastics and encapsulation by flexible plastics, which permits variations in specific volume of the paraffin without inducing significant mechanical stresses.

The encapsulation materials which could be used were selected following a variety of tests for compatibility and aging in the presence of paraffin by performing, on the one hand, heat cycles between ambient temperature and 70°C, and, on the other hand, prolonged periods of residence at 90°C.

11. Encapsulation with rigid plastics

Our work enabled us to select two rigid plastic encapsulation materials: polypropylene and chlorinated polyvinyl chloride (chlorinated PVC).

a) We showed that the mechanical properties under traction of test-pieces of these two materials did not change significantly following a period of residence of 3,000 hours in paraffin at 90°C, the values being as follows:

Characteristics at 20°C	Polypropylene	Chlorinated PVC
Stress at the flow threshhold	3 kgf/mm2	6.8 kgf/mm2
Elongation at the flow threshhold	22 %	7 %
Modulus of elasticity	40 kgf/mm2	250 kgf/mm2

Attention is drawn to the greater mechanical strength of chlorinated PVC compared with polypropylene, and to the fact that these two plastics have a good heat stability up to temperatures of the order of 100°C.

b) From the point of view of the permeability to paraffin, our tests showed that :

- polypropylene, which has a good impermeability up to 70°C, allows the diffusion, above this temperature, of some light fractions contained in paraffins. This phenomenon, which is also encountered at lower temperature with polyethylene, caused us to eliminate polyethylene for paraffin encapsulation,
- chlorinated PVC remains perfectly impermeable at 90°C.

c) We studied techniques for closing tubes at the ends and we showed that this could be done :

- by hot-pinching and welding between jaws in the case of polypropylene,
- by gluing or rotation welding of moulded plugs in the case of chlorinated PVC.

d) Because of the conjugate effect of the large variation in volume of paraffin during melting (15 %) and of the external pressure exerted by the water in the heating water loop (a few bars), the wall thickness of the tubes must be sufficient to prevent the tube from being crushed by buckling. In the case of chlorinated PVC, which has the higher mechanical properties, this makes it necessary to use wall thicknesses of 3 or 5 mm, according to the diameter of the tubes, 50 or 100 mm.

12. Encapsulation with flexible plastics

Following impermeability and aging tests carried out on several types of monolayer or multilayer film, we showed that the best results were obtained with a triple-layer complex film comprising :

- a 120 μ thick film of high-density polyethylene, which ensures weldability, on the inside,
- a 12 to 24 μ thick film of polyester, which imparts impermeability and improves the mechanical strength,
- a bonding adhesive chosen for its good heat stability (up to 120°C).

The impermeability, the cohesion, the mechanical properties and the chemical characteristics of a film of this type did not change after a period of residence of 5,000 hours at 90°C in paraffin. Flow tests under traction, which were carried out on dumb-bell test-pieces of the H1 type, show that, after 1,700 hours in paraffin at 55°C, a multi-layer film including 12 μ of polyester flows 2.5 % under a load of 1.15 kgf/mm2, and that the recovery after the load is removed is 40 %. As this load of 0.15 kgf/mm2 corresponds to the stress exerted by the hydrostatic pressure of the liquid paraffin contained in a flexible tube having a diameter of 50 mm and a height of 1 meter, placed vertically in the air, we see that the use of a multi-layer film containing a 24 μ polyester film would be desirable in the case of a paraffin heat store with air as the heat-transfer fluid. In the case of a paraffin heat store with water as the heat-transfer fluid associated to a heat pump, the external overpressure exerted by the water eliminates this problem.

These complex 3 layers thin films can be used in a simple manner with the aid of a heat-sealing machine, which makes it possible to produce elements having a bery wide variety of shapes : tubular "sausages", mattresses consisting of adjacent tubes, and the like. These advantages, together with the moderate cost of the film, namely 3 to 4 F per m2, make using a flexible film an attractive method of paraffin encapsulation.

2. Studies of the storage and discharge kinetics

21. Experimental studies of the storage and discharge kinetics

In the chain of heat transfers between the water and the core of the storage material, the heat exchanges with the water are high, namely of the order of 100 to 200 W/m2°C, by comparison with the other thermal resistances. This led us to direct the heat transfer study in the structures to the heat transfer within the paraffin and its encapsulation and at the interface of paraffin and encapsulation.

We studied experimentally the thermal behaviour of :

- structures comprising cylindrical tubes containing the paraffin, the heat-transfer fluid circulating outside these elements,
- structures comprising parallelepipedal elements containing the paraffin, the heat transfer fluid circulating outside these elements.

211. Experiments on tubes

a) The experiments carried out on single tubes placed in a water duct and cycled between 30°C and 60°C showed that, for a tube of Ø 116 mm, the insertion of internal heat drains, in the form of aluminium fins occupying 8.35 % of the volume, improves the exchange capacity with the heat-transfer fluid in the first hours (cf. figure 1) ; the average storage rate over 2 hours is thus increased by 17 %.

b) Experiments were carried out on a nest of 52 polypropylene tubes of 46/50 mm diameter and 1 meter length : they showed that the vertical or horizontal position of the tubes considerably changed the transitory distribution of the temperatures within the paraffin, because of the convection currents, but did not substantially modify the overall storage and discharge rates and capacities.

212. Experiments on parallelepipedal structures

We studied the fusion kinetics of paraffin contained in a tank of 500x500x60mm, receiving a constant energy flux over one of its large faces. It demonstrated the benefits of horizontal internal partitions, which permit a better distribution of the fusion of the paraffin between the top and the bottom of the element. From the energy point of view, the curves of figure 2 show that this effect results in a 20 % increase in the energy stored over 6 hours by changing the shape ratio $\frac{\text{height}}{\text{thickness}}$ of the paraffin cavities of 8 (initial tank) to 2 (tank with 3 partitions). A further decrease of this shape ratio to unity does not improve heat transfers. These expetiments were carried out with ARCOWAX paraffin, the characteristics of which are given in figure 3.

22. Modelling of the thermal behaviour of storage elements

Two models were developed : one related to encapsulating tubes and the other related to parallelepipedal elements filled with paraffin.

221. Model of tubes

This model, in pure conduction, is of "enthalpic" type, in which the heat accompanying a change of state is taken into account in the form of an apparent heat capacity varying with the temperature, such as is determined by differential enthalpic analysis measurements. The numerical determination is carried out by the "Crank Nicholson" method and by using a ratio of the intervals of axial space and of time which is close to the speed of the fluid.

This model was compared with the experiments carried out on single tubes and on a nest of 52 tubes. In both cases, we found a very good agreement as regards the change in the temperatures of the fluid passing through the store and in the paraffin temperatures in the central region of the tubes. Differences between the calculated temperatures and the measured temperatures were found at the ends of the tubes, because of the contraction on solidification of the paraffin in these regions. The overall energy behaviour of the nest of tubes is well simulated by the model.

222. Model of parallelepipedal elements

The Laboratoire de Thermocinetique de Nantes (CNRS) developed a model of a parallelepipedal paraffin element heated on one of its large faces. The model uses two units dealing successively with the different heat transfer conditions in the solid or liquid paraffin :

- one unit for pure conduction in the solid and in the liquid,
- one unit taking into account convection in paraffin. For boundary layers, a one-dimensional solution is based on an equivalent rectangular liquid cavity, the thickness of which is equal to the abscissa of the fusion front in the median part. The change in the fusion front is represented as the removal of the solid zone under the effect of a convection flux in the liquid.

The experiment showed that this model represented better than a purely conducting model the thermal behaviour in the presence of convection movements during storage.

3. Design studies of a paraffin-water store

31. Choice of the heat exchange area between water and paraffin

Using the simulation models developed, we compared the values of heat stored and the values of electrical energy consumed as a function of the exchange area between water and the paraffin, both in the case of tubes and in the case of parallelepipedal elements containing paraffin. This comparison was made when charging the store during 6 hours by means of a 2.5 kW heat pump operating at full power.

The following conclusions were made :

- in the case of the tubular structures, it is not advantageous to use tubes having a diameter of less than 50 mm, and the internal heat drains can only be justified for tube diameters greater than or equal to 100 mm. Figure 4 shows the effect of the exchange area and of the heat drains on the energy stored and on the performance coefficient of the heat pump,
- in the case of parallelepipedal elements, the thickness of paraffin between the exchange walls must be between 60 and 100 mm and that leads to 30 to 50 m2 of exchange area for a store of 1.5 m3. The gain in terms of heat stored is only 10 % when changing from 50 to 300 m2 (cf. figure 5).

32. Choice of the overall size of a store associated with a heat pump for a house space heating

We used a model of house space heating by a system including a heat pump and a heat store in parallel with the heat pump, the heat being supplied to floor heating pipes or to water radiators. This model was applied to the case of a 300 m3 house requiring 300 W/°C and located in the Paris region. We considered an external air/water heat pump which operates above -5°C outside temperature, with a water output temperature not exceeding 62°C.

The model takes into account, the solar contributions and the outside temperatures on an hourly basis as well as the regulation criteria of the heating system, the thermal inertia of the house and the dual electricity tariff :

- during 16 higher tariff hours : 0.378 FF/kWh,
- during 8 lower tariff hours : 0.220 FF/kWh.

We computed operating costs over a full heating season for several sizes of store and of heat pump and for several heat system regulation criteria. The conclusions are as follows :

- there is no significant difference in the gains in terms of operating costs between the heating systems with heated floors and the systems with radiators,
- with a 2.5 kW compressor heat pump, the gain is terms of operating costs provided by the installation of a store shows virutally no further increase beyond 1 m3 of store (cf. figure 6),
- an enlargement of the heat pump to 4 kW would lead to greater financial gains, but the additional annual saving made by enlarging the pump and the associated store does not appear to justify the corresponding extra investments,
- when considering a store of 1 m3 of paraffin melting at 47°C, associated with a 2.5 kW electrical heat pump, the annual gain in terms of operating costs provided by the store amounts to 436 F under these conditions, that is to say 19 % of the initial expenditure on electricity (cf. figure 6).

4. Design of a paraffin store and construction of a 1 m3 unit

41. Choice of the store structure

a) Analysis of various structures for the store led us to the following conclusions :

- the structures in which the storage material is encapsulated in elements which are immersed in a water tank implies constraints of a mechanical nature, which are reflected in the cost :

. the thank must withstand the pressure of the water circuit, which is a few bars,

.a rigid encapsulation must be sufficiently thick to withstand mechanically the differential pressure between the outside and the inside of the tubes, which leads to a substantial weight of encapsulating material,

. a flexible encapsulation requires supporting structures, because of the buoyancy of paraffin in water.

- the structures in which the heating water flows in exchangers immersed in the paraffin are better suited from a mechanical point of view ; the use of a thin walled parallelepipedal tank filled with paraffin is thus possible.

b) We compared from an economic point of view several possible designs of exchangers for stores with a paraffin tank. This analysis led us to choose elements made of roll-bond aluminium, the purchase price of which, when used on a rather large scale, would be less than 100 F/m2.

42. Design of a store and construction of a 1 m3 unit

a) We designed a store for a house of 300 m3 requiring 300 W/°C for space heating. This store, which contains about 0.9 m3 of paraffin, is shown in figure 7. Its main characteristics are as follows :

- volume of paraffin : 0.9 m3,
- exchanger consisting of eight 1,900 x 750 mm sheets of roll-bond aluminium, 61.5 mm apart, giving a wet area of 22.8 m2,
- parallelepipedal tank : width 860 mm, depth 540 mm, height 2,050 mm,
- thermal insulation by means of 50 mm polystyrene slabs.

b) A 1 m3 unit was constructed using this technology and filled with 690 kg of TOTAL Inter B paraffin. The photograph of figure 8 shows a view of this store, which was fitted and tested on a bench in parallel with an external air/water heat pump.

3. Tests of the paraffin/water 1 m3 store

The paraffin/water store, described above, which contains about 0.9 m3 of paraffin, was installed and tested on a bench comprising an external air/water heat pump with a 2.5 kW compressor.

51. Description of the test bench

The test bench makes it possible to reproduce charge/discharge cycles of the store under conditions representing those of an actual house. The heating water loop, of which the simplified diagrams corresponding to the heat storage and heat discharge phases are given respectively in figures 9 et 10, includes an exchanger with a cooling water loop, which makes it possible to simulate the heating needs of the house by removing a regulated amount of heat from the water loop to which heat is supplied by the heat pump and by the store. The modes of operation representing the different periods in a day are as follows :

- in the lower tariff period (8 hours), the pump covers the heat needs of the house and at the same time ensures the charging of the store with the excess heat power available.

The motorised "three-way" valve V1, controlled by a temperature regulator, distributes the flow created by the circulator C1 between the exchanger and the store, so as to keep the return temperature TR from the exchanger at a set value with varies with the external temperature according to the heating curve of the radiators for the house. The heat pump operates continuously up to the end of the lower tariff period or until the temperature of water leaving the heat pump reaches the limit of 62°C, when the store is full. In the latter case the heat pump covers the heat needs of the house, with control by the value of TR.

- in the higher tariff period (16 hours), the heat needs are covered by store discharge. The motorised valve V2 distributes the flow created by the circulator C2 between the store and a by-pass, so that the return temperature TR is maintained at the set value which varies with the outside temperature according to the heating curve of the radiators for the house. This heating by store discharge only is maintained as long as the discharge power can supply the needs. Beyond this, in order to utilise the store more completely the needs are covered with the aid of alternate sequences of heat pump operation and heat discharge from the store. As heat discharge is then insufficient to cover all the needs, fluctuations in the return temperature TR are allowed between two extreme values T_{lower} and T_{upper} on on either side of the set temperature. The values of T_{lower} and T_{upper} were chosen so as correspond to variations in the ambient temperature of the accomodation which were compatible with a good degree of comfort: ± 1°C relative to ambient temperature, which corresponds to ± 2 to 3°C relative to the heating water return temperature when the external temperature is 1°C.

52. Conditions of store testing

a) We carried out tests for three successive periods of 24 hours, 48 hours and 24 hours. By dividing the annual curve of the cumulative frequencies of the daily average of the external temperature into three regions, the following values of the external temperature were computed as representing one heating season in the Paris regions :

. 13°C for 24 hours,
. 8°C for 48 hours,
. 1°C for 24 hours.

b) During each of these three periods :

- the temperature of the climatic chamber simulating outside conditions was kept constant, as well as the thermal power removed from the water, which represents the needs of the accomodation, this being a known function of the external temperature for a given house,
- that part of the heat produced by the pump which does not have to be used immediately for heating the accomodation was stored during the lower tariff hours.

c) The values of the heat power representing the space heating needs and the values of the temperature of the water returning from the radiators are given for space heating needs of 300 W/°C for a house with an internal temperature of 19°C and a non-heating temperature of 15°C :

External temperature	Space heating needs	Temperature of water from the radiators
1°C 8°C 13°C	4.2. kW 2.1 kW 0.6 kW	48°C 33.5°C 23°C

53. Results of tests

The tests carried out showed that :

- for an external temperature of +13°C, the transfer of electricity consumption by the heat pump from the higher tariff hours to the lower tariff hours was 100 %,
- for an external temperature of +8°C, the store alone provided 2600 watts during seven hours, 2100 watts needed for space heating and 500 watts to cover water loop heat losses,
- for an external temperature of +1°C, the store covered the needs during higher tariff hours to the extent of 22 %,
- the average performance coefficient of the heat pump over the whole of these four days was 2.42,
- the distribution of the electricity consomption of 118 kWh was :
  . 73.7 % in lower tariff hours (87 kWh),
  . 26.3 % in higher tariff hours (31 kWh).

The electricity consumption of 118 kWh would cost 30.8 F for the four days in question, taking the higher/lower tariffs into account.

54. Comparison with heating by means of a heat pump without a heat store

We calculated the electricity consumption by the heat pump which, for the four typical days in question, would provide the same service in the absence of storage. To do this, we used the characteristics of the heat pump which we had measured on the bench for various conditions of external temperature, water return temperature and stops and goes of the heat pump. These calculations show that :

- the average performance coefficient of the heat pump would then be 2.35 ; the electricity consumption for the four days in question would increase from 118 to 121.5 kWh,
- the cost of the electricity consumed would increase to
  81 kWh x 0.378 + 40.5 kWh x 0.22 = 39.5F

The installation of the heat store therefore makes it possible, under the conditions of the experiment, to make a 3 % gain in the performance coefficient of the heat pump and to reduce the operating cost from 39.5 to 30.8 F . The gain extrapolated over a heating season of 210 days would be of the order of $210 \times \frac{8.7}{4} = 456$ F , very close to the computed value of 436 F (cf. 3.2).

6. Economics of paraffin heat stores

61. Paraffin store and water heat pump

We calculated that the factory cost price of a paraffin store of 0.9 m3, constructed with the technology described in paragraph 51, is about 5.000 F excluding the regulating system.

This cost is broken down in the following way :

- 700 kg of paraffin at 3,7 F/kg	2.590 F
- 8 exchange plates made of roll-bond aluminum	1.100 F
- tank	1.300 F
- heat insulation	80 F
- water collectors	90 F
- fitting	122 F
Factory cost price	5.282 F

The sale price of a store of this type would therefore be between 10.000F and 15.000 F, to which the cost of the regulating system would be added.

In view of the annual electricity saving, which in France, is only about 500 F with the current peak/off-peak tariff, this evaluation leads to a negative judgement as regards the economic advantage of a paraffin heat store in the case of a heating system with water heat pump for a 300 m3 house.

62. Paraffin store for space air heating with solar collectors

Based on the results obtained on paraffin encapsulation technologies and on modelling the thermal performance of a paraffin store, we have showed that a paraffin store on a air loop for solar heat collection and space heating should constitute, an advantageous alternative solution to the usual pebbles bed heat store . We showed that a store with 0.75 m3 of paraffin weighing 600 kg would provide the same service as a 1.5. m3 store of pebbles weighing 2.700 kg. Such weight and volume reductions make it possible, without mechanical reinforcement of the structures of the house, to install the store within the heated space of the house, the location being chosen to reduce the length of air ducts connecting the heat store to the solar collectors.

7. Conclusion

We have designed and tested a 1 m3 paraffin/water store associated with a heat pump for space heating. These experiments have led to positive technical results. Taking into account the performances of such a system over one heating season, such as they have been determined by computations and proved by experiments, we evaluated, with the electricity tariffs existing in France today, the economic advantage of such a heat store associated with a heat pump. It leads to a negative judgement as far as space heating of an individual house is concerned.

Based on the results obtained on paraffin encapsulation technologies and on modelling of the thermal performances of a paraffin store, we have shown that a paraffin store combined with an air loop for solar heat collection and space heating offers advantages over conventional pebbles bed heat stores.

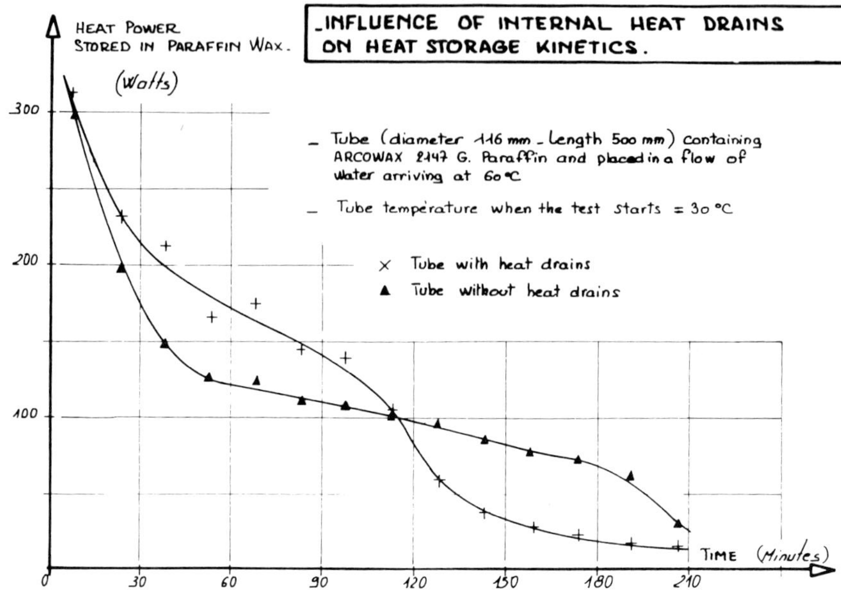
_INFLUENCE OF INTERNAL HEAT DRAINS ON HEAT STORAGE KINETICS.
HEAT POWER STORED IN PARAFFIN WAX.
(Watts)
_ Tube (diameter 116 mm _ Length 500 mm) containing ARCOWAX 2147 G. Paraffin and placed in a flow of water arriving at 60 °C
_ Tube température when the test starts = 30 °C
× Tube with heat drains
▲ Tube without heat drains
TIME (Minutes)

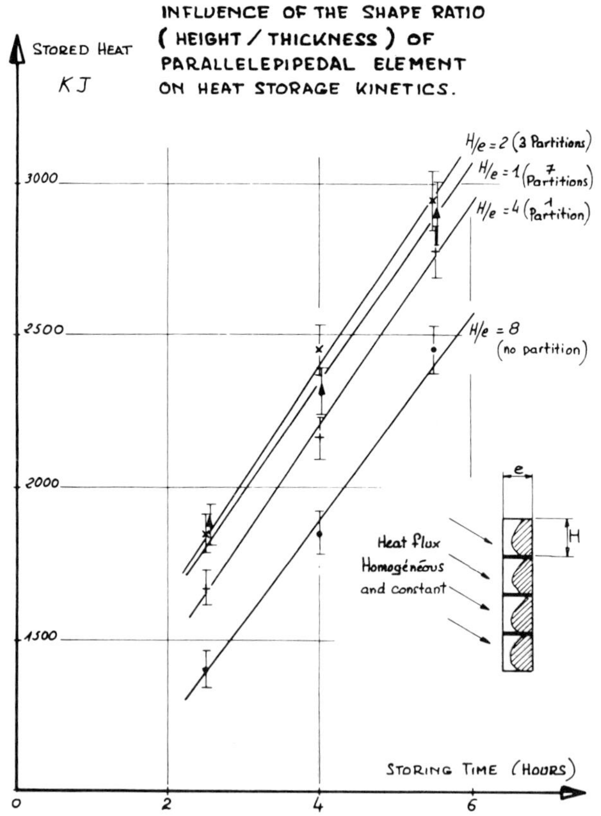
INFLUENCE OF THE SHAPE RATIO (HEIGHT / THICKNESS) OF PARALLELEPIPEDAL ELEMENT ON HEAT STORAGE KINETICS.
STORED HEAT
KJ
H/e = 2 (3 Partitions)
H/e = 1 (7 Partitions)
H/e = 4 (1 Partition)
H/e = 8 (no partition)
Heat flux Homogénéous and constant
STORING TIME (HOURS)

↑ Fig. 1

← Fig. 2

Figure 3

CARACTERISTICS OF PARAFFIN WAX

1) PARAFFIN ARCOVAX 2147 G USED FOR ELEMENTS TESTS

. Supplier : Societe Française des Petroles BP
10 Quai Paul Doumer - 92412 Courbevoie

. Caracteristics

Fusion peak temperature	47°C
Solidification peak temperature	46°C
Fusion enthalpy	36,2 $cal.g^{-1}$
Solid specific heat at 5°C	0,50 cal/°Cg
Liquid specific heat at 60°C	0,54 cal/°Cg
Solid density at 10°C	0,885 g/cm^3
Liquid density at 60°C	0,797 g/cm^3
Thermal conductibility	0,2 W/m°C

2) PARAFFIN INER B USED FOR THE 1 m3 STORE

. Supplier : Compagnie Française de Raffinage TOTAL
39 à 43 Quai André Citroën - 75739 Paris Cedex 15

. Caracteristics

Fusion peak	53,5°C
Solidification peak	51°C
Fusion enthalpy	39 $cal.g^{-1}$
Solid specific heat at 5°C	0,45 cal/°Cg
Liquid specific heat at 60°C	0,50 cal/°Cg
Density at 10°C (solid)	0,913 g/cm^3
Density at 60°C (liquid)	0,799 g/cm^3
Heat conductibility	0,2 W/m°C

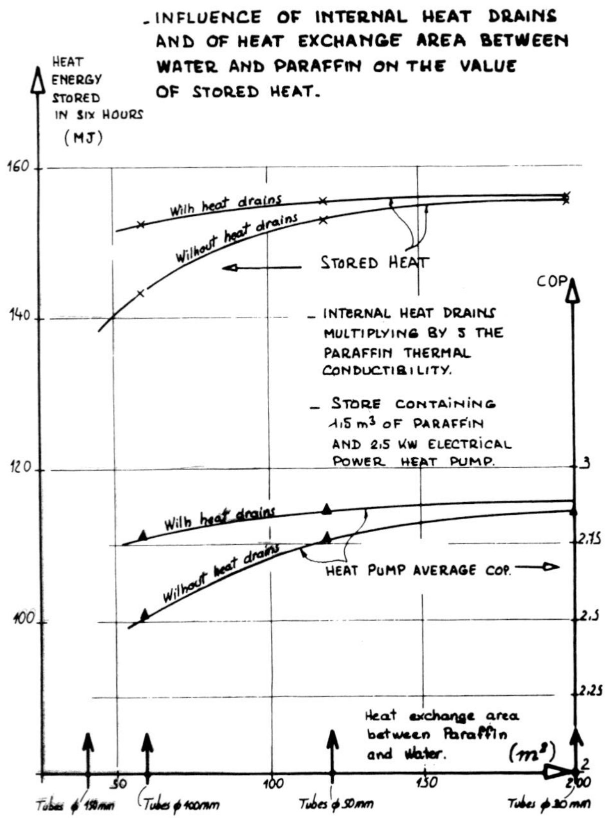

- INFLUENCE OF INTERNAL HEAT DRAINS AND OF HEAT EXCHANGE AREA BETWEEN WATER AND PARAFFIN ON THE VALUE OF STORED HEAT.
HEAT ENERGY STORED IN SIX HOURS (MJ)
160
140
120
100
With heat drains
Without heat drains
STORED HEAT
COP
- INTERNAL HEAT DRAINS MULTIPLYING BY 5 THE PARAFFIN THERMAL CONDUCTIBILITY.
- STORE CONTAINING 1.5 m³ OF PARAFFIN AND 2.5 KW ELECTRICAL POWER HEAT PUMP.
HEAT PUMP AVERAGE COP.
3
2.75
2.5
2.25
2
Heat exchange area between Paraffin and Water. (m²)
50
100
150
200
Tubes ϕ 150mm
Tubes ϕ 100mm
Tubes ϕ 50mm
Tubes ϕ 20mm

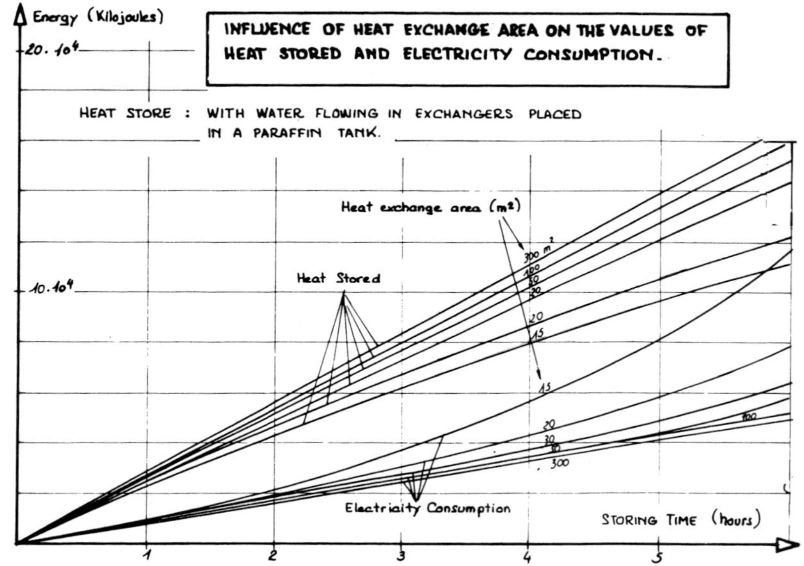

Energy (Kilojoules)
20.10⁴
10.10⁴
INFLUENCE OF HEAT EXCHANGE AREA ON THE VALUES OF HEAT STORED AND ELECTRICITY CONSUMPTION.
HEAT STORE : WITH WATER FLOWING IN EXCHANGERS PLACED IN A PARAFFIN TANK.
Heat exchange area (m²)
Heat Stored
Electricity Consumption
STORING TIME (hours)
1
2
3
4
5

↑ Fig. 5

← Fig. 4

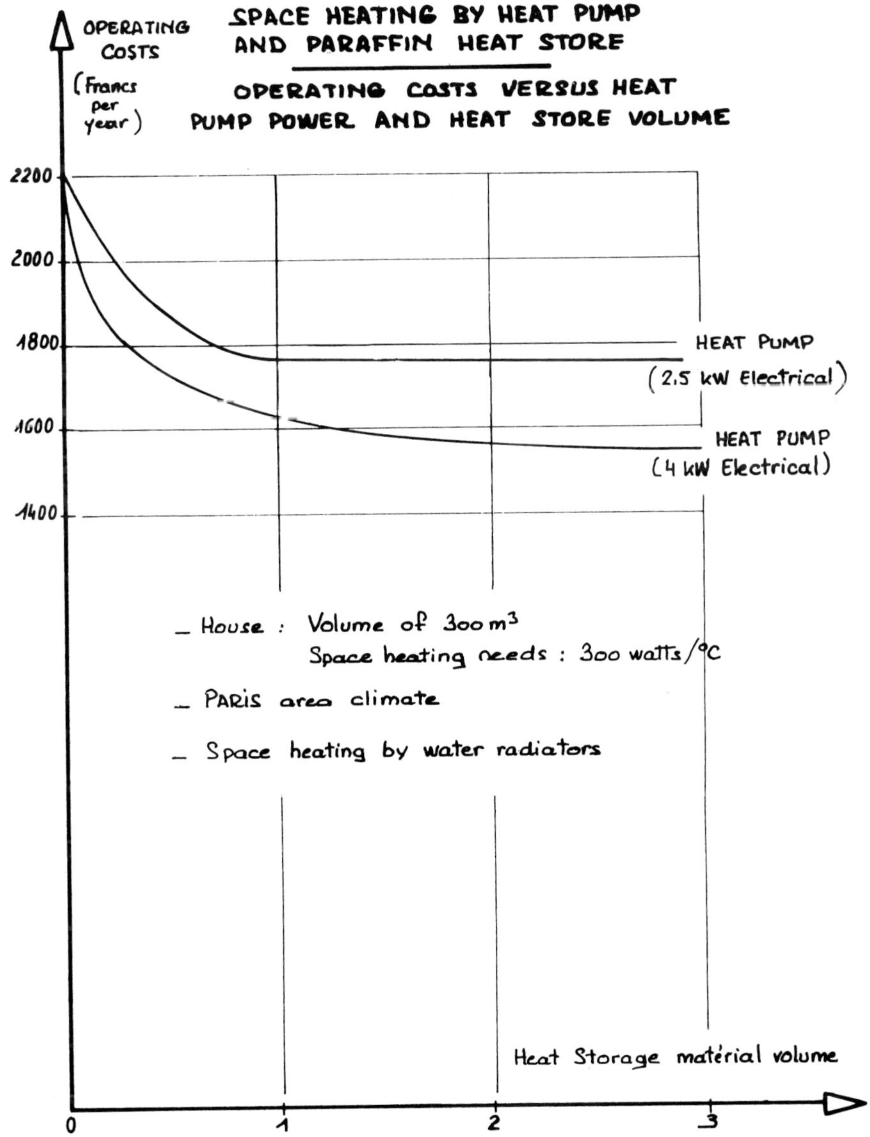
SPACE HEATING BY HEAT PUMP
AND PARAFFIN HEAT STORE
OPERATING COSTS VERSUS HEAT
PUMP POWER AND HEAT STORE VOLUME
OPERATING COSTS
(Francs per year)
2200
2000
1800
1600
1400
HEAT PUMP
(2,5 kW Electrical)
HEAT PUMP
(4 kW Electrical)
_ House : Volume of 300 m3
Space heating needs : 300 watts/°C
_ PARIS area climate
_ Space heating by water radiators
Heat Storage material volume
0
1
2
3

Fig. 6

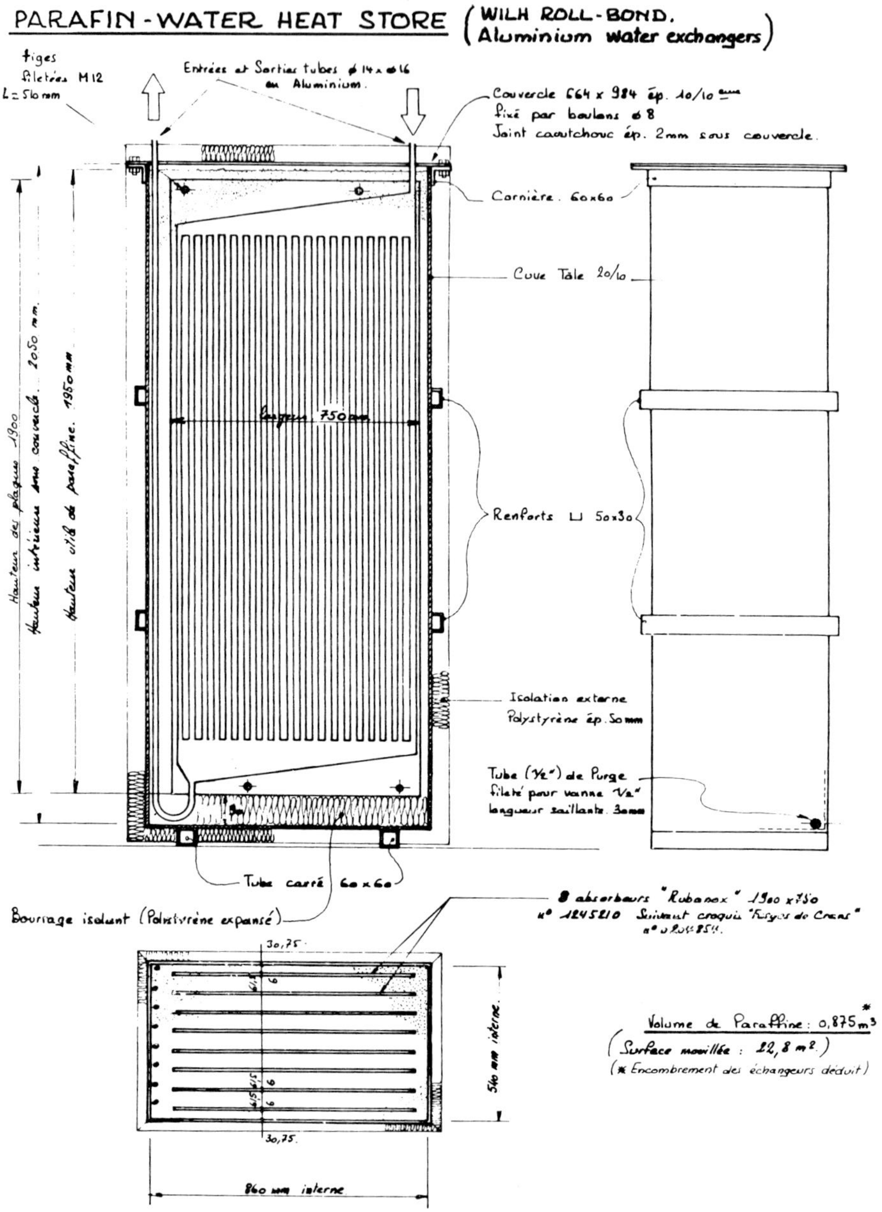
PARAFIN - WATER HEAT STORE (WILH ROLL-BOND. Aluminium water exchangers)
tiges filetées M12 L = 510 mm
Entrées et Sorties tubes ø 14 x ø 16 en Aluminium.
Couvercle 664 x 984 ép. 10/10 mm
fixé par boulons ø 8
Joint caoutchouc ép. 2mm sous couvercle.
Cornière. 60x60
Cuve Tôle 20/10
Renforts 50x30
Isolation externe
Polystyrène ép. 50 mm
Tube (½") de Purge
fileté pour vanne ½"
longueur saillante 30mm
Tube carré 60 x 60
Bourrage isolant (Polystyrène expansé)
Largeur 750 mm
Hauteur des plaques 1900
Hauteur intérieure sous couvercle 2050 mm
Hauteur utile de paraffine 1950 mm
8 absorbeurs "Rubanox" 1900 x 750
30,75
540 mm interne
860 mm interne
Volume de Paraffine: 0,875 m³
(Surface mouillée: 22,8 m².)
(* Encombrement des échangeurs déduit)

Fig. 7

A 1 m3 paraffin heat store on a space heating water loop with a heat pump

View of the test bench

Paraffin store seen from above showing the water heat exchanger made of sheets of roll-bond aluminum

Fig. 8

TEST BENCH WITH HEAT PUMP AND WATER - PARAFFIN HEAT STORE

HEAT STORAGE

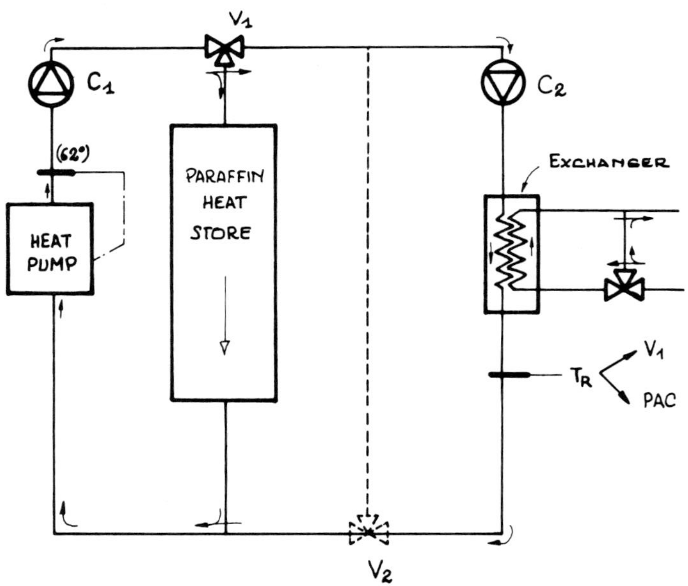

Fig. 9

TEST BENCH WITH HEAT PUMP AND WATER-PARAFFIN HEAT STORE.

HEAT DISCHARGE.

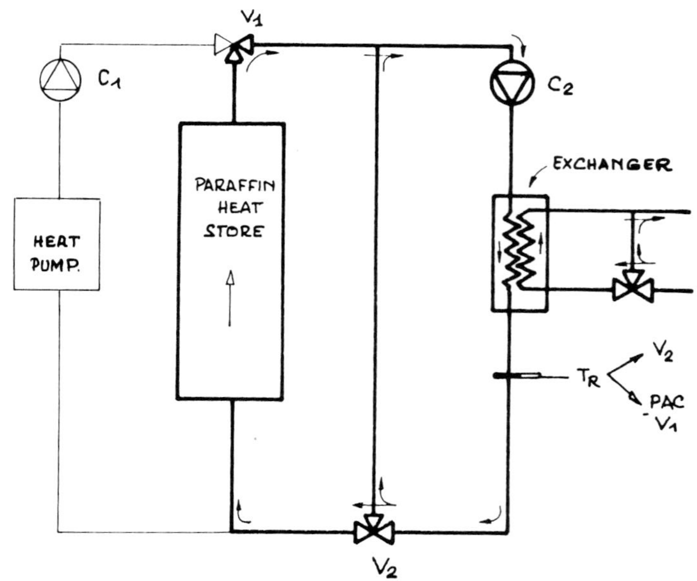

Fig. 10

QUESTIONS AND DISCUSSIONS

Pellegrini (CCR) : Assuming you could use a storage material with a narrow temperature range of fusion (e.g. solt hydrates at 64-65° C), would you expect a better performance in coupling your heat pump with the storage ?

Answer : a) As far as melting temperature is concerned, paraffin is well suited to the problem of space heating with heat pumps

- heat pump warm water to 60/62° C
- paraffin melts at 50° C

b) Other materials melting at about the same temperature, but with a higher fusion heat would lead to a smaller storage but the efficiency of the system would not be changed. This material should however behave properly under thermal cycling and it should not be corrosive to the packaging material. It should be noted that the hydrated mineral salts which have been tested, age under thermal cycling and are corrosive to the packaging material.

Prof. Ketelaar (Chairman) : 1. Without storage, but with heat pumps the annual cost is 2200 FF. Now what will be the cost of the same house under the same conditions with a conventional radiator heating system without heat pump ?
2. Over a larger temperature range the difference in specific heat content of paraffin and water come rather near.
3. Installing heat meters in apartment buildings leads to a rather considerable conservation. How is this in France ?

Answer : 1. As the heat pump COP has an average value of 2,42, the cost of electricity for space heating would go from 2200 FF per year for heat pump heating to 2200 x 2,42 = 5300 FF for resistor heating.

2. Yes. The difference between paraffin and water is zero for a temperature difference of 80° between the maximum storage temperature and the minimum temperature during heat release by the storage :

water $\Delta t = 80° \longrightarrow$ stored energy 80 cal/gr
wax $\Delta t = 80° \longrightarrow$ stored energy $[40 + 0,5 \times 80]$ cal/gr

Fusion heat storage is interesting when, as in the case of heat pump used for space heating, the temperature difference is rather small compared to 80° C.

3. Yes. It is not always possible in an existing apartment building. In new buildings it leads to savings of energy of 10 to 15 %. A problem is to find reliable heat meters.

Oudendal (Netherlands Railways) : Why did you use paraffin ? Have other materials been tested ? Is paraffin a fire hazard for the house ?

Answer :
1. The melting temperature of paraffin (~ 50° C) is well suited to heat pump maximum operating temperature for space heating (~ 60°) - paraffin is a rather cheap product which can withstand heat cycling without any change of properties. It is a "peaceful" product which can be packaged with plastics or aluminium.
2. For other applications, we have evaluated other products.
3. Paraffin can burn - like candles - But the fire hazard is much smaller with a paraffin tank than with a fuel tank.

Zegers (DG XII) : Your conclusions for economic feasibility is that daily heat storage with paraffin in combination with a heat pump is not attractive in France. In several other countries in the EC the electricity prices are twice as high as in France. Would you reckon that your heat storage system could be attractive in those countries ?

Answer : One has to make a calculation. It seems likely, provided that the electrical heat pump is close to economical feasibility with these high electricity prices.

Pilavachi (DG XII) : Did you make a sensitivity analysis related to the temperature in the heat storage system ?

Answer : We had a maximum temperature of 62° C due to the heat pump operating conditions. This temperature is smaller than the temperature usual with water space heating with fuel or gaz boiler (70 to 80°).
We did not investigate temperatures smaller than 62° C because it would increase the size of the storage and the size of the water/air heat exchangers in every room.

EVALUATION DU COMPORTEMENT D'UN STOCKAGE EN AQUIFERE CAPTIF UTILISE COMME RESERVOIR GEOTHERMAL

Auteurs : J. DESPOIS, H. CLAMENS, B. CONSEIL

Numéro de contrat : E-E-1-403-F

Durée : 12 mois 1er octobre 1980 - 1er octobre 1981

Chefs de projet : J. RASTOIN (CEA), M. RONC (SNEA)

Contractants : Comm. Energie Atomique (CEA) et Soc. Nat. Elf-Aquitaine

Adresses : CEA : Centre d'Etudes Nucléaires de SACLAY - DEMT/SEEN - 91191 Gif-sur-Yvette Cedex.
SNEA: Centre de Recherche Elf de Solaize - B.P. 22 - 69360 St-Symphorien d'Ozon.

Summary

Commissariat à l'Energie Atomique and Société Nationale Elf Aquitaine, have started studies to establish the feasability of underground heat storage (180-200°C) in captive ground water encountered at depths greater than 200 metres.
To demonstrate the feasability of this type of heat storage, CEA and SNEA have selected the site offered by the Commune of Thiverval-Grignon (30 kilometres west of Paris). A reconnaissance well was drilled in 1980 on the site of a huge garbage burning plant and equipped at a depth of 508 m in the aquifer of the Wealdien formation selected for the demonstration phase. The reconnaissance well permitted to know the important data concerning the hydrogeological situation of the formation. These data were then introduced in a numerical model, such calculations confirmed the feasability of the selected underground heat storage.

1. INTRODUCTION

Pour des stockages de courte durée et de petites dimensions, on peut envisager d'utiliser soit la chaleur sensible, soit la chaleur de changement d'état de corps appropriés. Compte tenu de la faiblesse des capacités volumiques, par rapport à l'énergie chimique d'un combustible classique, on aboutit vite à des réservoirs immenses et, le plus souvent, à un coût prohibitif du stockage lui-même.

L'utilisation de l'eau comme milieu de stockage est intéressante, car le coût de l'eau - même traitée - est faible et sa capacité calorifique volumique est élevée. Malheureusement, dès que la température dépasse notablement 100°C, les pressions croissent très vite et l'on devrait faire appel à des réservoirs très coûteux, voire impossibles à réaliser (pour les grands stockages) : on perd alors sur le réservoir tout ce que l'on a gagné par l'adoption de ce milieu de stockage économique. D'où l'idée de stocker la chaleur en souterrain, de manière à s'affranchir des problèmes de tenue en pression des réservoirs de surface. Deux options sont envisageables :

a) utilisation d'une nappe souterraine sous pression contenue dans un milieu poreux et perméable (nappe captive),
b) creusement d'une cavité en utilisant la poussée du terrain pour contenir la pression interne.

C'est ainsi que le COMMISSARIAT A L'ENERGIE ATOMIQUE (CEA) et la SOCIETE NATIONALE ELF-AQUITAINE (SNEA) associés pour la circonstance, ont entrepris des études pour établir la faisabilité d'un stockage thermique à haute température (180-200°C) dans un aquifère profond.

2. DESCRIPTION DU STOCKAGE SOUTERRAIN DE CHALEUR, ET PRESENTATION DU FORAGE DE RECONNAISSANCE

2.1. Principes généraux de fonctionnement

Un tel système de stockage comprend un puits central appelé "puits chaud" (température maximale 180°C), et plusieurs puits périphériques appelés "puits froids" (température environ 70°C) situés à près d'une centaine de mètres du puits central (Figure 1). Durant l'été, l'eau du terrain est extraite des puits froids et chauffée au travers d'un échangeur par les rejets thermiques disponibles sur place avant d'être réinjectée dans le puits chaud. L'eau dans l'aquifère circule alors lentement, et le front chaud se déplace suivant des cercles concentriques, qui graduellement s'éloignent du puits central. Durant l'hiver, le dispositif du stockage est connecté par l'intermédiaire d'un échangeur au réseau de chaleur alimentant l'agglomération concernée.

L'eau de l'aquifère circule alors en sens opposé. Elle est extraite du "puits chaud" et est refroidie dans l'échangeur placé sur l'eau de retour du réseau de chaleur urbain (qu'elle réchaufffe alors). Puis elle est réinjectée dans les "puits froids" périphériques. La zone chaude de l'aquifère diminue de taille, au fur et à mesure que le front chaud se rapproche du puits central jusqu'au printemps, où un nouveau cycle de stockage démarre.

Le système est conçu de façon que l'eau de l'aquifère circule en circuit fermé, et reste en permanence séparée de l'eau utilisée dans le réseau de chaleur (figures 2 et 3).

2.2. Choix du site de PLAISIR pour une opération de démonstration

Les résultats d'une première étude de faisabilité entreprise par le CEA et la SNEA ont paru suffisamment encourageants pour envisager une opération de démonstration portant sur 4 à 5 000 logements. Après une rigoureuse sélection parmi les nombreux sites possibles, celui de PLAISIR, dans la banlieue Ouest de Paris, a été retenu ; il présente de nombreux intérêts :

- du point de vue hydrogéologique : il existe plusieurs nappes bien séparées par des niveaux argileux limitant beaucoup le risque de drainance d'une nappe par l'autre, et offrant de plus de grandes possibilités dans le choix de la nappe. Parallèlement les nappes intéressantes du point de vue de la perméabilité sont situées à des profondeurs raisonnables (< 500 m).
- du point de vue du besoin de stockage : les conditions de surfaces sont assez favorables : d'abord une usine d'incinération fonctionnant toute l'année sur le site et pouvant être équipée de récupérateurs de chaleur (200°C), et ensuite plusieurs ensembles de logements, dont les réseaux de chauffage correspondants peuvent être fédérés, et connectés à l'usine d'incinération (et à son stockage de chaleur).

2.3. Rôle du forage de reconnaissance

Toutefois, suite à l'étude précédente de faisabilité, il était nécessaire de préciser les points suivants quant à l'emploi d'une température élevée de stockage (180°C) :

- effets de la température sur la perméabilité de la nappe,
- influence sur les équilibres physico-chimiques de l'eau du terrain,
- comportement de cette même eau dans les échangeurs de chaleur.

C'est pourquoi, avant d'engager l'opération pilote, il a été décidé d'effectuer des études complémentaires sur le site de PLAISIR. Ce programme de travail, qui a fait l'objet d'une aide à la recherche par la Communauté Economique Européenne (DG XII) avait pour objectifs :

- la détermination de l'aptitude du site au stockage,
- la vérification de la cohérence des hypothèses de faisabilité générale lors d'une confrontation à un cas réel,
- le dimensionnement du dispositif de démonstration et la prévision de son comportement.

Pour atteindre ces objectifs, il convenait alors d'exécuter un forage de reconnaissance, forage éventuellement réutilisable comme puits périphérique dans le dispositif de démonstration. Ce forage devait permettre des prélèvements de carottes et d'eau, des essais de pompages et des diagraphies. A partir de ces données, il convenait de mettre au point une procédure permettant de les exploiter et aboutissant à :

- la simulation du comportement physico-chimique du terrain et de l'eau
- la simulation du fonctionnement général du stockage,
- son dimensionnement optimum en fonction des données énergétiques de surface (usine d'incinération et taille du réseau de chaleur).

3. RESULTATS OBTENUS SUR LE PUITS DE RECONNAISSANCE

3.1. Coupe géologique du site de PLAISIR et équipement du puits au droit de la nappe envisagée pour le stockage de chaleur

Après une couverture de craie, puis de marnes jusqu'à 322 m, le forage de reconnaissance a rencontré successivement les trois ensembles de nappes captives attendues :

- l'albo-aptien de 356 à 428 mètres, protégé par 34 mètres de couverture argileuse,
- quelques petites nappes attribuées au barrémien,
- le wealdien de 478 à 530 mètres.

A 545 mètres, le forage a été interrompu dans les calcaires purbeckiens. A quelques petits détails près, ces résultats sont tout à fait conformes aux prévisions.

La nappe albienne, généralement très perméable dans les environs de PARIS, a été trouvée très argileuse et surtout très hétérogène. Des passées très minces de sables très perméables alternent avec des bancs presque imperméables.

Ces mauvaises caractéristiques, ainsi que la faible épaisseur des nappes barrémiennes, ont conduit à retenir finalement le wealdien comme réservoir.

Cet étage se décompose en 2 nappes de 21 et 15 mètres séparées par un banc d'argile de 5 mètres d'épaisseur, configuration déjà trouvée dans les anciens sondages les plus voisins, ce qui tend à montrer une bonne régularité des formations. Il y a peu de risque de faille ou de variation latérale du faciès. Pour des raisons de compatibilité géochimique et de différences de perméabilité, rapprochées de la quantité de chaleur à stocker, il n'est pas apparu souhaitable d'utiliser la totalité du wealdien.

L'analyse du comportement thermique montre que, dans ces conditions, il était préférable de retenir la couche supérieure. Cette option est confortée par le fait que la perméabilité de la couche supérieure, tout en étant largement suffisante, n'en est pas moins inférieure à celle de la couche inférieure, ce qui devrait limiter les convections. Ceci est d'autant plus net que c'est la partie supérieure qui est la plus perméable. Enfin, l'équipement de la première couche et la condamnation de la seconde est une opération plus simple que l'opération inverse.

L'ensemble de la couche retenue (wealdien supérieur) a été équipé pour la production, avec crépine JOHNSON et massif de gravier de 18" de diamètre.

Un essai de pompage à 30 m^3/h a été réalisé, avec étude des remontées de pression.

Il a montré une transmissivité intrinsèque d'environ 65 Darcy-Mètre.

L'existence d'un petit banc d'argile vers 490 mètres et la diminution de la perméabilité avec la profondeur permet d'interpréter de différentes manières cette transmissivité.

Pour les simulations, 2 cas extrêmes ont été envisagés :

- on considère la nappe dans son ensemble, en négligeant les variations internes, ce qui donne 3 Darcy sur 21 mètres,
- on considère que l'eau ne circule pratiquement pas à la base de la nappe. L'épaisseur totale est alors réduite à 14 mètres, mais la perméabilité à froid portée à 4,5 Darcy.

Il est vraisemblable que la réalité se situe entre ces deux hypothèses. De toute façon, cela importe peu pour le comportement hydraulique du système qui ne fait intervenir que la transmissivité qui est bien connue. Les erreurs ne peuvent concerner que le comportement thermique et le dimensionnement. On verra plus loin ce qu'il en est à ce sujet.

3.2. Mesure de la perméabilité et des caractéristiques thermiques de l'aquifère retenue pour l'opération de stockage

Les carottes extraites du forage ont été étudiées au laboratoire SNEA (P) de PAU. Elles ont fait l'objet de toutes les études et déterminations classiques et on insistera ici seulement sur quelques aspects originaux.

a) Un appareil a été spécialement mis au point pour replacer des carottes meubles dans des conditions de contrainte aussi voisines que possible des conditions réelles et d'assurer la percolation d'eau pressurisée, à des températures comprises entre 20 et 180°C, afin de mesurer les variations de perméabilité.
Conformément à ce que laissait supposer des travaux soviétiques et américains, ainsi que l'étude CEA-SNEA de faisabilité générale, il y a une baisse de la perméabilité intrinsèque, partiellement réversible, lorsque la température s'accroît. L'expérience a confirmé les 2 conclusions du rapport de faisabilité générale :

- la variation de perméabilité intrinsèque n'est pas liée obligatoirement à la présence d'argile et dépend au contraire de la teneur en quartz : le sable "propre" est plus affecté que le sable "sale" ;
- la baisse de perméabilité intrinsèque enregistrée ne compense pas la variation de viscosité de l'eau avec la température. Il s'agit d'une conclusion très favorable, car elle implique, moyennant certaines précautions :
  - . l'absence de risque de colmatage,
  - . la certitude de voir la perméabilité de Darcy (qui est exprimée en m/s et règle la circulation du fluide) s'accroître avec la température, la baisse de perméabilité intrinsèque n'ayant pour effet que de la voir s'accroître moins vite qu'on aurait pu le penser si l'on avait négligé ce phénomène.

b) L'étude des carottes et surtout des diagraphies a permis de mesurer (ou de calculer) les paramètres nécessaires à l'exécution de simulations spécialement conçues pour tenir compte de la précision effective des mesures physiques. On a ainsi déterminé la porosité, la masse volumique moyenne, les chaleurs massiques à différentes températures, ainsi que les conductibilités thermiques.
Les valeurs de perméabilité intrinsèque mesurées sur carottes avec l'appareillage de PAU n'ont pas été prises en compte directement, car on aurait introduit une erreur systématique, liée à une variance d'échantillonnage.
La valeur globale à froid, pour une hypothèse donnée d'épaisseur, est connue avec une grande précision. On a donc utilisé les mesures de laboratoire pour déterminer une loi de variation de cette perméabilité globale, en fonction de la température.

4. SIMULATION DU COMPORTEMENT DES DIFFERENTES CARACTERISTIQUES DU STOCKAGE

4.1. Simulation des équilibres physico-chimiques mis en jeu et leur analyse

Les carottes de forage ont été analysées minéralogiquement et chimiquement. Lors du pompage, des prélèvements d'eau ont été effectués avec de grandes précautions (pour éviter les variations de pH et le départ de CO_2 en cas d'exposition à l'atmosphère). Les analyses ont été effectuées par 2 laboratoires. Elles ont été complétées par des recherches bactériologiques. On a ainsi disposé de données suffisantes pour effectuer une simulation complète des phénomènes de dissolution et précipitation liés aux variations de température. Cette étude a été conduite avec l'aide de l'Université Louis Pasteur à Strasbourg (M. Fritz).

Pour cela un modèle spécial, dérivé des modèles existants à l'Université, a été mis au point, permettant de prendre en compte, dans une large mesure, la cinétique du système.

La description détaillée du modèle sortirait du cadre de cette brève communication. Il dérive des travaux de GARREL et CHRIST et du modèle KELLY-COB de HEGELSON, permettant de déterminer si une solution est sursaturée, sous-saturée ou en équilibre avec un solide, le modèle prenant en compte non seulement les ions simples détectables par l'analyse, mais aussi tous les ions complexes résultant des arrangements possibles des précédents. Les travaux de B. FRITZ ont conduit à corriger l'expression de la loi de DEBYE-HUCKEL, permettant de calculer le coefficient d'activité de la solution, afin de la rendre applicable à des solutions concentrées. En outre, un programme spécial prend en compte, par pas de 1°C, l'évolution des compositions des différentes phases, lors d'un échauffement ou d'un refroidissement (programme THERMAL de B. FRITZ).

Les résultats suivants ont été trouvés :

a) Dans le terrain :

Les transferts les plus importants ont lieu au niveau de la thermocline. Il y a, pendant le stockage, nourrissage des quartz et dissolution de calcite ; les phénomènes inverses se produisent lors de la récupération. Comme la thermocline balaye tout le stockage pendant la durée d'un cycle, les effets sur la porosité - et, indirectement, sur la perméabilité - sont très limités. Le calcul montre qu'ils sont insignifiants.

Les phénomènes les plus critiques concernent le voisinage des puits.

- Près du puits central, la dissolution de la silice est surtout importante au début du premier cycle, lors du stockage (environ 150 mg/kg d'eau injectée) et pourrait entraîner une certaine dégradation du massif filtrant. Il peut également y avoir un début de diagénèse des argiles avec formation de chlorites magnésiennes, mais cette réaction est fortement concurrencée par la dissolution de la kaolinite et est finalement peu critique.
- Près des puits périphériques, c'est la phase de récupération qui est la plus critique. Un certain nourrissage des quartz pourrait résulter de la réinjection d'une solution sursaturée en silice. Mais, aux températures envisagées à cet endroit, ce phénomène devrait être peu marqué. Il peut, par contre, en présence d'ions Aluminium libre, précipiter de la kaolinite. Au maximum, cette précipitation pourrait concerner 40 mg/kg d'eau. Cela serait suffisant pour entraîner un colmatage irréversible et l'échec de l'opération. Dans le terrain, c'est donc le comportement des puits périphériques lors de la récupération qu'il faut surveiller : une désaturation de l'eau en silice avant réinjection, par des procédés classiques et agréés pour les eaux potables peut éviter tout problème à ce niveau.

b) Dans les installations de surface :

Le chauffage de l'eau, lors du stockage, entraîne une sursaturation par rapport à la calcite, avec risque d'entartrage de l'échangeur, et accessoirement de la crépine du puits central (risque de colmatage). En évitant le dégazage du CO_2, le phénomène est réduit, mais non supprimé. Il affecte 50 mg/kg d'eau lors du premier cycle et 35 mg/kg lors des cycles suivants. Si l'on ajoute l'effet des oxydes de fer qui obéissent aux mêmes lois, on atteint une dizaine de tonnes de précipité par an, ce qui pose un réel problème. Il faut donc décarbonater l'eau avant chauffage.

Lors de la récupération, la sursaturation en silice n'est pas critique en surface, sauf si la température descend au-dessous de 50°C : il y a alors saturation non seulement par rapport au quartz (cinétique lente) mais aussi par rapport à la silice amorphe (cinétique rapide). Le précipité qui se formerait dans l'échangeur pourrait entraîner la ruine de celui-ci, mais cet accident peut être facilement évité, avec un réseau de chaleur à température de retour voisine de 70°C.

En résumé, il serait vain de sous-estimer ou de négliger les problèmes physico-chimiques, mais il y a des remèdes assez simples (précipitation à la chaux et à la magnésie, très facile à chaud, surtout quand, comme c'est le cas, on accepte une perte ionique élevée).

4.2. Simulation du comportement thermique du stockage - analyse du système

Un modèle numérique semi-analytique et semi-convectif a été mis au point (modèle STOICA).

En connaissant l'évolution de la production des rejets thermiques de l'incinérateur et des besoins thermiques du réseau de chaleur, et évidemment des caractéristiques du terrain données par le puits de reconnaissance, on peut ainsi prévoir le comportement détaillé du stockage. Le modèle calcule non seulement les températures de récupération, mais aussi les consommations électriques de pompage.

Il est volontairement rudimentaire : il est inutile de disposer d'une simulation sophistiquée, alors que les paramètres de nappe - et parfois de surface - sont mal connus. Il est préférable de disposer d'un modèle simple permettant d'effectuer de nombreuses simulations de longue durée. Le comportement d'un stock lors du premier cycle n'est en effet pas significatif.

Dans ces conditions, on a jugé correct d'effectuer 40 simulations portant chacune sur une durée de 4 ans. Ceci paraît un minimum indispensable pour se faire une idée du système.

L'originalité du procédé résulte du fait que l'on ne simule pas seulement la nappe mais l'ensemble du système source - stockage - utilisation. Ceci a abouti à des conclusions intéressantes et originales.

Dans toutes les hypothèses, le rendement est toujours médiocre ou mauvais la première année. Il s'améliore ensuite très vite pour se situer dans une fourchette de 60 à 70 % (selon les cas) au 4ème cycle. Les 50 % sont dépassés dans la plupart des cas dès la deuxième année.

a) Le résultat inattendu est le peu de sensibilité aux variations des paramètres du terrain autour des données moyennes fournies par le forage.
Il y a, à cela, une explication théorique : si le rendement est faible lors du premier cycle, une bonne partie de la chaleur demeure dans le terrain et constitue un coussin thermique améliorant les performances ultérieures. Il y a donc tendance à la convergence des résultats quels que soient les paramètres de terrain.
b) Au contraire, les conditions d'exploitation peuvent favoriser le stockage ou peser lourdement sur sa faisabilité.
c) Les énergies de pompage demeurent très acceptables (quelques % de l'énergie thermique récupérée, en prenant en compte stockage et récupération) à condition de choisir une pompe de récupération suffisamment performante.
d) Le système a pu être dimensionné. Pour 4 000 logements, on arrive à un rayon du stockage de 85 mètres.

Nota : On a également étudié en détail l'impact du stockage sur l'environnement : négligeable, ainsi que l'influence réciproque du stockage de chaleur et d'un stockage souterrain de gaz situé à 5 km de là, dans la même nappe.
Une simulation a été réalisée par Gaz de France, montrant que la cohabitation des 2 systèmes était possible.

5. CONCLUSIONS

Les différents problèmes susceptibles d'être envisagés lors de l'opération de démonstration ont été résolus :

- Comportement thermique et hydraulique de l'aquifère (la formation et son effluent) dans la gamme de température choisie).
- Stabilité du stockage de chaleur au sein de l'aquifère et de son environnement géologique.
- Détermination des paramètres techniques avec une précision convenable (nombre de puits, rayon du stockage, débit d'exploitation, caractéristiques dimensionnelles des équipements de fond et de surface).
- Simulation du bilan énergétique (avec un résultat satisfaisant).
- Etude d'impact (aucun inconvénient pour l'environnement).

Il est donc désormais possible et légitime d'entreprendre l'opération de démonstration prévue.

UNDERGROUND HEAT STORAGE

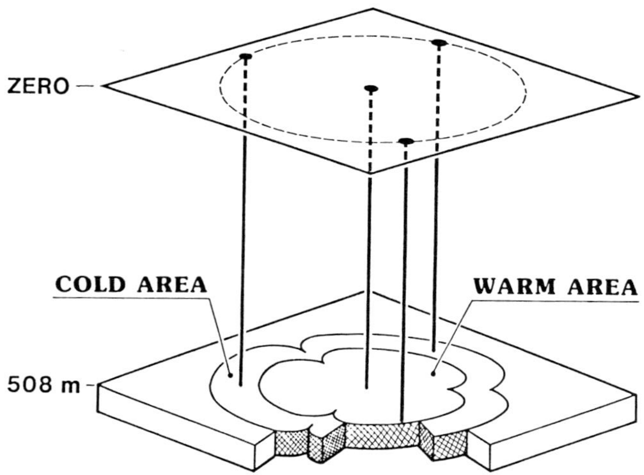

Fig. 1: SCHEMA GENERAL DU STOCKAGE SOUTERRAIN

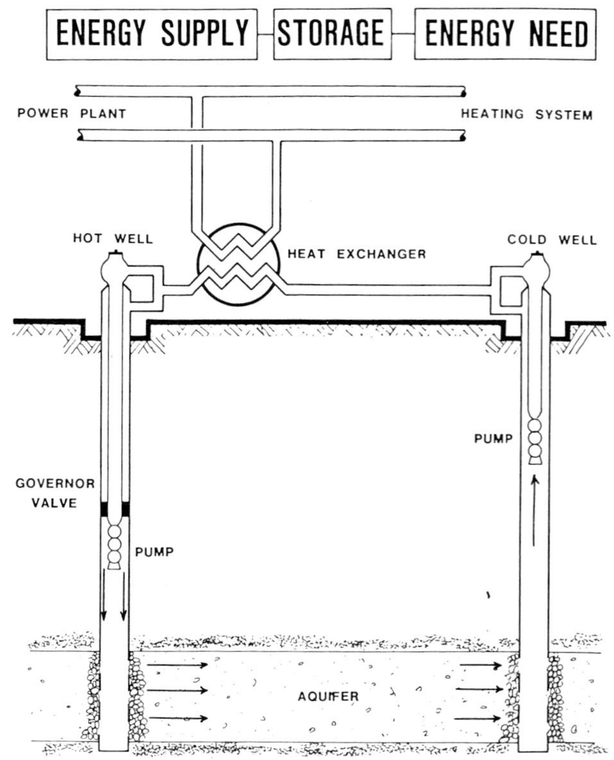

Fig. 2: SCHEMA DE FONCTIONNEMENT EN PHASE DE STOCKAGE (ETE)

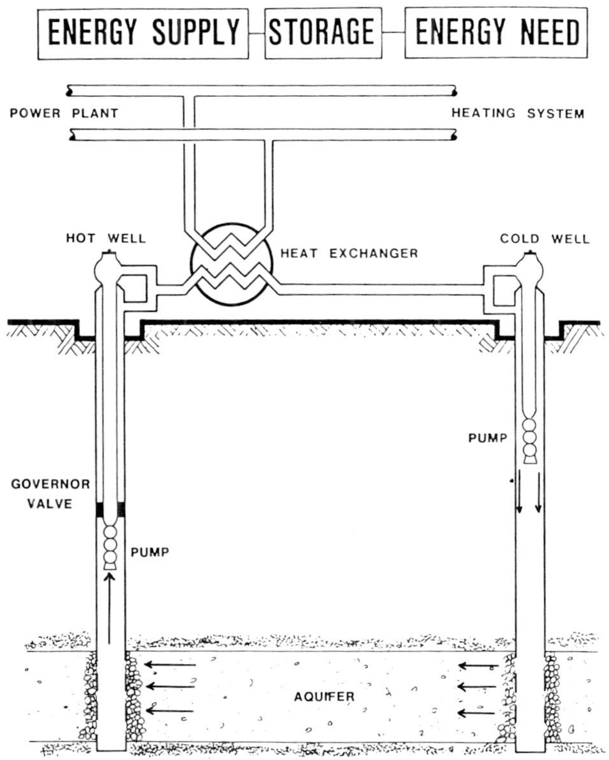

Fig. 3: SCHEMA DE FONCTIONNEMENT EN PHASE DE RECUPERATION (HIVER)

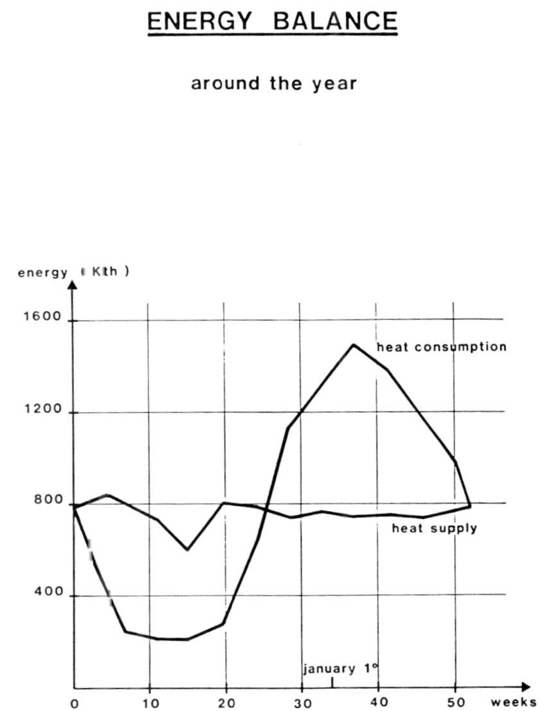

Fig. 4: EVOLUTIONS COMPAREES DURANT UNE ANNEE DES BESOINS DE CHAUFFAGE ET DE LA FOURNITURE DE DE CHALEUR

QUESTIONS AND DISCUSSION

Zegers (DG XII) : 1. What is the cost per m^3 (soil) of your heat system ? 2. What is the probability to find a suitable aquifer in an area with a radius of 20 Km from a location where a cheap heat supply and a heat demand exists?
3. Do you expect high salt concentrations in the water which could give corrosion ?

Answer : 1. It is very difficult to define such a price in our situation, but we could say less than 0,05 French francs per kWh (that will give a price of 5 FF per m^3.

2. In the sedimentary areas in France, it is easy to find such captive aquifers, but not always at a reasonable cost (for deep aquifers) or at a light feasability (due to grounds repartition).

3. Not in our present situation, the chloride distributions are not dependant of the temperature.

Pilavachi (DG XII) : What is the payback time for such system ?

Answer : We would be happy to give you an answer, but presently we don't know the efficiency of the storage system. Also data on the power of the garbage burning plant and the exact size of the heating district are not yet known.
Nevertheless we could hope a payback time less than 10 years.

Oudendal (Netherlands Railways) : Do you have estimates on the maintenance costs of the system for the whole period of 30 years ?

Answer : Yes, we did, but the hypotheses are not yet well precised

- lifetime of the pumps, especially for the one in the hot well
- necessity of work in the wells themselves (in fact no intervention has to take place before 30 years).

We hope that the average maintenance cost will be less than 5 % of the total initial one.

Ehringer (DG XII) : In the proposed system heat from a waste treatment plant is stored in the ground and used in winter for district heating. Would it be feasible to store the garbage in summer-time and to burn it in winter-time ?

Answer : It is quite impossible for reasons of public health. But it is possible to produce a low calorific solid fuel from urban waste. This transformation needs a lot of work, and the economical choice between this solution and over heat storage is not obvious.

Ketelaar (Chairman) : 1. Horizontal underground currents would lead to loss in heat (and treated water) from the region between pumps.
2. Industry as a source of heat can not be considered as it is an unreliable heat source over a long period (50 years).

Answer : 1. In a captive aquifer the underground water movements are very slow, and can, in our present demonstration plant, be compensated with the help of the peripherical wells.

2. It is true, for a single waste heat supply, but not if there is a local concentration of many industries.

SESSION IV - INSULATING MATERIAL

Chairman: C. BOFFA

Insulating foams with good fire resistance based on novolaque type phenolic resins in form of reinfored foams

Bituminous heat insulating materials

Heat accumulating building materials

Improvement of the functional thermal resistance of rigid polyurethane foams (PUR)

Research on buildings' general quality of insulation, especially with respect to developing and improving methods of postinsulation

Wall composed of different layers having high thermal value and aging of insulating

INSULATING FOAMS WITH GOOD FIRE RESISTANCE BASED ON NOVOLAQUE TYPE PHENOLIC RESINS IN FORM OF REINFORCED FOAMS

Authors	: Pierre DUBOIS, Christian REINAUDO (Laboratoires de Marcoussis) and Emile MOREL (IRCHA)
Contract number	: EEA-2-013 F
Duration	: 18 months 1 july 1980 - 31 december 1981
Head of project	: P. DUBOIS, Laboratoires de Marcoussis Centre de Recherches de la CGE
Contractor	: Laboratoires de Marcoussis Centre de Recherches de la CGE IRCHA
Address	: . Laboratoires de Marcoussis Route de Nozay - 91460 Marcoussis (France) . IRCHA BP 1 - 91710 Vert-le-Petit (France)

Summary

The research conducted by the Laboratoires de Marcoussis, the Research Center of Compagnie Generale d'Electricité and the National Institute of Applied Chemical Research (IRCHA) was related to phenolic resins panels for building thermal insulation. This research started from our previous results on the fabrication of foams from powdered Novolaque phenolic resins without acid catalyst.

We have developed the fabrication of 25x25 cm panels of reinforced phenolic foam with a density of 80 kg/m3. These panels possess a thermal conductivity equivalent to that of the most commonly used foams , polystyrene and polyurethane, and markedly better mechanical properties, thermal behavior and fire resistance.

The use of micro-wave heating instead of resistor heating has already provided encouraging results on samples : more homogeneous distribution of glass and foam and increased expansion of the phenolic foam in the presence of glass mat, leading to a density of 30 kg/m3.

O. Foreword

The research conducted by the Marcoussis Laboratories, the Research Center of the Compagnie Generale d'Electricite, and by the National Institute of Applied Chemical Research (IRCHA) was related to phenolic resins panels for buildings thermal insulation. These panels are made of reinforced foams possessing excellent fire resistance as well as good mechanical characteristics. This research started from our previous results on the fabrication of foams from powdered "Novolaque" phenolic resins without acid catalyst.

Phenolic foams possess excellent fire behaviour : inherent non-flammability and absence of toxic or corrosive droplets or fumes ; up to now their development was hampered by :

- the acid nature of the catalyst used with liquid resins, causing residual acidity in the finished product which could lead to corrosion of the substrates,
- mechanical weakness of the foams, which crumble.

We will successively present :

. the selection of the materials and the selection of the fabrication process for the panels,
. the characteristics of the reinforced foams thus obtained.

1. Materials and fabrication process

Novolaque thermo-setting formo-phenolic resins are based on three essential components : the basic resin, the blowing agent and the fillers.

1.1. The basic resins

Novolaque resins are characterized through three essential parameters :

- the melting temperature, which is a consequence of molecular weight ;
- the hexamethylenetetramine ratio, which characterizes how easily the resin sets ; in theory, 1.2 percent HMTA suffices to obtain a well-reticulated lattice but the amounts used in practice vary between 2.5 and 11 percent by weight ;
- the pouring behaviour which depends on the viscosity in the molten state.

We have made a comparison between three resins : PB 10, PB 15 and PB 17, whose characteristics are as follows :

Resin	Melting temp. (°C)	HMTA ratio (percent)	Pour at 125°C (mm)
PB 10	90 - 96	9.7 - 10.3	19 - 25
PB 17	77 - 83	10.3 - 10.9	54 - 66
PB 15	79 - 85	5.4 - 6.0	63 - 77

This comparison has led to the selection of resin PB 15 for the following reasons :

- low melting temperature,
- low viscosity.

Figure 1 shows that the density of the foam obtained varies inversely to the viscosity of the resin used. This is due to the fact that the released gas must exert a lower effort to overcome the viscosity forces within the molten resin when this latter is less viscous.

1.2. The blowing agents

Expansion of the Novolaque resin is obtained through release of a gas within the molten polymer. The expanded structure is preserved through hardening of the thermo-setting polymer. Novolaque being powdery, we have looked toward powdered solid expansion agents. The relative expansion and reticulation kinetics determine the quality of the foam. In addition to the nature of the resin and that of the blowing agent, the following parameters play a part: temperature, viscosity and superficial tension of the molten resin.

As blowing agent, we have selected dipentax, which can release 240 cm3 gas per gram. Figure 2 shows the influence of the amount of blowing agent used on the density of the foam obtained, at a constant curing temperature of 160°C ; when the blowing agent content increases, the foam density decreases and then stabilizes from a certain value of the blowing agent ratio, this being due to the bursting of pores when the pressure of the released gas becomes too high with respect to the resistance of the walls.

The results of figures 3 to 6 have been obtained through differential enthalpic analysis with an increase of the furnace temperature of 10°C per minute. Figure 3, relative to a mixture of resin and dipentax, shows two exothermal peaks corresponding to two successive reactions : resin reticulation between 120 and 185°C and chemical breakdown of dipentax at 200°C. When comparing this plot to those of figures 4 and 5, relative to resin and dipentax, one can see that plot 3 combines plots 4 and 5. The temperatures of the two peaks are quite different so that the reaction kinetics do not allow to obtain a satisfactory foam. It is thus interesting to try and modify the reaction kinetics so as to obtain better simultaneity between reticulation on the one hand and expansion on the other hand. For this purpose, we have successively incorporated four additives into the blowing agent : zinc oxide, alumina, carbon black and zinc stearate. Alumina and, mainly, zinc stearate (see fig. 6) allow lowering the dipentax breakdown temperature and bringing it closer to the resin reticulation temperature.

1.3. The fillers

Two types of fillers have been investigated :

- reinforcing fillers (glass mats or fibers) ; glass mat has been selected as mineral filler since it strengthens the material mainly by increasing its bending resistance,
- fillers which would lower the cost of the foam ; these changes lead to two opposite effects : they enhance foam expansion by lowering the breakdown temperature of the blowing agent (see fig. 6) and they make the foam heavier through the addition of a denser mineral product as shown by figure 7 in

the case of addition of alumina. It must be noted that the addition of alumina also leads to an increase of the foam oxygen index.

1.4. Obtention of low-density foams

a) The expansion of unfilled Novolaque powders can lead to low-density foams (up to 20 kg/m3) if the components of the mixture are well homogenized in the molten state in a mixing machine. The expansion process is however slowed when glass, mat or chopped fibres, is incorporated into the mixture. A density of 80 kg/cm3 has been obtained when the expansion took place in a oven. Observation of the expanded foam plates obtained through oven curing shows a color difference between the skin and the interior of the plates. The foam becomes all the more insulating as the foam expands and thus limits the thermal flow from the outside toward the center. The heat released by the reaction is too low to sustain expansion and hardening. The resin polymerization and blowing agent breakdown kinetics are disturbed. This observation is confirmed by a result : the density of thin plates is always lower than that of thick plates.

b) This has led us to investigate a heating mode allowing to obtain a higher in-depth temperature so as to obtain reinforced foams with lower density. The use of a micro-wave oven has already provided some encouraging results:

- a larger expansion, even in the presence of glass mat ; expansion of pre-impregnated products which led to 3 cm thick plates in an electrical resistor oven lead to 10 cm thick plates in a micro-wave open,
- the glass and foam are distributed in a more homogeneous manner,
- the cross-section of a panel of foam shows a dark-colored material at center, which points to good polymerization.

2. Characteristics of the reinforced Novolaque phenolic foam

2.1. Material composition and panel fabrication

The various components are powdery, except for one of the tensio-active agents which is liquid. Mixing is realized, in solid state, in a fast mixer whose blades rotate at 1800 rpm ; this powdery mixture is then mixed in a BUSS machine at 75°C. The mixing proportions (by weight) are as follows :

- Phenolic resin PB 15 = 100
- Dipentax blowing agent = 6
- Additives = 6.

We then realize a pre-impregnated glass mat as follows : 30 parts glass (b. wt) are incorporated into 100 parts (b. wt) of the above mixture. The glass mat is then pressed under pressure (3 kg/cm2) at 100°C for 4 minutes to obtain a 3 mm thick plate.

Finally, expansion is performed for 30 minutes at 200°C in a oven. The 250 x 250 x 30 mm panels thus obtained have a density of 80 kg/m3.

2.2. Mechanical characteristics

We have characterizedthe bending and compression mechanical behavior of panel elements subjected to constraints applied perpendicularly to the plate :

- . elasticity modulus — 23 kg/cm2
- . yield point — 1.2 kg/cm2
- . strength for 10 % crushing — 1.8 kg/cm2
- . strength for 50 % crushing — 5.2 kg/cm2

The values of the bending resistance show the beneficial effect of the skin present on the surface of the panels :

With skin	max	15 kg/cm2
	$E_{bending}$	1330 kg/cm2
Without skin	max	7.5 kg/cm2
	$E_{bending}$	55 kg/cm2

2.3. Porosity and thermal characteristics

a) Porosity

The measurements taken under an air overpressure of 400 mm Hg show a 24 % closed porosity.

b) Expansion coefficient

The plot of figure 8 shows the variation of the expansion coefficient according to temperature. This coefficient is equal to 1.6×10^{-5} at 20°C.

c) Thermal conductivity

The measured thermal conductivity is 0.04 $W.m^{-1}.K^{-1}$.

2.4. Ageing

2.4.1. Water absorption

a) Capillarity

The base of the sample is kept in 2 mm of water, the remainder of the sample being in the atmosphere of the enclosure at 20°C and 90 % relative humidity. After 1 000 hours ageing under these conditions, the ratio of moisture picked-up by the foam tends toward an asymptotic value of 1% by weight, which corresponds to the water picked-up by the resin.

b) Immersion

The specimens are immersed into water at ambient temperature. After 1 000 hours ageing, the ratio of water picked-up by the foam is equal to 24 % by volume, which corresponds to partial filling of the open pores.

2.4.2. Weather behavior

Phenolic resins are stable even at high temperatures. Specimens have been subjected to ageing tests in a weatherometer under conditions selected as typical of the extreme temperature, humidity and solar radiation conditions to be encountered by thermal insulating material placed externally :

. temperature 70°C
. relative humidity 90 %
1 kW/m2

No modification of the foam properties has been noticed after 500 hours of testing.

2.5. Fire resistance

a) Measurement of the limit oxygen index (LOI)

We measured the minimum oxygen ratio in a nitrogen/oxygen mixture required to sustain flame propagation in a vertical specimen.
Measured LOI = 39.

b) Measurement of flame propagation (UL 94)

We measured the ignition time and the flame propagation speed on a vertical specimen.
The material is classified V_o.

c) Epiradiator test

Three sets of 20 x 20 cm samples have been subjected to a thermal flux of 6 W/cm2 generated by a radiating panel, in the presence of a pilot flame ; this flux is roughly equivalent to that received by the ceiling of a room at the moment of general inflammation.

Set 1 : 30 mm thick reinforced Novolaque foam panels, directly exposed to the thermal flux.

After 30 minutes' expositions, photographs 1 show that the sample has been carbonized throughout its thickness ; the glass mat remains in place; the entire sample never ignited.

Set 2 : composites consisting of a 30 mm thick reinforced phenolic foam panel with a 10 mm thick protective plasterboard plate (Placoplâtre).

This sample exposed to the same 6 W/cm2 flux for one hour, never ignited. The plasterboard has split but the foam carbonized only over a limited thickness. Photographs 2 show the condition of both faces of the foam after this test.

Set 3 : composites consisting of a 30 mm thick expanded polystyrene bonded to a 10 mm thick protective plasterboard plate (Placoplâtre). After 12 minutes' exposition, the plasterboard has split and the polystyrene has sublimated in spite of the protection of the face exposed to the radiation. Photographs 3 show the condition of both faces after this test.

Figure 9 shows the evolution, during these tests, of the temperature of the face not exposed to the radiation for sets 2 and 3, these sets consisting of phenolic resin or polystyrene protected by means plasterboard. One sees that :

- the temperature has stabilized at 150°C after 1 hour testing in the case of phenolic foams,

- a sharp temperature rise occurs after 12 minutes' testing in the case of polystyrene ; this is when the polystyrene sublimates.

3. Conclusion

The cost of the materials used to compound the reinforced phenolic foams studied amounts to 9.40 FF/kg for a composite containing :

. 70 % by weight of phenolic resin costing 9 FF/kg,
. 22 % by weight of glass mat costing 9 FF/kg,
. 4 % expansion agent dipentax costing 13.65 FF/kg,
. 4 % additives costing 15 FF/kg.

This matérials cost of 9.40 FF/kg compares with 6.65 FF/kg for polystyrene and 10.5 FF/kg for polyurethane. The ex-factory cost of expanded plates is 12.85 FF/kg for polystyrene and 17.5 FF/kg for polyurethane.

We have developped the fabrication of 25 x 25 cm panels out of reinforced phenolic foam with a density of 80 kg/m3. These panels possess a thermal conductivity equivalent to that of the most commonly used foams : polystyrene and polyurethane, and markedly better mechanical properties, thermal behavior and fire resistance.

We have investigated a heating mode allowing to obtain more homogeneous temperature within the panel during polymerization so as to obtain foams with a lower density. The use of a micro-wave heating instead of resistor heating has already provided encouraging results on samples : more homogeneous distribution of glass and foam and increased expansion of the phenolic foam in the presence of glass mat, leading to a density of 30 kg/m3.

QUESTIONS AND DISCUSSION

Damsgaard Olsen (Technological Institute) Did you add fire resistant additives ?

Answer : No.

Pilavachi (DG XII) : As the materials resist temperatures up to 150° C there might be possible applications in industry. Did you consider this ?

Answer : We did not explore industrial applications.

Roux (CSTB) : What is the thermal conductivity for low density NOVOLAQUE phenolic resins and what is the relative humidity ?

Answer : We measured the thermal conductivity only for high density samples. The relative humidity was not measured.

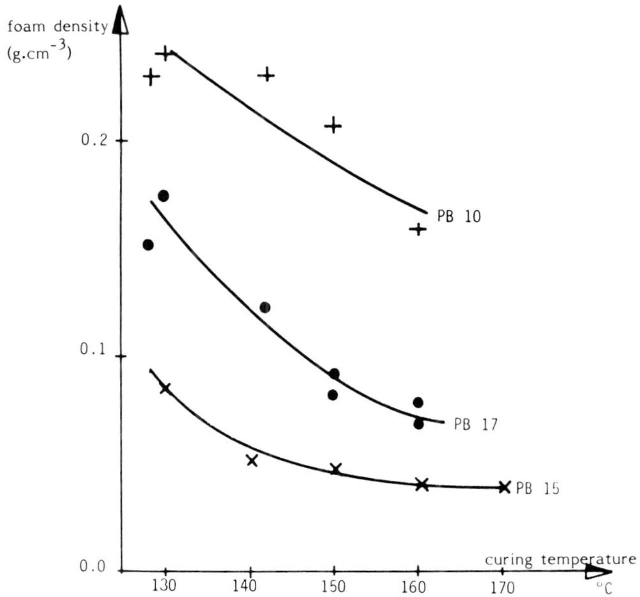

Fig. 1 Foam density versus curing temperature for 3 Novolaque resins
(Phenolic resin + 5% dipentax)

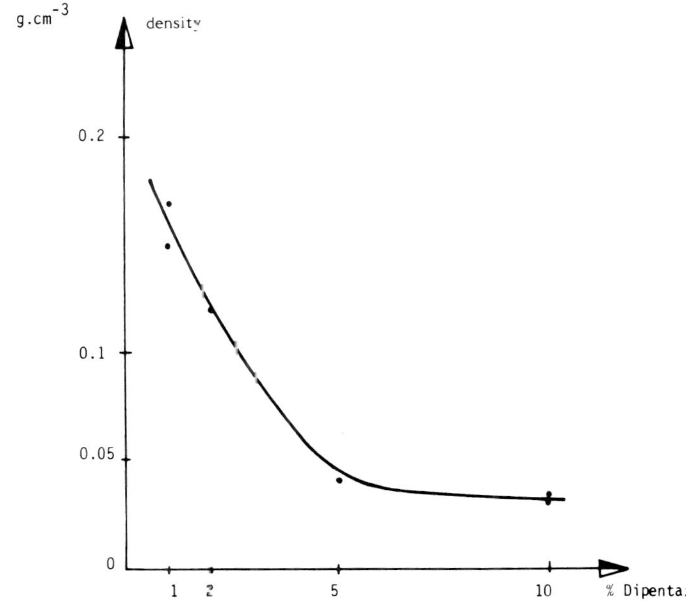

Fig. 2 Influence of the amount of blowing agent on the foam density
(PB 15 resin cured at 160°C)

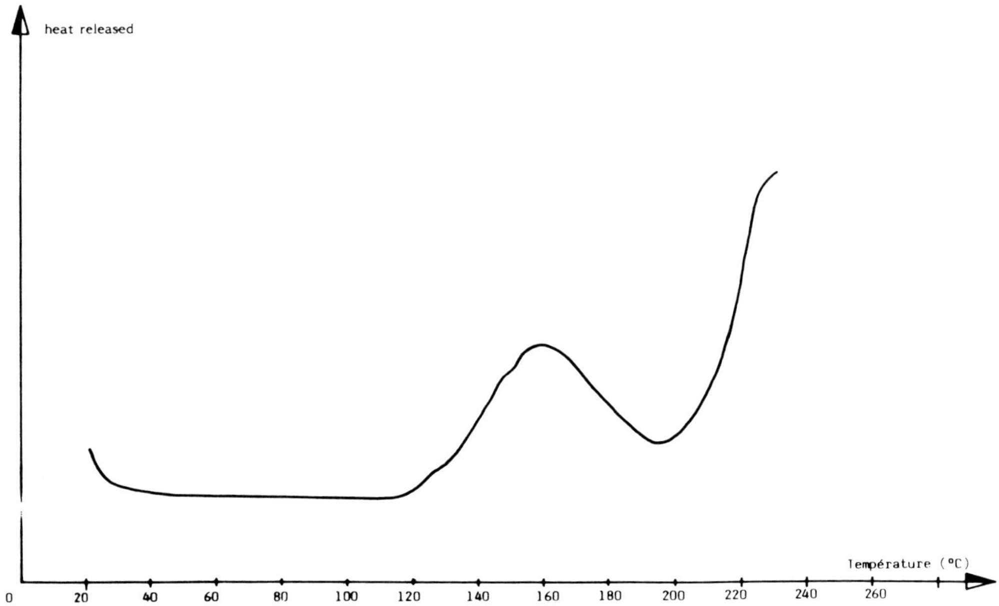

Fig. 3 Heat released by a PB 15 and dipentax mixture versus temperature

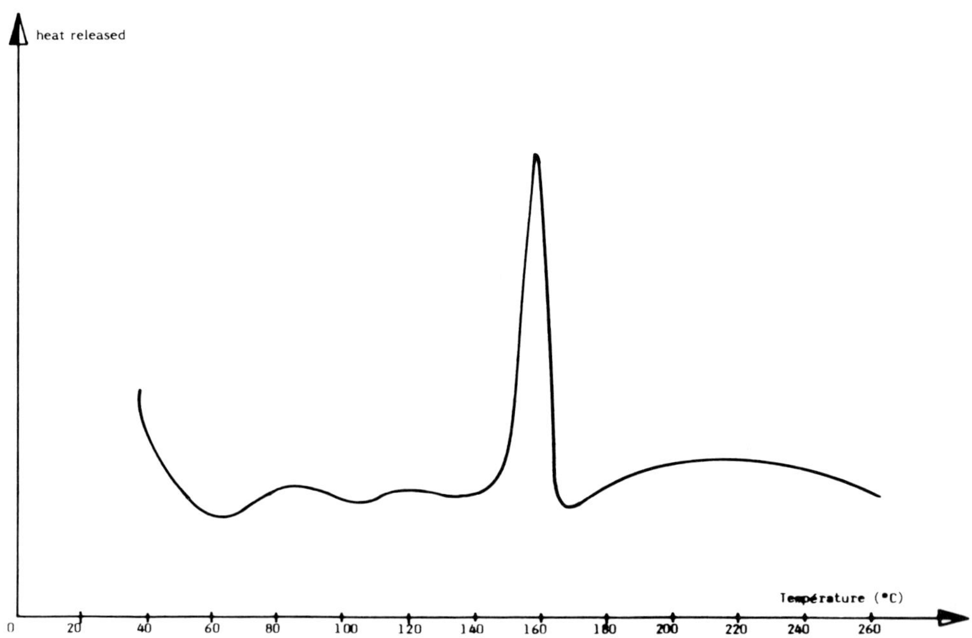

Fig. 4 Heat released by PB 15 resin versus temperature

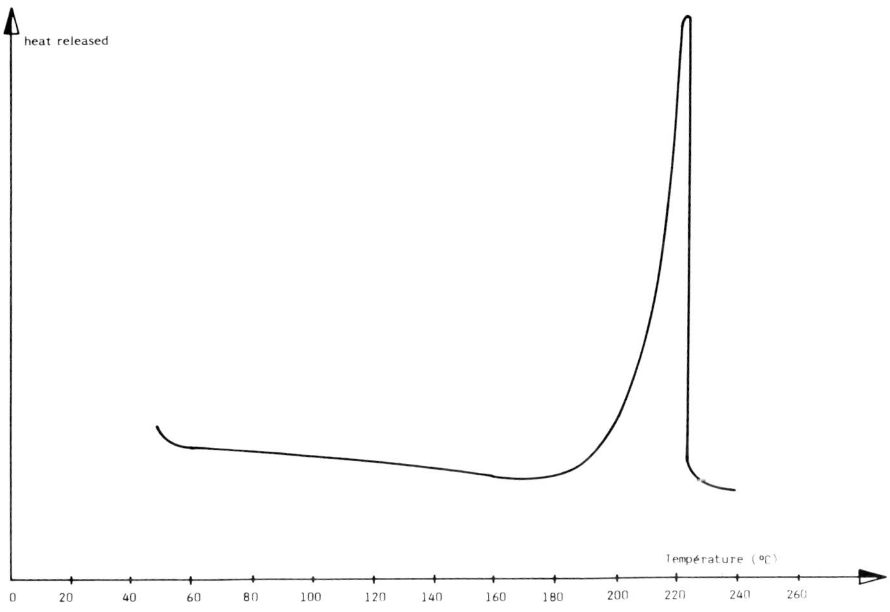

Fig. 5 Heat released by dipentax versus temperature

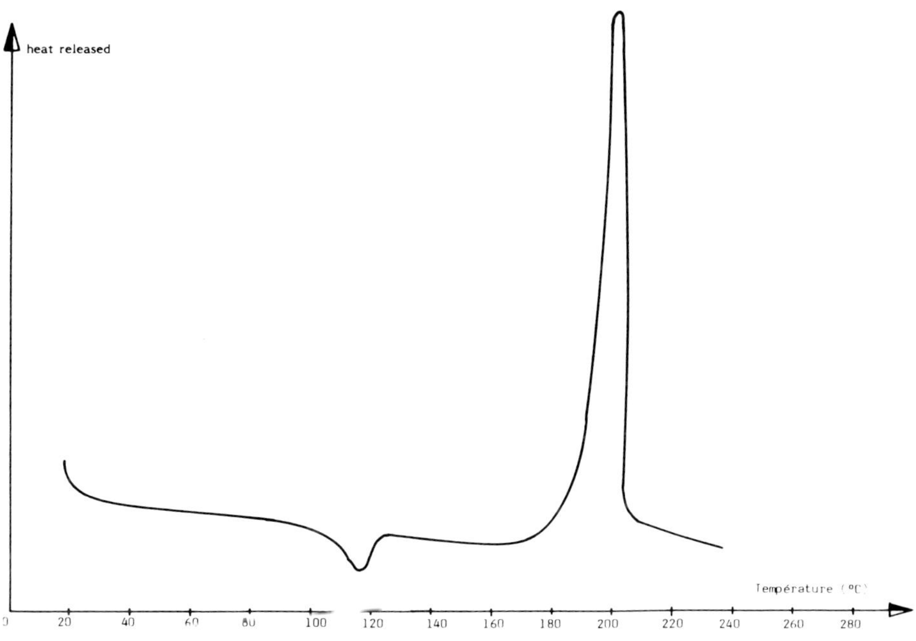

Fig. 6 Heat released by a mixture dipentax - Zinc stearate versus temperature

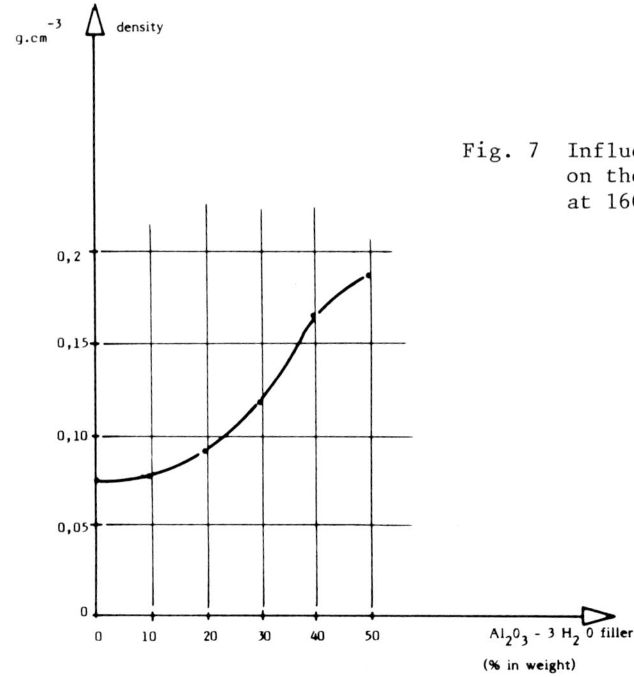

Fig. 7 Influence of the alumina content on the density of foam cured at 160°C

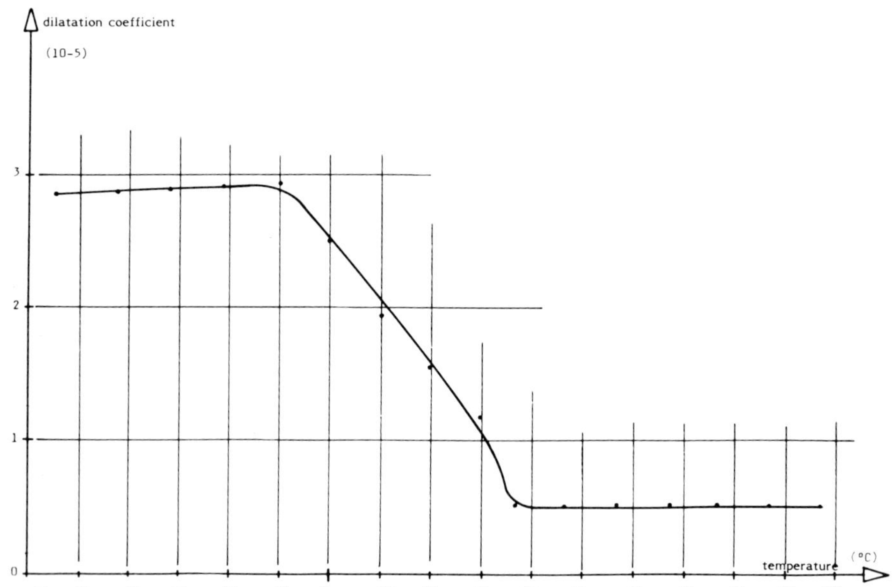

Fig. 8 Dilatation coefficient versus temperature

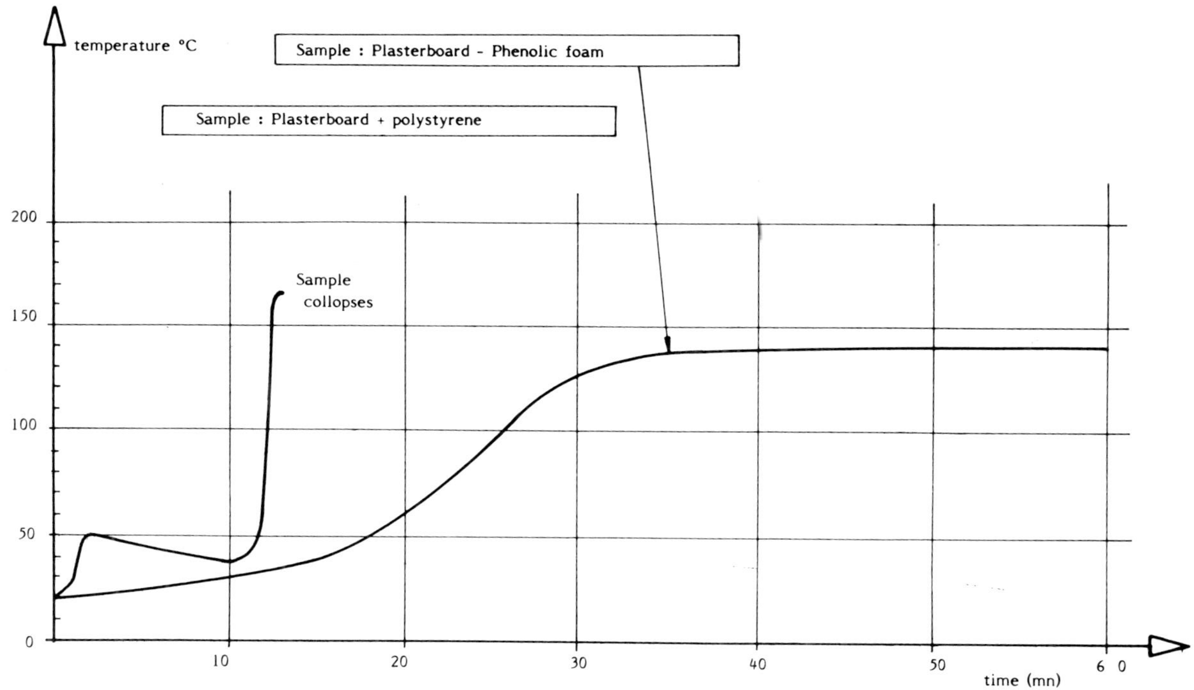

Fig. 9 20 x 20 cm samples under 6 W cm^{-2} heat flux - Temperature of the side opposite to heat flux

Side opposite to
heat flux

30 mm thick phenolic foam reinforced panel after
30 minutes under a heat flux of 6 w cm^{-2}

PHOTOS
1

Side under heat flux

Phenolic foam side

Phenolic resin + plasterboard plate
After one hour under a 6 W cm^2 heat flux

PHOTO3
2

Plasterboard side

Polystyrene side

Polystyrene + Plasterboard

After 12 minutes under a 6 W cm^{-2} heat flux on the plasterboard side

Plasterboard side

PHOTOS
3

BITUMINOUS HEAT INSULATING MATERIALS

Authors : L. BOCCHI, V. CASTAGNETTA

Contract number : EE-A-2-017-I

Duration : 10 months (1 March 1982 - 31 December 1982)

Head of project : Dr. Ing. V. Castagnetta, Industria Italiana Petroli (Italian State Oil Group ENI)

Contractor : Industria Italiana Petroli S.p.A.

Address : Piazza della Vittoria 1, 16121 Genova, Italy

Summary

The project envisages energy saving in the heating of civil and industrial buildings with the use of improved insulating materials, e.g. waterproofed against humidity and water vapour and protected against the attack of polluting agents in the atmosphere, by means of bituminous products. The research aims at the utilization of any kind of granular or fibriform insulating material, where each individual element is completely coated with an impermeable and protective bituminous binder. The latter, as it also binds the elements together, makes possible the manufacture of insulating panels. The coating and manufacturing process can be performed by thoroughly mixing the insulating material with a relatively hot bitumen or with a cold bituminous emulsion and by pressing the mixture so as to form suitable panels. The report covers both the hot and the cold process, as well as the technical characterization of the bituminous heat insulating panels obtained.

1.1 Introduction

Our project comes within the general framework for economizing on energy consumption for the heating of civil and industrial buildings. It envisages a better utilization of the properties of the insulating materials through the use of the waterproofing and protective properties of bituminous binders.

In fact, the thermal conductivity coefficient λ values of insulating materials and the normal techniques for their application are known, but some comments have drawn attention to the deterioration of the λ due to the presence of humidity and/or the degradability of certain insulating materials.

As the waterproofing and protective properties of bituminous binders are also known, it is believed they can be exploited with a view to eliminating or at least reducing the λ deterioration referred to above.

Our research covers the use of any insulating material in granules or fibres and its objective is to create - by using bituminous binders - the protection and waterproofing of each individual insulating element against humidity and water vapour, protecting it, moreover, against the attack of any polluting agent in the atmosphere and thus against any degradation capable of reducing its efficiency and life.

Thanks to the impermeableness and indegradability of bituminous insulating materials, energy consumption in the civil and industrial building heating field can drop to and stabilize itself at considerably lower levels.

After preliminary laboratory tests and controls, the new type of bituminous heat (and sound) insulating materials has been patented in Italy by IP (Industria Italiana Petroli of the ENI Group) and the patent was then extended to other European countries.

1.2 Work programme

The work programme considers the two alternatives of the coating process of the insulating granular material (at present expanded polystyrene) with hot bitumen or cold bituminous emulsion.

It has been deemed opportune and expedient to introduce the cold process alternative in as much as it allows the saving of energy for the heating of the bituminous binder and does away with any possible pollution aspect in the processing phase.

1.2.1 Hot process

The research programme for the hot process includes the following phases:

a) laboratory study of the mix composition and preparation of initial samples of bituminous insulating panels;
b) basic characterization of the initial samples (measure of the thermal conductivity λ);
c) experimental preparation of a number of panels (prototypes) covered with aluminium paper, bitumen impregnated boards or felts, kraft paper, or uncovered;
d) basic characterization of the prototypes (λ measurement), also for acoustic insulation;
e) complete characterization of the best prototypes;

f) ageing tests of the best prototypes (compared with traditional insulating materials) in presence of different types of atmosphere:
   - normal (unpolluted) ambient
   - marine (sea environment) ambient
   - urban ambient ($CO - NO_2 - CO_2 - Cl_2$)
   - industrial ambient
   - hydrocarbon vapours (xylene, n-decane)

g) characterization of the aged prototypes.

1.2.2 Cold process

The research programme for the cold process includes the following phases:

a) laboratory study of the mix composition and preparation of initial samples of bituminous insulating panels;
b) basic characterization of the initial samples (λ);
c) laboratory preparation of a number of panels (prototypes) covered with kraft paper or uncovered;
d) complete characterization of the prototypes according to UNI 7819 Specification.

Ageing tests are not included in this programme, as the finished products (panels) obtained with the cold process are expected to have the same performance as the panels obtained with the hot process.

1.3 Progress of work - Preliminary results

1.3.1 Hot process

Initial laboratory samples were prepared with a mixture of polystyrene granules (two sizes), hydrated lime filler and 80/100 pen. bitumen, and the mix was pressed to form panels.

Thermal conductivity, measured by the Padua University, showed a result of 0.040 kcal/h°Cm.

Experimental prototypes were then produced at a mixing plant, having care to heat bitumen at the lowest possible temperature, such as to allow coating of the polystyrene granules without deteriorating them.

Prototype panels, covered with aluminium paper or with bitumen impregnated light board, were tested for thermal conductivity (ASTM C 177) at Assoreni (ENI Group Laboratories) with the following results:

- prototype with aluminium paper : 0.044 kcal/h°Cm
- prototype with bituminous board : 0.046 kcal/h°Cm

which confirm the results obtained with the initial samples prepared in our laboratory.

Prototypes were also tested by Assoreni for sound insulation, in comparison with some traditional insulating materials, with the following acoustic attenuation results:

- our prototype with aluminium paper : + 39.0 dB
- our prototype with bituminous board : + 39.8 dB
- normal expanded polystyrene : + 22.7 dB
- compressed cork covered with flannel : + 31.0 dB
- bitumen and rock wool layers : + 39.5 dB

The tests were carried out under the following conditions:

- test frequency : 1000 Hz
- loudspeaker tension : 2 Volt
- octave selectivity : 20

The above results confirm the good sound insulation characteristics of our prototypes.

The following ageing tests in polluted atmosphere were carried out at IP Central Laboratory (20 days at 30°C):

- urban atmosphere:
  gas: CO - NO_2 - SO_2 - Cl_2
  respective concentrations (desumed by reports on air quality in a number of towns, p.p.m.): 60 - 0.1 - 1 - 1
- hydrocarbon vapours (xylene, n-decane): hydrocarbons were added on water surface with a 50/50 ratio, so that - according to their vapour tension curves (in vapour saturated ambient) at 100°F - a concentration of 4000 p.p.m. of hydrocarbons in gas phase might be achieved.

The other ageing tests were executed by ISTEDIL (Istituto sperimentale per l'edilizia) under the following conditions:

- normal (unpolluted) atmosphere : ASTM G 23 Method (500 hours)
- marine (sea environment) atmosphere : ASTM B 117 Method (500 hours)
- industrial environment atmosphere : ISO 3231 Method (5 cycles).

All samples of prototypes aged at IP Laboratory and at ISTEDIL have been sent to ASSORENI for retained density and thermal conductivity controls; the results are expected before the end of the year.

ISTEDIL has also been involved with the following characterizations of our original prototypes:

- density (UNI 6349 Method)
- compression strength (UNI 6350 Method)
- response into heat and flame (ASTM D 635 Method)
- thermal conductance and transmittance (ASTM C 236 Method).

1.3.2 Cold process

The cold process study is developed by TECNOCASA in collaboration with the University of L'Aquila.

Initial laboratory samples have already been prepared with a mixture of polystyrene granules (two sizes) and a special bitumen emulsion added with colloidal mineral. The mix has been compressed to form panels, either covered with bitumen impregnated kraft paper or uncovered.

After curing, panels have shown good strength and stiffness, even better than with the hot process.

At the moment the study is being developed for the optimization both of the mix composition and of the compression and curing conditions. A number of optimized prototypes is being prepared and will be completely characterized by the University according to UNI 7819 Standard (main tests: density - compression and tensile strength - thermal conductivity - resistance to flame progress).

All results will be available by the end of the year.

1.4 Conclusions

Our research aims at the production of bituminous heat (and sound) insulating panels which - owing to the use of granular or fibriform insulating elements coated with a waterproofing and protective bituminous product - cannot be affected by humidity and/or by normal or polluting atmospheric agents, so that a higher level of insulation and a longer duration of that level can be expected with consequent energy saving in civil and industrial building heating.

Both a hot and a cold process is being studied for the production of the above improved insulating panels and the characteristics of the final products are being assessed.

At the moment it is not yet possible to draw the conclusions, but the work in progress looks encouraging: from preliminary results the characteristics of the bituminous insulating panels seem to be reliable and, in any case, the cold production process seems to be preferable.

The final results and the conclusions will be available by the end of the year.

QUESTIONS AND DISCUSSION

Prof. Boffa (Chairman) : Did you vary the temperatures during the aging tests ?

Answer : Yes these tests were done at different temperatures.

Prof. Boffa (Chairman) : What is the cost of these bituminous heat insulating materials ?

Answer : 3000 Lire/m2.

Zegers (DG XII) : Did you do accelerated aging tests ?

Answer : Yes we did standard accelerated aging tests. For example the aging process was accelerated by increasing the concentration of polluting material.

Ehringer (DG XII) : Is the material resistent against heating of the wall by solar radiation ? How is the material applied ?

Answer : Yes it is resistent against solar heating. The material can be injected in the wall cavities but can also be used in the form of panels.

Damsgaard Olsen (Technological Institute) : We use this material and performed thermal conductivity measurements. This material (DRAINPLAST) is used in Denmark and Germany for cellar walls.

Prof. Fontana (Instituto per l'Edilizia) : Do you intend to investigate fibrous material ?

Answer : In this project we will only study granular material.

Zegers (DG XII) : Are there problems with the inflammability of the material ?

Answer : Bituminous material is of course inflammable but we use only very little of it. Moreover it is possible to add additives which decrease the inflammability.

HEAT ACCUMULATING BUILDING MATERIALS

Authors	: G. Rossi, G. Solaini, A.M. Tonelli
Contract number	: EE-A-2-016-I
Duration	: 36 months 1 April 1981 - 31 March 1984
Head of project	: Dr. A.M. Tonelli, Luigi ROSSI-ARCO s.r.l.
Contractor	: Luigi ROSSI-ARCO s.r.l.
Address	: Luigi ROSSI-ARCO s.r.l. Viale Romagna 14 I-20133 MILANO

Summary

New building materials with high transverse thermal diffusivity have been conceived and proposed, with the aim of increasing the amount of heat which can be stored in the buildings' walls. The proposed building structure consists in a combination of materials characterized by differentiated thermal properties, such as bricks or concrete walls in which metallic elements are disposed in the sense of thickness. In buildings' perimetral walls external insulation is required. Heat can be supplied to the walls by radiation or convection; in this last case, warm air must be supplied by means of heating elements disposed all along the wall's basis.

This system allows to reduce the demand of energy and to improve physiological comfort, by means of a better distribution of the room temperature, of a better radiative exchange among walls and persons, and of a lower operating plant temperature.

The research is concerned with the study of the thermal behaviour of the proposed structure, by means of theoretical and experimental investigations. Up today, two different theoretical models have been conceived, by means of which it has been possible to obtain some results concerning walls' charge times and temperature distribution, for different sizes and geometrical distribution of the thermal conductors.

The high heat storing capability of the structure is confirmed by these data. Furthermore, four large concrete walls, with iron cylinders disposed horizontally, in the sense of thickness, have been built inside two greenhouses. Measurements of internal and external air temperatures, and of surface temperature of the walls, heated by solar radiation and by a conventional heat supply system, have been carried out. The experimental data show the positive action of walls on the room temperature behaviour.

1.1 Introduction

Although this paper is presented in a meeting concerning insulating materials, it may be useful to point out that the study of heat accumulating materials is complementary with respect to the research in the field of thermal insulation. There is no doubt, however, that heat storage and thermal insulation may be seen as two aspects of the research concerning a proper thermal behaviour of buildings.

The present proposal takes into account any situation in which it is convenient to apply an external insulation and to make the best use of the heat storage capability of internal and external walls. On the basis of these assumptions it has been conceived a model of building structure in which thermal diffusivity has been increased by the insertion of metallic elements and the insulation of external walls is provided by an external cover.

Being the heat storage and the charge and discharge times functions of the transversal thermal diffusivity, the best distribution of the metallic elements is rather complex to be determined, and consequently mathematical models and analysis are required. The experimental phase is devoted to check some theoretical results and to verify industrial feasibility and energy saving possibilities.

1.2 Description

1.2.1 Description of the model of the heat accumulating material.

In the proposed model (patented) the conducting elements are disposed perpendicularly, with respect to the wall's surface, in order to create high conductivity "channels" and to increase heat diffusion.

In Fig. 1 the schematic drawing of an externally insulated wall, with thermal conductors, is shown. Heat is propagated both "directly" along the normal to the wall's surface, and "indirectly" along the directions determined by the cylinders' radii. The same model is suitable for internal walls. In this way transversal diffusivity is increased without any variation of vertical conductivity.

The proposed heat supply system is based on natural or forced convection from floor to ceiling, and operates along the wall's perimetral basis (Fig. 2). As a consequence, the wall's surface temperature gradient becomes negative, going from the bottom to the top. At the end of the charging phase, when the heating system is disconnected, the walls give back heat to the room by natural convection and by radiation. In some cases (greenhouses, solar passive buildings) walls may be heated by radiation. In the summer time, cold can be supplied by means of a cooling system, disposed at the top of the wall, acting simmetrically, with respect to the heating system.

The presence of heat storing walls allows to reduce the demand of energy and to increase the physiological comfort by means of a better distribution of the room temperature, of a better radiative exchange among walls and persons, and of a lower operating plant temperature. Furthermore, the possibility of operating at low temperature makes it possible the installation of non-conventional plants such as solar systems, heat pumps, etc. Finally, the presence of a large heat storing mass allows to use energy when it is available (solar radiation) or when it is cheaper (electricity in the night time, for electrically-driven heat pumps).

1.2.2 Description of a thermal plant with heat storing walls installed in a greenhouse.

Four prototypes of the proposed structure have been realized inside two greenhouses, built near Castagneto Po (Torino, Italy), on the slope

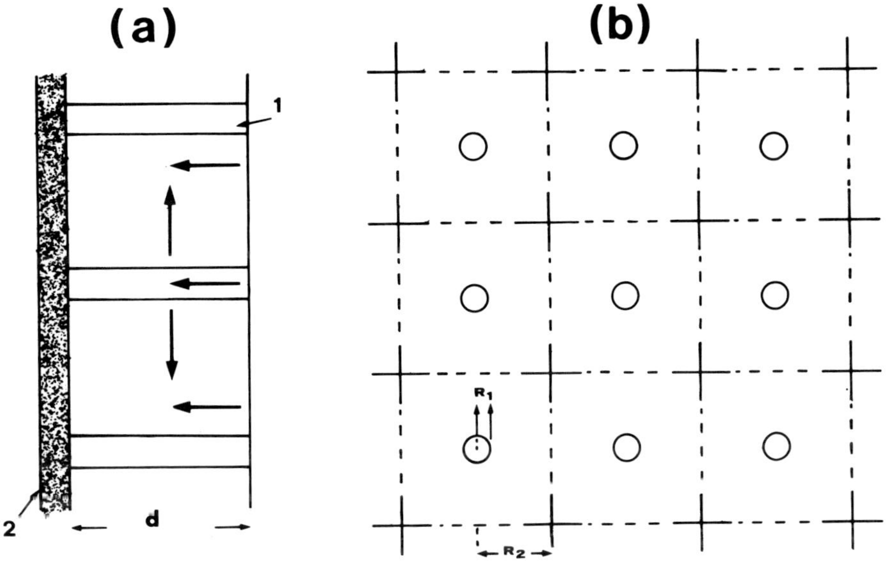

Fig. 1 - Scheme of a heat storing wall, externally insulated.
(a) section; (b) front (1 = thermal conductors,
2 = external insulation).

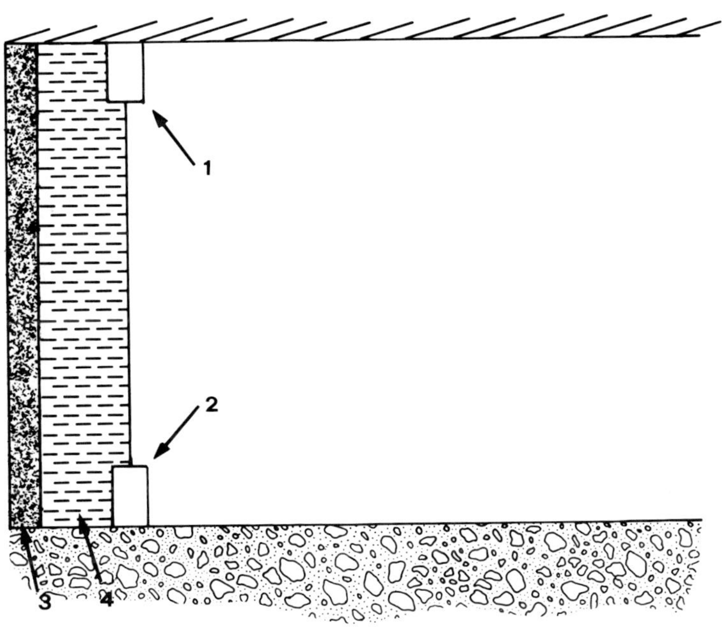

Fig. 2 - Scheme of heating (1) and cooling (2) systems.
(3 = external insulation, 4 = heat storing wall).

Fig. 3 - Greenhouses external view.

Fig. 4 - Internal view with heat storing walls.

Fig. 5 - Detail of the lower wall with heating system.

Fig. 6 - Temperature recording apparatus.

of a hill, facing South (Fig. 3). In each greenhouse there are two walls sustaining the terrane and forming two steps (Fig. 4). The walls are 44 m long, 2.25 m high, whereas thickness varies continously from 0.25 to 0.35 m. Iron cylinders(Ø = 16 mm, lenght = 200 mm) are disposed horizontally along the thickness direction at a relative distance varying in a range of about 0.2 - 0.3 m. All along the wall's basis tubolar radiators are disposed, to supply heat during the coldest periods of the year (Fig.5). Heat is supplied by a conventional plant. Systems for the recycling of warm air accumulated immediately below the ceiling will also be installed.

The experimental apparatus consists, at the present, of a six-tracks temperature recorder, and of several thermistors of different types(Fig.6). Having the greenhouse been completed in the spring of 1982, measurements carried out are not concerning the winter season.

1.3 Progress of work and results.

1.3.1 A theoretical model for the analytical determination of charge and discharge times has been proposed.

The thermal behaviour of the heat storing material has been investigated for an externally insulated wall, charged by air convection, under the assumptions of a uniform initial temperature, and of a very high value of metallic elements conductivity, with respect to the medium.

By means of this study it has been possible to find the values of the ratio τ_1/τ_0 between the charge time τ_1 of the proposed material and the charge time τ_0 of the homogeneous medium with same thickness, and same charging conditions (the charge time is defined as the time required for reaching, in a given point, a temperature value equal to 63% of the asymptotical temperature).

The values of τ_1/τ_0 have been obtained by means of the following methodology:

- writing Fourier's equation with simplified boundary conditions;
- expressing the solution in terms of a series of the eigenfunctions of Laplace operator - ∇^2.
- finding the minimum eigenvalue α_{min} (taking into account the given boundary conditions), by means of Dirichlet's variational principle;
- finding the value of charge time $\tau_1 = \tau_{max} = 1/\alpha_{min}$ D (where D is the thermal diffusity of non conducting material); values of τ_0 can be found in literature.

Several diagrams have been obtained taking into account different geometrical distributions, different sizes of conducting elements, and different values of thickness. The diagram of τ_1/τ_0 versus R_1 is shown, for different values of R_2/R_1, keeping constant d = 10 cm (Fig.7a), and for different values of d, keeping constant R_2/R_1 = 10 (Fig.7b). R_2,R_1 and d are defined in Fig. 1. The diagrams have been plotted under the assumptions of h(liminar coefficient) = 7 W/m^2 K and of λ_m(conductivity od medium) = 0,84 W/m K.

1.3.2 In order to find the values of temperature and heat flux at different times and positions inside the wall, a computer program has been developed on the basis of the following assumptions(thermal nets model):

- the wall's volume has been divided in several isothermal elementary parts (nodes),considered, from a geometrical point of view, as points;
- the thermal capacity of each node has been calculated, as well as the thermal conductance between two adjoining nodes;
- a set of linear equations has been obtained, considering the heat balance

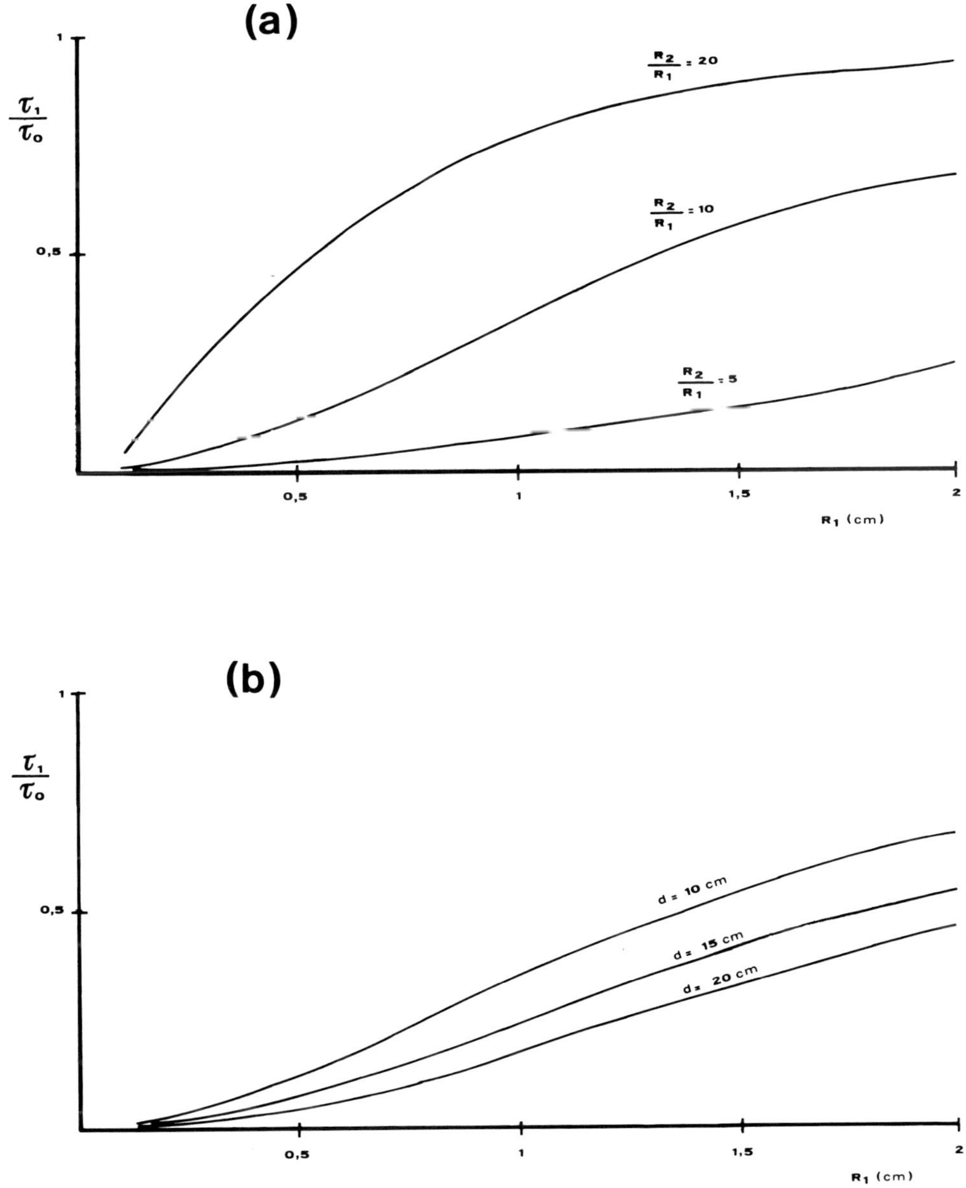

Fig. 7 - Ratio τ_1/τ_0 versus R_1 at constant d = 10 cm (a), and at constant R_2/R_1 = 10 (b)

of each node during a given time Δt.

In Fig. 8 a scheme of nodes distribution and connections is shown. Taking into account the evident simmetry conditions, it is possible to limit the investigation to the half of the parallelepiped $R_2 \times R_2 \times d$. This volume has been divided in 30 parts (nodes from 1 to 30), whereas the conducting element has been divided in 3 parts (nodes 31 to 33). Nodes 34 and 35 represent external and internal air, respectively. It is evident that the nodes number can be increased (or decreased), and that a higher number of nodes corresponds to a higher precision of results. The same in true for time intervals.

Several data have been obtained by means of the computer program, concerning temperature and heat flux inside the wall. In order to give an example, a diagram of stored heat versus time is shown in Fig. 9, whose data refer to concrete and iron, and have been determined on the base of the following assumptions: T_{ext} (external temperature) = T_o (initial wall temperature); $\Delta T = T_{ext} - T_{int} = 30°C$; $h = 12\ W/m^2\ K$; $R_1 = 7{,}5$ mm.

1.3.3 Some experimental data have been obtained by recording the surface temperature of the two walls of a greenhouse, as well as the temperature of internal and external air, and of the water of the heating plant. Six thermal probes have been used: two for the walls, two for internal air (placed in the upper and lower part of the greenhouse), one for the external air and one for the plant. All the internal probes have been placed in the greenhouse central section. Measurements have been carried out in the course of April 1982.

During sunny days walls have been charged by solar radiation, whereas during cold cloudy days they have been charged by the heating plant. In Fig. 10 and 11 recorded data are shown, referred to a cloudy and to a sunny day, respectively (1 = upper internal air; 2 = external air; 3 = upper wall; 4 = heating system).

1.4 Analysis of results and comments

Although results obtained by the two theoretical models are quite different, as far as numerical values are concerned (being diffent the boundary conditions assumed in the two cases), the results themselves agree in showing that the heat storing capability of the proposed materials is higher with respect to the conventional ones, and that the ratio τ_1/τ_0 becomes lower when, keeping constant the metal volumetric percentage, the thermal conductors are thinner and their number is greater (Fig. 7 an 9).

In particular, Fig. 9 shows that for a metal volumetric percentage of 3% the proposed wall can store, during its charge time τ_1, twice as much the heat stored in the same time by the conventional wall.

Better results can be achieved by increasing the volumetric percentage of thermal conductors, and by enlarging their head section exposed to air convection. In this way the conductors shape would be similar to nails' one.

As far as the weight increase of the proposed material is concerned, with respect to conventional ones, it is to be noted that a volumetric percentage of 6% (double of the one considered above) corresponds to a weight increase that, for iron, is less than 20%.

Measurements performed on the heat storing walls built in the greenhouse (Fig. 10 and 11) show that walls are able to store and give back heat in a relatively short time, and to keep a considerable thermal stability inside the greenhouse, notwithstanding the low volumetric percentage of iron (less than 0.2%). Recorded data show that walls are

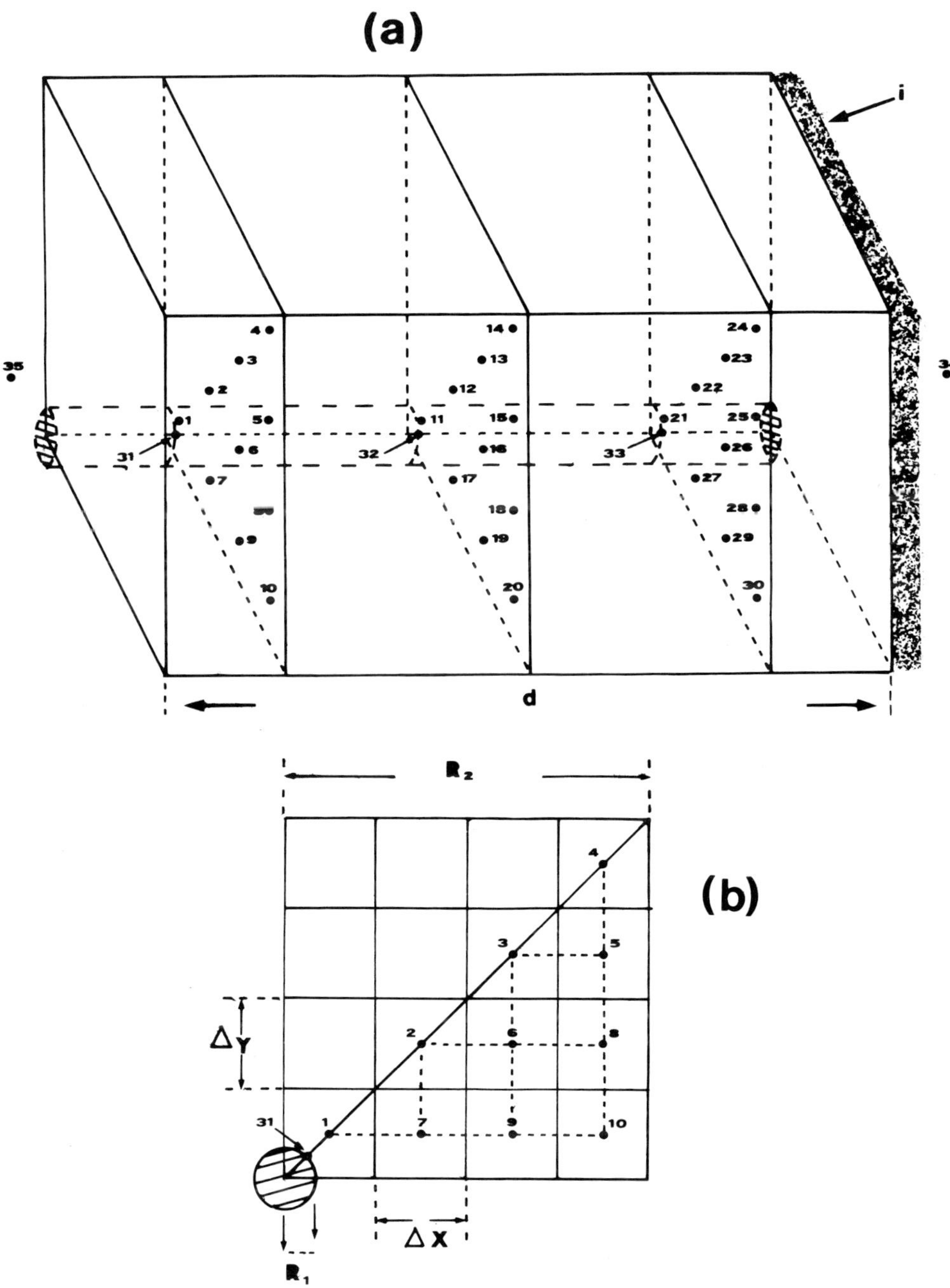

Fig. 8 - Scheme of nodes distribution: (a) assonometric view, (b) section (i=external insulation). In (a) node j is joined to node j + 10. Nodes from 10 to 20 and from 20 to 30 are connected with 35 (internal air) and 34, respectively. The thermal conductor is divided into three parts (31,32,33) with connections 1-31, 11-32, 31-33.

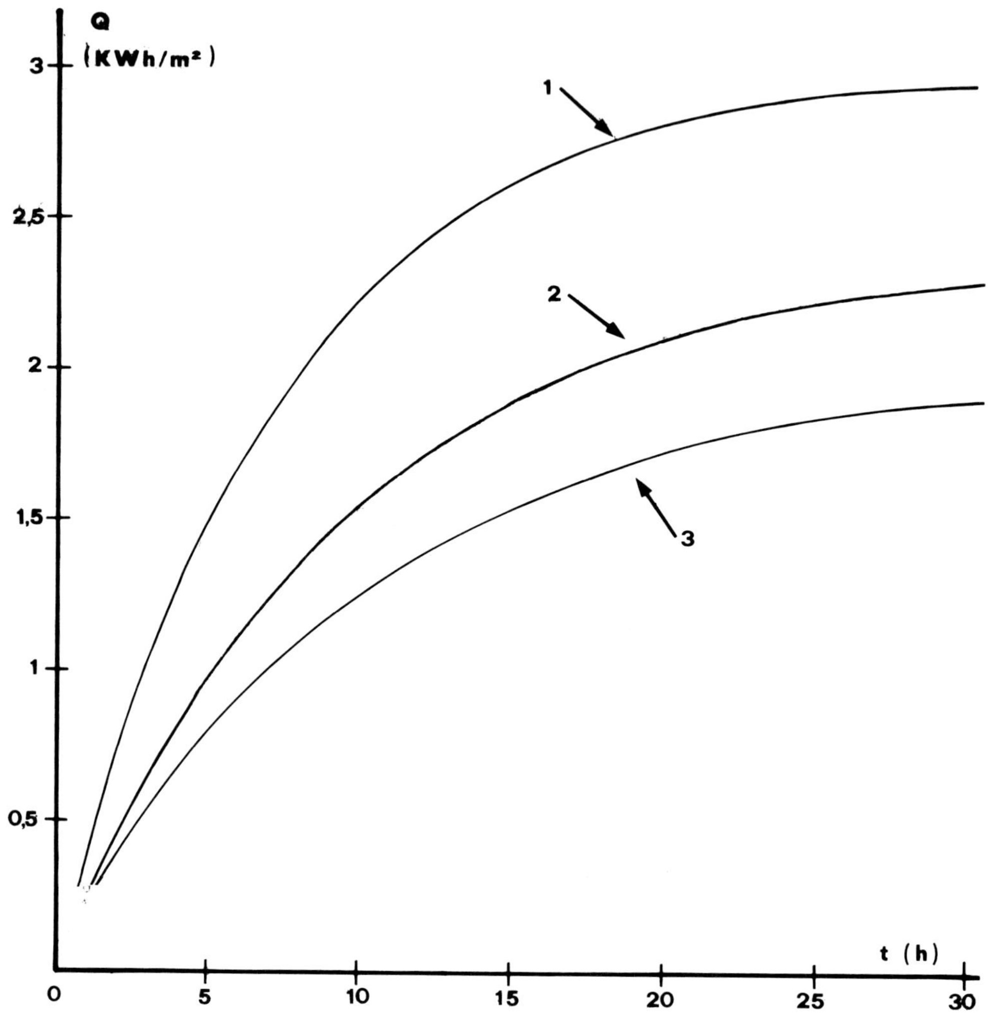

Fig. 9 - Stored heat Q versus time t. Curves 1 and 2 refer to the proposed material, for $R_2/R_1 = 5$ and $R_2/R_1 = 10$, respectively. Curve 3 refers to a conventional wall.
In all cases d = 20 cm. Thermal insulation is provided by a polyurethane layer 5 mm thick.

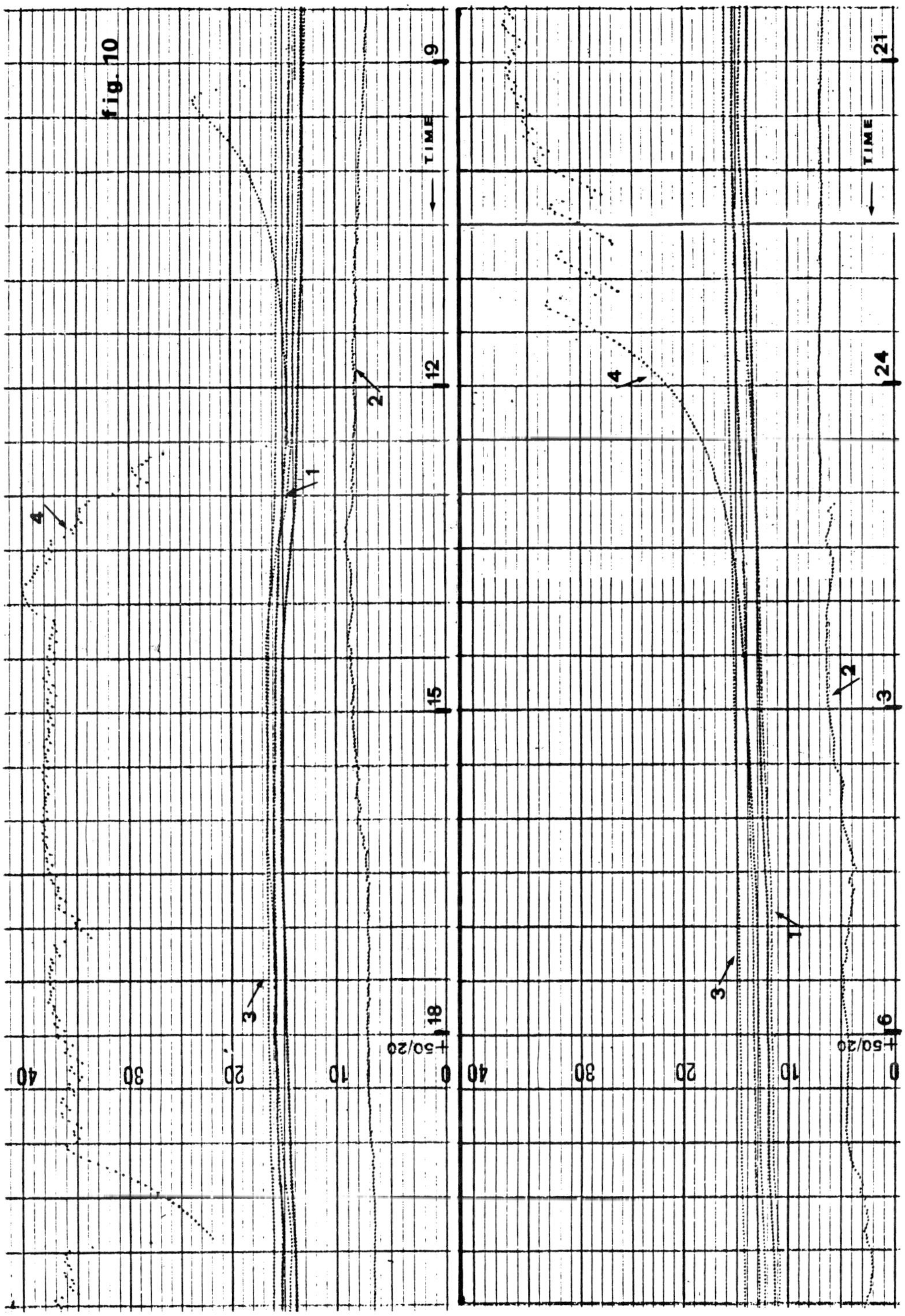
fig. 10
TIME
9
12
15
18
21
24
3
6
40
30
20
10
0
+50/20
1
2
3
4

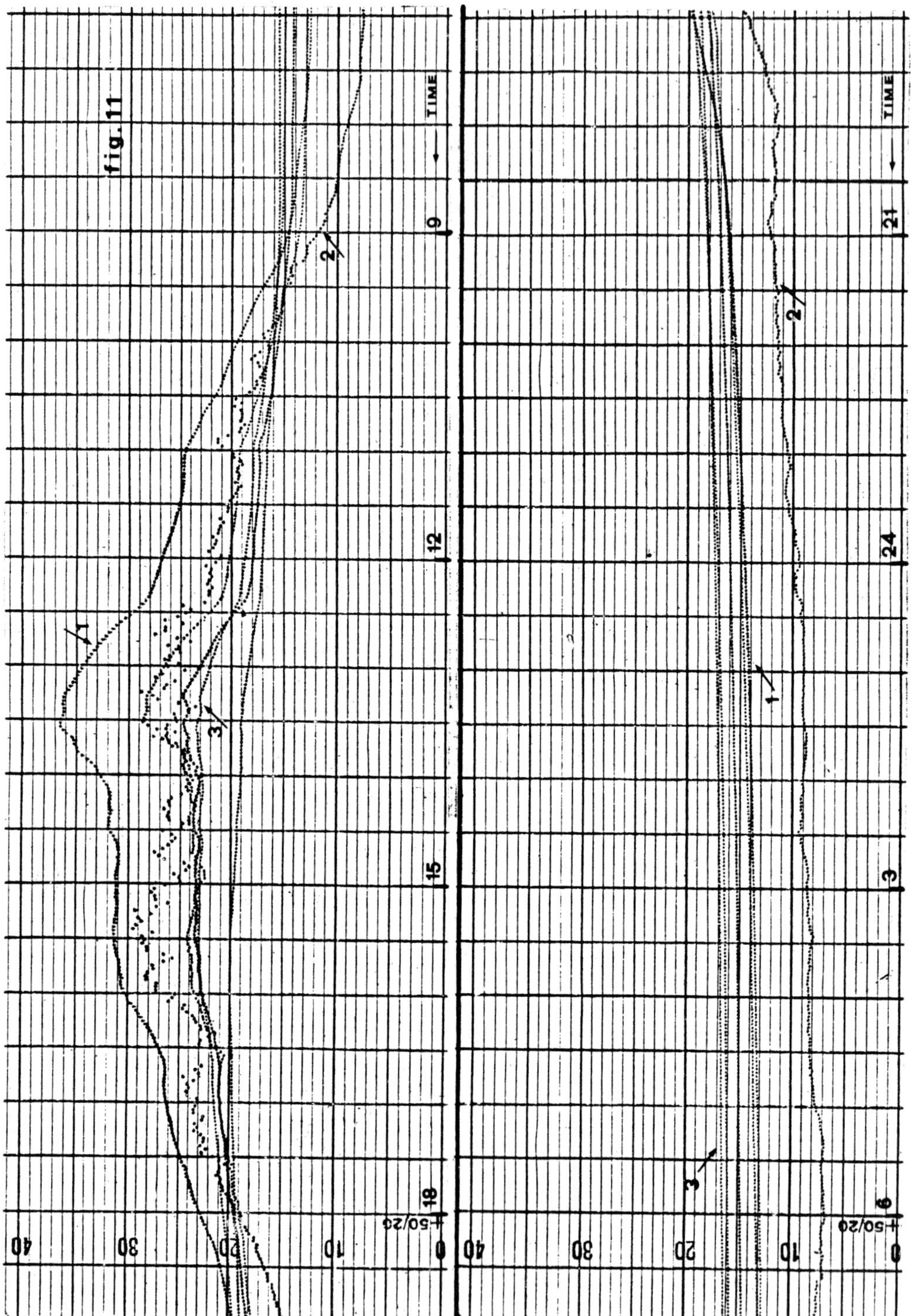
fig.11
TIME
9
12
15
18
21
24
3
6
40
30
20
10
0
50/20
1
2
3

particularly reactive when charged by solar radiation, and that satisfactory results can be achieved also when the conventional heating system operates at low temperatures (less than 40°C).

2. Conclusions

Transverse thermal diffusivity of conventional walls can be increased by inserting metallic elements in the sense of tickness: on the basis of this assumption, new building materials have been proposed, in order to increase the amount of heat that can be stored in building's walls.

A theoretical research concerning the thermal behaviour of the proposed structure has been carried out. Several results have been found, concerning thermal charge times, temperature distribution and stored heat, for different sizes and geometrical distribution of the thermal conductors. From these data it turns out that, keeping constant the volumetric percentag of metal, heat is stored faster when the number of conducting elements is greater. Furthermore, the proposed theoretical model allows for the optimization of geometrical features and distribution of thermal conductors, when boundary conditions are given.

Four large heat storing walls (made of beton and iron) have been built inside two greenhouses, in order to obtain experimental results, and to investigate (in a first approach) on the technical aspects concerning the proposed materials. Some experimental data have been found, which show that walls are able to give a noticeable contribution to the thermal stability of internal air.

QUESTIONS AND DISCUSSION

Pilavachi (DG XII) : How did you determine the heat content of the walls ?

Answer : With the computer model and simulated data.

Zegers (DG XII) : What are the applications for these heat storage walls ?

Answer : For the storage of solar heat and possibility of electric heat during the night.

Prof. Fontana (Instituto per l'Edilizia) : I found that the accumulated heat in the wall increases when the wall contains a metallic matrix.

Answer : Yes we agree, we will investigate this.

Zegers (DG XII) : What is the size of that increase ?

Answer : 20 %.

Damsgaard Olsen (Technological Institute) : Did you use the wall to calibrate the computer model ?

Answer : Yes.

IMPROVEMENT OF THE FUNCTIONAL THERMAL RESISTANCE OF RIGID POLYURETHANE FOAMS (PUR)

Authors : G. ROUX, A. BALME, A. SILVENT, A.M. ANDREANI, B. FOGG

Contract number : EEA-2-015-F

Duration : 15 months 1 March 1981 - 30 May 1982

Contractors : Centre Scientifique et Technique du Bâtiment
Centre d'Etudes Nucléaires de Grenoble

Address : CSTB
24, rue Joseph Fourier
F. 38400 Saint Martin d'Hères

CEN-G
Avenue des Martyrs
BP 85
F. 38041 GRENOBLE CEDEX

Summary

The K value of rigid polyurethane foams increases from about 18 mW/m°.K after processing to nearly 30 mW/m°.K after weathering. The aim of this study is to reduce this asymptotic value to about 20 mW/m°.K, by manufacturing foams with different formulations and following the K value evolution in hot-humid conditions. It was shown that the asymptotic value could be decrease to 25 mW/m°.K. Preliminary tests have shown that the introduction of reactive additive in polyol could decrease this value. That will be checked in the prolongation of this study.

1. Introduction

Due to gaz diffusion through the cell walls of rigid polyurethane foams, the K value of these products increases with time from the original value. The asymptotic value taken in account for thermal calculation is in France 0,029 W/m.°K.

The aim of this research is to achieve a significant improvement in this functional thermal resistance of foams to 0,020 W/m°.K by inhilition the process described above without incurring a serious increase in manufacturing costs.

This study is undertaken jointly by the CSTB and the CEN-Grenoble in collaboration with ICI-France SA as raw materials producer and SECMER as manufacturer of equipments for processing rigid polyurethane foams.

2. Experimental procedure

All foams tested in this study were manufactured in our laboratory. Small blocs (30 x 30 x 30 cm) were foamed using a two components SECMER machine with a capacity of 10 kg/minute. After a curring of one week in the laboratory these blocs were cutted in samples of 10 x 10 x 2,5 cm. Each sample was marked to know the expansion direction and the foam formulation. After that, the reference value of the thermal conductivity λ_o was measured using a heat flow meter ANCON 88 according the french standard NFT 56.124. In the same way dimensions, compressive strength, density was measured. Then samples was weathered in a hot and humid oven at 70°C 95% RH for one, seven, fourteen and twenty eight days. After weathering all these caracteristics were measured again and so we can follow their evolution versus aging.

We have applied this experimental plan on different formulated foams. So we have study the influence of the nature and the amount of catalyst, surfactant on the evolution of λ. We have correlated this evolution with the shape of foam cells. More over we have studied the influence on λ evolution of the rate of water, R_{11}, of the density and of the equipment for processing the foam.

In an over way we have, by preliminary tests, studied the influence of some reactive additives on λ evolution. This new way will be taken in account in the prolongation of this contract in 1983.

3. Results

The first part of this program concern the classification of 12 surfactants and 6 catalysts. All them are commercial products. The results are given on table I and table II.

The second part concern the study of the influence of the amount of surfactant and catalyst on λ values. In this way we increase from 1 to 5 the rate of each additive. The values are given on table III and table IV.

The correlation between the asymptotic value of λ and of foam cells geometry is shown on fig. 1.

For a given formulation, the influence of the rate of water on λ evolution is shown on fig. 2. We can see that λ_o is almost independant of the rate of water where as the asymptotic value moves with an optimun for 2% of water. We have seen too, that dimensional stability decreases when the rate of water goes from 0% to 4%.

By following λ evolutions versus density we have checked that the optimun is arround 35 kg/m^3 (fig. 3).

The influence of change in processing system is given on fig. 4 wich shows that our fabrication is representative of an industrial one. However it is evident that a high pressure system gives a best foam.

The influence of the R_{11} rate in the formulation between 30% to 50% is given on the fig. 5. It shows that the thermal conductivity is not clearly connected to the R_{11} amount in the formulation when this rate is up to 40%.

Finally, by preliminary tests, we have evaluated the feasability of decreasing the gaz diffusion through the cell walls using a reactive additive. It is a chemical compound having hydroxyl groups which react with isocyanate. It is neither a polyether nor a polyester but a low molecular weight polymer.

The fig. 6 gives the influence of such a compound and shows that the initial value is higher, due to water presence in the compound, but the kinetic of the λ evolution is slower than with a typical foams. This way of investigation will be taken in account in the future.

4. Analyse of results and comments

After studying the influences of nature and rate of catalyst and surfactant, of the rates of water, R_{11} in formulation, the kind of process, on the foam properties, we can explain the following facts.

4.1. Small effect of surfactants (tables I - III)

The asymptotic value λ_∞ does not clearly depend neither from the nature nor from the rate of surfactant and particulary if we compare with the catalyst influence. More over, surfactant modifications do not affect the foaming reactivity exept for non silicon surfactants.

4.2. High effect of catalysts (tables II - IV) fig. 1

After the evident effect on reactivity, increasing the catalyst rate gives a more regular and a better cells configuration. The direct result is a consecutive reduction of λ_0 and λ_∞. An over effect is an increasing of the dimensional stability of the foam.

On the over hand, changing the catalyst acts realy on the foam friability and the reaction kinetic. Moreover, the catalyst classification on the base of the thermal resistance, depends strongly on the catalyst rate.

4.3. Varying effects of water (fig. 2)

The modification of the water rate in the formulation of foams have no real effect neither on cells geometry nor on the thermal conductivity. However there is a real effect on the foaming reactivity, on the foam density and on the dimensional stability. That is an important point because a foam must have a low λ value and a high dimensional stability.

4.4. Limited effect of molding machine (fig. 4)

It seems that the change of low pressure molding machine does not induce a real change in thermal conductivity, for a given formulation. On the over hand the air flow during the foaming process have a greater influence. Changing from a low pressure to a high pressure machine seems to give better results. Never theless all these observations correspond to tendancies more than strong effects.

4.5. Non uniform effect of air

For low pressure system, introducing air during the mixing have a good effect on the cells shape and on thermal conductivity, but air in excess gives a contrary effect.

An optimal flow rate must be adjust. With our experimental configuration, this air flow is 40 l/h for a resin flow of 5 kg/mn.

4.6. No real effect of R_{11} amount (fig. 5)

It is clear that up to 40% neither the initial value λ_o nor the asymptotic value λ_∞ are realy dependant of the R_{11} rate. So it is no interesting for cost reasons to increase this rate up to 40%.

All these results was obtain with the same sorbitol initiated polyether polyol with OH n° 490 having a functionality of 5,6. The same experimental procedure was repeated with two other polyol :

. TEOA initiated polyether, OH N° 550 f = 3

. Sucrose " " , OH N° 450 f = 6

The conclusion is that the type of polyol does not change the observations presented above.

4.7. Relation with gaz diffusion

For several specimens having different formulations, before, during and after weathering, the foam gaz composition was measured using a high resolution mass spectrometer VG. This volume composition must be converted into partial pressure which is proportional to the mass of each gazeous constituent. This conversion entail the open cells average measure for each foams specimen. The accurancy of this measure is not good enough and so the conversion is not yet possible. However using the total gaz pressure measure and the volumic composition it is possible to give some interesting conclusions. Whoever the polyol type, it is possible to correlate the kinetics of the λ and the total pressure evolutions (or the air concentration in the gazeous solution). So the gaz kinetic diffusion can be correlate to the foam formulation (rate of surfactant and particullary). That point is actually important for the continuation of this study.

5. Conclusion

The thermal conductivity taken in account for insulating calculation, λc, is about 30 mW/m°.K in Europ countries. The finality of this study is to decrease this value, corresponding to the asymptotic λ evolution, near 20 mW/m°.K.

It was shown firstly that using commercial raw materials for polyurethane foams and particulary catalysts and surfactant it is possible to get an asymptotic value higher than 35 mW/m°.K for the thermal conductivity, value which is not in respect with λ_c.

On the over hand, it was shown that it is possible, by ajustment of the formulation and density, to get an asymptotic value near of 25 mW/m°.K.

Moreover it was shown that the diffusion kinetic of gaz through the cell walls could be correlated with the foam formulation. As preliminary tests have shown that suitable reactive additives (having hydroxylfunction) mixed with the polyol may reduce the asymptote value λ_∞,it seems possible to got a λ_∞ value lower than 25 mW/m°.K.

TABLE 1 - SURFRACTANT EFFECT -

SURFACTANT		1	2	3	4	5	6	7	8	9	10	11	12
cells cotation (de 1 à 5)		3	3	3	3	3	3	3	4	3	4	5	
λ // (mW/m°.K)	0 day	25	26	26	26	25	32	26	26	27	26	28	no foam possible
	1 -	30	32	31	31	32	34	33	32	32	31	33	
	2 -	31	33	32	32	32	35	35	33	32	32	33	
	4 -	31	32	32	34	33	34	35	33	32	31	34	
λ ⊥ (mW/m°.K)	0 day	21	21	21	21	21	21	23	22	24	24	26	
	1 -	26	26	24	26	25	25	29	25	28	28	31	
	2 -	27	27	28	29	25	27	29	27	28	29	31	
	4 -	28	29	28	29	26	29	29	29	28	29	31	

TABLE II - CATALYST EFFECT -

Sample		1	2	3	4	5	6	1	2	3	4	5
Surfactant		1	1	1	1	1	1	5	5	5	5	5
Cream time (sec)		25	25	12	25	45	8	25	25	10	20	45
String time (sec)		120	135	135	95	130	115	135	135	135	100	135
end of rise (sec)		180	225	255	155	185	220	210	240	255	150	195
Cells cotation (1 à 5)		3	3	3	2	4	3	3	3	3	2	4
λ// (mW/m°.K)	0 day	21,7	24,4	22,7	22,0	24,0	23,0	24,4	24,6	23,1	22,0	25,0
	1 -	23,3	25,4	23,9	23,1	26,9		25,2	25,8	24,1	23,6	28,3
	14 -	27,4	28,6	28,3	27,4	30,5		29,1	29,3	28,3	27,9	31,8
λ ⊥ (mW/m°.K)	0 day	20,6	22,0	20,8	19,5	19,8	21,6	20,5	21,6	20,7	19,4	19,6
	1 -	21,1	22,8	21,7	20,1	20,4		20,7	21,9	21,4	19,6	20,1
	14 -	23,4	25,6	25,5	23,2	24,7		23,8	25,4	25,5	23,0	24,6

Polyol (Sorbitol 96% - water 4%) P. oxyde = 100
Surfactant
Catalyst
H_2O
Index NCO/OH = 105

TABLE III - SURFACTANT -

Ref.	POLYOL Nature	POLYOL (%)	SURFACTANT Nature	SURFACTANT (%)	CATALYST Nature	CATALYST (%)	FOAMING Nature	FOAMING (%)	WATER (%)	CREAM TIME (s)	STRING TIME (s)	CELLS COTATION	FREE RISE density (kg/m³)	FREE RISE λ_o (mW/mK)	BLOCS density (kg/m³)	BLOCS λ_o oven (mW/mK)	BLOCS λ_1 day oven (mW/mK)	BLOCS λ_{14} day oven (mW/mK)
52	(96% SORBITOL, 4% WATER) P.O.	100	1	1	1	2	R_{11}	40	2	27	128	4	22,0	24,4	24,5	23,8	26,3	30,2
50		"	"	2	"	"	"	"	"	20	110	3	21,0	25,0	24,0	23,1	23,4	28,4
51		"	"	3	"	"	"	"	"	23	120	2	22,0	25,4	24,5	22,8	25,8	29,5
60		"	"	4	"	"	"	"	"	21	109	2	22,0	24,4	24,5	21,2	24,9	28,0
53		"	7	1	"	"	"	"	"	23	123	4	20,4	24,3	24,5	25,2	28,4	35,8
54		"	"	2	"	"	"	"	"	26	121	3	20,1	25,2	24,5	23,2	26,1	31,0
55		"	"	3	"	"	"	"	"	23	127	2	22,5	24,1	25,0	21,8	24,9	29,9
61		"	"	4	"	"	"	"	"	20	120	2	21,0	24,0	25,0	21,4	24,0	27,6
65		"	"	1	5	"	"	"	"	18	60	1	24,0	20,6	26,0	20,1	22,2	26,6
66		"	"	5	"	"	"	"	"	17	51	1	24,5	20,3	27,0	18,8	21,2	25,6

TABLE IV - CATALYST EFFECT -

Ref.	POLYOL Nature	POLYOL (%)	SURFACTANT Nature	SURFACTANT (%)	CATALYST Nature	CATALYST (%)	FOAMING Nature	FOAMING (%)	WATER (%)	CREAM TIME (s)	STRING TIME (s)	CELLS COTATION	FREE RISE density (kg/m³)	FREE RISE λ_o (mW/mK)	BLOCS density (kg/m³)	BLOCS λ_o oven (mW/mK)	BLOCS λ_1 day oven (mW/mK)	BLOCS λ_{14} day oven (mW/mK)
37	(96% SORBITOL, 4% WATER) P.O.	100	1	1	4	1	R_{11}	40	2	45	230	4	22,5	27,7	24,0	27,5	29,5	32,6
38		"	"	"	"	2	"	"	"	25	137	3	22,0	25,4	25,5	24,8	26,3	29,7
32		"	"	"	"	3	"	"	"	20	100	2	23,5	23,6	25,5	22,6	23,7	28,0
39		"	"	"	"	4	"	"	"	16	71	2	22,0	23,4	26,5	21,7	22,4	27,1
62		"	"	"	"	5	"	"	"	15	60	1	22,0	21,6	24,5	19,7	21,8	26,6
65		"	"	"	5	"	"	"	"	18	60	1	24,0	20,6	26,0	20,1	22,2	26,6
64		"	"	2	5	1	"	"	"	56	199	4	23,0	24,8	23,5	26,9	31,4	38
58		"	"	"	"	1,9	"	"	"	38	160	4			24,0	25,1	29,3	33,6
59		"	"	"	"	3,5	"	"	"	24	73	2	23,5	21,8	26,0	19,5	21,8	26,8
63		"	"	"	"	5		"	"	16	54	1	24,0	20,5	27,0	18,7	20,9	25,1

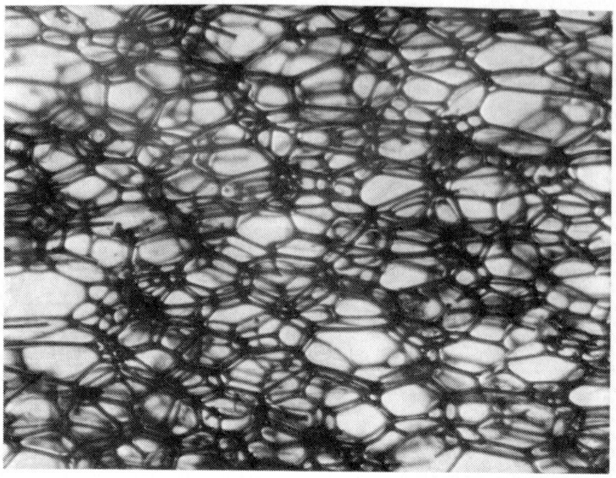

λ_o 18.8

λ_∞ 25.6

cells size 1

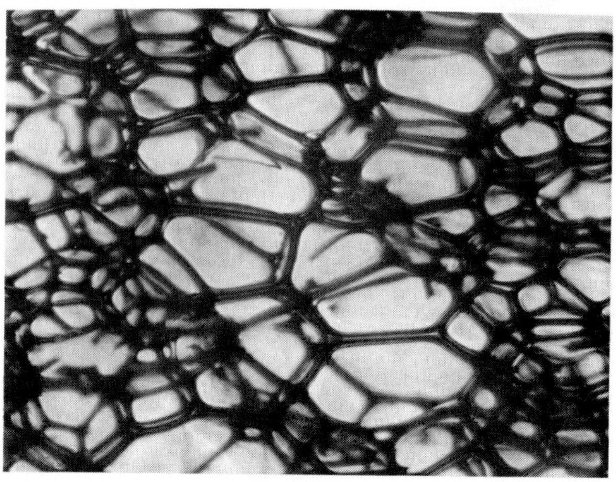

λ_o 28.5

λ_∞ 38.5

cells size 4

Fig. 1 - Cells macrographies for two foams having different cells size and λ values -

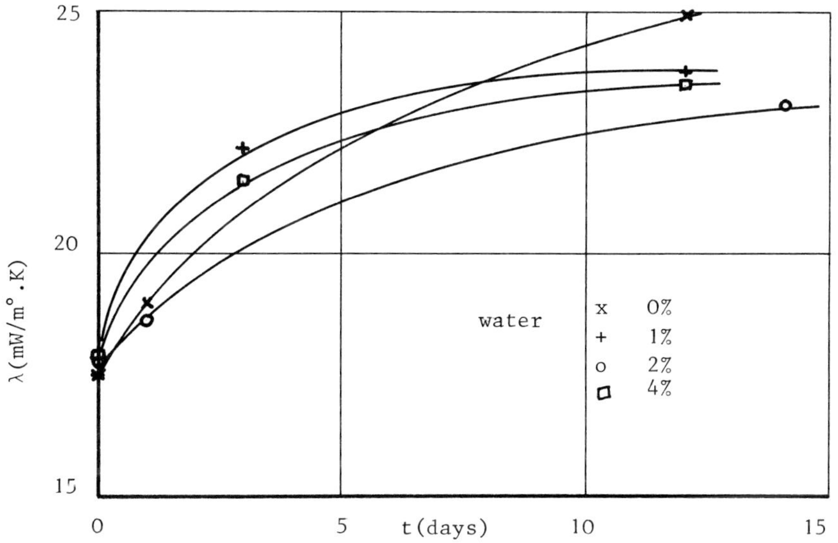

<u>Fig. 2</u> – λ evolution versus weathering duration, influence on the rate of water –

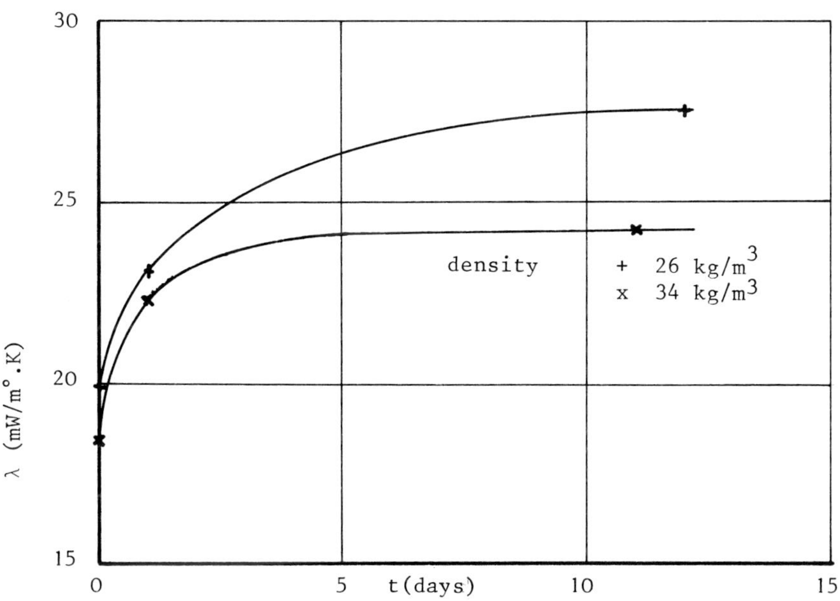

<u>Fig 3</u> – λ evolution versus weathering duration, influence of foams density –

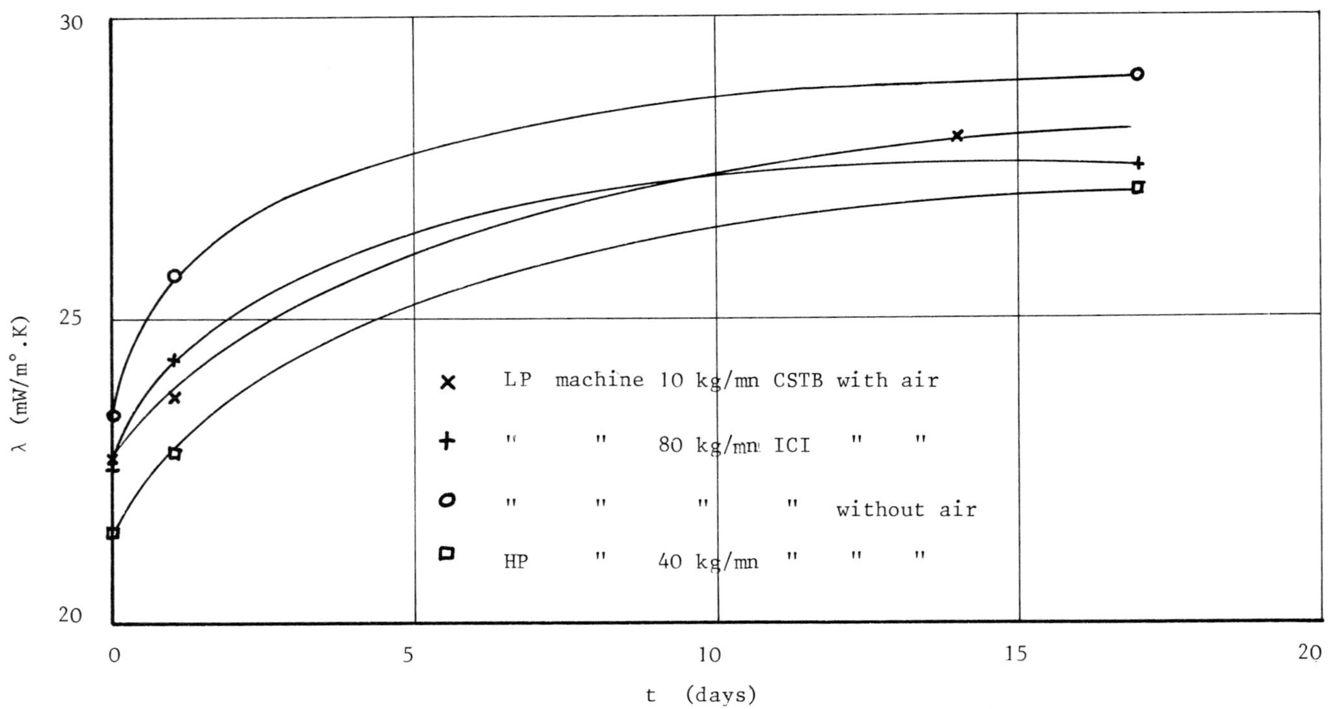

Fig. 4 - λ evolution versus weathering at 70°C 95% HR. Influence of molding machine -

QUESTIONS AND DISCUSSION

Zegers (DG XII) : What is the time constant for the increase of thermal conductivity of this insulation material due to aging ?

Answer : For the investigated insulation material this is roughly 15-20 years.

Pilavachi (DG XII) : How did you measure the pressure in the cells of the insulation material ?

Answer : Air from the cells is sucked into a bottle which initially is at vacuum and the quantity of air is measured. Also the volume of the cells from which air has been extracted is measured. With these data the pressure can be determined.

RESEARCH ON BUILDINGS' GENERAL QUALITY OF INSULATION, ESPECIALLY WITH RESPECT TO DEVELOPING AND IMPROVING METHODS OF POST-INSULATION

Part-project Ia and Ic

Development of a simple model of calculation and a measuring equipment for estimating the energy consumption in buildings.

Part-project IIIe

External control method of cell plastic insulation.

Authors	:	Tommy Nielsen, Gregers Chr. Larsen, Anne Abildgaard
Contract No.	:	EE-A-2-023-DK
Duration	:	24 months, January 1, 1982 - December 31, 1983
Head of Project	:	Arne Damsgaard Olsen
Contractor	:	Dept. for Building Technique, Technological Institute
Address	:	Gregersensvej, DK-2630 Taastrup, Denmark.

Summary

The energy consumption in a building varies throughout the day and night and throughout the year as a complex function of the construction of the building, its use, and the weather. The most important parameters that affect the energy consumption are discussed. It has been estimated which parameters it will be necessary to measure in order to estimate the energy signature of a building. The energy accounts of a building is: Energy supply = energy accumulation + energy loss. All important parameters from the energy accounts are measured with intervals of ten minutes and for each measuring interval the balance entry of the energy accounts must be zero. Special measurements are made continuously of the insolation through windows and of the air infiltration. Both these methods of measurement have been developed by the Technological Institute. The variation of the energy balance in relation to time will make it possible to locate possible insulation defects. The mobile measuring system is based on a computer that controls the measurements and collects and treats the data.
As regards the cell plastic materials a method has been developed for the determination of the structure of the material. The thermal conductivity depends on the structure, f.inst. the size of the cells and the total air content. The method is based on measurements made on a thin cut of the material and executed in an automatic picture analyzer. Air content, area index, and index variations are measured directly from the cut. The method will be used for both quality control and control of the finished insulation work.

Part-project Ia and Ic

1. Introduction

The energy consumption in a building varies throughout the day and night and throughout the year as a complex function of the construction of the building,its use, and the weather.

In the following the influence of the most important parameters on the energy consumption will be discussed. It has been estimated what parameters must be measured in order to determine the energy signature of a building and how to measure these parameters.

2. Description of a theoretical model for measuring and calculating the energy consumption of a building and the measuring equipment

In order to define the meaning of energy loss and energy supply the concept of energy reference envelope is introduced:

By the energy reference envelope of a building is meant an envelope that surrounds the house completely and is placed in such a way that the energy brought in through the envelope may be transformed into heat energy in the building, whereas the energy brought out through the envelope can not be brought back again.

Normally the energy reference envelope is identical with the outer surface of the building, i.e. the outer surface of walls, windows, roofs, etc.; it might, however, be situated elsewhere in the construction, if special models of calculation or special placings of the measuring equipment require it.

The energy reference envelope may be considered as the physical background for the energy accounts of a building following this equation:

Energy supply = energy accumulation + energy loss.

Once the energy has been brought in through the energy reference envelope, it is normally transformed into heat, before it again is brought out through the energy reference envelope. During this transformation of the energy, it may be transformed into some intermediate forms, just as some energy is normally stored, so that variations in the amount of stored energy within the energy reference envelope cause phase shifts between energy supply and energy loss.

Table No. 1 gives a survey of the most important parameters included in the energy accounts of a building. All these parameters must be measured continuously in order that the accounts balance at any time.

A building is divided into sections.

Fig. No. 1 gives an example of energy flows to and from a single section of a building as well as the parameters to be measured in order to determine the energy flows to and from this section.

The total energy accounts of the building is given as the sum of the accounts covering the various sections.

Energy Flows	+	÷
A. Supply from:		
A1 Heating plant	X	
A2 Electric Installation	X	
A3 Persons	X	
A4 Insolation (light energy)	X	
B. Losses through:		
B1 Transmission to outer air		X
B2 Transmission to ground		X
B3 Transmission to neighbours		X
B4 Air flow, by infiltration		X
B5 Air flow, loss through chimney		X
B6 Water flow, hot waste water		X
C. Accumulation in:		
C1 Constructions	(X)	X
C2 Furniture	(X)	X
C3 (Air)	(X)	X
C4 Water	(X)	X
D. Balance	(X)	X
Control:	SUM =	SUM

Table No. 1. Survey of model of calculation.

The individual energy flows can be registered as follows (see Table No. 1):

A1 Heating Plant

Measurement of the supplied quantity of oil by means of a flowmeter and of the smoke temperature by means of thermocouple.

A2 Electric Installation

Measurement of the total electricity consumption by means of an automatic counter (Landis & Gyr).

A3 Persons

The number of persons in the building is registered by means of a system of pushbuttons.

A4 Insolation/light energy

Measurement of the difference between the incoming and the reflected radiation through the windows by means of a differential light energy transducer (developed by the Technological Institute).

B1-B3 Transmission loss to air, ground and neighbours
Measurement of air temperatures and surface temperatures indoors as well as outdoors by means of thermocouples.

B4 Air flow, by infiltration
Measurement of the air infiltration in the building by means of the tracer gas method (constant concentration of SF_6). The equipment for this measurement has been developed by the Technological Institute.

B5 Air flow, loss through chimney
Measurement of the smoke temperature by means of a thermocouple.

B6 Water flow, hot waste water
The volume of water from washing machines, etc. is known from the technical data sheet. Consequently, the energy loss can be estimated by measuring the water temperature.

C1-C2 Accumulation in constructions and furniture
This is calculated on the basis of the registered surface or air temperatures. The absorption of heat in relation to time constitutes a square root function until the moment $\tau_1 = C_1 \cdot M_1 \cdot \pi/2$. C_1 is the heat capacity per m^2 of the constructions, and M_1 is the thermal resistance from the air or the surface, where the temperature is registered, to the layer where the heat accumulation is supposed to take place.

C3 Accumulation in air
Is not measured, because of the poor heat capacity of the air.

C4 Accumulation in water
Is registered directly by measuring the temperature of water at different places in the central heating system.

D Balance
All the measurements are made with intervals of ten minutes, and within each measuring interval the balance entry of the energy accounts must be zero. If this is not the case, one or more of the following types of errors might have occurred:

1. Incorrect assumptions on the heat producing system, f.inst. the efficiency of the central heating system.

2. Incorrect assumptions on the insulating system, f.inst. thermal conductivity, surface resistances, thermal bridges, etc.

3. Insufficient knowledge on the way the air flow follows through the building.

4. Uncertainty on the determination of water consumption and the temperature of the waste water.

5. Incorrect assumptions on the heat capacity of the construction.

The variation of the balance in relation to time will make it possible to locate the error and thus demonstrate possible defects and insufficiencies in the construction of the building or in its use.

When the balance is constantly at zero, the model of calculation is sufficient during the measuring period. The model may then be used to demonstrate in which respects the building can be improved and to estimate the effect of an improvement, possibly in combination with a renewed measurement after the improvement has been carried out.

In order to measure the above parameters the following system has been developed:

The measuring system is based on a computer system where all the measured parameters are converted into digital values that are stored on the diskette, see Fig. 2.

The computer prints the data or plots them as tables, curves, etc.

The measuring of the parameters is controlled by the computer that records one measuring point at a time.

In order to obtain the most flexible system, the system has been designed with a number of branch and scanner boxes, which may be placed close to the various measuring points. The individual boxes may be connected mutually to each other and to the computer by a single 7-core cable. This cable is used for control and transfer of analog as well as digital measuring signals.

The computer may also control an equipment for measuring the air infiltration.

Computer and equipment for measurement of the air infiltration is shown in Fig. 3.

3. Results of the project

So far the following results have been obtained:

1. The model of calculation has been worked out.
2. The measuring system has been designed on the basis of the model of calculation.
3. The various components of the measuring system have been specified and have been bought, ordered or put into production.
4. The detailed planning has started.
5. The data recording programme is being worked out.

4. Analysis of results and comments

In order to investigate the possibility of describing in a simple way the energy loss from a building, a well-insulated house of 125 m^2 with 12,1 m^2 glass area in windows and doors, and a volume of 337 m^3 has been calculated:

The U-values are:		
	Outer walls :	$U = 0{,}13\ W/m^2\ ^oC$
	Windows and doors:	$U = 2{,}6\ W/m^2\ ^oC$
	Roof :	$U = 0{,}10\ W/m^2\ ^oC$
	Floor :	$U = 0{,}17\ W/m^2\ ^oC$

The results are shown in Fig. 4.

The full-drawn line (N=0) illustrates the transmission loss as a function of the outdoor temperature. To this loss is added the loss by the air infiltration, which is also a function of the outdoor temperature, resulting in the dashed curves N=0,2, N=0,5, and N=1,0, corresponding to low, normal, and high air infiltration.

The insolation through the glass area is not a function of the outdoor temperature, but is subject to seasonal variations. In the figure is shown how the insolation on sunny days in Denmark varies according to the mean temperature of the day. The average energy consumption of the house at a given time of the year can at a given air infiltration rate be registered as the difference between the air infiltration curve and the insolation.

Example: It is assumed that the energy consumption has been measured during the periods November-December and December-January. In both periods the insolation is assumed to be half the insolation of sunny days, and the air infiltration rate is constant, N= 0,5 h^{-1}.

The curve now shows the following:

Month	November		December		January	
t_u	4,9		2,2		0,1	°C
Transmission + air infiltration N=0,5	2,40		2,75		3,15	kW
Insolation/2	-0,65		-0,48		-0,78	kW
Energy consumption	1,75		2,27		2,37	kW
Difference		-0,52		-0,10		kW
Temperature difference		2,7		2,3		°C
The variation of the energy consumption in relation to the outdoor temperature ("the energy signature")		-193		-43		W/°C

This example clearly shows that in Denmark the insolation has a significant influence on the size of the energy consumption during the seasonal variations. Consequently, it would be inconceivable in Denmark to estimate the energy signature of a house by measuring only the oil consumption and the outdoor temperature.

The equipment mentioned in the project description: "Monitoring unit for Building Energy Signature type DAU 16/2 of the V.N.R. Electronique S.A., Lausanne" is capable of measuring only the working period of the central heating system and the outdoor temperature as a function of time, and it may already now be said that his equipment will be insufficient for estimating the energy signature of a house. Consequently, the Technological Institute has decided to develop an equipment that can measure so many energy parameters that the following can be obtained:

1. Detailed knowledge on energy supply, energy loss, and energy accumulation, and
2. a prompt result, because no long periods of measurements are required to calculate average results.

5. Conclusion

A model for calculating the energy balance of a house has been worked out, a model that takes into account all essential parameters:

1. Energy supply
2. Energy loss
3. Energy accumulation.

Further, a computer controlled measuring system has been designed. The system can carry out a Continuous, Automatic Registration of the Energy Balance (CAREN).

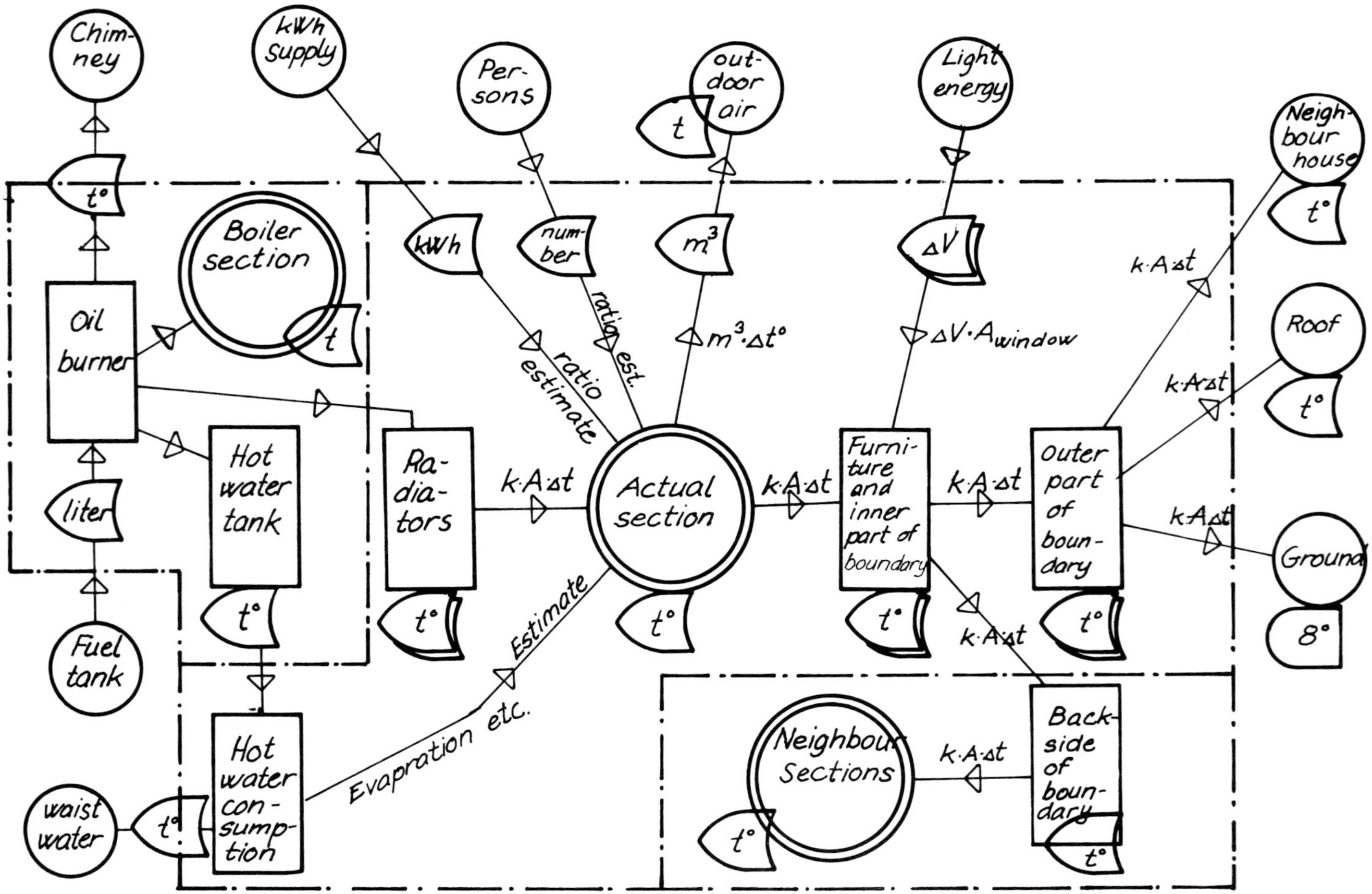

Fig. 1 Energy flow to and from one section of a building

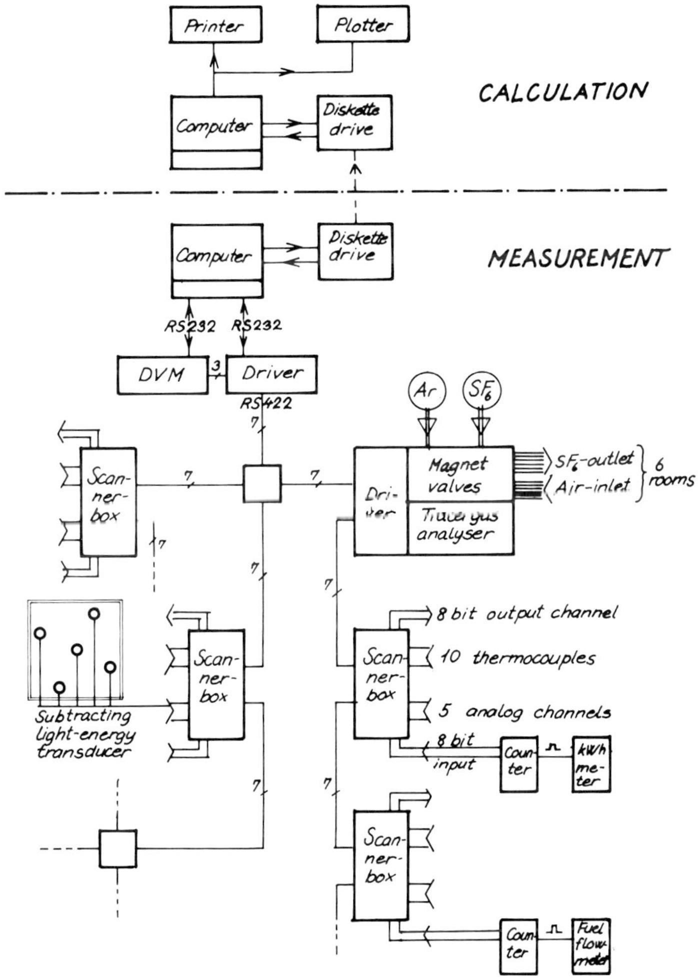

Fig. 2 System design

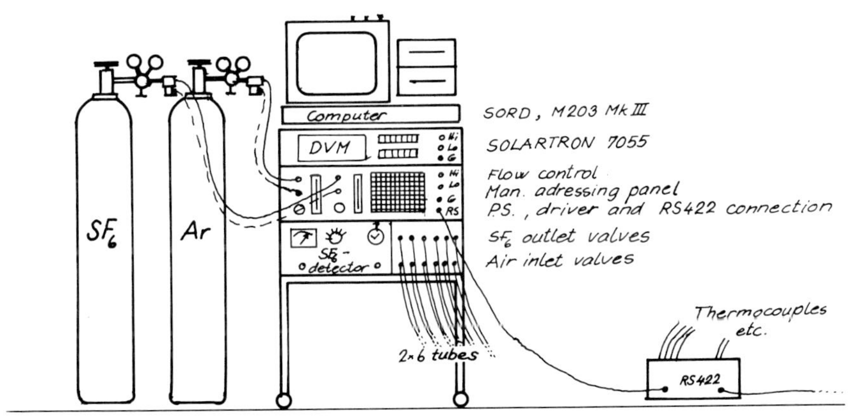

Fig. 3 System view

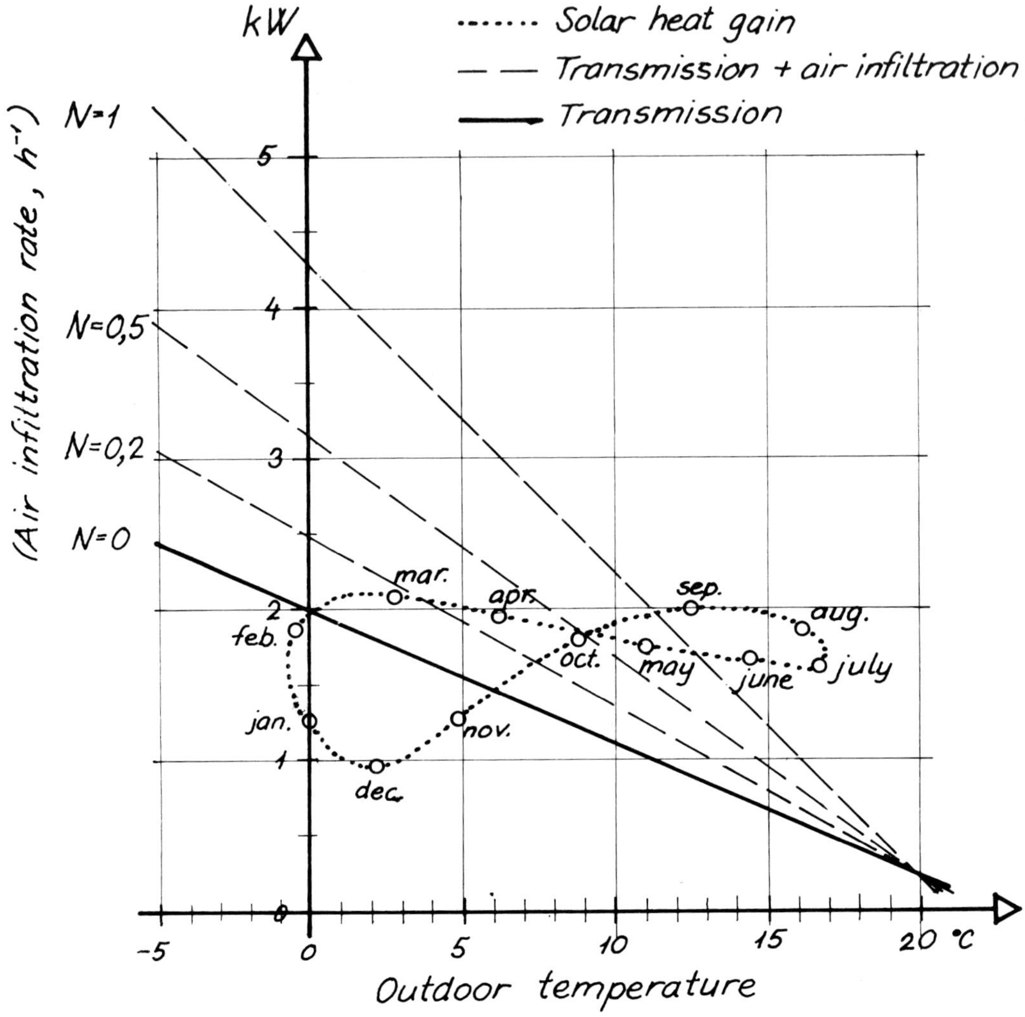

Fig. 4 Powerconsumption and solar heat gain in a well-insulated house:

Area	125 m^2
Glass-area	12 m^2
Volume	337 m^3
Indoor	t = 20°C
Ground	t = 8°C

U-values	
Outer walls	0,13 W/m^2 °C
Windows, doors	2,6 -
Roof	0,10 -
Floor	0,17 -

Part-project IIIe

1. Introduction

The increasing demands on the efficiency and life of the insulating materials have been followed by a need for estimating and controlling the heat insulating qualities of the materials. This part-project treats a description of a method to estimate the structure of foamed insulating materials.

Foamed plastic materials as f.inst. polyurethane and polystyrene consist of three-dimensional cells in a matrix of plastic. The cells may be open or closed (air/gas filled) depending on the production method.

The cell structure of the finished product is decisive of the thermal conductivity of the material. The cell structure can be determined by an optical analysis of small samples of the material having been carefully prepared in order to avoid damaging the cell walls.

2. Measuring method

The most important parameters needed for determing the cell structure are the average cell size and total air per cent. The thermal conductivity depends on the size of the cells, as big cells have a higher conductivity than small ones. The total air per cent indicates the part that the cells (gas/air) constitutes in relation to the matrix of plastic. It is thus of importance to the insulating capacity. The total air per cent may also be used for estimating the efficiency of the different processes and the capability of the various materials to form an appropriate pore structure.

These parameters can be measured by quantitative microscopy on the following principle. The prepared sample is placed in the microscope in transmitted light and an optical magnification of 20x. In the microscope a raster system is placed on the optical picture and the number of intersections between the cell walls and the raster system is registered. On the basis of these measurements the above parameters are calculated.

The analysis is carried out manually or automatically. In this case it is automatic by means of an automatic quantitative microscope (Leitz TAS).

3. Preparation

The above measuring method requires that the sample to be investigated is plane and without defects. For the preparation is used a high-frequency microtome consisting of a razor blade mounted on a vibrator. The amplitude and the frequency can be regulated to produce thin plane parallel cuts of the material with only few damaged cell walls. The result is shown on the photo overleaf.

4. Results and analysis

The measurements made on the above principle gave the following parameters:

- total air content %
- area index and index variation.

The area index indicates the average area of the cell walls in the cut, and the index variation is a measurement of the variation between these areas.

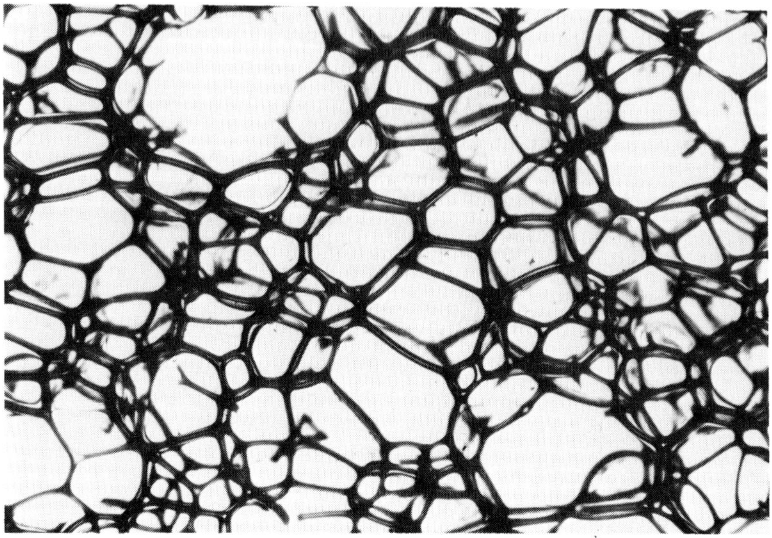

Below is shown examples of measuring results from prepared samples:

	1	2	3	4
L air content %	69	67	57	62
Area index mm^2	0,019	0,022	0,011	0,012
Index variation	0,0018	0,0023	0,0009	0,0009

So far, measurements have been carried out on materials with unknown thermal conductivity. The purpose of these measurements is partly to examine whether the preparation is sufficient and partly to estimate which parameters it would be appropriate to measure. Consequently, these preliminary results can not be used to estimate the thermal conductivity of the material, but only to estimate the method of preparation and measuring.

5. Conclusion

On account of the geometric shape (three-dimensional structure) of the cells, it is not possible to define a specific pore area, because the cut mainly shows the limits of the cell in the form of top, bottom, and side walls. The preliminary measurements seem to indicate that it would be appropriate to measure an area index which indicate the average area of the cell wall. That parameter, combined with the variation in the areas, is a measure of the homogeneity and fineness of the material.

The measurements are carried out on samples of the same thickness, so that it will be possible to compare the parameters of the different samples.

QUESTIONS AND DISCUSSIONS

Dubois (Marcoussis) : How did you measure the air loss of the house ?

Answer : We used two methods. With the first method, tracer gas was brought into a house and the decrease of the tracer gas concentration with time was measured. This is a time consuming process.
The second method consisted of keeping the tracer gas concentration constant. The air exchange rate was calculated from the amount of gas which had to be added.

Dubois (Marcoussis) : What was the initial tracer gas concentration ?

Answer : 5 ppm.

Prof. Boffa (Chairman) : It would be useful it you could get in contact with Mr Dupagne from the University of Liège, Belgium, who in the E.C. contractors group "characteristics of houses" is working along the same lines. A drawback with models for buildings is that the behaviour of the occupants is very different and unpredictable. Simulation of this behaviour is therefore impossible.
Finally it is important to know what the aging effect will be on insulation.

Damsgaard Olsen (Technological Institute) : On the point of aging of insulation material, 300 firms in Denmark are working together in order to investigate this problem.

WALL COMPOSED OF DIFFERENT LAYERS HAVING HIGH THERMAL VALUE AND AGING OF INSULATING

Authors : D.M. FONTANA - R. ZARI - E. GIRAUD

Contract number : EE-A2-014-I

Duration : 18 months 1 January 1982 - 30 June 1983

Head of project : Prof. Ing. D.M. Fontana

Contractor : Istituto Sperimentale per l'Edilizia S.p.A.

Address : C.P. 7237 00100 Roma Nomentano - ITALY

Summary

Objects

- Variation test of thermic characteristics in consequence of thermic stresses.
- Variation test of thermic characteristics in consequence of vibrations.

Obiect of the experiments: building walls to be largely used in Europe.

We intend to determine the transmittance values in stationary conditions, those of damping and phase lag in sinusoidal stabilized periodical conditions.
The test equipment is that of calorimetric type used in steady temperature with a protecting cover chamber (thermostatic chamber type ACS).
For the execution of the experiments in periodic conditions we determine the wanted range to the thermostatic regulator set point of the type P.I.D.
The air temperatures are detected by thermistors, those superficial by thermocouples.
The measured quantities are directly recorded and worked out by an data acquisition control system model HP 3054 DL.
The ageing is simulated:

- by thermic stresses, obtained by sudden climb of the tested walls temperature up to 70°C, this temperature is kept steady for 3 hours, followed by cooling using a stream of water which falls on the walls.

- By vibrations having the same spectrum and energy typical of those we find in road traffic.

We describe the equipment installed and preliminarily tested for the determination of thermic parameters in question.

1. Introduction

The main objects of the research are constituted by:

- Variations test of the thermic characteristics in consequence of thermal stress.
- Variations test of the thermic characteristics in consequence of vibrations.

The experiments are particularly about building walls, of light, medium, and heavy type, to be largely used in the European Community.

1.1. Quantities and method of measure

The quantities are defined:

- Over-all heat transmission coefficient $K = \frac{Q}{A\ (T_1 - T_2)}$ (1)

Where Q = Heat flux through area A

A = Area of the specimen

$T_1\ T_2$ = Temperature of the environments separated by the sample

(By temperature of the environments we mean the average value of the air temperature in it and of the walls which delimit the environments, excluding thet to be tested).

- Damping σ and phase lag φ

It is defined by referring to the constant values T_2 and ranges of T_1 of the type

$$T_1 = T_1^m + \theta \operatorname{sen}(\omega \tau) \quad (2)$$

If T_2' is the superficial temperature of the wall from the side of the environment 2, we can place in steady periodic regimen

$$T_2' = A + \sigma\theta \operatorname{sen}(\omega\tau - \varphi) \quad (3)$$

relation wich defines σ and φ.

To determine K, we must measure in stationary condition, Q, T_1, T_2, A and then we can deduce K.

To determine σ and φ we fix the constancy of T_2 and a sinusoidal range of T_1: the ranges of T_1 and T_2 are recorded and once reached the steady periodic regime (3), we are able to find σ and φ .

1.2. Production of the thermal stess

Thermal stress, in the present work, is a thermal story conventionally defined as follows:

a) a sudden climb of temperature on the wall surfaces to be test, T_2 being kept constant. This rise is characterized by:

- final value of T_1 equal to 70°C;
- initial condition $15°C < T_1 = T_1' = T_2' = T_2 < 25°C$
- climb time $\tau_s < 0,5$ h
- specific thermal flux on the wall surfaces: $< 0,5$ KW/m^2

  This power can be exceeded only if we don't obtain climb times inferior to 0,5 h with it.

b) Maintenance of this temperature for three hours.

c) Flow irrigation of wall surface until return to environment temperature.

1.3. Vibration test

The vibration will be produced in vertical direction on the walls placing on a vibration mounting. The oscillations will be of a aperiodic type, with a continous spectrum, in absence of pure tones.

It has already been prepared the motion equipment of the specimen, from its place of construction and assembling to that for thermical experiments and vibrations test.

The mounting carriage will be of course modified with elastic suspensions so that the specimen-mounting system natural frequency will be sufficiently far away from those contained in excitation spectrum.

We are carefully examining the problems relative to the choice of the frequency spectrum and power of vibration.

2. Description of the equipments

We have up to now manifactured the equipment to determine the following parameters: over-all heat transmission coefficient K, damping σ , phase lag φ .

The thermoregulation group consists mainly in the two following sections:

a) the first section is formed by a 5 walls, called guard box, inside to which a second 5 walls cell, called metering box is contained.

b) The second section is formed by a single 5 walls cell called thermostatic box.

Both the sections are setted up on a rail-travelling frame (fig. 1 & 2, photo 1 & 2).

For measuring the thermical transmittance the calorimetric test method at constant temperatures ASTM C 236 is used. It needs two thermoregulated cells between wich the sample is placed. The two cells have different temperatures, which constance is controlled for a long period of time to ensure steady conditions of temperatures and thermal flow.

All data are stored, recorded and periodically tabulated by a computer controlled data logger (photo 3 & 4).

The sinusoidal range of T_1, has been obtained by a heating device wich can obtain a pre-determined value of T_1.

The set-point will be varied by a periodical law obtained by cinematic devices.

The regulation system is feed-back controlled with proportional, derivative and integral action.

The temperature T_1' and T_2' are measured by thermocouples.

We are now planning the equipments for the production of thermal stress and vibration.

3. Progress of work

Whit reference to their composition, the types of walls to be tested have been defined as follows:

- Heavy concrete with included insulating material (fibrous or expanded plastic materials plates).
- Lightweight concrete with included insulating material (fibrous or expanded plastic materials plates).
- Clay brick with included insulating material (fibrous or granulated or expanded plastic materials plates) or without insulating material.
- Porous clay brick included insulating material (fibrous or granulated or expanded plastic materials plates) or without insulating material.
- Two different types of insulated metal paneling.

The equipment mentioned in number 2 has been tested and various experimental determinations have been carried out in stationary conditions on different kinds of walls in order to calibrate the equipment in the best way (fig. 3).

The software relative to the simulation on a computer of the mathematical model of the walls is to be tested yet.

4. Conclusions

The theoretical bases of the work and the employed equipments are described.

The choise of the walls commonly used in the European Community, which represents the first research pourpose, has been carried out in collaboration with the field operators.

Although the equipment calibration has been problematic, now all runs satisfactorily and the experimental tests are starting.

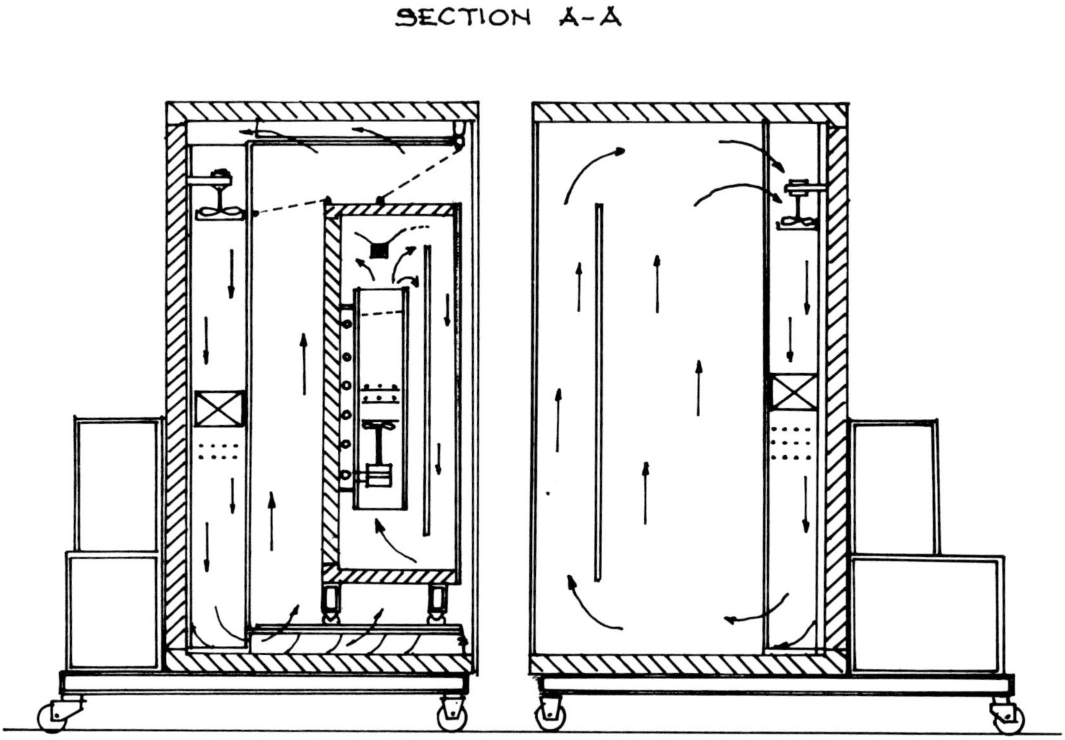

Figure 1 Thermoregulation group

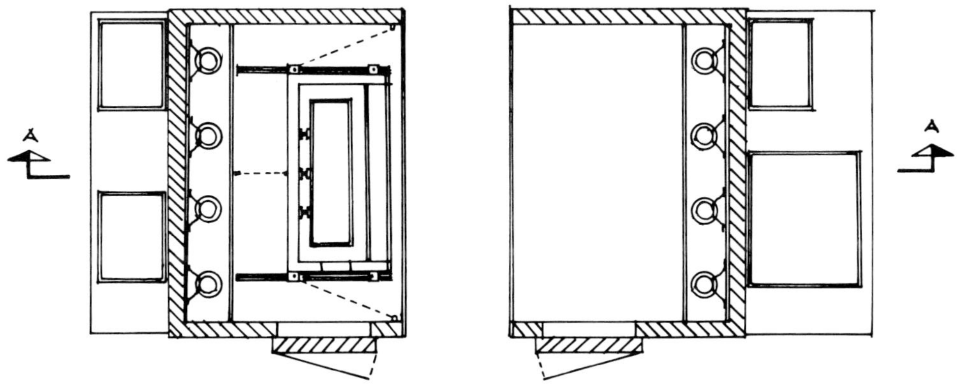

Figure 2   Thermoregulation group

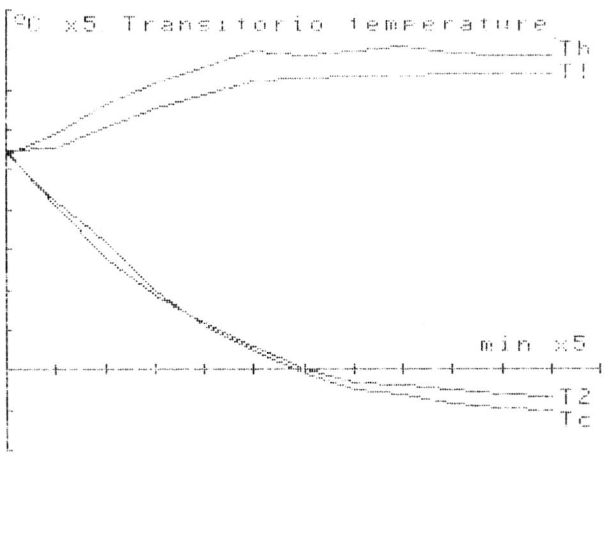

Figure 3   Transient of the temperatures in the conditioning initial stage

Photo 1 Thermoregulation group

Photo 2 Thermoregulation group

Photo 3 Computer-controlled data logger

Photo 4 Energy meter

SESSION V - INSULATION

Chairman: J. UYTTENBROECK

THE PRODUCTION OF SELECTIVE OPTICAL COATINGS ON PLASTIC SHEET

Authors	:	R.P. HOWSON, M.I. RIDGE, C.A. BISHOP
Contract number	:	EE-A-2-021-GB
Duration	:	36 months 1 July 1980 - 30 June 1983
Head of project	:	Dr. R.P. Howson, Loughborough University of Technology
Contractor	:	Loughborough University of Technology
Address	:	Loughborough, LE11 3TU, Leicestershire, U.K.

Summary

A technique of vacuum deposition is described for producing heat reflecting films onto sheet plastic for application in domestic windows. Thin metal films of silver sandwiched between dielectric films and electrically conducting oxide coatings of indium, indium doped with tin and cadmium/tin are shown to have the properties required for a heat reflecting coating for visible windows. Durability measurements suggest that conducting oxide films have more promise. Indium oxide coatings on polyester sheet have been made which have been produced onto plastic transferred from roll-to-roll at a rate of 0.25 m min^{-1}. They are highly transparent and have a low temperature emissivity of about 0.3.

It is shown that their use as a secondary window can improve the performance of the system to a level previously unknown and this can be done with a small weight penalty and, what is most important, can be projected to be able to be manufactured at a cost which makes the unit a very efficient use of energy resources. It should have a pay-back period of considerably less than one year, both in cost and in energy saved, and has the potential to be made in large areas to make a significant impact on natural energy use.

1. Introduction

Energy costs have been rising more rapidly than the cost of incorporating insulation into buildings, a trend which is confidently predicted to continue, and a situation has been created where existing buildings are insulated to a level below that which is justified by the cost of heating. This is not only a current problem but one which will be with us for along time; it is estimated that 80% of the domestic buildings which will be used in the year 2000 have already been built and to inferior insulation standards than those that will be demanded by the price of fuel at that time. Insulation standards of existing buildings have to be raised with methods which are properly effective in that the cost of their installation is justified by the energy that is saved. Solar energy is free and it is important this this be utilised as far as possible. All existing houses have solar collectors installed which can be very effective on occasions, but can also create problems. These are the windows. It is important to consider these elements of a building as they affect energy capture and energy loss.

Domestic windows play a large part in determining the living environment for most of the population of the world. They fill the apertures through which natural light is admitted to a building, which prevents air movement and, hence, the loss of heat. The essential material and design of such windows has changed little from medieval times. This is because of the properties of glass, a strong stable economical material with high resistance to mechanical scratching, which has a good visual transmittance. Unfortunately, it has a high thermal conductance and the thermal insulation it offers is principally obtained from the layers of air at either side and is, as a consequence, quite low and dependent upon air movement. The insulation effect of air, or some other gas, can be obtained by enclosing a layer within the window with two glazing elements and this has been the most obvious thermal development in window design in recent times. Windows, however, have a great advantage as thermal elements in a building in that they allow the energy from the sun to penetrate into the interior. The energy is trapped inside because the glass is absorbing in the infra-red, the region at which the interior re-radiates the heat generated by absorbed solar energy. This is known as the "greenhouse effect". Even for fossil fuel heated housing at Northern European latitudes in winter there is very nearly an energy balance for south facing vertical double glazing elements; they gain almost as much energy from the sun as they lose due to heat loss from the warm interior. There remains the radiation interchange between an individual and the outside environment, which takes place with the window because it is opaque to heat radiation. If the surface of the window can absorb and radiate heat radiation very easily then discomfort will be felt by an individual due to radiation heat transfer with a cold window.

The thermal performance can be improved with a surface coating which gives visual transmittance with a high infra-red reflectance. The high infra-red reflectance has two purposes; to reflect heat radiation and to provide a surface with a very low emissivity which limits radiated heat transfer. For example in a double glazed window, two thirds of the heat transfer is by radiation between the two panes. This is so high because of the high emissivity of glass.

Such a surface has been created economically by the addition of a plastic sheet coated with aluminium to the window surface. The aluminium is deposited onto polyester film, which is opaque to infra-red radiation, but the protective laminated layer is polypropylene which is transparent and allows the aluminium film to utilise its high infra-red reflectance.

Unfortunately, these films may only be utilised where solar control is also required because good infra-red reflectance is only maintained with considerable loss in visible transparency. In these circumstances the winter solar gain is also lost so that window covered with this film may still be worse than one without in terms of winter energy conservation.

There exists a strong need to create surfaces on windows which have low emissivities and maintain the solar gain which can be applied to improve the thermal performance and comfort levels and remain cost effective.

The covering of a window, be it double or single glazed, by one layer of plastic coated with a suitable material would be expected to decrease the thermal conductance equivalent to creating the effect of another glazing element; this is projected to be achieved at one tenth the cost of the glass that would be required as well as providing the desired solar control. It is this fact that prompts the search for a technique and material for coating plastic rather than to coat the glass directly.

The prime requirements to manufacture, by vacuum coating, a film having the desired optical properties are that it should be able to be produced at a very high rate at as high a pressure as possible. These requirements are set by the high capital cost of the producing equipment and its influence on the final cost of the product. The process must meet the requirements of the plastic film substrate which will not withstand high temperature operation.

The basis of the methods and materials that need developing is that they provide a technique which can be readily adapted to the equipment used in the vacuum metallising of polyester sheet in the present packaging industry.

2. The Properties required for Window Films

The difference between the wavelengths at which energy is received by a window and those at which it is re-emitted is illustrated in Figure 1. The control of light and energy from the sun is obvious; what is not so clear is the influence of the presence of low emissivity surfaces on the thermal properties of the structure. A plastic sheet with various emissivities can be adhered directly onto a single pane window. The difficulty with this system is that a lower glass temperature will result which is liable to cause condensation on the inner surface. Water is opaque to heat radiation so that the thermal advantage will be lost. In double pane windows, two thirds of the heat transfer is by radiation and considerable improvement can be made to the properties by using a low emissivity surface on the inner pane and facing the outer. This is seen in Figure 2. This can be further improved if the plastic sheet is freely suspended between two panes of glass.

The properties required for window films can be specified in terms of the two spectra.

The solar energy input must be maximised whilst the insulation provided by a low emissivity surface is improved by using multiple elements.

The filter must be designed to allow all the energy from solar radiation to penetrate the building which requires a transition point where the film goes from completely transmitting to completely reflecting at about 3 microns. This is illustrated in Figure 1.

Comfort for a person is dependent both upon air and radiant temperatures. The radiant temperature is determined by the interchange of energy between the individual and his surroundings by infra-red radiation. If the surroundings are at a low temperature and have a high emissivity, then considerable heat loss will be experienced and the air temperature will have to be raised to maintain adequate comfort levels. This effect is commonly experienced next to a single pane window on cold days and can be specified by an equivalent perfect absorber at a temperature which represents the energy exchange. The effect can be minimised by raising the temperature of the surroundings or using a low emittance surface to prevent the radiation exchange.

For a temperature difference between the room and the outside of 22°C the equivalent black body temperature seen by an occupant is 14°C below ambient for a normal single pane window. This reduces to 5°C when a coating with an emissivity of 0.3 is put on the inside of the window and 1.7°C for one with a value of 0.1.

3. Materials for Simple Optical Films

The material has to be non-toxic, cheap, easily available and require to be used in as thin a layer as possible consistent with durability. As few as possible layers should be used to obtain the desired properties. These requirements exclude conventional dielectric interference coatings and has lead to there being two possible systems.

(a) A metal film, with low intrinsic absorption, used in a sandwich between two high refractive index dielectrics. Suitable materials are: copper, gold and, best of all, silver.

(b) An electrically conducting metal oxide film generally of indium oxide doped with tin.

Both these systems require the production of an oxide film at a high rate with carefully controlled properties, especially in the case of the electrically conducting variety.

4. Deposition Techniques

Our original hypothesis was that the addition of energy to a growing film by exposure to bombardment by neutral heavy ions should lead to properties being attained which were normally seen only at higher temperatures. This was tested with evaporation and sputtering methods. Only reactive methods using pure metal sources were considered in order to achieve the simplest realisable system which could achieve high rates. This gave good results with both systems, but has proved to be symptomatic of a more important factor, which is the careful control of the oxygen to metal ratio. With the high rate used it was apparent that it was the oxygen gas admission rate that determined the properties obtained.

The precision control of rate, combined with a high sputtering rate capability,and operation at a relatively high pressure to provide the reactive gas, has led us to the system of d.c. planar magnetron sputtering of the metal or alloy in an atmosphere of argon and oxygen. The ion bombardment of the substrate provided by a radio frequency discharge activates the oxygen reacting of the substrate surface to give higher rates of deposition. In order to achieve optimum properties in the films we have found it necessary to control the power to the sputtering source and the oxygen gas admission rate to within 0.1%. The conditions were adjusted in accordance with the properties of the film continuously monitored in a roll coating apparatus.

Planar magnetron sputtering is a technique in which a magnetic field is created parallel to the surface of the sputtering cathode which enhances the discharge by trapping secondary electrons. The advantages of such a system are :

1. A high rate of sputtering is obtained because of the confinement of the plasma close to the target surface. The limit is determined by a factor involving the thermal conductivity of the target, the efficiency of the water cooling, the melting point of the sputtering material and its sputtering yield.
2. The heating of the substrate by stray electrons is low. This is the limiting factor for high rate conventional sputtering in many cases.
3. Sputtering can take place at a low gas pressure because of the increased path length of electrons within the plasma and prevention of their escape. This is achieved by the crossed electric and magnetic fields.

When this sputtering source is used in the presence of a reactive gas to create a compound, further advantages ensue:

1. The rate of deposition can easily and accurately be controlled to allow balanced processes to be maintained. In particular the high surface ion current density allows sputtering of the metal from the metal target even in a reactive gas, with the compound formation taking place on the substrate surface rather than on the target surface.
2. The gas pressures required for sputtering and for producing reactive processes are compatible.
3. Alloy metal targets can easily be sputtered to give mixed metal oxides of similar metal ratios.

If an additional field is applied to the substrate surface, a discharge can be created which bombards the growing film surface with ions. The advantages of "ion plating", used in conjunction with sputtering, are :

1. The reactive gas is activated by ionization and driven towards the substrate surface by the ion plating radio frequency field.
2. The addition of surface energy by the reactive and neutral gas atoms to the growing film results in control of structure and nucleation phenomena without using high temperature substrates.
3. The gas pressures needed for the sputtering and ion plating discharges are compatible.

A roll coating system offers advantages for the development of coatings, especially with the reactive process. These are :

1. High rate equilibrium processes are possible. Long substrate lengths allow lead-in times which eliminate shuttering disruption.
2. Rapid preparation gives pure deposits. The ratio of depositing species to residual impurity gas is high.
3. High experimental productivity is obtained; many samples can be produced on one roll in one evacuation. In-situ monitoring facilitates parameter optimisation in a short time.
4. The results are applicable to commercial processes.

5. Deposition Apparatus

Two separate pieces of equipment, including the magnetron source, have been designed and manufactured by ourselves for the deposition of oxide and metal films onto plastic sheet, transferred from roll-to-roll over a drum capable of being taken to high radio frequency potentials.

The design in both cases was conditioned by the initial equipment that was modified. Roll Coater 'A' was constructed to use an external mount magnetron source, 'B' used the motor drive mechanism available in a Balzers 510 Coating Unit, together with the existing electrode feed-throughs for the internal magnetron sources.

The apparatus of 'A' is illustrated in Figure 3. Figure 4 shows a section of a round planar magnetron that was built.

6. Theoretical

To allow the prediction of the properties of filters utilising conducting oxides with anti-reflecting dielectric oxides, a computer programme for multi-layers has been developed using the properties of the conduction carriers as the variables of the optical constants of the films, relating them through the theory of Drude. Values of the plasma frequency ω_p can be obtained; measurement of the refractive index in the visible η_ℓ, gives ε_ℓ the relative electrical permittivity from $\varepsilon_\ell = \eta_\ell^2$

$$\lambda p = \left\{\frac{\varepsilon_\ell}{\varepsilon_\ell - 1}\right\}^{\frac{1}{2}} \lambda_o$$

where λ_o is the wavelength where the reflectance reaches a minimum. Drude theory gives the refractive index, η and extinction coefficient, κ, from :

$$\eta^2 - \kappa^2 = \varepsilon_\ell \;\{ 1-\omega_p^2/\{\omega^2 + \tau^{-2}\}\}$$

$$2\eta\kappa = \varepsilon_\ell \{\omega_p^2/\{\{\omega\;\tau\}\{\omega^2 + \tau^{-2}\}\}\}$$

τ is the electron relaxation time related to the electrical mobility, μ, through $\mu = q \dfrac{\tau}{m^*}$

where m^* is the effective mass.

Applying thin film interference theory and information obtained from measurements of sheet resistivity, Rs and film thickness, we have shown that excellent agreement can be obtained with experiment with these films when a value of τ obtained from $\tau = 1/\{\omega_p^2 \times \varepsilon_o \times \varepsilon_\ell \times d \times Rs\}$ is used.

The way is now open to consider how possible changes in the relaxation time and the plasma frequency can determine the relationship between electrical and visible optical properties; the plasma frequency is controlled by the free carrier density, N, through

$$\omega_p^2 = \frac{Nq^2}{m^* \;\varepsilon_o\;\varepsilon_\ell}$$

and $Rs = \dfrac{1}{\sigma\; d} = 1/Nq\;\mu.d$

Metal films can initially be treated by using published values of their optical constants.

Programmes have been developed for the two cases in BASIC on a HP45 computer. They have been developed from first principles using a generalised form for the series of reflectances and transmittances with successive calculations along the series of layers. They give reflectance and transmittance as a function of wavelength and allow the calculation of average values for the visible spectrum or any other region.

7. Results

The measured spectra of various oxide films are shown in Figure 5 and the related optical and electrical parameters are given in Table 1. Samples of indium oxide were taken from a roll of material which was made in a continuous process where 10 metres of the roll of polyester sheet was coated at about 0.25 metres per minute. The rate of sputtering was over 0.5 μm min^{-1} kW^{-1} and the power used was about 3.5 kW into the one target. The remaining material is being made into a 1 metre square window for detailed testing. Measurements made with similar systems suggest that a value of 1.25 Wm^{-2} ^{o}C^{-1} will be obtained if the sheet is used as a centre element in an existing glass double pane structure or 2 Wm^{-1} ^{o}C^{-1} if used as a suspended element in addition to a single pane window.

Durability testing has been made using temperature and humidity cycling in the presence of simulated sunlight. These tests were severe because they exposed the film side of the sheet to the 'exterior' environment. Indium tin oxide showed slight signs of crazing after considerable exposure; indium oxide was worse, but when examined under a microscope showed that in fact the film was lasting very well except for dimension changes which caused the film to buckle, but not to break. Indium oxide has been observed to cause more bending of the plastic sheet and so has greater compressive stress.

PARAMETER	ITO	CTO	IO
Relative electric permittivity	4	4	3.5
Plasma wavelength (x 10^{-6}m)	1.3	2.5	2.0
Plasma frequency (x 10^{15})	1.45	0.75	0.94
Sheet resistance ($\Omega/\square$)	17	24	30
Film thickness (x 10^{-9}m)	330	280	336
Carrier density (x $10^{26}m^{-3}$)	6.1	3.4	2.1
Electrical mobility (x $10^{-4}m^2V^{-1}s^{-1}$)	17	30	29
Carrier relaxation time (x 10^{-15}s)	2.8	5.8	2.6
Effective mass	0.29	0.34	0.16

TABLE 1

Our work on the metal films is not so well developed. Dielectric oxide films have been prepared together with metal films. These have only recently been put together to give the characteristics shown in Figure 6. Experience with films of silver made in a 'static' apparatus has lead us to suspect that durability will be a considerable problem with such films as they deteriorate quickly in a normal atmosphere and with handling.

8. Discussion

The results have confirmed our predictions for the advantages of an experimental method for the production of electrically conducting oxides which are good heat reflectors whilst remaining visibly transparent and are durable. It is important to realise that conventional laboratory methods have not given properties that were useful, unless the film was heated well above temperatures that plastic will tolerate, or have been made very slowly and then only with indium oxide doped with tin. We can produce a variety of oxides with suitable properties at high rates onto cold plastic substrates.

9. Application

An approximate cost for an oxide or metal filter can be computed from the data we have acquired.

Calculation soon reveals that the cost of the coated plastic sheet will be determined by the capital cost of the coating equipment and, hence, the rate of production. A machine installation could well amount to £1m which will cost about £1,000 per day in charges. With a multiple source machine a rate of 20 metres/min. of 2 metres wide material might be expected. This is ten times slower than the rate obtained with current conventional metallising machines using aluminium. With such a slow rate most of the time will be spent in sputtering so that run times of 10 hours per day should be obtained after development. Then 20,000 m^2 should be produced per day. This may well be halved with machine maintenance but a cost of only 10p per m^2 of coating is then predicted. The cost of the plastic substrate is about 25p per m^2 and material costs would amount to about 10p per m^2. It is expected that problems will arise in the distribution and use of the volume produced and also the availability of source materials such as indium. One machine will double the U.K. use of indium and the price may rise, but the principal problem will remain that of guaranteeing the supply of the material.

We have investigated how coated film could best be used. We believe that a plastic sheet, coated both sides, and stretched and mounted in an aluminium frame is the most promising system. We have bought a sophisticated frame which is available for approximately £8.00 to create a secondary window of one metre squared. Coated plastic sheet will be durable, resistant to scratching and prevent static build-up, which causes dust to stick to the window, and give the window a performance which will fully justify the use of a good quality frame. The addition of this secondary window will take a single pane window from above 6 to less than 2 Wm^{-2} $^{o}C^{-1}$. A saving of up to 5 Wm^{-2} $^{o}C^{-1}$.

For a moderate climate of 2,000 degree days, this gives a saving of 250 kW hours per square metre of window per year. This is approximately twice the energy content of the coated plastic sheet.

With electrical heating at 4.7 pence per unit, a payback period of considerably less than one year will be obtained.

This crude estimate of cost and energy saving does indicate the considerable potential for such a system, without consideration of the other advantages of light-weight, enabling large glazed areas to be treated without any significant extra weight load on the structure. There is the potential of providing windows and light-weight wall structures having a thermal performance considerably in excess of anything that can be achieved today, using multiple elements which still allow advantage to be taken of solar input. We believe that this application is a very productive use of our diminishing fossil resources.

10. Conclusions

A vacuum coating technique using reactive magnetron sputtering has been developed which has been shown capable of producing coatings onto plastic sheet at a high rate. These coatings have the property of reflecting heat radiation whilst remaining transparent to visible radiation and when used in a window structure suppress the radiation transfer which forms the main component of heat loss in a double pane structure. It is estimated that if a commercial coating was created and the coated sheet was used as a lightweight secondary window system, the cost could be repaid in one year, with most of this arising from the aluminium frame used to mount the film. The large area capability of such a process could well be used to create a real reduction in the national use of energy for domestic, public and commercial building heating.

Acknowledgements

The author is pleased to acknowledge the members of the Thin Film Group in the Department of Physics at Loughborough University, without whose work this report would have been impossible.

J.N. Avaritsiotis helped initiate this work and did some of the earlier experimentation with ion plating.

M. Stenlake did most of the construction. Later measurements on windows were made by G.P. Cottman.

References

I have not referred directly to sources of information but have chosen to give an indication of the major sources which allow the subject to be pursued if desired.

1. Fan, J.C.C. and Bachner R.J.
   Transparent heat mirrors for solar-energy applications
   Applied Optics Vol. 15, pp 1012 - 1017 (1976).

2. Glazer, H.J.
   Highly transparent low emissive coatings on the basis of ZnO/Ag/ZnO, ZnO/Au/ZnO, $Bi_2O_3/Au/Bi_2O_3$, ZnO/Cu/Zno and $Bi_2O_3/Cu/Bi_2O_3$.
   Proceeds of the 8th International Vacuum Congress, Cannes 1980 (Suppl. "Le Vide les Couches Minces No. 201). Vol 1, pp 723-726.

3. Frank, G, Kaner, E and Kostlin.
   Transparent heat reflecting coatings based on highly doped semiconductors. Thin Solid Films Vol 77, pp 107-117 (1981).

4. Howson R.P. and Ridge M.I.
   Deposition of transparent heat reflecting coatings of metal oxides using reactive planar magnetron sputtering of a metal and/or alloy.
   Thin Solid Films Vol 77, pp 119-125 (1981).

5. Mahdjuri, B.
   Highly insulated window glazing.
   Energy Research Vol 1, pp 135-142 (1977).

6. Ridge, M.I., Howson R.P. and Bishop C.A.
   New techniques for the roll coating of optical thin films.
   Proceedings of the SPIE Los Angeles Technical Symposium, 25-29th January, 1982, Los Angeles, California, USA.

7. Schiller S, Heisig V, Steinfelder K, Strumpfel J.
   Using the d.c. magnetron for the reactive deposition of oxides.
   Proceedings of the International Conference on Ion Plating and Allied Techniques, London, U.K. July, 1979, p. 211.

8. Howson R.P., Ridge M.I.
   Heat mirrors on plastic sheet using transparent oxide conducting coatings.
   Proceedings of the SPIE Los Angeles Technical Symposium, 25-29th January, 1982, Los Angeles, California, USA.

9. Lampert C.
   Heat mirror coatings for energy conserving windows.
   Lawrence Berkeley Laboratories LBL Report No. 11446, August, 1981.

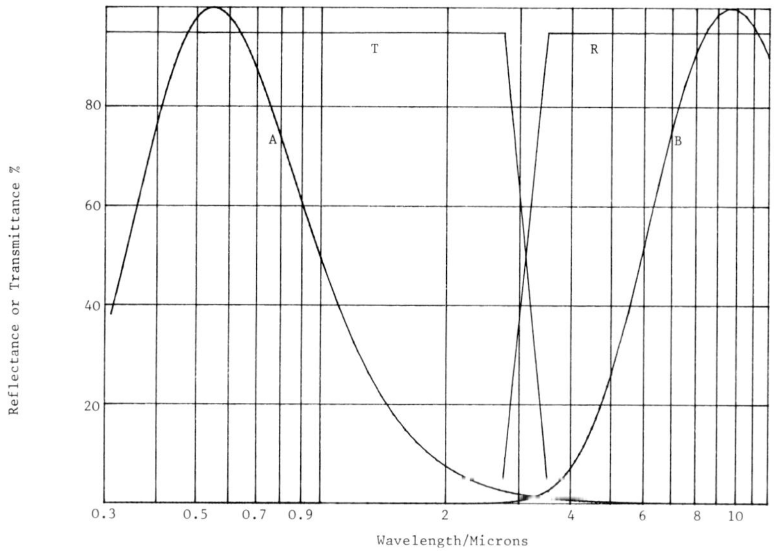

Figure 1
An approximate normalised emission spectra is shown for the sun (A) and for room surfaces (B) together with the selective properties required for a selective optical surface for energy conservation. R refers to the reflectance and T to the transmittance for such a filter.

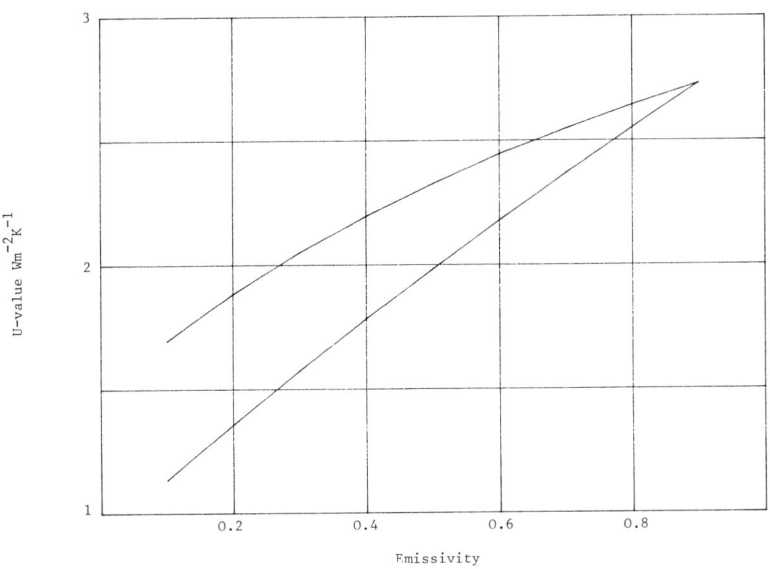

Figure 2
The U-value for a double pane window with the inside pane having a coating (A) on the side facing the other pane only and (B) on both sides.

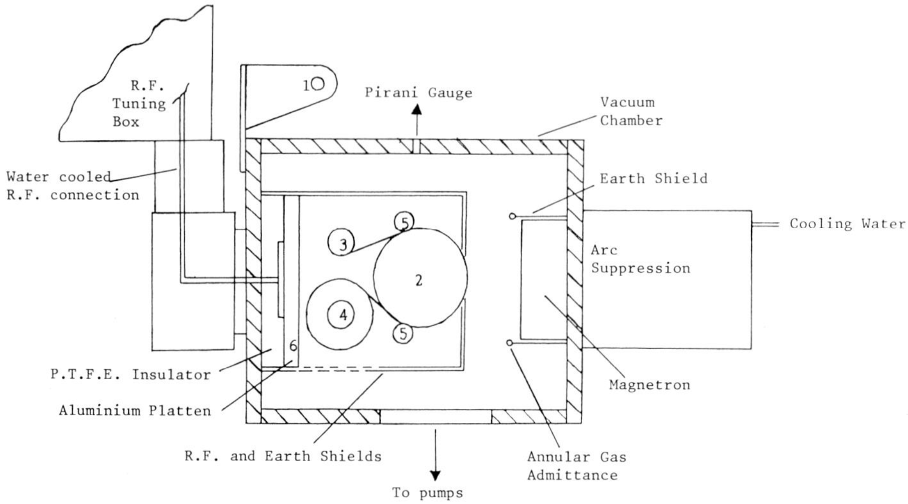

Figure 3

The roll-to-roll coating apparatus detail and vacuum system showing the magnetron source in position.

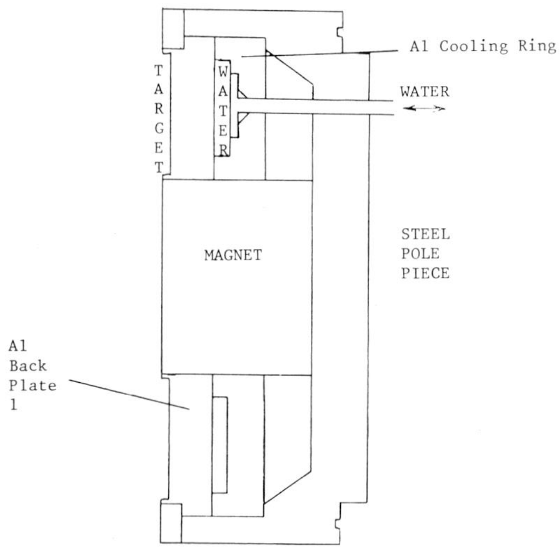

Figure 4

Magnetron source detail.

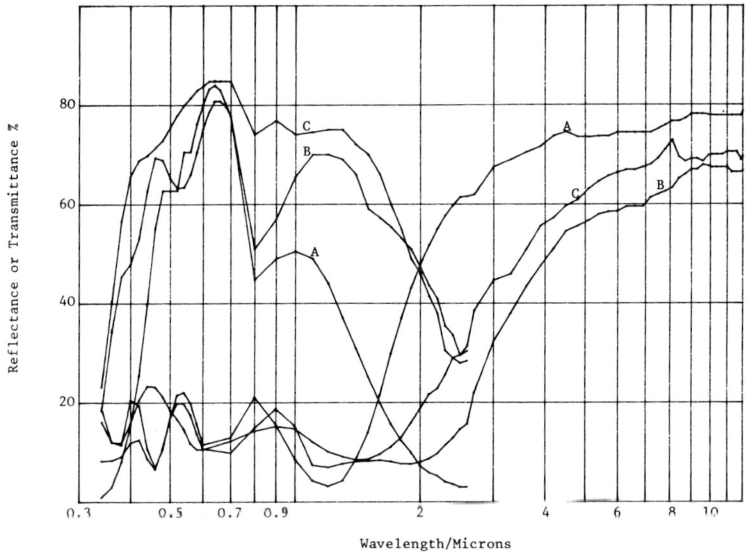

Figure 5

The selective optical properties obtained with (A) indium 10% tin oxide, (B) indium oxide, and (C) cadmium 2:1 tin oxide.

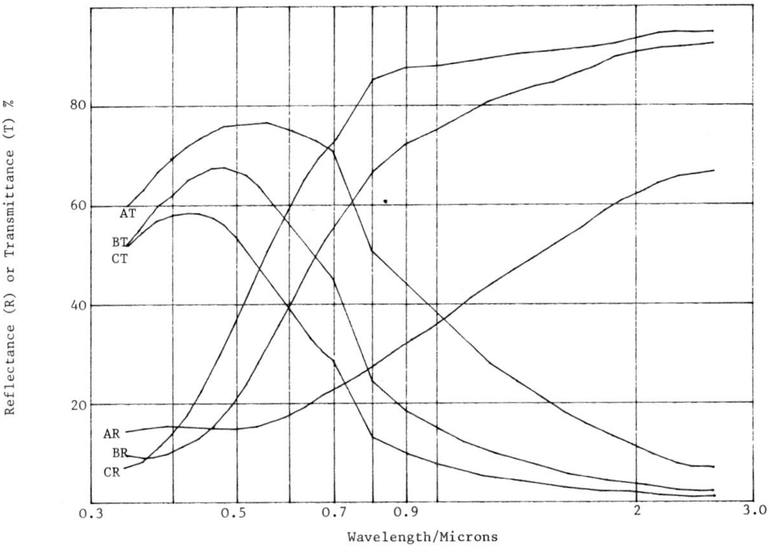

Figure 6

The experimental spectra of silver films sandwiched between dielectric layers of reactively sputtered tin oxide of an approximate thickness of 22 nm with a refractive index of 1.92 at 633 nm. The low temperature heat emissivity of A is 0.19, of B 0.08 and of C 0.05. The films were deposited on plastic sheet.

QUESTIONS AND DISCUSSION

Lagarde (CEA-CENG) : What is the average life time of the plastic substructure ?

Answer : This is not known in detail but the evidence suggests that this will be at least five times the payback period (which is less than one year), the aluminium frame will last much longer so that the replacement cost of coated plastic will be much lower than a new installation.

Ruyssevelt (PCL) : What is the theoretical "U" value with a sheet of coated plastic with an emissivity of 0,3 and which is framed at a distance of 20 mm from the window ?

Answer : About 2 $Wm^{-2}\ {}^{o}k^{-1}$.

Ruyssevelt (PCL) : Since the plastic does not give support to the frame is the frame therefore more expensive than other frames of secondary double glazing ?

Answer : It can be quite inexpensive $\approx$ £1 a metre run.

Dubbeld (TNO) : When a firm would want to built a machine to produce these coated plastic sheets. Would you support these firms and do they have to pay rights for the process ?

Answer : We believe that it is still too early to create an industrial process. Problems remain with the supply of indium in large amounts, e.g. one machine will double the U.K. consumption of that material, and with the toxicity of cadmium. New materials remain to be investigated and a pilot plant is needed to discover the problems in the new process. We will support any manufactures with the process. The rights would have to be negociated with the supporters of the R+D programme who include the Commission.

DEPOSITION OF SELECTIVELY TRANSMITTING FILMS BY THE PYROSOL PROCESS

Authors : G. BLANDENET, M. COURT, Y. LAGARDE

Contract number : EEA-2-022-F

Duration : 24 months 1 January 1981 - 31 December 1982

Head of project : G. BLANDENET, Laboratoire d'Etudes des Matériaux Minces

Contractor : Commissariat à l'Energie Atomique
Centre d'Etudes Nucléaires de Grenoble
Département de Métallurgie de Grenoble
Laboratoire d'Etudes des Matériaux Minces

Address : CEA - CENG
85 X
F-38040 GRENOBLE

Summary

The aim of this research is to decrease the cost of selectively transmitting films coated on glass by the Pyrosol Process with an in-line machine by technical improvements (tin oxide coatings) and economic improvements (indium oxide coatings). The protection of the glass by means of an under-coating, composed of SiO_2 (deposited by CVD or dipping) or SnO_2 (deposited by Pyrosol Process from organo-tin compounds) is the solution. On the other hand, the production cost of indium oxide coatings is reduced by the optimization of the indium extraction-line. Good results are obtained for double glazings (coated with $SnO_2(F)$: emissivity = 0.2, coefficient of heat transmission K_T = 2.2 W/m^2/° C ; coated with I.T.O. : emissivity = 0.15, K_T = 1.9 W/m^2/° C).

1.1 Introduction

In traditional buildings, the window area is equal to 10 per cent of the walls on an average and thirty to forty per cent of the heat losses are from the windows. However, window insulation must not involve a reduction of penetrating solar light whereas architecture has to search for a solar energy captation as natural as possible. The real problem is to conceive glazings with a great resistance to internal-external heat flow and an important solar factor when they are exposed to the sun.

One of the envisaged solutions is the use of a single or double glazing with a selective coating, transparent in the solar spectrum and reflecting in the infra-red range. Theoritical calculations show that a double glazing coated with an infra-red reflective layer of emissivity near 0.3, for example, has approximatively the same heat insulation as an uncoated triple window (table I).

Table 1 : Calculated values of coefficient of heat transmission

Glazings (8 mm air-spacing)	Uncoated $W/m^2/^\circ C$	Coated with a reflective layer ($\varepsilon \simeq 0.3$) $W/m^2/^\circ C$
single	5.8	
double	3.2	2.26
triple	2.3	1.4

Thereby, the field of application of conducting transparent oxides (In_2O_3, SnO_2, Cd_2SnO_4) is expanding rapidly at the present time, in particular in the area of selective glasses. These oxides are deposited on glass by various methods, such as cathode sputtering, chemical vapour deposition, and reactive chemical spraying. This last deposition technique is based on the pyrolysis of an aerosol obtained by pneumatically spraying a solution of a mineral salt or an organo-metallic compound (Chemical Spray Method). Based on the same principle, the Pyrosol Process, a deposition technique developed and applied in the Laboratoire d'Etudes des Matériaux Minces of the French Atomic Energy Commision, uses an aerosol produced by ultrasonic spraying. Prior research work has resulted in the successful deposition on glass, by means of Laboratory equipment (carousel furnace) of thin films of tin-doped indium oxide and fluorine-doped tin oxide presenting optical and electronic properties equivalent to or better than the deposits obtained using other methods (EEF Contrat N° 201.77). In addition, it has led to the development of a conveyor-type (run-through) Pyrosol machine for coating flat glass plates of relatively small size but capable of being easily extrapolated. (EEF Contrat N° 334.78.1). The aim is now to lower the cost of the selective glasses obtained by the Pyrosol conveyor coating process by improving both the technical and economic data, working in the following main directions :

- enhancing the properties of the fluorine-doped tin oxide coating (which is comparatively less expensive than the tin-doped indium oxide but gives lower performance in particular with respect to transmission in the visible region).
- better control over aerosol production.
- reducing the cost price of the tin-doped indium oxide deposit by optimising the preparation of indium acetylacetonate and by recovering unused indium-bearing products.

1.2 Description of the equipment

Two type of equipment are used : the first is intented for laboratory work, the second (the conveyor-type machine) is suitable for the continuous coating of glasses size 300 x 400 mm. The laboratory apparatus, of simple design, is used to study the feasibility of different coatings. It mainly comprises (fig. 1) :

- an aerosol generator : this ultrasonic atomizer comprises a glass vessel containing the liquid to be spraying and fitted at the bottom with a ceramic transducer excited by a H.F. generator (frequency : 800 kHz).
- a cylindrical reaction chamber.
- an 800 W heating plate set vertically.
- an aerosol injection nozzle (it consists of a pipe having a diameter sufficiently large to avoid coalescence of the droplets).
- an exhaust to evacuate the gases produced by pyrolysis.

The conveyor-type machine comprises (fig. 2) :

- an in-line furnace into which the substrate are conducted by means of a conveyor system. Its primary role is to ensure the heating of the glass substrates without breaking up to the deposition temperature, then to promote vaporization of the solution.
- an aerosol injection nozzle (its directs the aerosol into the pyrolysis zone. Its temperature is regulated to prevent premature vaporization of the sprayed solution).
- one or more aerosol generators (ultrasonic atomizers).
- a gas extraction system to evacuate the gas effluents outside, for safety reasons due to the use of chemicals that may be toxic and to ensure the gas mixture circulates near the substrates to be coated.
- a washing tower in which the intact indium-base products and the by-products of pyrolysis can be recovered.

1.3 Description of activities

To reduce the cost of selective glasses, on solution considered is to replace the tin-doped indium oxide deposited on the glass by a cheaper fluorine-doped tin oxide. However, thin films of this second oxide deposited from a solution of $SnCl_4$ in methanol onto a sodium-calcium type glass result in a milky white appearance along with a deterioration of the physical properties of the coating, in particular a slight decrease in transmission in the visible region. A protective layer must first be deposited on the glass to act as a barrier to prevent any contact between the glass and the halogenated products generated by the pyrolysis of $SnCl_4$. The materials considered for this layer are oxides such as SiO_2, TiO_2 or ZrO_2.

The optimization of indium acetylacetonate preparation is in progress. The Minemet Company is responsible for this part of the research work which consisted first in doping the indium acetylacetonate with certain impurities to simulate the composition of the raw materials in the indium plant extraction process. Effectively, depending on the influence of the impurities contained in the indium acetylacetonate on the physical properties of the oxide deposited on the glass by the Pyrolysis Process, this compound may be obtained from the so-called "rich" solution, from unpure metal, from purified metal, or from intermediate products. The chemical composition of the raw materials in the acetylacetonate production line is given in the table II.

Table II

X	"Rich solution" g.l^{-1}	$\frac{X}{In}$ ppm	Unpure metal p.p.m.	Purified metal p.p.m.
In	150			
Zn	200		30	10
Pb	0.6	4000	4000	10
Sn	0.3	2000	2000	10
Tl	0.07	470	100	5
Cd	0.02	134	100	5
Al	0.1	670	30	10
Cu-Ca	0.002	15	10	5
Fe	0.4	2680	10	10
Si	0.1	670	10	10
Na	0.1	670	10	10
Sb	0.01	67	100	10
Ga	0.1	670		

The most troublesome elements from the point of view of quantity seem to be zinc, lead, iron, whereas the quantity of tin must be controlled since this element is used to dope the indium oxide layer. However, particular attention must be paid to any aluminium, thallium, gallium, sodium and iron present, since these elements can entail a deterioration of the electronic properties of the indium oxide coating (by increasing the number of traps and thereby decreasing electrical conduction and infra-red reflection) as well as of its transmission in the visible region (by increasing absorption of the coating , especially due to the presence of iron. So the elements that form a complex compound with acetylacetonate, the chief reagent used to prepare indium acetylacetonate, must be identified. Batches of indium acetylacetonate representative of the extraction line were prepared and coatings of indium oxide were obtained by the Pyrosol Process. The characterization of these coatings gives the possibility of optimizing one stage of the indium extraction line (fig. 3) and to decrease the preparation cost of indium acetylacetonate and consequently to reduce the production cost of the glazings.

Before preparing the batches representative of the extraction line, the partition curves of each of the impurities were plotted as a function of the impurity content in the starting product. The operating mode used to produce indium acetylacetonate is the method employed by the Laboratoire d'Etudes des Matériaux Minces of the CEA for the work carried out in the scope of the EEF contrat N° 201.77. In this operating mode, impurities can be voluntarily introduced in the form of a nitrate or chloride. The indium acetylacetonate is then precipitated at a pH of 5.5-6.5 then recrystallized in petroleum ether after being dissolved in benzene.

It might seem that some impurities would be eliminated at the pH of precipitation but at this pH the indium is kept in solution in the form of soluble acetate, just as some impurities form soluble acetic complex compounds (Pb-Cu-Al).

The quantity of impurity added is measured in the dissolving solution then in the indium acetylacetonate.

1.4 Characterization of the glazings

Deposited materials are identified by X-ray diffraction. The thickness of the deposits is measured by the stylus method (Talysurf apparatus by Rank Xerox), after chemical etching. The surface resistances are measured by the in line four point method and the resistivities calculated from the formula :

$$\rho_{\Omega\ cm} = R_{\Omega} \times e_{cm}$$

(R = surface resistance, e = thickness of the coating).
The optical properties of these thin films are determined by measuring the optical transmission in the visible region of the spectrum and the infrared region (Beckman Acta PIV spectrophotometer). The determination of coatings emissivity is carried out at room temperature, by an optical device made in our laboratory and based on the reflectivity measurement with a black body at 150° C. In addition, the heat losses of a double glazing in which the external glass is coated by a selective layer on the internal side, are measured, and this with different air-spacings.

1.5 Results and discussions

- Deposition of a protective layer on the glass.

Comparative tests were carried out simultaneously on sodium-calcium type glass, Pyrex glass and sodium-calcium type glass previously coated by CVD with a silicon oxide layer (thickness 500 Å). When deposited on a glass previously coated with SiO_2, the layers consists of a single phase (quadratic SnO_2 ASTM Sheet 21.1250) identified by X-ray diffraction (fig. 5) ; no trace of chlorine is detected by X-ray analysis with a scanning electron microscope and the layer adheres well on the substrate. When the oxide is deposited on Pyrex glass or on sodium-calcium type glass, X-ray analysis detects a higher content of chlorine when the substrate is a sodium-calcium type glass (richer in sodium). In addition, the X-ray diffractograms (fig.5) show a line that does not belong to the quadratic SnO_2 phase but corresponds to the NaCl line (200) (ASTM Sheet 5.628). A comparison of the intensity ratios of the NaCl line and the (101) SnO_2 line indicates the higher content of NaCl when the substrate is a sodium-calcium type glass. On this same glass, after deposition of tin oxide, cubic shaped crystals were found and examined with the scanning electron microscope (fig. 4). The formation of sodium chloride is due to the reaction of hydrochloric acid, produced by the pyrolysis of tin chloride ($SnCl_4$) with the sodium in the glass. The presence of a silica layer (500 Å thick), acts as a barrier and prevents the hydrochloric acid vapour from contacting the glass during deposition of the oxide layer. It is worth noting that the presence of sodium chloride not only gives a milky white appearance, probably due to the diffusion of light by NaCl micro-crystals, but is accompanied by a deterioration in the adherence of the deposit on the glass. To avoid such phenomena, the gas produced due to pyrolysis must be prevented from contacting the surface of the glass substrate. This can be done by depositing a layer to act as a protective barrier. Different materials were selected to form this layer : SiO_2, TiO_2, ZrO_2. In each case, when the conditions of deposition are defined, the effectiveness of the layer is tested simply be depositing SnO_2 from stannic chloride, $SnCl_4$. The deposition temperature of the protective layer must not be above 500° C, which is the limiting temperature for the glass.

Among these materials, the deposit of SiO_2 (500 Å) gave the more effective barrier. However this thin film must be deposited by chemical vapor deposition or by dipping. On the other hand a good protection of glass is made by the Pyrosol Process, using a SnO_2 coating (1000 Å) deposited from a solution of an organo-tin compound.

- Optimising the preparation of indium acetylacetonate

It must be noted that the results presented here were obtained using synthetic solutions.

Zinc is only detected in indium acetylacetonate when its content in the dissolving solution is at least 2 % of the initial indium content. Since the zinc content is at most 100 p.p.m. in the unpure metal, this element will not be detected in the indium acetylacetonate prepared in this manner. An experiment carried out with a solution having a zinc content equivalent to that of the so-called "rich" solution (Zn/In ratio 200/150) showed that 2.7 % of the initial zinc is precipitated with the indium acetylacetonate.

The lead content of indium acetylacetonate increases very gradually with the ratio of added impurities ; whether prepared from unpure metal or solution, the chelate always contains about 100 p.p.m. of lead compared with indium.

When the tin content in the initial solution increases, the partition coefficient reaches a maximum. However, for an initial content corresponding to that of the "rich" solution or that of the unpure metal, tin is not detectable.

Aluminium may be accidentally present in the unpure metal, and for this reason high aluminium contents were added to the dissolving solution. When present, it precipitates out almost entirely with the indium acetylacetonate.

Reagents and acetylacetone in particular may contain stray iron (depending on production batches an iron content of up to 30 mg/l may be found). A blank test with indium containing less than 10 p.p.m. of iron was carried out. The mean iron content of the indium acetylacetonate obtained is found to be about 50 p.p.m.. It is found that the indium acetylacetonate will never be without iron unless the acetylacetonate is purified. If the raw materials have an iron content of not more than 50 p.p.m., the indium acetylacetonate will contain at least 50 p.p.m. A solution containing copper only will give with acetylacetonate an insoluble acetylacetonate. But when indium is present, the indium acetylacetonate precipitates prefenrentially without involving the copper. For copper contents corresponding to the rich solution or to the unpure metal, copper is not detected in the indium acetylacetonate. As for copper, for thallium, cadmium, calcium, sodium, antimony contents corresponding to the so-called rich solution or to the unpure metal, these elements are not detected in the indium acetylacetonate. Consequently, lead, zinc, aluminium and iron are the only impurities detected in the indium acetylacetonate. Starting from the unpure metal, only lead (100 p.p.m.) and iron (due to the reagents) are detected in the end product. Starting from the solution as raw material, the acetylacetonate will contain zinc as well as lead and iron.

Batches of indium acetylacetonate representative of the extraction line were prepared and deposits of indium oxide were obtained by the Pyrosol Process. The results show that the most harmful impurity is iron which causes spontaneous burning of the gases in the deposition furnace and results in an excessive emissivity (0.35) and overhigh resistivity ($1.27 \times 10^{-3}\Omega.cm$). The other impurities in very small quantity seem to have little effect on the properties of the indium oxide coatings. Consequently, it would appear that in the indium extraction process, the so-called "rich" solution cannot be used as raw material for the synthesis of indium acetylacetonate.

However, good results (table III) may be obtained from other intermediate stages of the indium extraction line (fig. 4) which are the "unpure metal" or the "purified solution". In these cases, the production cost of indium acetylacetonate is reduced. An estimation is in progress.

Table III

Raw materials	Rich solution	Unpure metal	Purified solution	Pure metal
Thickness of coating Å	4500	4000	4500	4500
Emissivity	0.35	0.18	0.16	$\leqslant$ 0.1

Selective coatings with a good reflectivity in the infra-red range (fig. 6) and so a low emissivity ($0.1 \leqslant \varepsilon \leqslant 0.2$) have been used in the elaboration of double glazings. They have also an high optical transmission in the visible region of solar spectrum (fig. 7), thus a good energy transmission and a welcome heat contribution in winter. The experimental results, depending on the nature of the coatings and the thickness of the air-spacing are shown in table IV. (The selective layer is coated on the internal side of the external glass).

With regard to an uncoated single glass window a double glazing (thickness of air-spacing = 12 mm) with a selective coating of SnO_2 (F) presents a saving of $4W/m^2/°$ C. The same comparison for a double glazing (thickness of air-spacing = 12 mm) with a selective coating of I.T.O. presents a saving greater than $4W/m^2/°C$.

2. Conclusion

In the immediate future, the optimization of indium acetylacetonate preparation must allow a reduced cost of indium tin oxide coatings and enable the production of less expensive glazings. The selective glazings with SnO_2 (F) coatings cheaper than I.T.O. are however just worse in performance In the industrial programme, a licence agreement is being investigated in two ways with a french manufacturer :

- on the one hand, the short-term industrialization of the Pyrosol Process

Table IV

Thickness of air-spacing mm \ Nature of coating	Coefficient of heat transmission $W/m^2/° C$	
	SnO_2 (F) emissivity = 0.2	In_2O_3 (Sn) I.T.O. emissivity = 0.15
6	2.7	2.6
8	2.4	2.2
10	2.2	1.9
12	2.0	1.8
visible light transmission of glazing %	69	75

for producing glazings (40 cm x 30 cm) used in electric domestic applications (ports of cooking-stoves : for energy savings and safety). This production is already practical as shown by the studies which have been carried out with the in-line prototype machine,

- and on the other hand, the industrial development of the selective glazings (1 m x 1 m) manufactured for buildings.

Moreover, agreements will signed between the Frenc Atomic Energy Commission and different manufacturers of chemical products to insure the supply of cheaper products. In another connection, at the request of another manufacturer in the European Communauty, a technical collaboration to compare different techniques of selective glazing production is going to begin.

During the last five years, the Pyrosol Process has, in addition, found now important applications in electronics and related fields (microelectronics, optoelectronics ...).

Figure 1 : Laboratory apparatus

Figure 2 : In-line machine

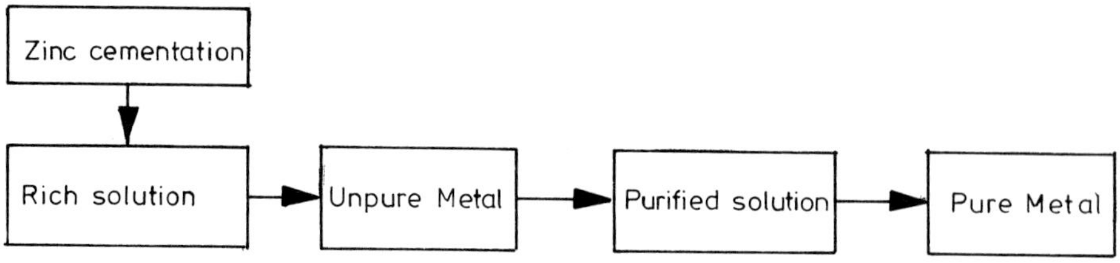

Figure 3 : Indium extraction line

Figure 4 : SEM examination of $SnO_2(F)$ layers deposited on float glass

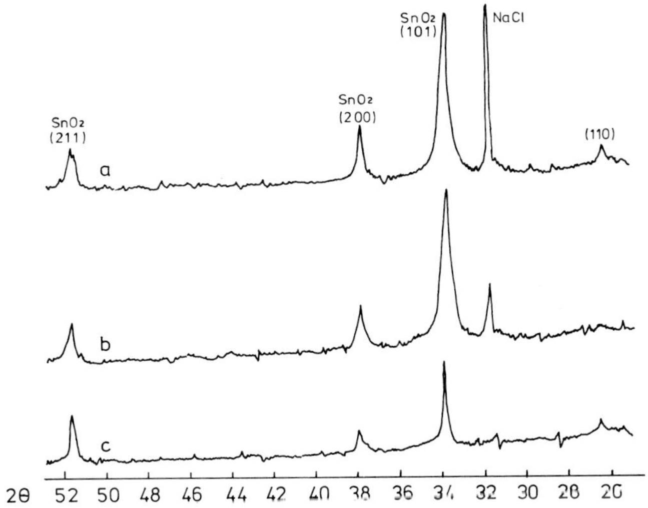

Figure 5 : X-ray diffractograms of SnO_2(F) coatings.
a - on sodium calcium glass
b - on borosilicate glass
c - on SiO_2 pre-coated sodium calcium glass

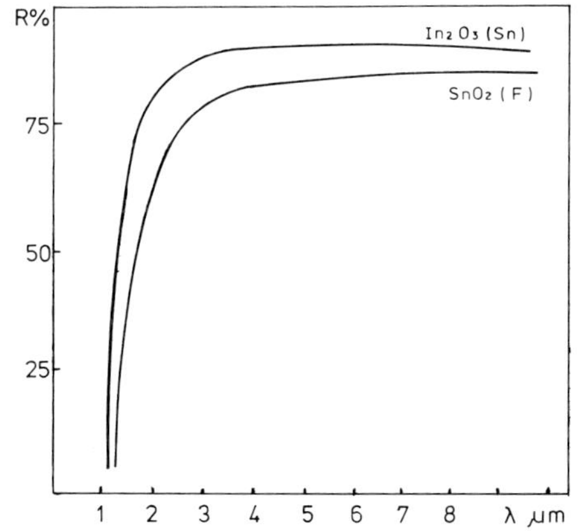

Figure 6 : Infra-red reflection of selective coatings

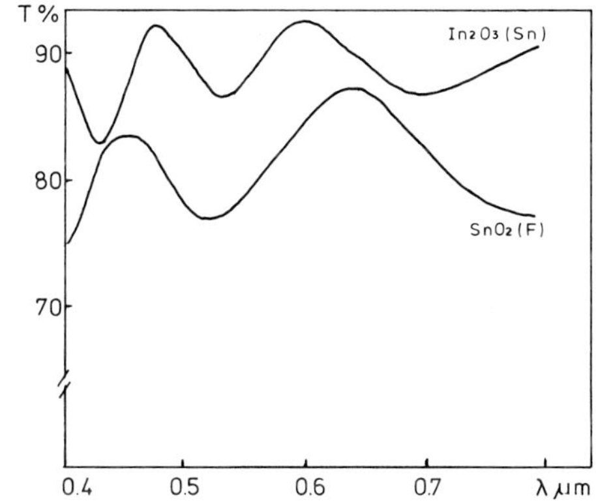

Figure 7 : Selective coatings optical transmission in the visible range

QUESTIONS AND DISCUSSION

Uyttenbroeck (Chairman) : What is the maximum width of the glass surface you are able to coat ?

Answer : Presently 30-40 cm. For industrial production this width will have to be increased.

Uyttenbroeck (Chairman) : Is it possible to give for each solution you have worked on, the values for

Transmission factor as a function of wavelength
Emission factor as a function of wavelength
Cost in the case of industrial production
Evaluation with time of the properties under real conditions ?

Answer :

	I.T.O. In_2O_3 (Sn)	SnO_2 (F)
Thickness	4000 Å	4000 Å
Transmission in visible range	85 - 88 %	80 - 83 %
Emissivity	~ 0.15	~ 0.2

The cost price of these coatings (yield 50 %, coating thickness 4000 Å) which result from laboratory production is given below :

I.T.O. Emissivity = 0.15			SnO_2 (F) Emissivity = 0.2		
Acetylacetonate d'indium	Solvent	Coating	$SnCl_4$	Solvent	Coating
43.5 F (2500 F/Kg)	13.4 F (16 F/l)	56.9 F/m^2	32 F (69F/100 ml)	6.4 F (20 F/l)	38.4 F/m^2

However, if we want to estimate the cost price of these coating for an industrial production, we must consider the cost price of metal in the starting compounds, especially in the case of I.T.O. coatings. Actually the cost price of indium acetylacetonate is equal to ten times the cost price of indium contained in this compound.
We think that it is possible to obtain a cost price of indium acetylacetonate equal to two times the cost price of indium for an industrial production. On the other hand, we must try to decrease the cost price of the solvent (16 FF $\rightarrow$ 2 FF/l) by using other cheaper solvents. A study is necessary.
After these considerations, the cost price of coatings for an industrial production may be as follows :

In_2O_3(Sn)	I.T.O. Emissivity : 0.15		SnO_2 (F) Emissivity : 0.2		
indium Acetylaceto-nate	Solvent	Coating	$SnCl_4$	Solvent	Coating
8.7 FF (500 F/kg)	1,7 FF (2 FF/l)	10.4 FF/m^2	3.2 FF (7 FF/100 ml)	0.64 FF (2 FF/l)	3.84 FF

In the case of I.T.O. coatings, recent results show that the same emissivity can be obtained with I.T.O. coatings 2000 Å thick, the corresponding cost price should be devided by a factor 2 (5.2 FF/m^2), and the output speed of the machine can be increased.
The behaviour in time in real conditions has not been tested. However, after heating at 200° C in air during 1000 hours, the properties of I.T.O. coatings are not changed. The adherence is always very good on glass. In the point of view of chemistry, these coatings are etched only by hydrochloric acid and hydracids. We think that the behaviour in time is the behaviour of the glass substrate.

Howson (Loughborough University) : What is the output speed of your 14 metre machine, how thick is the oxyde film ?

Answer : The output speed will be between 10 and 20 cm/min for I.T.O. coatings of 4000 Å thick.

Littler (PCL) : I have three questions :

1. a) I find the figure of 88 % transmission for daylight very high for coated glass when there is not anti-reflection treatment;
   b) It would be helpful to have τ at various angles of incidence.
2. What is the value of the emissivity ?
3. What measure is available of the homogeneity of coating across wide pieces of glass ?

Answer : 1. a) The transmission in the visible range of a pane of float glass (5 mm thick) coated with indium tin oxide is 88 %, limited by interferences phenomenon (see figure below).
b) The measurements of the transmission at various angles of incidence have not been carried out.

2. The values of emissivity are the following :

	Thickness	ε
I.T.O. In_2O_3 (Sn)	4000 Å	~ 0.1
SnO_2 (F)	5000 Å	~ 0.2

3. Homogeneity of coatings (40 cm x 30 cm) is good. The thickness measurements show variations lower than 10 %.

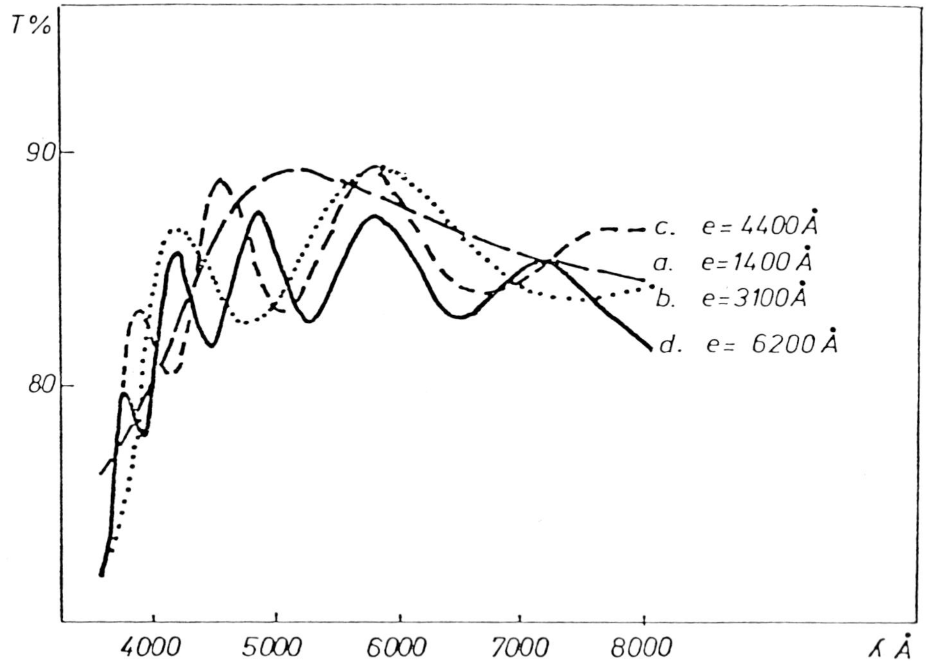

Transmission as a function of the light frequency and coating thickness.

INSULATED SHUTTERS

Authors : Thermal Insulation Laboratory

Contract number : EE-A-2-018-DK (G)

Duration : 19 months, 1 September 1981 - 31 March 1983

Head of project : Mr. Mogens R. Byberg, Thermal Insulation Laboratory

Contractor : Technical University of Denmark

Address : Building 118
DK - 2800 Lyngby
Denmark

Summary

During the first ten months of the project period alternative ways of design and placing insulating shutters have been analysed. The work resulted in two different types of shutters - one for existing buildings and one for new buildings. The solutions try to satisfy different kinds of demands - well insulating, easy to operate, flexible in application, architecturally good and economic.

The shutter for the existing building has internal placing and is in daytime parked in a shutter case. The shutter for the new building is placed between two layers of glass in a specially designed window element of which the shutter is an integrated part. In daytime the shutter - which is sidehung - is parked along the casing.

The shutter material is for both shutter solutions Polyisocyanurate cellular plastic (PIR). This material has physical qualities which satisfy the demand for low density, low thermal conductivity, sufficient stiffness and low moisture permeability.

At the end of the paper the influence of the energy consumption and the rentability of the shutters are briefly discussed.

Introduction

This energy conservation project deals with the problem of increasing the insulation value of window constructions by using thermally insulating shutters during the night.

As the Danish Building Regulations [1] claim a U-value for windows not exceeding 2.9 $W/m^2 {}^oC$ the basic window construction for the project will have at least two layers of glass, either as a hermetic double pane or as two separate layers.

The increasing oil prices have initiated window constructions with three or four layers of glass, the three-layer construction being common in Denmark and the other Scandinavian countries.

To improve the insulation value of these windows it is necessary to add insulation material in thicknesses not less than 30 mm (referring to mineral wool). Insulated shutters have already been used in several conservation houses for the last eight years, for the most part mounted on the outside of the windows.

Figure 1. South aspect of Conservation House at Hjortekær near Copenhagen

As you will realize from the photo, figure 1, the external shutters will demand sufficient distance between the windows to enable parking of the shutters in the daytime. Further the architectural aspect of the house will be influenced considerably by the shutters.

The purpose of this project is to develop, to design and to test insulating window shutters and we are aiming at shutters with interior location for existing buildings and shutters located between two layers of glass for new buildings.

The choice of shutter construction

As indicated in the introduction external shutters will influence the architecture of the house and demand regularly positioned windows. Further, the daily opening and closing of the shutters will demand either access from the outside to the shutters (as obtained in single story houses) or a mechanical device for handling the shutters from the inside of the house. Automatic systems are supposed to be too expensive to be a realistic alternative to the manual systems.

External shutters have earlier been investigated (e.g. B. Saxhof: External insulating shutters in energy conservation houses, descriptions and experiences from three low-energy houses at Hjortekær, Denmark [2]). All of them have been applicated to new houses.

For the existing buildings the internal shutters - giving the same energy savings as the external shutters - seem to be the more realistic solution.

For new buildings with more than one story the shutter location between external and internal glasses is chosen for reasons mentioned below.

If insulating shutters - a relatively new building component in Scandinavia - shall have a possibility to be commonly accepted, they will have to be:

- easy to operate
- flexible in application
- economic
- well insulating
- architecturally good

Internal shutter

The internal shutter opens the possibility of:

- direct manual operation from the inside
- a lightweight construction, as the shutter is protected from the weather
- assembly and maintenance from the inside

but it causes:

- risk of condensation on the inside of the pane

To avoid moisture condensation on the pane or to minimize the condensation it is necessary to secure that both the shutter and the joints are moisture tight. During the night the panes, now situated on the cold side of the thermal insulation, will reach surface temperatures below the dew-point of the inside air.

When the shutters are dismounted in the morning it is inevitable that a certain amount of moisture will condensate on the pane.

The problem has been analyzed during pilot tests in a cold/hot box in which measurements have been carried out for shutters located on windows with 2 and 3 layers of glass.

Shutter: PIR (Polyisocyanurate) cellular plastic with a thickness of 45 mm.
Height and width: 1200 mm x 1200 mm

Placement of the Shutter: The shutter is located on the interior as shown below.

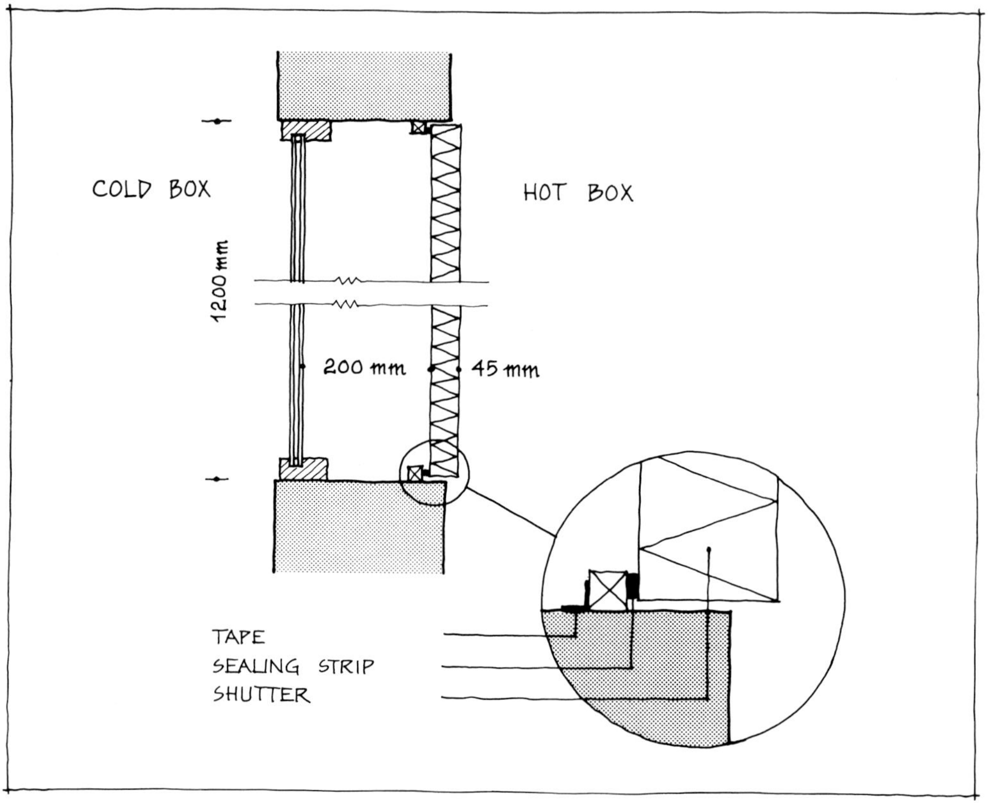

Figure 2. Test in cold/hot box - shutter located on the interior.

The measurements were carried out in the cold/hot box with the following parameters:

- exterior temperature: -5 to -11 °C
- interior temperature: +20 °C
- interior relative humidity: 30 to 55%

The test series were spread over several days and nights. The shutters were put up in the afternoon and removed next morning.

No condensation was found on the panes with the shutters in position due to the aimed favourable tightening between shutter and frame.

When the shutter was removed condensation was produced all over the pane. If electric resistance heaters were located below the window it took half an hour before the condensation vanished. If the electric resistance heaters were located on the wall opposite the window it lasted about an hour before the condensation vanished.

The condensation vanished faster from a pane with 3 layers of glass than from one with 2 layers of glass due to the higher pane surface temperature.

The pilot tests show that the risk of condensation is small when the shutters are located on the interior and the joints are tight. There will be no condensation when the shutters are in position. The condensation which is produced when the shutters are removed will be temporary.

The shutter material (polyisocyanurate cellular plastic) is found to be excellent, the features are:

- low thermal conductivity ($\sim 0.03\ W/m^{\circ}C$)
- low density ($\sim 30\ kg/m^3$)
- sufficient stiffness (no frame is needed)
- low moisture permeability due to the material itself and to the coating film necessary for the production.

The sawn edges of the material have to be protected. This problem will be solved in connection with the jointings.

Figure 3 shows the principle of the internal shutter with two adjoining shutters.

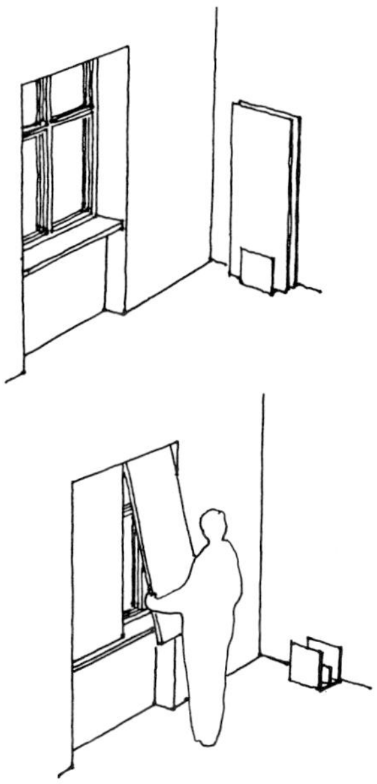

Figure 3 - Shutters for the existing building

Location of the shutters takes place manually - each shutter unit weights about 2 kg. The shutters are placed in a shutter case during the day. Independence from curtains and the mutual distance between the windows will be achieved by this shutter solution.

Joints between internal shutters

Much attention has been paid to the joint problem and many different solutions have been tried and rejected. It is essential that the joint should be air-tight so that no moist air can circulate from the indoor climate to the space between shutter and pane. Further the jointing material must be soft and with very little friction to the casing.

Figure 4 shows details of the edge of the shutter in the solution of to-day. The PVC profile with plastic foam covered by nylon cloth is continued along all four edges of the shutter allowing the foam to tighten either to the casing or to the foam of the ajoining shutter.

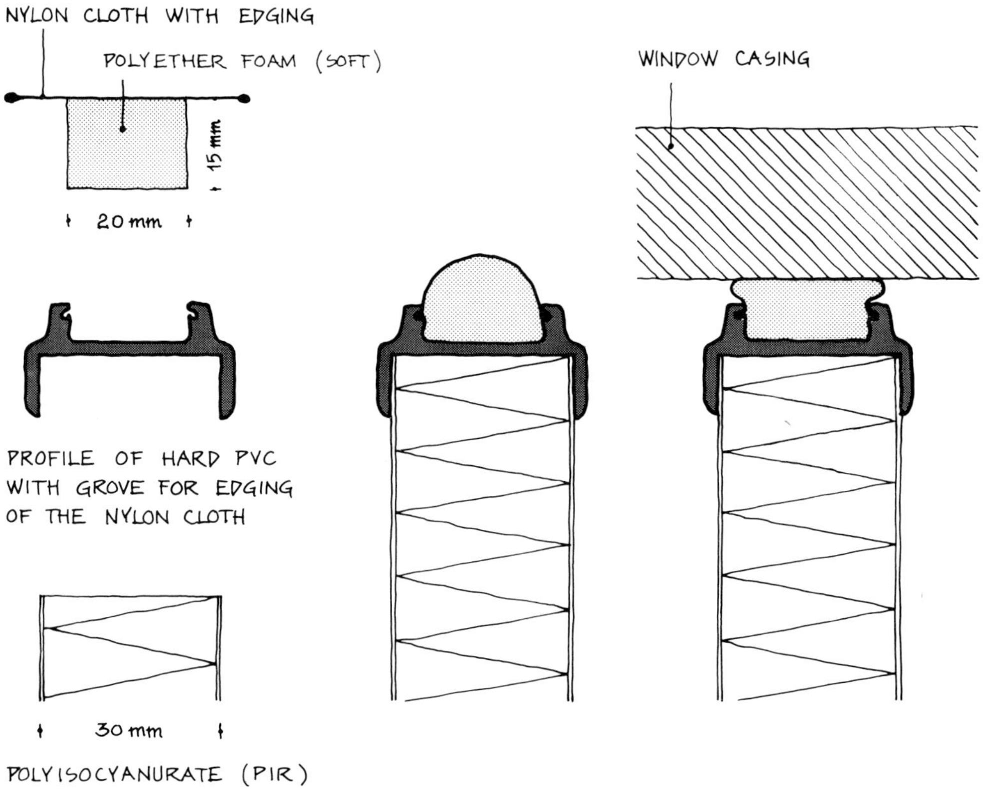

Figure 4. Details of the edge-profile for the internal shutter

Although figure 4 describes a PVC-profile we have only until now prepared profiles made of wood and steel plate for the prototypes. These prototypes have shown that also the corners are very well filled with foam, which is not the situation for normal types of weatherstrips.

In the near future the prototypes will be tested in the cold box/hot box facility. If no condensation problems will arise the PVC-profile will be produced.

The window construction with shutter between two layers of glass

Figure 5 shows the principle of the special window construction.

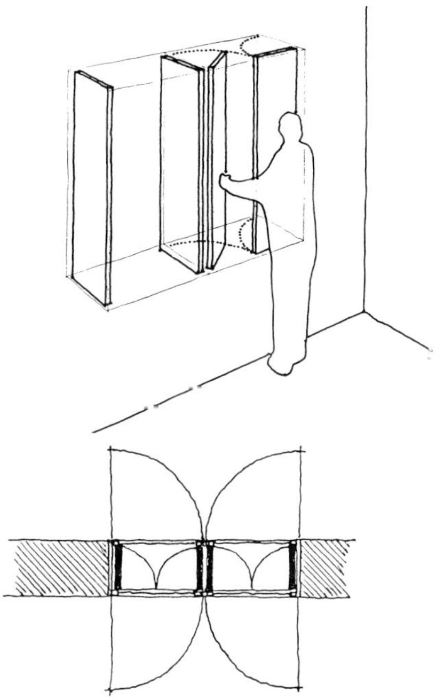

Figure 5 - Shutters for new building

The insulating shutters are part of the window construction and are placed between the two layers of glass. In the passive position the shutters will be parked along the casing. The interior windows are opened and the shutters operated manually. The window elements can be located side by side. Above are shown two elements placed side by side.

As a result of the Danish Building Regulations (BR-77) [1] the thickness of the outer walls is to-day about 0.35 m or even more as the layer of insulation material will be minimum 0.15 m to satisfy the maximum limit for the U-value, 0.3 $W/m^{2\,o}C$. In the future the wall thickness may be even greater.

For the prototype of this shutter construction we have chosen a distance between the glasses corresponding to the wall thickness. This allows two sidehung shutters to be located between the glass with the possibility of day-parking along the casing.

Both glasses are of the hardened type, they are sidehung as indicated in figure 5, the internal one with weatherstrip to secure air-tightness and the external one with weatherstrip along three of the edges to secure tightness against driving rain but allowing for a small amount of ventilation (like a normal two-stage joint).

The shutters are boards of the PIR polyisocyanurate foam like the internal shutter already mentioned. In the prototype of to-day, which has just been finished, the shutters are supplied with an edging of wood with a tightning of brush-type strips.

This prototype will also be tested in the cold box/hot box facility in the near future.

Why this peculiar construction

There are several good reasons for this special window design. Some of the problems, which this construction solves, are mentioned earlier:

- parking in the day-time
- architectural influence
- weather resistance
- application and maintenance

To these reasons can be added:

- two separated layers of glass solve the problem of puncture of hermetic two-layer panes
- no risk of condensation
- no need for sophisticated mechanical device for opening and closing
- the shutters can be used in multistorey as well as one storey houses.

Unfortunately, the width of the window is limited to max. 0.6 m corresponding to a wall thickness of 0.35 m. This allows the regulatory emergency opening of 0.5 m.

Energy savings

The maximum utilization of the insulation value is obtained for both types of shutters as the joints will be air-tight. In this case it is very easy to calculate the energy saving based on U-values.

Two layers of glass will normally have the U-value 2.9 $W/m^{2o}C$. With 30 mm of PIR-foam the thermal resistance will be increased by

insulation	$0.03\ m/0.03\ W/m^oC$	$= 1.00\ m^{2o}C/W$
air space between shutter and pane:		0.16 -
		$1.16\ m^{2o}C/W$
double pane:	$1/2.9\ W/m^{2o}C$	= 0.34 -
		$1.50\ m^{2o}C/W$
Which gives the U-value:		$0.67\ W/m^{2o}C$

So, in the night with closed shutters the U-value is reduced from 2,9 to 0,67 $W/m^{2o}C$, but in the day-time it is unchanged 2.9 $W/m^{2o}C$. Thus an estimate of 50% "shutter-time" gives a mean U-value of 1.79 $W/m^{2o}C$. The energy saving per m^2 based on 72000 degree-hours per year will be

$$(2.9 - 1.79) \times 72.000 \times 10^{-3} = 80\ kWh/m^2\ year.$$

For a single-family house in Denmark you will typically find window areas about 20 m^2. Thus the energy savings will amount to 1600 kWh/year. This is a conservative estimate as the shutters are used in the coldest part of the day and as the solar gain in the daytime is disregarded.

Economy

1600 kWh corresponds to 700 DKR or 1360 DKR depending on wether you use oil or electricity for the heating system respectively which indicates an investment of 9.000 DKR or 16.000 DKR respectively for an Exact Payback period of 10 years. 9.000 DKR corresponds to 450 DKR/(m^2 of shutter). Fuel cost is assumed to escalate 4% above the inflation rate and the discount rate is assumed to follow the inflation rate.

This is the reason for selecting the very inexpensive internal shutter solution. A more detailed calculation of investment and payback period will be shown in the final report of the project.

Conclusions

The purpose of the project is to develop and to test insulating window shutters for both existing and new buildings. Two separate shutter constructions have been selected: An internal shutter for the existing buildings and a shutter between the internal and external layer of glass for the new buildings. The use of external shutters has been analysed very closely and is found of minor interest especially for multistorey buildings.

Both shutter constructions are based on a stiff board of PIR-polyisocyanurate foam with very low thermal conductivety. Special attention has been paid to the joints around the shutters to secure a high level of air-tightness.

Pilot experiments have been carried out concerning the condensation problems in connection with the internal shutter. The shutter construction for the new buildings consists of a combined window and shutter construction with the internal/external glasses separated by a distance corresponding to the thickness of the outer wall. For both types of shutters the U-value of the two layer pane is reduced to a mean U-value of 0.67 $W/m^2 {}^oC$ corresponding to an annual energy saving of approx 1600 kWh for a typical Danish single family house.

In the project the laboratory tests of the thermal performance of the shutter prototypes still remain as well as tests of the condensation risk regarding the internal shutter.

Although the investigations have not yet been completed the project group finds that it is possible to design insulating shutters for both the existing and the new buildings presenting excelent insulation values combined with low investment.

References

[1]: Danish Building Regulations, BR-77, Ministry of Housing, 1977.

[2]: Bjarne Saxhof, "External Insulating Shutters in Energy conservation Houses, description and experiences from three low-energy houses at Hjortekær, Denmark". Thermal Insulation Laboratory, 1982. Report No 82-30, 1982.

ENERGY SAVING BY USING ROLLING SHUTTERS

Authors : Ing. M. Dubbeld and Ing. B. Knoll

Contract number : EE-A-2-019-NL

Duration : 12 months 1 October 1981 - 1 October 1982

Head of project : Ing. M. Dubbeld

Contractor : TNO

Address : TNO
Research Institute for Environmental Hygiene
P.O. Box 214
NL-2600 AE DELFT

Summary

For the Ministry of Housing and Physical Planning and the Commission of the European Communities a research project is carried out called "Energy saving by use of rolling shutters".
The aim of the rolling shutter project is to make a cost-benefit calculation of the supplied results of a reflective layer, edge seals and second layer and combinations.
The U-values measured up till now are reported together with the total progress of the project.

1. INTRODUCTION

This investigation is a part of the National Program which is coordinated by the Steeringgroup for Energy Saving in Buildings and partly founded by the Ministry of Housing and Physical Planning and the Commission of the European Communities. The parts of the investigation are:

1) Inhabitants' acceptation of rolling shutters carried out by Bouwcentrum;
2) Insulation and climate effect measurements, Bronswerk and TNO;
3) Energy saving calculations, TNO;
4) Cost-benefit calculations, Bouwcentrum;
5) Possible integration in the building construction, Bouwcentrum;
6) Summary report investigations E.C., TNO.

The first point is not carried out under contract of the European Commission. The research program is a follow up of the work for energy saving with curtains and windowsill. It can be divided in two parts:

1) rolling shutters carried out in the Bronswerk testroom,
2) windowsill and parapet by IMG-TNO.

The aim of the rolling shutter program is to make cost-benefit calculations of the supplied results of a reflective layer, edge seals and second layer and combinations. The results of the insulation and the climate effect measurements can be seen in the paper of Mr. J. Bekker [3].

In the testroom of the TNO Research Institute for Environmental Hygiene the following experiments will be carried out:

- effect of the disturbing of the cold draft along the window surface,
- catch of cold air from the window with the windowsill to prevent draft,
- effect of reflective paint on the parapet and the radiator.

2. MEASUREMENTS

The principle of the measurements is the same as used by TNO for these and earlier investigations for the effect of curtains and windowsill [1].

The heat loss through the façade is the summation of the heat loss through the parapet and through the window. Because the heat flow through the window cannot be measured directly by heat flow meters it must be calculated as the balancing item of a heat balance. Two heat balances may be set up for the test.

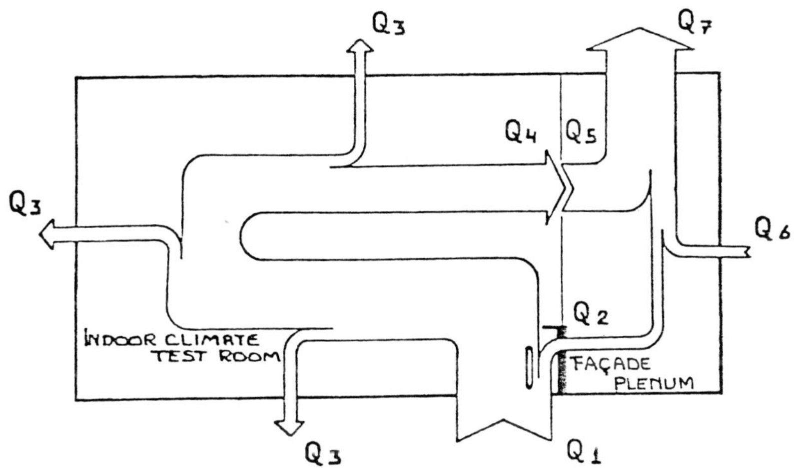

HEAT FLOWS

2.1 Heat balance of the room

The heat flow through the window, Q4, equals the heat flow supplied by the radiator, Q1, diminished by the heat flow through the parapet, Q2, and that through the walls of the room, Q3.

2.2 Heat balance of the façade plenum

The heat flow through the window, Q5, equals the heat flow discharged through the façade plenum, Q7, diminished by the heat flow through the parapet, Q2, and the heat flow from the outside, Q6. The heat loss through the window can be determined by these two ways. The difference between Q4 and Q5 is a measure for the precision of the measurement. The limits are 4 % of Q4 or 25 W. The hall round the test facility can be controlled in temperature. The U-value will be measured 33 times per hour. The mean value of these measurements is one measuring point. The data are collected by a data acquisition system HP 9825. In total 100 measuring points are used.

3. PROGRESS OF THE WORK AND RESULTS

From the total program the progress is as follows:

1) Inhabitants' acceptation is ready;
2) Rolling shutters is still going on, windowsill and parapet are ready;
3) Energy saving calculation is waiting for results of 2;
4) Cost-benefit calculation after 3;
5) Integration in the building construction will start in the next three months;
6) Summary report E.C. after finishing all the participated contracts.

With a new planning we can finish the contract by the end of June 1983. Therefore continuation of the contract till that date has been asked for. The results of the rolling shutters can be found in the paper of Bronswerk [3]. The results of the windowsill and parapet are mentioned here.

3.1 Progress of the work and results TNO

The measurements in this testroom are started with the facilities which are used for the investigation of energy saving by curtains with single glass. After calibrating the testroom measurements are carried out with single glass, a windowsill of 0,15 m and

1) horizontal disturbance on the window,
2) special windowsill to catch the cold air from the window.

The following results are measured:

1) Horizontal disturbance on the window.
   The U-value of the window will be 8 % higher with horizontal disturbance, so that the heat losses increase (compare figure 2 with reference situation in figure 1).
   Horizontal strips : height 10 mm
   width 30 mm
   Distance between strips: 0,46 m
   The sharp-edged strips increase turbulence near the window beyond the strips. This gives a better heat exchange. The vortex under the strips also keep the flow near the window after passing the strips (figure 2, flowpattern).
2) Windowsill catches the cold air from the window.
   Different situations are investigated. In general they all give the same results (figures 3 to 10). In the case the cold air from the window is streaming to the parapet, the mixed air from the radiator and the window streams closer to the window. These situations cause lower

heat losses through the parapet but higher losses through the window and the total façade.

3) When there is a crack between window and parapet and a curtain is used, the heat losses will be higher than without crack (figures 11 and 12).
   Conclusion: With horizontal disturbance, catching cold air from the window and a crack between the windowsill no energy saving can be reached.
4) The use of a reflective paint on the parapet U-value ≈ 1 W/m^2K (absorption ε = 0,35) has the following effect on the U-value: parapet: -21 %, window: -4 %, façade: -6 %.
5) The use of a reflective sheet on the parapet U-value ≈ 1 W/m^2K (absorption ε = 0,1) has the following effect on the U-value: parapet: -25 %, window: 0 %, façade: -3 %.
6) The use of a reflective paint on the backside of the radiator (absorption ε ≈ 0,40) has the following effect for a parapet of U-value ≈ 1 W/m^2K: parapet: -18 %, window: -4 %, façade: -5 %.

Supplemental velocity measurements are carried out for better understanding of the physical phenomena. An example of the measurements is given in figure 13.

4. CONCLUSIONS

1) With horizontal disturbance on the window or catching cold air from the window between the windowsill no energy saving can be reached.
2) A crack between the window and the windowsill gives a higher heat-loss when the curtains are closed. A crack must be avoided.
3) The use of reflective material on the parapet (U-value ≈ 1 W/m^2K) gives lower heat losses through the parapet:
   - reflective paint on parapet (ε = 0,35) -21 %,
   - reflective paint on the backside of the radiator (ε ≈ 0,4) -18 %,
   - reflective sheet on parapet (ε = 0,1) -25 %.

5. REFERENCES

[1] Bossers, P.A. e.a. — Energy saving by the use of curtains and windowsill. Report C 422, IMG-TNO (Dutch).

[2] Crommelin, R.D. and P.M. Maaskant — Determination of possible energy saving in a room by infrared reflective finishing of the parapet behind a radiator. Report C 396, June 1977.

[3] Bekker, J. — Energy saving by use of rolling shutters. Contractors' meeting on insulation, 28th September 1982.

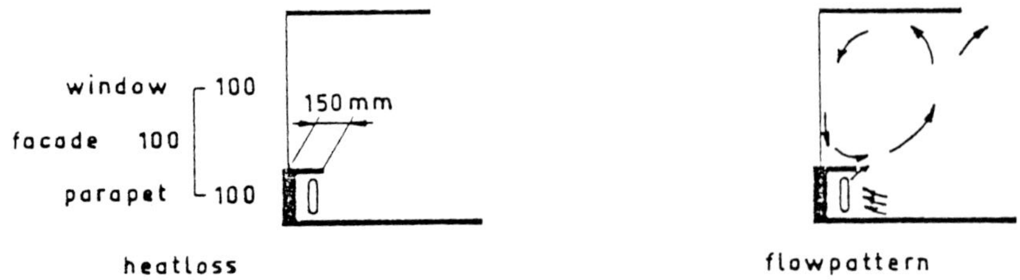

Figure 1 : Reference situation

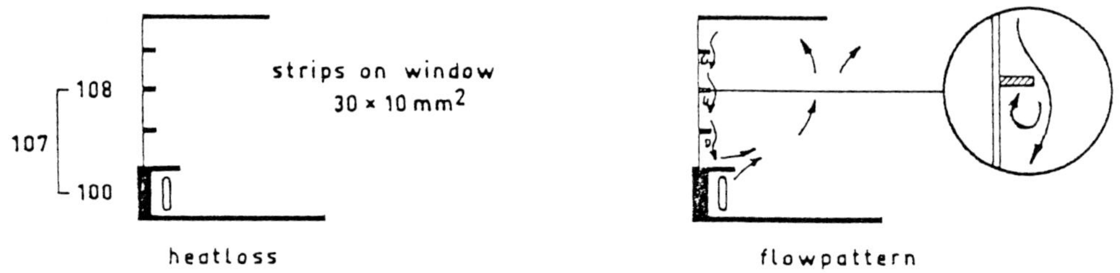

Figure 2 : Horizontal disturbance on the window

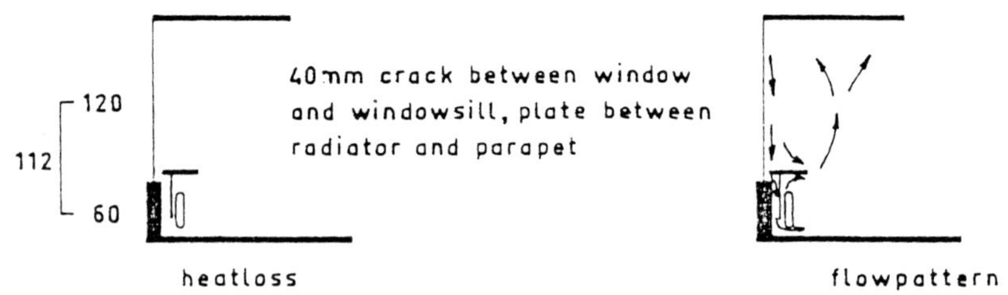

Figure 3: A part of the cold air from the window flows along the parapet and the radiator

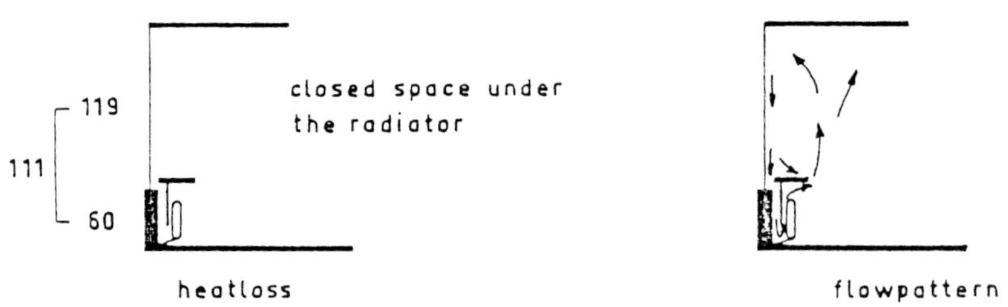

Figure 4 : as figure 3, no flow of cold air under the radiator

Figure 5: as figure 3, with windowsill 250 mm

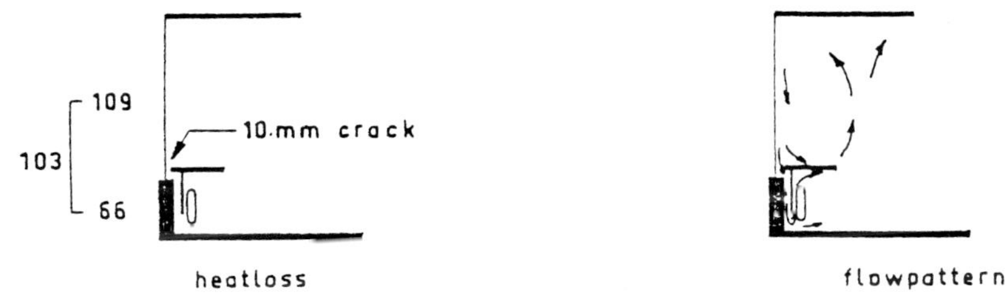

Figure 6: as figure 3, with 10 crack between window and windowsill

Figure 7: as figure 3, no crack between window and windowsill

Figure 8: Windowsill as figure 1 with crack

Figure 9 : as figure 8, with 40 mm crack

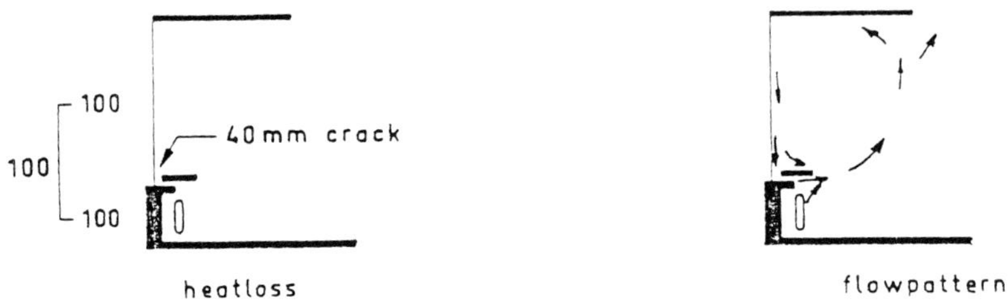

Figure 10 : double-plated windowsill with 40 mm crack

Figure 11 : as figure 9, with a curtain

Figure 12 : as figure 11, windowsill without crack

◯ = > 18 cm/s
○ = 15 – 18 cm/s
o = 11 – 14 cm/s
ᴑ = 8 – 10 cm/s
• = 5 – 7 cm/s
= < 5 cm/s (near the radiator not measured)

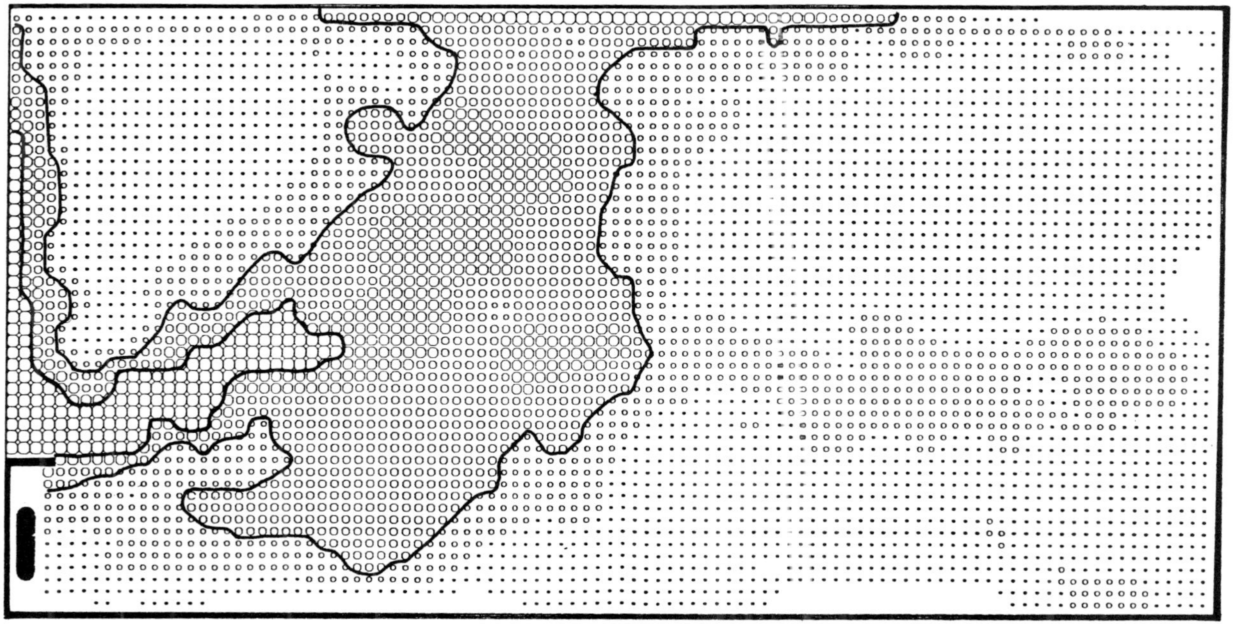

Figure 13 Measured velocities in a cross-section of the testroom with ref:ective point on the pararet (facade at the left).

QUESTIONS AND DISCUSSION

Littler (PCL) : Do you intend to do cost-benefit calculations and when will the results be published ?

Answer : The cost-benefit calculations will be started after finishing the U-value measurements and energy saving calculations. The results will be available in March 1983.

Littler (PCL) : We use a U-value for a single glazed window of 5,8 $W/m^2\ {}^oK$ at a wind velocity of 5 m/s. TNO/bronswerk gives 4,8 $W/m^2\ {}^oK$. Could you comment on this ?

Answer : The U-value will also be calculated at a wind velocity of 5 m/s, the result will be reported in the final report. Presently we are only interested in the supply heat resistance of different devices and the supplied layers or seals.

ENERGY SAVING BY USE OF ROLLING SHUTTERS

Paper to be read on the contractor's meeting on "Insulation" in Brussels on September 28th, 1982.

Author : ir. J. Bekker

Contract number: EE-A2-019-NL

Duration : 12 months

Head of project: ing. M. Dubbeld

Contractor : IMG-TNO

Sub-contractor : Bronswerk Airconditioning Research Centre

Address : Bronswerk B.V.
Postbox 28
3800 HC AMERSFOORT
Netherlands

Summary

In the total set-up of the research "Energy saving by use of rolling shutters" the majority of the measurements has to be carried out by the Bronswerk Airconditioning Research Centre. A total of 26 different shutter-combinations has to be measured, of which 30% has been realized now. Special attention has been given to the construction of the testroom and to the measuring process, to achieve a reliability in the results, better than 4%. Except the requested U-value measurements, some additional parameters have been measured:

- the Fanger PMV-value
- extra temperature-measurements on and between the shutter-combinations, for a better physical understanding. (These figures have not yet been worked out).

1. Introduction

In 1979, the Bronswerk Airconditioning Research Centre was invited to enter a discussion for the planning of a research on energy saving by use of rolling shutters. The research should be a continuation of the research programme "Energy saving by the use of curtains and window-sill" carried out by the TNO Research Institute for Environmental Hygiene (Delft) in 1978 and reported in Report C422. Partners in the discussion were the TNO Institute, the Centre for Building Research "Bouwcentrum" (Rotterdam) and Bronswerk, as the only institute of an industrial origin in this set-up. The tasks in the coming research were to be divided as follows:

- TNO: Co-ordination of the research and editing the final report. Calculation of the energy savings. A part of the measurements.
- Bouwcentrum: Research to social acceptance of rolling shutters. Cost/profit analysis. Research to the incorporation of rolling shutters in (existing) buildings.
- Bronswerk: Carrying out of U-value measurements of various rolling shutters.

The research has to be done by order of the (Dutch) Ministry of Housing and Physical Planning and by the Commission of the European Communities. In December 1981 Bronswerk, as a subcontractor to TNO, was instructed to prepare and start the measurement programme.

Some historical details about the relation between TNO and Bronswerk might be interesting. In the 1960's Bronswerk, as a contractor and manufacturer in the field of heating and airconditioning, was a customer of IMG-TNO, and many problems were solved in the TNO indoor climate testroom. With the increasing interest of airconditioning, and specially with problems of air-distribution, Bronswerk decided to build a testroom of its own. This decision was made in 1967. In the following two years TNO gave a generous support during the building process; TNO was specially in charge as the main-contractor and supplier for all the measurement devices (Figure I). Since the realization in 1969 of the Bronswerk Research Centre there is a permanent relation and exchange of information between the institutes, especially during improvements of laboratory facilities and instrumentation.

The U-value measurements programme at Bronswerk includes:

facade	rolling shutter	outside temperature ^oC
single pane	normal	+ 5
	normal	+ 12
	normal	− 3
double pane	normal	+ 5
	reflecting	+ 5
	reflecting	+ 12
	reflecting	− 3
	normal, with brushes between shutter and window frame	+ 5
	double	+ 5
	double	+ 12
	double	− 3
	double, with brushes	+ 5
	curtain (cotton)	+ 5
	curtain and rolling shutter	+ 5
double pane, no window-sill	no shutter; radiator at enlarged distance from parapet	+ 5
double windows	normal	+ 5
	reflecting	+ 5

facade	rolling shutter	outside temperature °C
double pane	double double (variant) horizontal alum. blinds, reflecting shutter outside shutter outside, reflecting	+ 5 + 5 + 5 + 5 + 5

2. Description of the installation and the measuring equipment.

2.1. The installation of the B.A.R.C., as far as this research is concerned.

A testroom, corresponding to the E.C.-specifications is constructed on the testfloor (figure II):

length: 4 m — glass area : 5 m2
width : 3 m — parapet : R > 1 m2 K/W
height: 2,5 m — other walls: R ≈ 3 m2 K/W.

On both sides of the permanent glass pane a wooden frame, of 0,1 m thickness, is mounted to obtain a facade of 0,2 m with the glass in the centerline. The facade plenum length is limited to 3 m by means of 2 closing panels. Air of the required temperature is supplied through circular apertures, and by means of a panel in the middle, an increased air flow is realized around the central panel and along the glass pane.

The supply of cold air in the facade plenum is indicated in figure III. A 29 KW-chiller chills the capacity of a 0,7 m3 storage vessel, filled with R11 as a heat transport medium. Two hot-gas-bypass valves serve to "kill" the compressor capacity to a great extent. From the storage vessel, and by means of a three-way-valve mixing system, the R11 medium is supplied to a fin-tube heat exchanger in the air circuit. The air system with a 4000 m3/h ventilator and 3 dampers, supplies the facade plenum with cold air. The air quantity is measured with a DIN-venturi and a BETZ micro-manometer. The pressure in the facade plenum is zero to the surrounding.

The supply of hot water to the radiator in the testroom is indicated in figure IV: from a 116 KW boiler the water circulates through a 1,5 m3 storage vessel. From the storage vessel and via a three-way valve system, the water is supplied to the radiator. The flow is determined by means of a rotameter.

2.2. The measuring equipment

In 1981, the 12 years old set of measuring equipment which was obtained from IMG-TNO, has been fully replaced by a new set of equipment, bought from Hewlett Packard. The present set consists of:

- a HP-85A micro computer with 32 K-byte memory.
- a 3497A data acquisition system with 3498A extender, giving a max. of 300 channels.
- a 9872C plotter, A3-size, for on-line presentation of results.

To the set, the following probes can be connected:

- thermo couples of the copper-constantan T-type.
- heat-flux meters of the TPD-type WS31 (106 mm Ø), with a calibration factor of about 4,5 W/m2.mV.
- low-velocity anemometers of TNO-type (and other types). Not used in this research.

In the testroom a total of 143 (147) channels is connected, 23 (27) heat-flux meters and 120 thermo-sensors. See figure V as an example of a complete experiment.

According to the E.C.-specifications and own requirements the measuring channels are divided as follows:

A. heat-flux

4 flux meters in each wall, roof and floor	20
3 flux meters in parapet	3
additional in calibration panel	(4)
	23
	(27)

B. temperature

radiator	water, in		3
	water, out		3
room temperature	4 x 4		16
wall temperature	5 x 4		20
facade plenum			4
air-supply facade plenum		air in:	4
		air out:	4
hall, outside the testroom		6 x 4	24
globe-temperatures			6
glass-surface		outside	6
		inside	6
surface parapet			3
surface radiator			2
air-temperatures between rolling shutters,etc.			
various distributions. max. channels:			18
			120

Apart from these 147 channels the water flow through the radiator and the air flow through the facade plenum are determined "personally" from the rotameter and the venturi, and fed into the computer.

3. Determination and calculation of the U-value. Reliability check of the result obtained.

The determination of the requested U-value cannot be measured directly. The heat flow through the glass pane with rolling shutter appears as a difference in a heat balance.

From the testroom and from the facade plenum, two separated heat balances are calculated (see figure VI). The heat balance of the testroom shows as: $\emptyset_1 + \emptyset_2 + \emptyset_3 + \emptyset_4 = 0$

Heat flows from inside to outside are calculated as positive.

$\emptyset_1$ - heat flow from the radiator
$\emptyset_2$ - heat flow through the parapet
$\emptyset_3$ - leak flow through floor, roof and other walls
$\emptyset_4$ - heat flow through glass pane with rolling shutter.

The heat flow $\emptyset_4$ is derived as: $\emptyset_4 = - \emptyset_1 - \emptyset_2 - \emptyset_3$
And from this the requested U-value as:

$$U = \frac{\emptyset_4}{A_4 \cdot (t_{globe} - t_{plenum})}$$

A4 : glass area
t-globe : globe temperature
t-plenum: outside air temperature.

The heat balance of the facade plenum shows as: $\emptyset_2 + \emptyset_5 + \emptyset_6 + \emptyset_7 = 0$

$\emptyset_5$ - heat flow through glass pane with rolling shutter.
Has to be equal to $\emptyset_4$.
$\emptyset_6$ - leak flow through "other" walls of facade plenum.
$\emptyset_7$ - heat flow from cold air supply.

The heat flow $\emptyset_5$ is derived as: $\emptyset_5 = - \emptyset_2 - \emptyset_6 - \emptyset_7$

Following the E.C.-specifications, the reliability of the measurement is calculated as:

$$\left| \frac{|\emptyset_4| - |\emptyset_5|}{\emptyset_4} \right| \leq 0{,}04$$

In other words: the relative difference between $\emptyset_4$ en $\emptyset_5$ has to be smaller than 4%.

4. Additional parameters to be measured.

4.1. In the E.C.-specifications only the determination of a U-value is asked. We considered it as useful (necessary) to get more detailed information about the behaviour of the model (glass/shutter/curtain/blind-combination). For this purpose a maximum of 18 temperatures in the model is registered: (see figure VII).

4.2. The U-values to be determined have a close relation, not only to the energy savings, but also to the human comfort to be obtained. With this background we installed the sensor of a "Comfy-test" (Reci) in the testroom at an equal distance to the facade as the E.C.-reference globe, to measure directly the Fanger-PMV (Predicted Mean Vote) at different values of clo-value and activity-level ("metabolism"). For a check of the "Comfy-test" some extra temperature measurements (such as the radiator surface) are made to be able to calculate the PMV with a computer programme. With the "Comfy-test" also the effective temperature (the "red scale") is measured directly for a check with the globe.

5. First results obtained in the present research.

Each test is performed after a waiting period of normally 36 to 48 hours after the last adjustments on the installation (which is longer than foreseen). After the waiting period, a perfect stationary situation is reached, and then the test is repeated 3 times with 1 hour intervals. The obtained stability can be checked by comparison.

The presented U-values are the mean values of the 3 executed measurements.

		U-values in W/m2K			R-values in m2K/W	
test no.	title	U_4 (glass + shutter)	U_2 (parapet)	U_f (facade)	R_4 (glass + shutter)	improvement Δ R
1	single glass no shutter + 5°C outside	4,83	0,479	3,40	0,207	-
2	single glass standard normal rolling shutter + 5°C outside	3,43	0,417	2,44	0,292	0,085
3	single glass standard normal rolling shutter + 12°C outside	3,43	0,448	2,45		
4	single glass standard normal rolling shutter - 3°C outside	not yet	executed			
5	double glass no shutter + 5°C outside	2,31	0,414	1,69	0,433	-

6	double glass standard normal rolling shutter + 5°C outside	2,07	0,396	1,52	0,483	0,050
7	double glass rolling shutter reflecting outside + 5°C outside	1,74	0,450	1,32	0,575	0,142
8	double glass rolling shutter reflecting outside + 12°C outside	1,73	0,459	1,31		

6. Analysis of the work performed.

On starting up the testroom and the measurement programme some additional provisions had to be made. The two main reasons to require these provisions were:

- The required accuracy of 4% in the heat balances, which is very strict (but useful).
- Notwithstanding a test area (glass) of 5m2 the heat flows in the balance show very low absolute figures (under 100 Watt, for example), which can change a small absolute error in a measurement to a high relative error!

As a consequence of these two items, we made the following provisions:

6.1. Physical properties of the testroom.

In the normal practice of the Bronswerk laboratory, checking and improving airconditioning installations, the testroom is built with standard partitions (R-value about 0,3 m2K/W). As the temperature in the lab. hall is kept constant and equal to the testroom-temperature, the "leakage" through these partitions is mostly under 100 W, which gives a satisfying heat balance with 2 or 3 KW power in the installation to be tested. The same 100 W-difference would render the present test useless. Furthermore, the airthightness could not be granted. Therefore we decided to design new partitions. The testroom, built up with these new partitions, is air-thight and has a R-value better than 3 m2K/W. The "leakage" is now reduced under 10 W, and can be determined with good accuracy.

6.2. New calibration of the facade plenum.

The first testruns resulted in disappointing heat balances. As we felt sure about the testroom, the facade plenum had to be guilty! See chapter 3 and figure VI for the heat balance of the facade plenum:

$$\emptyset_2 + \emptyset_5 + \emptyset_6 + \emptyset_7 = 0$$

The "leakage" $\emptyset_6$ is determined as:

$$\emptyset_6 = A_6 \cdot U_6 \cdot \Delta t$$

A_6 total area of "other" surfaces, except glass area and parapet.
U_6 known as 0,69 W/m2K, as a "mean" value for the total 10 m-length facade plenum.
Δt = t hal- t plenum.

We decided for a new calibration of this specific 3 m-length facade plenum with 2 closing panels. Therefore, a <u>measurement</u> of $\emptyset_5$ was necessary. Special panels to 5 m2 surface were constructed for a full cover of the glass area, and 4 extra heat flux meters were built in the panel. Now $\emptyset_6$ could be determined as the single unknown difference in the heat balance. The new determined U-value came out as U_6=0,63 W/m2K, a reduction of 8,7%. This value improved the heat balances in a sufficient way.

6.3. The thermal stabilization of the testroom.

The heat balances came out as still not fully satisfying. Where in our normal practice a time span of 4 to 12 hours is sufficient for a good balance, in the actual calibration of the facade plenum it came out that we needed 36 to 48 hours for a perfect stability ("to get the last 0,1 ^{o}C out of it"). This was against our expectance, our hall temperature being constant. The now followed procedure is that we check 10 important temperatures in the test set, every hour, and watch their stability. If there are no longer changes in the readings (36 to 48 hours after the last adaptions on the installation), three complete experiments are carried out, and calculated, automatically by the computer. The results are presented (on-line) at the plotter (see figure V, which is reduced from A3 to A4-size). See figure VIII for an illustration of the stabilization-process.

6.4. Rotameter for water-flow in radiator.

To improve the exactness of the rotameter, we chose a different floating element in the same tube. This reduced the maximum flow from 250 liter per hour to 100 liter per hour, thus decreasing the error in the measurement.

6.5. A heat balance under 100 W; consequences for the installation.

6.5.1. Cold air supply to the facade plenum. (see figure III)

A cooling power less than 0,1 KW had to be created. Practically the total power of the compressor is "killed" by the hot-gas-bypass valve. Following to the compressor, a 0,7 m3 storage vessel stabilizes the temperature, and the three-way mixing valve determines the resulting cooling power.

6.5.2. Hot water supply to the radiator. (see figure IV)

Again a power less than 0,1 KW has to be created with boiler of 116 KW normal capacity (supplying 90 ^{o}C water). The boiler is provided with a non-standard maximum thermostat (going as low as 25 ^{o}C) to provide a water temperature slightly higher than the requested temperature. The temperature is again stabilized in a 1,5 m3 storage vessel, and then a three-way mixing valve determines the water definite temperature.

7. Conclusions.

7.1. In relation to the measurement process:

- The determination of the thermal power of the facade plenum is restricted by the measurement of the Δ t of the air supply, which is small. With a measuring accuracy of 0,1 K in the Δt, a relative error of 10% or more can be derived. This also causes a relativation of the value
$$\frac{\phi_4 - \phi_5}{\phi_4}$$
- The determination of the requested U-value, via the heat balance of the testroom, has an accuracy better than 4%.
- A perfect thermal stability of the testroom, which is a necessity for a good heat-balance and reliable result, can be maintained over a long period.

7.2. In relation to the results obtained so far:

- A standard normal rolling shutter, mounted inside a single pane facade, causes an increase of the R-value of 0,085 m2K/W.
- The same, but mounted inside a double pane facade, causes an increase of the R-value of 0,050 m2K/W.
- The same, but with double pane and a rolling shutter reflecting on

the outside, causes an increase of R-value of 0,142 m2K/W.

- There is no significant difference in R-value with different outside temperatures (+ 5°C and + 12°C).

Figure I: The Bronswerk Airconditioning Research Centre.

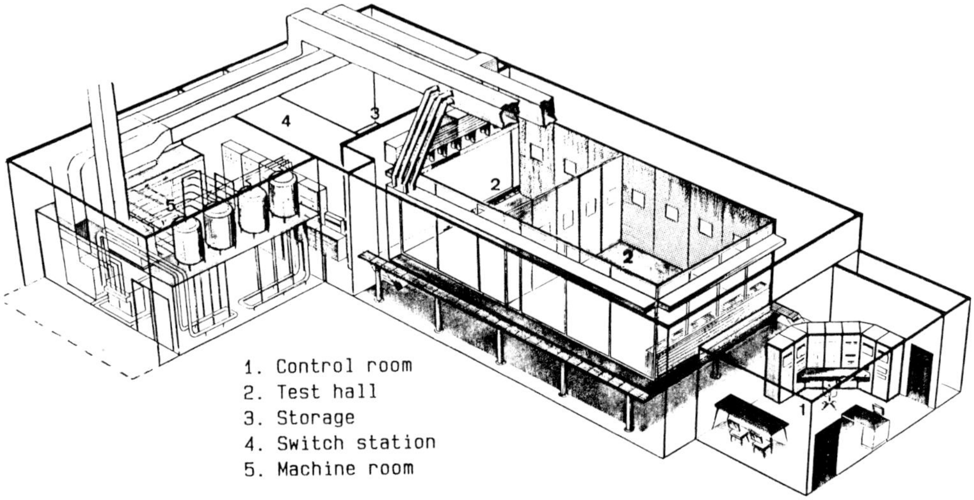

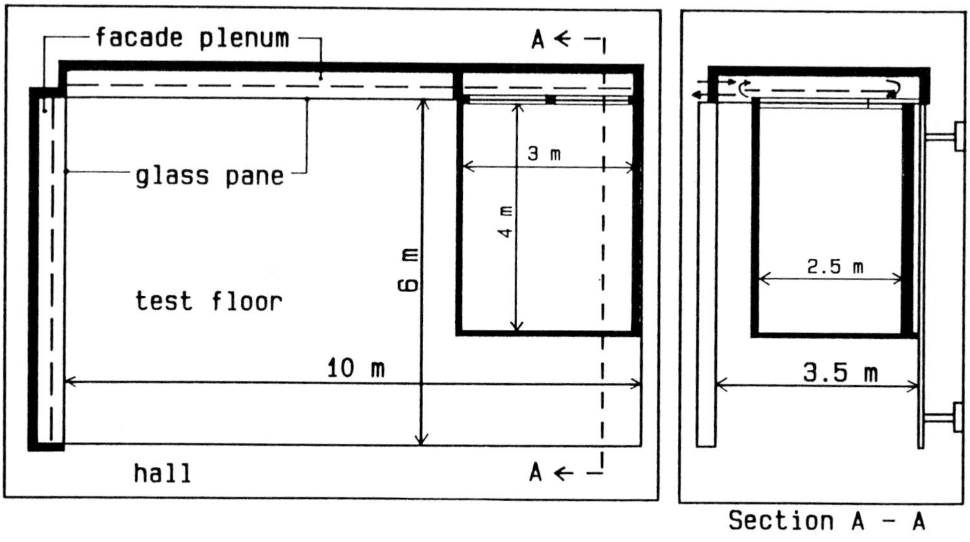

Figure II: Facilities of the Bronswerk Airconditioning Research Centre.
(testroom for the present research in red)

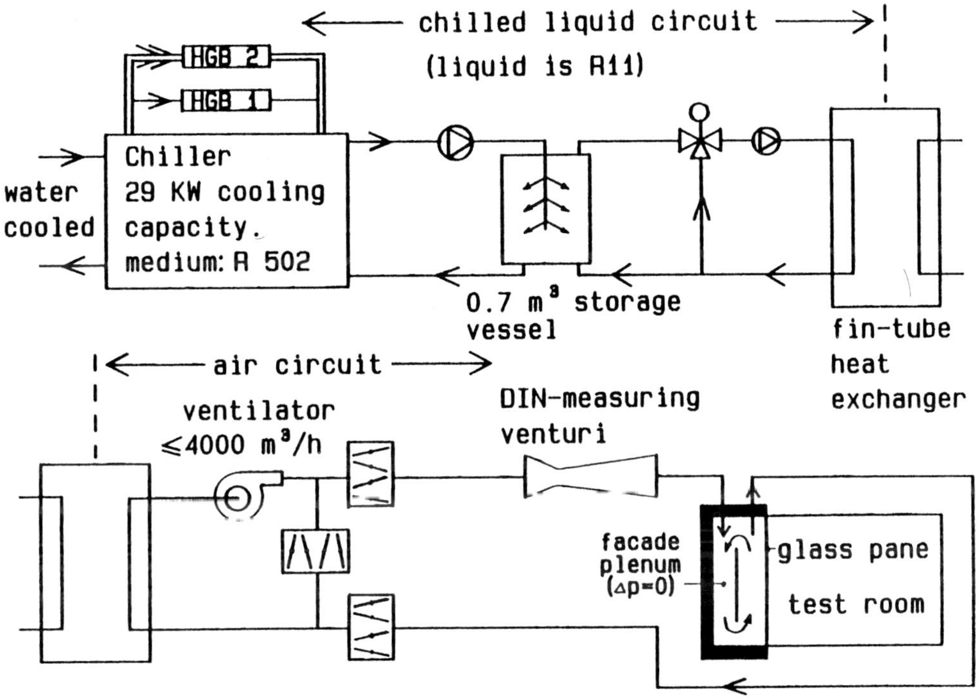

Figure III: Cold air supply.

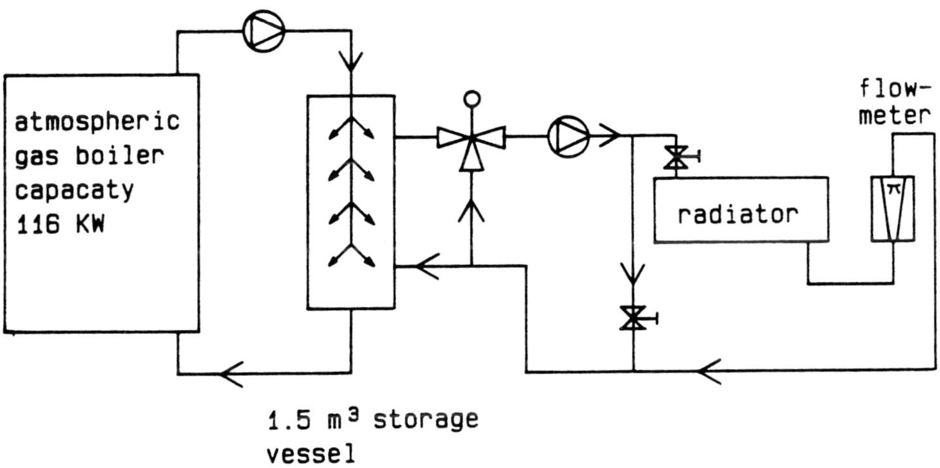

Figure IV: Hot water supply.

BRONSWERK AIRCONDITIONING RESEARCH CENTRE

ASSIGNMENT: ENERGY SAVING BY USE OF ROLLING SHUTTERS

TEST : Double glass - normal shutter - outside temperature + 5 deg. C

TEST NO : 6 DATE : 21-07-1982 TIME : 10:35 h EXP. NO : 17

description	chan.	values run	values mean	result
water temperature radiator in	66 88 89	37.1 37.0 37.1	37.1	Δt_w = 2.6
water temperature radiator out	60 65 72	34.5 34.4 34.5	34.5	
flow water l/h		54		
heat flow radiator				$Ø_1$ = -162.9 W
heat flow parapet (wall 2)	161 160 162	5.96 5.21 6.80	5.99	$Ø_2$ = 14.7 W W/m²
heat flow floor (wall 3,1)	143 142 141 140	-0.13 -0.13 -0.31 -0.29	-0.21	$Ø_{3,1}$ = -2.62 W W/m²
heat flow right wall (wall 3,2)	156 157 158 159	-0.19 -0.18 -0.37 -0.42	-0.29	$Ø_{3,2}$ = -2.94 W W/m²
heat flow left wall (wall 3,3)	154 152 155 153	-0.14 -0.09 -0.22 -0.20	-0.16	$Ø_{3,3}$ = -1.67 W W/m²
heat flow ceiling (wall 3,4)	150 149 148 151	-0.06 -0.09 0.15 -0.17	-0.04	$Ø_{3,4}$ = -0.53 W W/m²
heat flow back wall (wall 3,5)	144 146 147 145	-0.13 -0.17 -0.14 -0.11	-0.14	$Ø_{3,5}$ = -1.03 W W/m²
total heat flow other walls				$Ø_3$ = -8.8 W
heat balance test room				$Ø_4$ = 156.9 W
room temperature test room A	16 11 46 44	20.5 20.5 20.3 20.1	20.3	t 1 (200 / 150 / 75 / 10)
B	58 54 57 56	20.5 20.5 20.4 20.2	20.4	t 2 (200 / 150 / 75 / 10)
C (plan: A E B / C D)	51 50 45 48	20.4 20.4 20.3 20.1	20.3	t_K=20.3 t 3 (200 / 150 / 75 / 10)
D	40 53 52 49	20.4 20.4 20.3 20.1	20.3	t 4 (200 / 150 / 75 / 10)
temperature facade plenum	100 111 112 116	5.3 5.1 5.1 5.4	5.2	t_{rk} = 5.2
U-value tested shutter				K= 2.0 W/m² K
hall temperature (wall 6) behind fac. plenum	114 117 21 22	22.5 22.1 22.1 22.2	22.2	$t_{hal,6}$ = 22.2
heat loss fac. plen.				$Ø_6$ = -208.8 W
air temperature facade plenum in	103 106 109 120	4.0 4.0 4.0 4.2	4.1	Δt_{rk}=1.2
air temperature facade plenum out	101 102 104 115	5.3 5.3 5.3 5.3	5.3	
flow fac.plen. kg/h		1152		
heat supply facade plenum				$Ø_7$ = 385.1 W
heat balance facade plenum				$Ø_5$ = -161.6 W
inaccuracy heat balance				3.0 %

description	chan.	values run	values mean	result
globetemperatures A	41	20.3		t_{z1} = 20.3
B	43	20.3		t_{z2} = 20.3
C	47	20.3		t_{z3} = 20.3
D	55	20.4		t_{z4} = 20.4
E	1 13	20.3 20.4	20.4	$t_{z,ref}$ = 20.4
wall temperature 3,1	83 86 87 88	20.2 20.2 20.2 20.3	20.2	$t_{0,1}$ = 20.2
wall temperature 3,2	81 82 85 92	20.4 20.3 20.2 20.4	20.3	$t_{0,2}$ = 20.3
wall temperature 3,3	88 89 91 95	20.4 20.3 20.4 20.4	20.4	$t_{0,3}$ = 20.4
wall temperature 3,4	84 90 93 42	20.3 20.4 20.4 20.3	20.3	$t_{0,4}$ = 20.3
wall temperature 3,5	80 94 96 97	20.3 20.4 20.4 20.4	20.4	$t_{0,5}$ = 20.4
surface temperature parapet	20 26 29 37	21.7 21.0 19.9 19.4	20.5	t_b = 20.5
surface temp. rad.	2 5	32.3 33.8		
temperature glass surface inside	23 31 32 33 35 36	14.2 12.9 11.7 12.3 12.6 13.5	12.9	t_g = 12.9
temperature glass surface outside	105 107 108 110 113 118	7.1 7.2 6.8 8.0 7.6 7.8	7.4	t_g = 7.4
hall temperatures	70 77 27 28	20.6 21.2 21.1 20.6	20.9	$t_{hal\,1}$ = 20.9
	24 25 34 38	20.6 20.7 21.5 21.5	21.1	$t_{hal\,2}$ = 21.1
	62 63 67 75	20.3 20.7 20.6 20.8	20.6	$t_{hal\,3}$ = 20.6
	30 61 71 76	20.8 20.3 20.6 19.6	20.3	$t_{hal\,4}$ = 20.3
	64 73 74 78	20.4 20.3 20.7 20.4	20.5	$t_{hal\,5}$ = 20.5
temperatures in test model	0 3 7 9 10 14	13.7 18.3 15.0 14.1 13.4 12.7		

Figure V: Presentation of the test results.

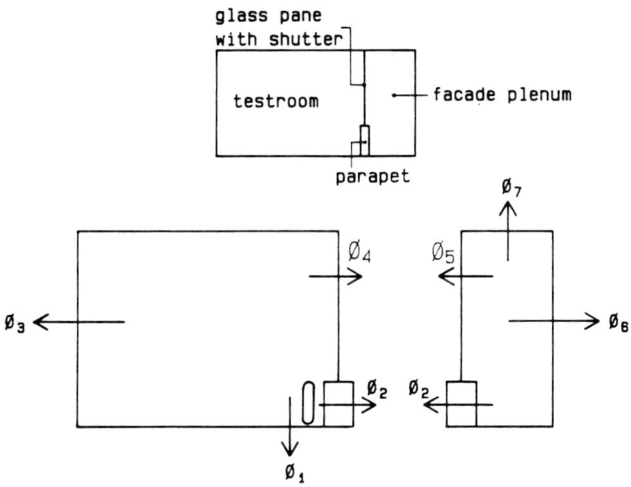

Figure VI.

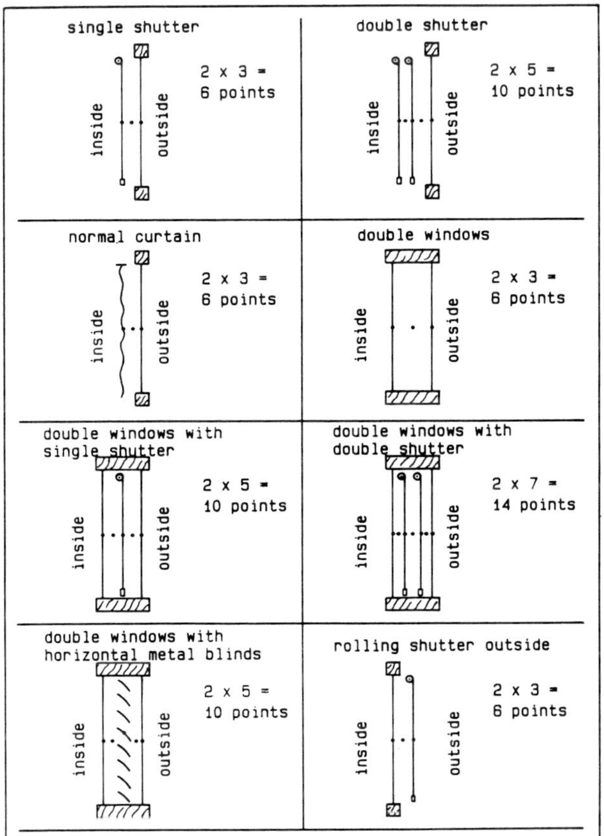

Figure VII : Additional measurements in model.

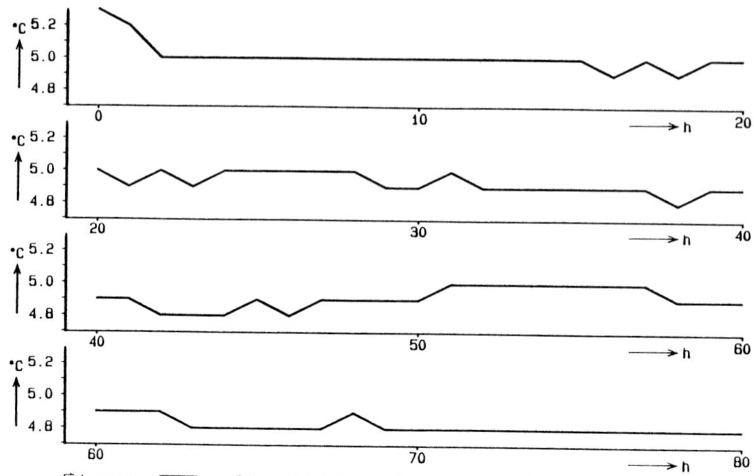

Figure VIII: Stability of temperatures
(Example: The air temperature in the facade plenum)

BENEFITS OF INSULATING BLINDS AND THE IMPORTANCE OF EDGE SEALS

Authors : J.G. LITTLER, P.A. RUYSSEVELT

Contract number : EEA-2-020-82-UK

Duration : 24 months May 82 - April 84

Head of project : Dr. J.G. Littler

Contractor : Building Unit, Polytechnic of Central London

Address : Polytechnic of Central London
35 Marylebone Road
GB - London NW1

Summary

The aim of this project is to research and develop a form of window insulation for retrofit to existing buildings. After preliminary investigation it was established that a roller blind system would be the most suitable for this application for the following reasons:

- It could easily be adapted to different shapes and sizes of window.
- The whole of the window aperture could be enclosed.
- The storage space required when not in use is minimal.
- It would be easy to operate.
- It could be cheap to produce.
- If well designed the system would be an attractive addition to the window.

Having decided upon a roller blind system it was necessary to evaluate the energy saving potential of various constructions before progressing to a prototype.

A significant source of heat loss through windows in the U.K. is by air infiltration. Testing was carried out to assess the ability of various edge seals to reduce air infiltration through windows. The results of the air infiltration testing were used in the computer simulation of a 3-bedroomed, semi-detached house with different types of insulating blinds applied to the windows. From the results of the simulations a comparison was made between the various types of insulating blind and secondary double glazing. It was found that one blind in particular proved to be substantially more effective than secondary double glazing. This blind is now being developed to the prototype stage for air infiltration and thermal testing.

1.1 Introduction

The need to reduce heat loss through windows is well established. The main sources of heat loss through windows are radiation, conduction and convection, their combined effect is represented by the U-value of the window. However in the U.K. heat loss due to air infiltration through the window and around the frame is a significant factor. To be effective an insulating blind must reduce not only the U-value of the window, but also the heat loss due to infiltration.

This paper analyses various methods of edge sealing insulating blinds and shows that effective edge sealing is worthwhile.

1.2 Air Leakage Testing

To assess the effectiveness of various methods of sealing a single layer insulating blind an air pressure chamber built to test air leakage to BS 4315:Part 1:1968 was used. However the procedure involved in BS 4315:Part 1:1968 was designed to represent very severe conditions and was not suited to this work. The chamber consisted of a well sealed box with an aperture in which components were fixed. The chamber was fitted with rotameters to measure air flow (lm^{-1}) and a manometer to measure the pressure drop from inside to outside (mm water).

The chamber was modified in two ways. Firstly a completely sealed frame was added to the aperture to allow various seals to be applied to both faces of a thin plastic film which was used to represent an impervious blind material (Plate 1 and Fig.1). The top and bottom of the frame were designed to seal the plastic film completely whilst adjusting to the thickness of the seals being tested (Fig.2). Secondly it was necessary to increase the accuracy of reading on the manometer, this was achieved by focussing a travelling microscope with a built-in scale on the column of liquid thus allowing readings to be taken to 0.05 mm.

The following procedure was adopted for the tests. In each case the pressure was increased to 2mm water and then reduced to zero. Air flow readings were taken at 0.25, 0.5, 0.75, 1.0, 1.5, 2.0 mm water, this range is equivalent to 0 to 19.2 Pa and is representative of pressures commonly experienced in a UK suburban situation. The first test was performed with the window aperture completely sealed, the readings taken were the 'blank' readings to be subtracted from all subsequent test readings. Further tests were carried out on the components described in table 1 and the results can be seen in Figure 3.

1.22 Friction Testing

These tests were performed to assess how much various seals would impede the free movement of the roller blind. A 25mm diameter, aluminium spring roller was used in conjunction with a plastic film .13 mm thick and a plastic coated, aluminium bottom rail 3 mm thick. The blind was tested in 715 mm heights of the seals set out in table 1. The roller was first pre-tensioned until the blind returned freely through the whole height of the seals and the number of revolutions was noted. Then the blind was loaded up and notes were taken of the forces required to move the blind a further 10mm when 150, 300, 450 and 600 mm were trapped in the seals.

1.23 Computer Simulation

To compare the energy saving potential of various insulating blinds and different methods of edge sealing with secondary double glazing a computer model was employed. This model, called Suncode was developed

see fig. 1
see fig. 2
Plate 1

Fig 1. Section at head of frame (base detail similar)

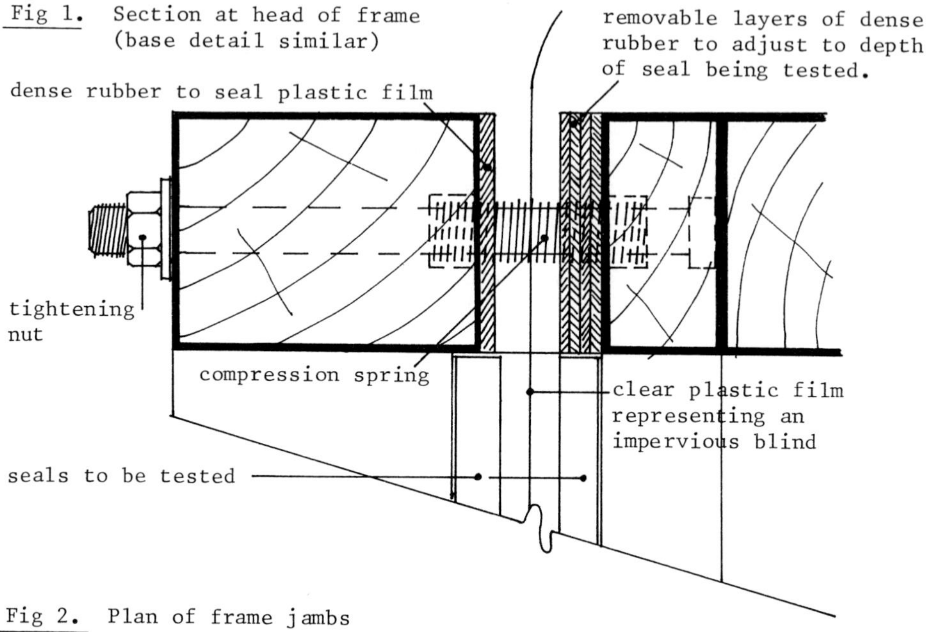

Fig 2. Plan of frame jambs

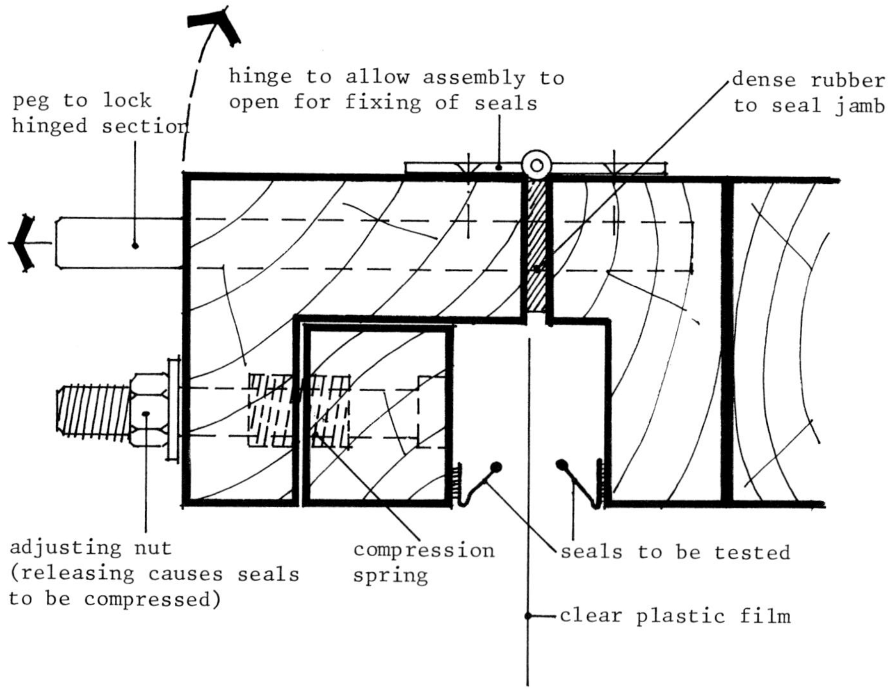

by the Ecotope Group, Seattle, USA and has the backing of ETSU in Great Britain. The hourly simulation allows building materials to be represented in both conductance and capacitance, it takes account of solar radiation and the overshadowing effect of the surrounding environment. Suncode has the facility to schedule various elements such as the temperature setpoint of the heating system and the internal heat gains, so that the occupancy pattern of the building can be accurately represented. This scheduling facility was used to represent insulating blinds with edge seals being drawn, this was done by reducing both the U-value of the windows and the air change rate of the building for given time periods.

Suncode was used to model the performance of a typical 3-bedroomed, semi-detached house with a total floor area of 115m^2 and a glazed area of 24m^2. The construction of the house was taken to be pre 1982 U.K. Building Regulations giving the following U-values: walls = 1.0 Wm^{-2}K^{-1}; roof = 0.7 Wm^{-2}K^{-1}; floor = 1.3 Wm^{-2}K^{-1}; single glazing = 5.6 Wm^{-2}K^{-1}. The house was assumed to be on a suburban estate and therefore overshadowed to varying degrees in all orientations. Weather for the London area was used in the simulation.

The first computer run (Fig.4 line a) was performed on the house as described so far, with a winter air change rate of 1.5 h^{-1}. The second run (Fig.4 line b) was performed with the introduction of secondary double glazing, reducing the window U-values to 2.8 Wm^{-2}K^{-1} and the winter air change rate to 1.2 h^{-1}. The air change rate was reduced on the basis that the closed windows of the house contribute 40% of the total air change rate of the house* (footnote) and the edge seals in this case are 50% effective. The next three sets of runs were performed with insulating blinds applied to the windows through the hours of darkness during the winter, the resultant U-values were varied from 2.8 Wm^{-2}K^{-1} to 1.0 Wm^{-2}K^{-1}. In the first set of runs (Fig 4. line c) the insultating blinds were assumed to have no edge seals and the winter air change rate remained at 1.5 h^{-1}. The second set of runs (Fig 4. line d) had insulating blinds with edge seals that were 50% effective thereby reducing the night-time, winter air change rate to 1.2 h^{-1}. The insulating blinds in the third set of runs (Fig 4. line 3) had seals which were 85% effective and the night-time, winter air change rate was reduced to 1.0 h^{-1}.

1.3 Results

1.31 Results of Air Leakage Testing

The results of the air leakage tests are given in Fig 3 and Table 1. The leakage rates in m^{3}h^{-1} per m crack length were calculated by dividing the total air leakage rate for the components by the crack length. In the case of the window the crack length was 2.34 m and the area of the window as 0.39 m^2. For the combinations of the window and seals the shorter crack length of the seals, 1.43 m was used.

1.32 Results of Friction Testing

All the seal types set out in Table 1 were tested. The seals type B allowed the blind to be drawn and freely returned only when the roller was pre-tensioned by 4 revolutions. At a point where 600mm of blind was

* The figure of 40% of the total air change rate is based upon the air leakage around the openable sections of the window and between the frame and the wall. Both these paths of infiltration were assumed to be sealed by the addition of secondary double glazing and the insulating blinds. 40% is quite a common figure due to the poor standard of house building in general and window manufacture in particular in the U.K.

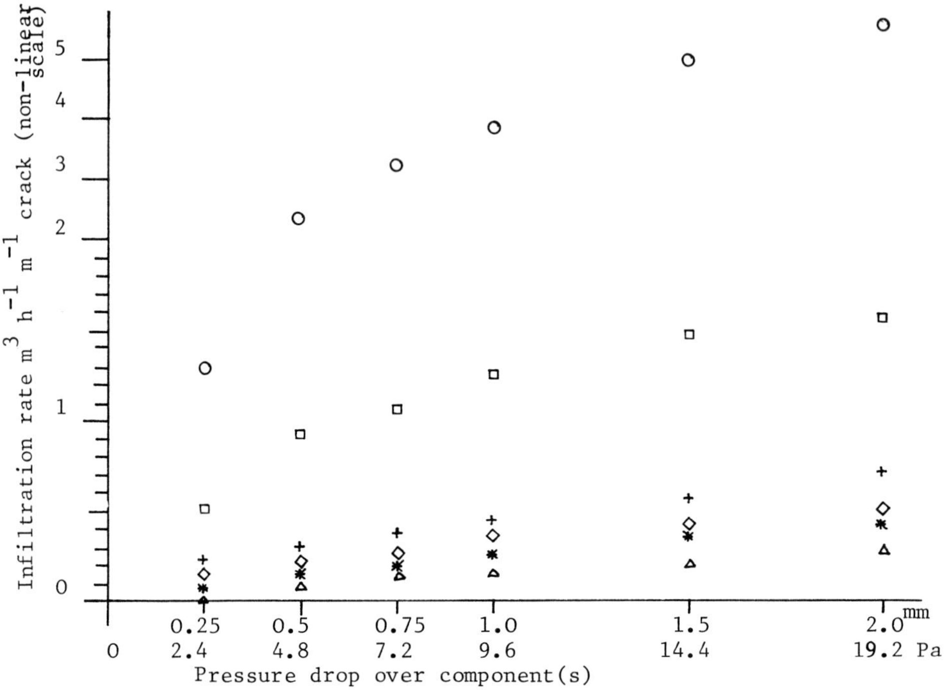

Fig 3 Results of air leakage testing

Table 1 Components tested in air leakage testing

			% reduction on A in the range 0-19.2 Pa	Deviation +	Deviation −
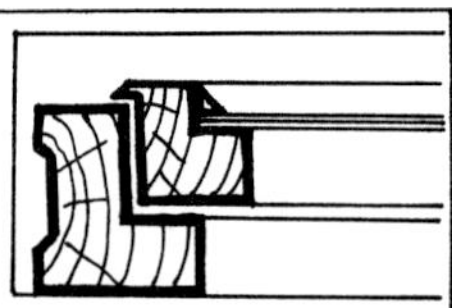	○	A side hung, timber, casement window, non-weatherstripped			
Components B-E were tested in series with A.					
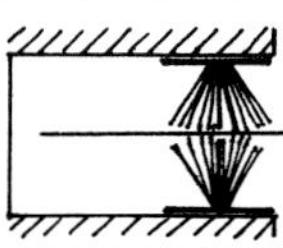	□	B Brush seals, 5mm deep with polythene core, self adhesive, compressed to 9mm.	66.6	5.8	5.7
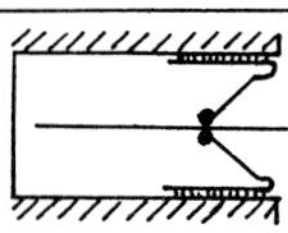	+ ◇	C P.V.C. seals, 7mm deep, self 1. adhesive, compressed to 1. 13.5mm 2. 2. 12 mm	87.8 91.2	1.7 1.1	3.8 1.4
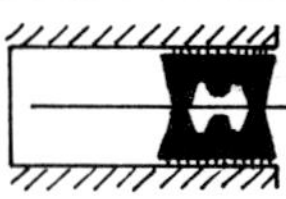	△	D E.P.D.M. synthetic rubber seals, 4mm deep, self adhesive, compressed to 7mm.	96.2	3.8	1.5
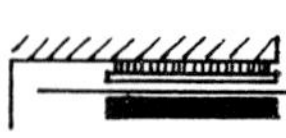	✱	E Flexible magnetic strip, 14 x 1.5mm, over plastic film, attracted to self adhesive ferrous strip	93.6	3.3	1.2

trapped in the seals a force of 2 Kg. was required to move the blind a further 10mm. The seals type C1 allowed the blind to be drawn and returned when the roller was pre-tensioned by 6 revolutions. At a point where 600 mm of blind was trapped in the seals a force of 3.5 Kg was required to move the blind a further 10mm. The seals types C2, D and E did not allow the blind to move at all.

1.33 Results of Computer Simulation

The results of the computer simulations can be seen in Fig 4 and Table 2.

1.4 Analysis of Results and Comments

The results of the air leakage tests show that quite significant reductions in the air leakage rates through windows can be achieved by various types of seals applied to a simple roller blind. The most effective seals tested were the magnetic and E.P.D.M. types, however these did not allow the free movement of the blind and therefore could only be applied once the blind was fully drawn.

From Table 2 it can be seen that blind c2.8 which has no edge seals may result in quite significant reductions in energy consumption. Based on recent work at Cardiff University (1) the U-value of 2.8 $Wm^{-2}K^{-1}$ could be achieved by a single layer material with one low emissivity surface facing the window creating a freely ventilated air-space approximately 20 mm wide. If side and top brush seals together with a bottom magnetic seal were added to blind c2.8 the air leakage rate of the windows could be reduced by 50%. The U-value of blind c2.8 would also be reduced to $\sim 2.3\ Wm^{-2}K^{-1}$ because air movement between the blind and the glass would be reduced by virtue of the edge seals. Blind c2.8 with edge seals would therefore correspond to blind d2.3 from Table 2 which achieves greater savings, but would be more expensive to produce and not as easily operated.

The most effective blind simulated was blind e1.0 from Table 2, this achieved greater savings than secondary double glazing.

1.41 Construction of Blind Type e1.0

Fig 5, and following calculations show how a blind with a U-value 1.0 $Wm^{-2}K^{-1}$ when applied to single glazing can be constructed.

With reference to Fig 5 the U-value of the blind

$$U = \frac{1}{r_{so} + r_{a1} + r_{a2} + r_{si}}$$

$$U = \frac{1}{0.056 + 0.180 + 0.60 + 0.123} = \frac{1}{0.959} \sim 1.0\ Wm^{-2}K^{-1}$$

To achieve the 85% edge sealing required a combination of EPDM and magnetic seals could be used but these may only be applied once the blind is drawn. A means of applying these seals using no action other than that of fully drawing the blind, is being developed at PCL. Both layers of material would roll down from one roller under the weight of the bottom rail, controlled by a ball chain and cog.

1.42 Benefits of Blind Type e1.0

The blind type e1.0 achieved greater saving than secondary double glazing according to the results of the computer simulation. In order to compare like with like the secondary double glazing must be easily openable i.e. hinged or sliding. It is very likely that blind e1.0

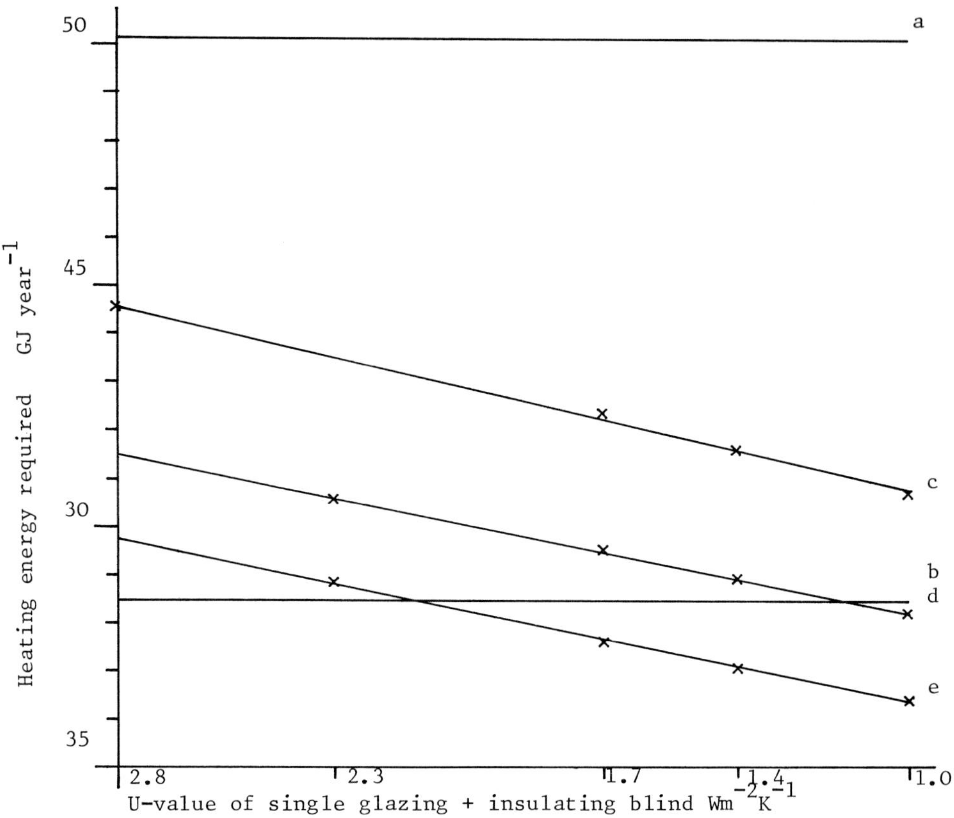

Fig. 4 Results of computer simulation (refer to table 2)

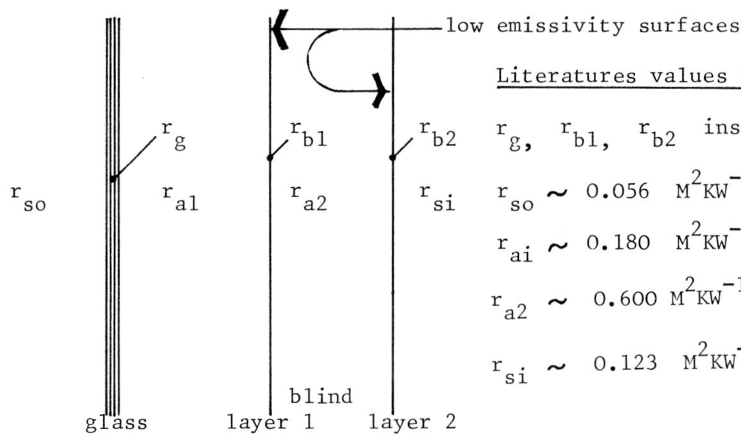

Literatures values for resistances

r_g, r_{b1}, r_{b2} insignificant

$r_{so} \sim 0.056$ M^{2}KW^{-1}

$r_{ai} \sim 0.180$ M^{2}KW^{-1}

$r_{a2} \sim 0.600$ M^{2}KW^{-1}

$r_{si} \sim 0.123$ M^{2}KW^{-1}

Fig.5 Construction of insulating blind e1.0

Line Type and No.	Window construction in 3-bedroomed semi-detached house	Window U-value $Wm^{-2}K^{-1}$ day	Window U-value $Wm^{-2}K^{-1}$ night	Winter air change rate h^{-1} day	Winter air change rate h^{-1} night	heating energy GJy^{-1}	reduction on (a) $GJ\ y^{-1}$	energy saving per m^2 glazing $GJ\ y^{-1}$	cost saving per m^2 glazing £15.30 GJ^{-1} £
a	single glazing	5.6	5.6	1.5	1.5	50.036			
b	a + secondary double glazing	2.8	2.8	1.2	1.2	38.422	11.614	0.484	7.41
c 2.8	a + insulating blinds	5.6	2.8	1.5	1.5	44.416	5.62	0.234	3.58
c 1.7	"	5.6	1.7	1.5	1.5	42.272	7.764	0.324	4.95
c 1.4	"	5.6	1.4	1.5	1.5	41.690	8.346	0.348	5.32
c 1.0	"	5.6	1.0	1.5	1.5	40.916	9.120	0.380	5.81
d 2.3	a + insulating blinds	5.6	2.3	1.5	1.2	40.626	9.410	0.392	6.00
d 1.7	"	5.6	1.7	1.5	1.2	39.469	10.567	0.440	6.74
d 1.4	"	5.6	1.4	1.5	1.2	38.852	11.184	0.466	7.13
d 1.0	"	5.6	1.0	1.5	1.2	38.091	11.945	0.498	7.62
e 2.3	a + insulating blinds	5.6	2.3	1.5	1.0	38.764	11.272	0.470	7.19
e 1.7	"	5.6	1.7	1.5	1.0	37.615	12.420	0.518	7.92
e 1.4	"	5.6	1.4	1.5	1.0	37.043	12.993	0.541	8.28
e 1.0	"	5.6	1.0	1.5	1.0	36.282	13.754	0.573	8.77

Table 2 Results of Computer Simulation

could be produced at a lower cost than hinged or sliding secondary double glazing. The blind would not reduce the winter solar gain and in a building with more south facing windows or less overshadowing than the one simulated this would be more significant. Because the seals are only applied when the blind is drawn the free movement of the blind is not impeded at all. The surface temperature of the blind would almost certainly be higher than that of double glazing or single glazing. In theory if a room has a large area of glazing drawing the blind on a cold evening would allow a reduction in the ambient temperature whilst still maintaining thermal comfort. The risk of night-time condensation on the windows is greatly reduced as the edge seals would prevent warm moist air reaching the cold surface of the glass.

2. Conclusions

Having analysed various methods of edge sealing and calculated their effect when applied to different insulating blinds the following conclusions may be drawn:

(i) The most effective forms of edge sealing are EPDM and magnetic, however these may only be applied once the blind is fully drawn.

(ii) The application of effective edge seals to a two layer, low emissivity, insulating blind would drastically reduce not only the U-value of the window but also the heat loss due to air infiltration. This particular insulating blind would be more beneficial than secondary double glazing and possibly cheaper.

References

1. An Energy-Efficient Roller Blind, A.R. Bates et al, Department of Physics, University College Cardiff, G.B. Paper presented at the Las Vegas Window Energy Show, August 1982.

QUESTIONS AND DISCUSSION

Helcké (CCR) : Have you considered the effect of sudden pressure changes (such as due to a door being opened) on the side seals of your blind ?

Answer : We have tested our seals up to 20 mm water and only the magnetic seal failed (at 12 mm water). However this test does not accurately represent the effect of a sudden increase on pressure. Gusting tests will be performed on the prototype blind when it is constructed. We believe that the combination of effective seals and a side channel of adequate depth should prevent the blind material from blowing out.

Byberg (DTH) : What is the width of your test facility ?

Answer : The overall width of our test facility is 1200 mm. However the window described in this paper was 650 mm wide. When the prototype blind is constructed it will be tested on a wider window.

SESSION VI - NATURAL VENTILATION

Chairman: J. UYTTENBROECK

THE MEASUREMENT OF AIR INFILTRATION RATES IN LARGE ENCLOSURES AND BUILDINGS

Authors : C.J. BRAYSHAW, J. DEWSBURY

Contract number : EE-A-5-051-GB

Duration : 24 months 1 September 1981 - 31 August 1983

Head of Project : Mr. I.N. Potter

Contractor : The Building Services Research and Information Association

Address : Old Bracknell Lane West,
Bracknell,
Berkshire RG12 4PU,
England.

Summary

The aims of the project are to develop a method of measuring the rate of air infiltration in large single cell enclosures and buildings an to use the method to collect data on actual infiltration rates. A survey has been made of existing and novel methods of measuring air infiltration rates. Two methods which have been judged to merit further consideration have been identified. The first method is based on the use of methane as a tracer gas, measuring its concentration with an infrared laser technique. The second method is based on the use of ethanol vapour as a tracer gas, measuring its concentration with a fuel cell detector. Initial trials of the latter method are now in progress.

1 INTRODUCTION

1.1 General

The effect of air infiltration as a ventilation source and its contribution to energy losses in buildings has been recognised as an important research topic in recent years. Since the early definitive work of Dick (1) in 1949, research in the field of air infiltration has rapidly gained momentum. The current awareness of the finite nature of primary energy sources, in particular fossil fuels, has given rise to a climate of energy consciousness which has injected additional impetus to the research of air infiltration in buildings.

A survey carried out by the UK Building Research Establishment (2) has shown that buildings in the United Kingdom, in both the commercial and domestic sectors, account for at least 40% of the total consumption of prime energy sources. Furthermore, the space heating requirement in the domestic sector consumes an average of 64% of the buildings energy requirement. As the standard of building insulation is improved due to regulatory requirements and better construction techniques, the effect of heat losses due to air infiltration effects can become the dominant source of heat loss.

The bulk of research regarding air infiltration carried out in the UK and elsewhere has concerned domestic dwellings. Several reviews of the current status of research into air infiltration phenomena have been carried out (3, 4, 5). Ross and Grimsrud (4), and Liddament (3) have both highlighted the absence of any significant practical work in progress in large commercial or industrial buildings, although procedures predicting air infiltration and air movement in such buildings have been developed (6) but these require substantiating. Some studies relating to air infiltration in office block type environments have been, and are currently being undertaken e.g. (7, 8, 9), but none of these are in large open plan buildings, and in particular those relying solely on natural ventilation. It has been suggested that the absence of such work is, in part, indication of the practical difficulties involved. The purpose of this report is therefore to critically examine the alternative proven and also novel techniques of measuring air infiltration in order to seek to find the optimum solution to the problem of measurement in large enclosures. The term large in the context of this programme of research is considered to include buildings of internal volume of 1000 m^3 or more.

An excellent review of air infiltration measurement techniques was published some years ago by Hitchin and Wilson (10), in which the currently available and proven methods were detailed, and the shortcomings and difficulties associated with each method highlighted. Since that time, significant advances in the field of instrument technology have been made. However, the fundamental physical problems associated with the measurement of air infiltration remain.

1.2 Air infiltration in large buildings

Air infiltration is a natural phenomenon which occurs in all types of building or enclosures unless stringent actions are taken to eliminate the paths by which adventitious air flows can take place. It is known to be a function of the parameters detailed below, although the list is by no means exhaustive, since the exact physical processes by which infiltration may take place within the macroenvironment of the structure are not clearly understood.

i) Building geometry and orientation

ii) Standard of building construction.

iii) Prevailing wind conditions; wind speed and direction.

iv) Internal to external temperature difference.

v) Internal pressure distribution.

In order to measure air infiltration, some method of differentiating between the internal and external air must be sought. Conventionally, this may be achieved by altering either the chemical or physical qualities of the internal air, since to achieve similar results with the outside air would not be technically or economically viable due to the problems of scale.

Numerous methods of altering the properties of the air have been used in the past, using some form of tracer gas to 'seed' the air, with varying degrees of success. The type of tracer gas chosen for individual applications has depended on two factors:

i) the environmental suitability of the gas

ii) gas detection techniques

The participants in this meeting will be familar with the standard measurement methods. These are the concentration decay method,the constant concentration method and the constant emission method. However, because of their volume, large buildings, particularly those of single cell construction, exhibit a number of fundamental problems.

i) The local ventilation rate may vary significantly from one part of the building to another.

ii) The high ceilings associated with commercial buildings may create substantial temperature gradients within the space, thus creating recirculation.

iii) The large enclosed volume coupled with normal construction practices of commercial buildings resulting in high ventilation rates may require extremely large quantities of tracer gas to be injected in order to achieve reliable measurements. Mixing of the tracer gas may also be a problem.

iv) A large enclosed space will require a multiplicity of sampling points, which will require long sampling tubes if a central detector unit is utilised. The distribution of sampling points within the test space will be a crucial factor, and may be influenced by the method of analysis adopted. The number of sampling points selected has depended historically on constraints imposed by the detector system. For example, an infrared absorption detector for tracer gases would normally require up to two minutes to complete a measurement, hence in a nodal system of ten discrete measuring stations, each could be serviced at a rate of once per twenty minutes. One of the schemes proposed in this paper enables multipoint sampling to be carried out at the same moment. In this context, the distribution of sampling points will therefore depend on technical considerations and not instrumentation constraints.

The extent of these problems is not known at this time, although certain research groups are known to have carried out some encouraging preliminary studies. It is therefore envisaged in this present study that trials will be conducted in a controlled space in order to investigate these problems before full scale measurements are implemented.

1.3 Tracer gases

The participants in this meeting will be familar with the properties desired of a tracer gas, and the present popularity of nitrous oxide and sulphur hexafluoride.

Novel tracer gases which are being considered are ethanol vapour and methane.

Ethanol has a molecular mass of 46.07, compared with 28.97 for the mean molecular mass of dry air. The vapour has a lower explosive limit of 43000 ppm and an upper explosive limit of 190000 ppm. The liquid is highly inflammable. The threshold limit value (ACGIH 1980) for ethanol vapour is 1000 ppm. It is not believed to have any long term harmful effect when inhaled in concentrations below the threshold limit value.

Methane has a molecular mass of 16.04. It has a lower explosive limit of 53000 ppm and an upper explosive limit of 140000 ppm. It is a simple asphyxiant, so the acceptable concentration in air is determined by the availability of oxygen. The minimal oxygen content should be 18%, compared with somewhat over 20% in fresh air, which gives a maximum acceptable methane concentration of about 2% or 20000 ppm.

1.4 The problem of mixing

Many researchers have expressed their concern towards the problems of mixing of tracer gas with air in the test space (10, 11, 12, 13, 14).

The standard tracer gas techniques assume perfect mixing both initially and throughout the measurement process. In practice, perfect mixing is not achieved and the air infiltration rate is usually approximated. Essentially there are three basic types of mixing problems associated with tracer gas measurements:

i) the mixing of tracer gas with the internal ambient air,

ii) the mixing of infiltrating air with the internal air/tracer gas,

iii) the circulation of air within the test space.

It is common practice to install a tracer gas measurement system with the injection and sampling points spatially separated, particularly in the case of multi-node systems. Consequently there is an inherent time delay between the time the tracer gas is injected into the test space, which may be a sub-compartment in the test space and the time the additional tracer volume is detected as an increase in concentration at the sampling point. This results in an apparent increase in infiltration rate until the tracer is well mixed.

The mixing of the infiltrating air into the test space may result in two effects. When infiltrating air enters the test space it may not be evenly dispersed, which results in a variation in tracer gas concentration from point to point. The use of a multi-node sampling system is often used to minimise this effect. The second effect that may occur concerns infiltrating air which exfiltrates out again without mixing. This phenomena will not result in a change in tracer concentration and hence ths calculated infiltration rate. Since this 'short-circuiting' does not affect the indoor air quality or building heat load, the phenomena may not be a significant problem.

The circulation of air within the test space may also constitute a problem when administering air infiltration measurements.

It is usually assumed that the whole of the physical volume of the test space participates in the infiltration phenomenon. If however, there are parts of the test space which do not participate in the process of infiltration, then the effective volume of the test space will be reduced. Many of these volumes may be easily recognised and some compensation made when determining the volume of the test space, e.g. closed cupboards, packing cases, etc.

Other areas which may not be participating in the air infiltration process may not be so easily recognisable. For example, in large industrial buildings with high internal ceilings, stratification vaused by temperature effects may isolate a significant volume adjacent to the ceiling.

Gale (13) has discussed the use of mixing fans to minimise the homogeneity problems. Two techniques are usually employed:

i) to inject the tracer gas into the test space, mix it with a fan, and then turn off the fan before undertaking the analysis, and

ii) to mix the air continually with a fan.

The former approach avoids disturbance of the natural paths or air movements due to convection whilst the latter approach gives a better average result but may affect the actual air infiltration rate. Some researchers have adopted the use of central ventilating/heating system to distirbute the air-tracer gas mixture, which is initially well mixed using the ventilation system fan or blower, into various points within the test space (8, 15, 16, 17).

Dick (18) and Honma (19) found that tracer concentrations in single rooms were relatively uniform if auxiliary fans were used, while Kelnofer et al (20) performed spot checks during their infiltration trials and found minimal concentration variation throughout the building under investigation.

As summarised by Gale, the advantages of mixing fans are that the tracer gas concentration is uniform and stratification is minimised whereas the disadvantages are that the natural air flow paths are disrupted and streams of air may impinge upon room openings.

1.5 The application of air infiltration measurement techniques to large enclosures

None of the standard techniques for the measurement of air infiltration are directly suitable for use in the application of measurement in large enclosures without additional development.

Modern advances in instrumentation techniques have removed the fundamental problem associated with early tracer gas measurements, viz the detection of the tracer gas in very low volumetric concentrations. Indeed, the detection of concentrations in the order of 1 ppm are now commonplace. The critical problem in the adoption of tracer gas techniques to the measurement of air infiltration rates in large spaces concerns mixing, both the initial mixing of the tracer within the confines of the test environment, and also the mixing of the incoming infiltration air.

It is known that some work, as yet unpublished, is in hand in the UK to investigate this problem. It is envisaged, however, that considerable effort will be required during the laboratory phase of the current project to investigate fully the problems associated with this phenomena.

2. PROPOSED MEASUREMENT TECHNIQUES

2.1 Detection of infrared absorption using laser techniques

The characteristic peaks in the infrared spectrum exhibited by most gases has been utilised in the field of air infiltration measurement for many years, and many infrared gas analysers are currently available. However, the use of laser detection equipment is novel in this field, and the methods proposed offer several unique advantages over conventional infrared gas analysis equipment. One possible method of detecting the infrared absorption of a tracer gas using a tuned laser is detailed below.

Detection of backscatter radiation

The basic instrument, detailed schematically in Figure 1, comprises two laser units whose light outputs are combined to form a single collimated beam. The primary or measurement laser unit (laser A) is tuned to radiate infrared light at a frequency which is coincident with the peak absorption frequency of the tracer gas. The secondary or reference laser (laser B) is slightly off tuned from the peak absorption frequency so that it is not absorbed by the tracer gas and thus provides a reference. The collimated output beam may be traversed through the test space and scattered light reflected from the boundary surfaces measured by the associated detector units. The logarithm of the ratio of the two detector units represents an indication of the concentration of tracer gas along the beam path length.

When there is no gas present in the test space, both detectors will have similar outputs. Moreover, if during a beam traverse the boundary surface changes from say brick to metal cladding, with a corresponding change in reflectivity, both the measurement and reference beam will be equally affected, and the rates of the detected signals will remain unaltered. The system is similarly immune to the effects of changes in atmospheric pressure, temperature, humidity, path length, opacity and other gaseous contaminants in time or along the beam length. The instrument therefore has the major advantage of being immune to the majority of sources of extraneous error. With a small quantity of tracer gas present in the test space, the output of the measurement detector will fall due to infrared absorption. However, any additional variation caused by other factors will create an equal change in both detector systems providing such variations are linearly dispersed in the 3.38 - 8.40 μm spectrum. The infrared absorption spectrum for ambient air reveals no significant variations in transmission at these wavelengths and should not therefore present a problem. However, the effects of air contaminants in the industrial environment must be considered, particularly if a peak absorption coincides with one laser frequency and not the other. Providing the concentrations of such contaminants are not time dependent, a reference measurement taken before tracer injection will provide a zero datum.

The parameter measured by the laser detector system is the effective mean tracer gas concentration integrated over the optical path length, i.e. units of ppm metre. In order to rationalise the units of measurement, it may prove necessary to incorporate an optical range finder instrument to determine the optical path length, and hence reduce the unit of measurement to mean ppm.

An instrument which would be suitable for air infiltration measurements would be a derivative of the 'Eagle' system patented by ICI and designed and marketed by G.P. Elliot Electronic Systems. This instrument was primarily developed for the surveillance of oil rig drilling and production platforms and petrochemical plants. However, the technical and environmental specification for a unit suitable for the application of indoor measurement would not be as stringent. In order to take advantage of development work already conducted and also the cost advantage of helium/neon lasers in the 3.4 μm spectrum, the proposed system would be used in conjunction with methane as a tracer gas. Methane (CH_4) has a lower explosive limit (LEL) of 5.3% by volume, which is well in excess of the maximum concentration of 0.1% by volume likely to be required for infiltration measurements. In addition, methane is relatively inexpensive and in pressurised containers would present no serious injection problems.

The advantages of the system are seen to be:

i) immunity from extraneous sources of error due to mixing and other constituents of internal air

ii) speed of response; the production unit has a response time of 10 msecs, although the projected response time for the infiltration system would be approximately 10 seconds

iii) the output from the system yields a continuous measurement of tracer gas concentration

iv) the system provides effective averaging of tracer gas concentration at source

v) ease of setting up

vi) portability and flexibility

Major drawbacks of the system would be the high initial capital cost of the equipment, which would be in the order of £50,000, the proving of the system for infiltration studies and the need for the space where the infiltration rate is to be measured to be free from obstructions. The cost is based on preliminary negotiations with the supplier, inclusive of tracer gas injection equipment and sundry incidentals.

Preliminary specification for the equipment would be as follows:

Measurement laser	Helium/Neon
Reference laser	Helium/Neon
Measurement range	0-500 ppm meter
Measurement accuracy	1% of reading
Topographical range	up to 150 metres
Time constant	10 seconds

Construction

Lasers, laser ejector mirrors and telescope with detector fixed to mounting plate. Signal conditioning equipment mounted in a rack.

2.2 Fuel cell analytical detector

The possibility of using fuel cells as a primary source of energy by converting chemical energy directly into electrical energy has been the subject of extensive research, particularly in the field of space exploration. However, under certain conditions, the electrical power from a fuel cell is dependent on the gaseous fuel it receives, so that the use of such devices as analytical gas sensors is possible. This subject has been extensively researched in the UK, to the point where analytical fuel cell detectors for ethanol and formaldehyde are now on the market and an adoption of such a device is proposed as a suitable detector for air infiltration measurements.

Principle of the fuel cell detector

The fuel cell detector, shown in simplified form in Figure 2, is constructed from porous PVC discs (30mm ø x 1mm) which have been gold plated on each surface to form the electrodes of the cell. After electroplating, the discs are immersed in phosphoric acid electrolyte. The cell is then mounted in a sealed plastic moulding with the necessary electrical contacts. The plastic moulding provides a small deadspace volume above each electrode, together with the inlet and outlet ports above the anode plate.

The general principle of the analytic fuel cell is quite simple although the exact chemical reactions which take place are not clearly understood. At one electrode, the anode, one chemical compound is reduced and at the other, another chemical compound is oxidised. Therefore, the essential requirement for the fuel cell to operate as an analytical sensor is that the chemical to be detected must be either oxidisable or reducible, at the particular electrode surface, the former applying to most situations. The electrical response of the cell to the presence of an oxidisable gas introduced through the sample port is proportional to the quantity of gas which is absorbed onto the face of the electrode, and hence which is present in the gas sample. Since the absolute quantity of the oxidisable gas is small, the oxygen requirement to complete the reaction at the cathode is correspondingly small, so that diffusion from the air in the deadspace is sufficient. The fuel cell may be used to measure ethanol vapour in concentrations as low as 1 ppm. The following reactions are thought to occur:

At the anode (working electrode)

$$CH_3CH_2OH \rightarrow CH_3CHOH + H^+ + e$$

$$CH_3CHOH \rightarrow CH_3CHO + H^+ + e$$

At the cathode (reference electrode)

$$H_2O + \tfrac{1}{2}O_2 + 2e \rightarrow 2OH^-$$

Electrons generated by the oxidation process cause the working electrode to become polarised with respect to the reference electrode. The working electrode is depolarised by connecting an external resistance across the cell electrodes. The small potential difference, in the order of millivolts, generated across this resistor by the electron flow is an indication of the quantity of gas at the working electrode.

A typical response curve is shown in Figure 3. The external resistance gives rise to an exponential discharge curve, and the combination of this with the charge curve due to the ethanol oxidisation results in the overall response curve characterised by the rapid increase in output to a peak followed by an exponential decay. The level of the peak voltage is directly related to the concentration of oxidant in the cell.

The main advantages of the fuel cell transducer are:

1) Low threshold value of detection, about 1 ppm

2) Long life, the only consumables being atmospheric oxygen and the gas being analysed. The active life of the cell is therefore independent of the number of samples taken.

3) No power requirement

The fuel cell does suffer some minor disadvantages, viz:

1) The fuel cell is not an absolute detector and therefore requires calibration from time to time. This is normally achieved using laboratory prepared standards.

2) The fuel cell response is not temperature independent. This effect can be compensated over a limited temperature range by incorporating a thermistor device in the voltage amplifier normally required for instrumentation purposes. Alternatively, the fuel cell may be maintained at a constant temperature of 60°C, at which point the thermal coefficient response is 0.2%/°C, and which has the additional advantage of improved response time.

Fuel cell integrated system

The ethanol fuel cell described above forms the basis of a remote detector unit currently under development by the manufacturers of the equipment, Lion Laboratories Ltd. The unit comprises a temperature stabilised fuel cell detector unit complete with remotely actuated sampling system and inbuilt calibration facility. A series of ten such units would form the fundamental hardware of a microprocessor based detection system detailed schematically in Figure 4.

Ethanol vapour would be used as the tracer medium. The injection of the tracer into the test space could be based either on the apparatus used to generate the calibration standard vapours as presently used by the manufacturer, i.e. passing compressed air through a known water/alcohol mixture or on direct injection of metered quantities of liquid ethanol. The quantity and concentration of tracer ejected from the headspace above the water/alcohol mixture depends on the gas flow rate and alcohol concentration respectively, and can be very accurately controlled. The proposed system has several advantages over existing multi-sampling gas analyser techniques:

1) Analysis of the gas sample is carried out at the point of measurement, negating the requirement for long lengths of tubing to connect the sample port to the gas analyser.

   This is particularly important in the case of large structures where lengths of tubing up to 50 metres in length may be required.

2) Measurement of the gas concentration at each sampling point may be taken at the same time, at a period of once per two minutes, compared with approximately once per twenty minutes for a conventional multiplexed input sample gas analyser with an equivalent number of sensors.

3) Additional sensors may be added without reducing the system response time.

4) Ease of tracer injection.

5) The proposed system would be extremely versatile, enabling infiltration measurements to be made using almost any method, from the simple decay method to a fully automatic unattended data gathering system.

Several drawbacks to the system should be considered:

i) Although fuel cell detectors have been used for many years to measure ethanol concentrations, the application to air infiltration measurements is novel.

ii) Siting of the remote detector units may pose a problem, though no worse than the siting of sample tube inlets.

iii) Each remote sensor requires individual calibration.

The projected cost of the system would be as follows:

Computer plus peripherals	£3,500
Input/output device	£2,500
Remote sensors (10)	£4,000
Tracer injector units	£3,000
Sundry consumables	£2,000
	£15,000

3. PROGRESS OF WORK

Two possible methods of determining the rate of air infiltration in large enclosures have been presented. Neither of them have yet been proved to be suitable, but their major advantages and disadvantages appear to be as set out below:

System	Advantages	Disadvantages
Laser	Only one instrument required	High initial cost of equipment High power requirement (4KW) Enclosure must be empty or nearly so.
Fuel cell	Can give multi-point measurement of tracer concentration.	Labour involved in positioning and calibrating large number of sensors

Work has now begun on small scale laboratory trials of the fuel cell detectors, to verify that they can be used to measure low concentrations of ethanol vapour, and to develop a suitable method of ethanol injection.

REFERENCES

1. DICK J.B.
Experimental Studies in Natural Ventilation of Houses
JIHVE Vol 17 1949 pp 420 - 466.

2. BAKKE P. et al
Energy Conservation: a study of energy consumption in buildings and possible means of saving energy in housing
BRE CP 56/75 1957

3. LIDDAMENT M.
A survey of current research into air infiltration in buildings
Air Infiltration Centre Tech Note AIC TN 2.80 1980.

4. ROSS H. and GRIMSRUD D.
Air Infiltration in Buildings : Literature survey and proposed Research Agenda
LBL-W7822 1978

5. KRONVALL J.
Air leakage of buildings - A literature list
Lund Institute of Technology Report 77 1970.

6. MOULT A. and DEAN R.B.
CAFE - A compute program to calculate the flow environment.
CAD 1980. 4th International Conference on computers in design engineering, Brighton 1980.

7. ELECTRICITY RESEARCH COUNCIL
Infiltration tests at Ringway House, Basingstoke
Electricity Council Report /R1088 1977.

8. GROT R.A., HUNT C.M., HARRJE D.R.
Automated Air Infiltration Measurements in Large Buildings
Air Infiltration Centre Conference, Windsor 1980.

9. SHAW C.Y.
Methods of Conducting small scale pressurisation tests and air leakage data of multi-storey apartment buildings
Air Infiltration Centre Conference, Windsor 1980.

10. HITCHIN E.R., WILSON C.B.
A review of experimental techniques for the investigation of natural ventilation in buildings.
Building Science Vol 2 pp 59-82 1967.

11. HUNT C.M., KING J.C., TRECHSEL H.R.
Air Infiltration, a review of some existing measurement techniques and data.
Building Air Change Rate and Infiltration Measurements
ASTM STP 719 pp 3.23 1980.

12. SHERMAN M., GRIMSRUD D.T., CONDON P.E. and SMITH B.V.
Air Infiltration Measurement Techniques
Air Infiltration Centre Conference, Windsor 1980.

13. GALE R.
Ventilation Heat Loss outside in
Gas Engineering and Management 19 No. 11 1979.

14. GRIMSRUD D.T.
Case Studies in Air Infiltration
DoE Report, LBL, University of California 1978.

15. HUNT C.M. and TREADO S.J.
Air Exchange Measurements in a high rise office building
ASHRAE/DoE Conference Proceedings, Florida Dec 1979.

16. HARRJE D.T. and GROT R.A.
Instrumentation for monitoring energy usage in buildings at Twin Rivers
Energy and Buildings 1 (3) April 1978.

17. HARRJE D.T. and COPPER J.B.
Instrumentating energy audits
Princeton University, Centre for Environmental Studies, July 1979.

18. DICK J.B.
Measurement of ventilation using tracer gas technique
Heating Piping and Air Conditioning May 1950.

19. HONMA H.
Ventilation of dwellings and its disturbances
TEKNISKA MEDDELANDEN Vol 3 No. 2 1975

20. KELNOFER W.J., HUNT C.M. and DIDION D.A.
Determination of combined air exfiltration and ventilation rates in a nine storey office building
Conference Proceedings Vol II Purdue University April 1976.

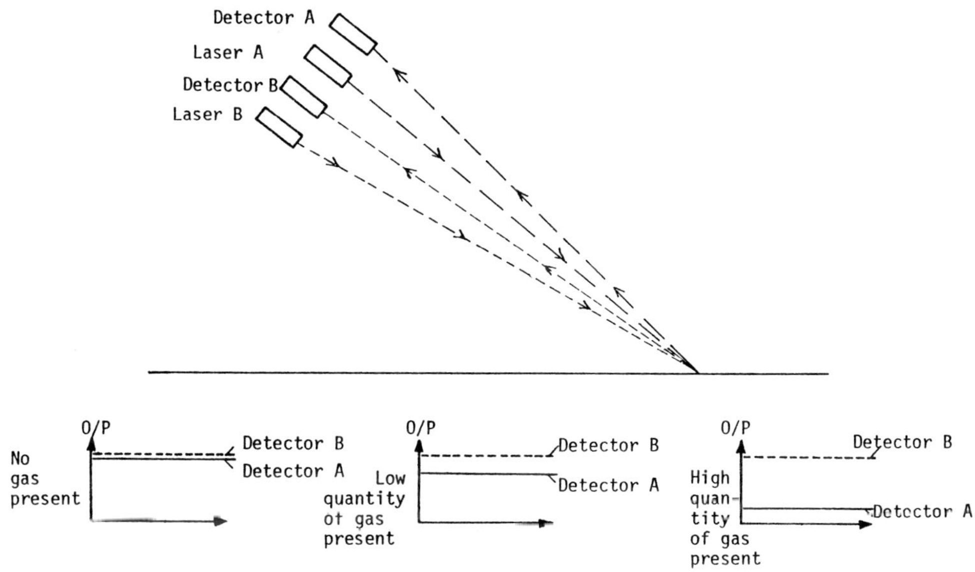

Figure 1 Principle of laser gas detection technique

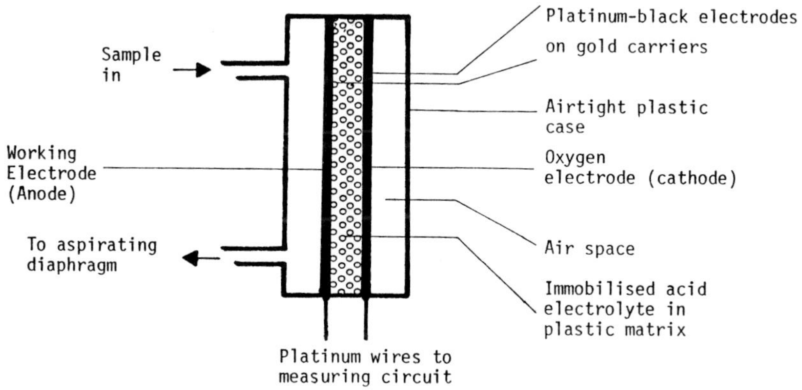

Figure 2 Schematic of fuel cell sensor

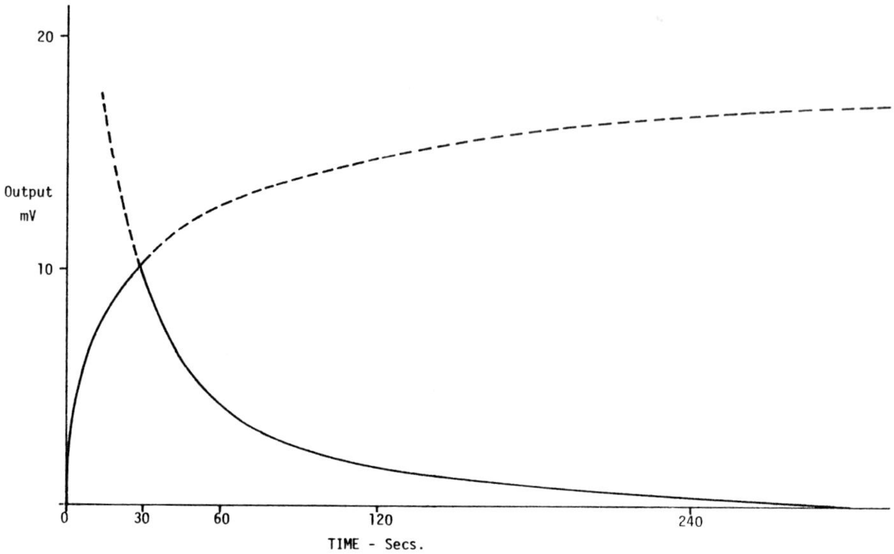

Figure 3

Response curve of a fuel cell

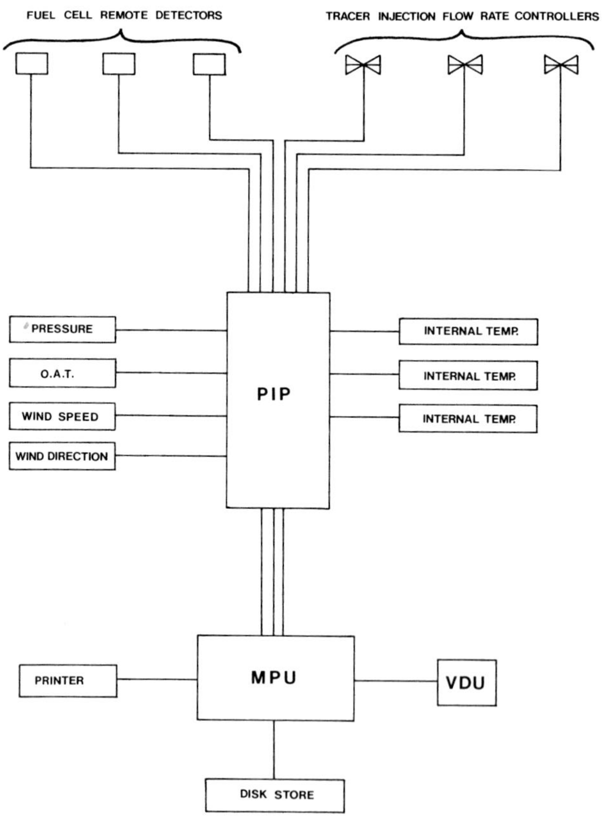

Figure 4

Fuel cell integrated detection system

QUESTIONS AND DISCUSSION

Uyttenbroeck (Chairman) : The laser technique gives an idea about the concentration of the tracer gas on the total length of the emitted and reflected beam. To have an idea about the ventilation rate is it not necessary to suppose or to make sure that the concentration of the tracer gas is homogeneous along the total length of the path ?

Answer : - Yes it is necessary to have a good mixing of the tracer gas, but by scanning the beam across the space a non-uniform distribution will be revealed. But it is possible with an "artificial wall reflector" to decrease the length op the path.

- We dit not measure with the laser technique. We only measured the concentration in the testroom with fuel cells.
- The laser technique is very expensive. The fuel cell technique is cheaper and can be tested with one cell.

Ehringer (DG XII) : You are using small quantities of tracer gases; could they be absorbed by the walls, the floor etc. of the room ?

Answer : This has to be considered. We do not know the answer for ethanol. Water is a very bad tracer for exactly the reason you suggest.

Ehringer (DG XII) : Could the difference of density of tracer gas and air variation give problems ?

Answer : In practice once you have made the initial mixing it is all right.

Uyttenbroeck (Chairman) : Should it not be useful to come to a standardised method to control if tracer gases react with building materials ?

Answer : Such a method should be interesting.

Collet (BTI) : 1. It is not necessarily a question of measuring the average of the tracer gas concentration, but it is a question of measuring the ease of the air leaving the enclosure. Did you measure the concentration in different directions .
2. How do you avoid the disturbances of the low voltage signal from the fuel cell ?
3. Have you tried with a conventional unit to measure a.c.r. in large enclosure to see what sort of problems there are in measuring a.c.r. ?

Answer : 1. The laser method relies on the concentration of tracer gas in the space being approximately uniform. By scanning the beam across the space in various directions, it will reveal whether the tracer gas concentration is or is not uniform.

2. We will probably amplify or convert the signal to digital before transmitting it to the central data collection unit.

3. No. We began by looking at novel types of instrumentation because there is a general feeling among researchers that existing techniques are too cumbersome or expensive for large single-cell buildings. We have disregarded SF_6 as unappropriate for industrial premises due to sensitivity of Freon type refrigerants. In addition we have found the analysers to be rather unreliable. NO_2 presents some difficulties in occupied spaces due to recent links with some medical conditions. Additionally the quantity required can be quite significant. We therefore looked at more abundant cheap tracers and hence new novel detectors.

Uyttenbroeck (Chairman) : The new techniques especially the fuel cell method is certainly also very interesting for normal rooms.

REVIEW OF TECHNIQUES FOR MEASURING VENTILATION RATES IN MULTI-CELLED BUILDINGS

Author : M D A E S PERERA

Contract No : EE-A-5-050-GB

Duration : 30 months 1 July 1981 - 31 December 1983

Head of project : N O MILBANK, Head of Building Services and Energy Division

Contractor : Building Research Establishment's Building Services and Energy Division

Address: : Building Research Establishment
Building Research Station
Garston
Watford WD2 7JR
Herts, UK

Summary

The overall objective of this project is to broaden the area of knowledge of natural ventilation to include types of buildings other than dwellings, in particular larger and more complex buildings such as offices. This report satisfies the first of the three phases of the project by (a) presenting the underlying theoretical basis for measuring airflows in complex, multi-cellular buildings and, (b) by reviewing the existing experimental procedures which allow us to carry out these measurements.

It was shown that inter-cell airflows could be determined only by generating N sets of tracer gas conservation equations, where N is the number of cells. These equations are then algebraically manipulated to form N sets of the equation $^{(1)}\{Y\} = [A].^{(1)}\{X\}$ where 1 refers to the cell number, $\{Y\}$ is the time variation of the concentrations, and $[A]$ is the intercell concentration matrix. This equation can then be solved for $\{X\}$, the intercell flow rate matrix.

Various experimental procedures to carry out multi-cell ventilation measurements have been reviewed and on the strength of this report, three systems have or are being designed. These systems will be used for (a) 'grab' sampling, (b) multiple tracer measurements and (c) measuring averaged airchange rates over long periods.

1 Introduction

Infiltration and natural ventilation are among the least well understood aspects of building design, although most buildings, whether commercial, public or domestic, are naturally ventilated. In recent years, research into air infiltration and ventilation has increased, but mainly with respect to dwellings rather than bigger, more complex buildings like offices. Problems of scale and the lack of appropriate techniques have deterred investigation of bigger buildings.

In 1980 the heating and lighting of public administration and commercial buildings accounted for 15% of the total primary energy consumption, of the United Kingdom, at a cost of £3,320M (1). Over a quarter of this may be attributed to ventilation and infiltration emphasising the need for better control of ventilation and for the reduction of infiltration.

This is further illustrated by a recent report (2) dealing with a programme of retrofit measures on Government office buildings. Although significant reduction in annual fuel costs was achieved over the programme as a whole, particular difficulty was found in reducing consumption in a number of high energy using buildings. A common feature of these was that occupants judged them to be generally cold and draughty, suggesting excessive ventilation and infiltration. Subsequent calculations for a typical building indicated that a reduction in ventilation rate from 3 to 1 air changes per hour could reduce energy consumption by 17% and plant operating time by 18%. A field trial undertaken on a large office building in London demonstrated a reduction of 22% in energy consumption by improving the tightness of windows by weather-stripping. However, the cost-effectiveness of measures such as this can only be assessed if methods are available for determining existing infiltration and ventilation rates in such buildings. It is the purpose of the programme of research, of which the work described in this report forms a part, to develop such techniques to enable a better understanding of the ventilation of large complex buildings and to provide those concerned with energy management with tools to determine the effectiveness of remedial measures.

As part of the work commissioned by the EC, this report reviews the techniques and sets out the underlying theoretical basis for measuring ventilation rates in complex buildings taking into account the following:

(i) the use of multiple, as well as single, tracer gases,

(ii) the use of different injection and sampling strategies, in particular:

- (a) controlled sampling and injection to give ventilation rate and internal exchange rates on a continuous basis,
- (b) intermittent sampling and injection, including 'simple' grab-sampling methods for off-site analysis,
- (c) long-term averaging methods, including the use of passive emitters and samplers.

The work described in this report provides the basis for field measurements of ventilation rates in naturally ventilated office buildings which can then be related to energy consumption and the effectiveness of remedial measures like weatherstripping.

2 Theoretical basis of tracer gas techniques

2.1 Governing equations

The basic equations of all tracer gas methods can be derived by considering the general equations governing tracer gas concentrations in multiple-connected spaces (ie cells) using an approach similar to that given by Sinden (3) and Sherman et al (4).

The following assumptions are made:

(a) the building is composed of N cells in which the air and tracer is perfectly mixed,

(b) Q_{ij} and Q_{ji} are the volume air flow rates between cells i and j where the two flow rates are not necessarily equal,

(c) $C_i(t)$ is the concentration of the tracer in the ith cell at time t,

(d) the volume of each cell is V_i,

(e) production of the tracer in any cell i is $f_i(t)$.

'o' suffix refers to the air space outside the building.

The conservation of the mass of tracer gas gives the following set of equations:

$$V_i \cdot \frac{dC_i}{dt} = f_i + \left[Q_{oi} \cdot C_o + \left\{ \sum_{j=1}^{N} Q_{ji} \cdot C_j \ (1 - \delta_{ij}) \right\} \right] - \left[Q_{io} \cdot C_i + \left\{ \sum_{j=1}^{N} Q_{ij} \ (1 - \delta_{ij}) \ C_i \right\} \right] \tag{1}$$

for i = 1, 2, ..., N and j = 1, 2, ..., N. The independent variable t has been dropped from $C_i(t)$ and $f_i(t)$ for clarity. The Kronecker delta,

$$\delta_{ij} = 1 \text{ for } i = j$$

and $\quad = 0$ for $i \neq j$.

The conservation of mass of air gives another set of equations,

$$Q_{oi} + \sum_{j=1}^{N} Q_{ji} \ (1 - \delta_{ij}) = Q_{io} + \sum_{j=1}^{N} Q_{ij} \ (1 - \delta_{ij}) \tag{2}$$

(Inflow) (Outflow)

If S_i is defined as the outflow of air from i to the outside, then from Eqn (2),

$$S_i = Q_{io} + \sum_{j=1}^{N} Q_{ij} \ (1 - \delta_{ij}) \tag{3}$$

If the tracer gas concentration,

$$C_i(t) \gg C_o$$

then Eqn (1) reduces to,

$$\{K\} = [Q]\,\{C\} \tag{4}$$

where the elements of column matrix $\{K\}$ are given by $k_i = V_i\dot{C}_i - f_i$ and those of the square matrix $[Q]$ by $q_i = S_i\,\delta_{ij} + Q_{ji}(1 - \delta_{ij})$.

2.2 Direct solution of the tracer gas equation

Given the initial conditions

$$\{C\}_{t=o} = \{C(0)\} \tag{5}$$

and knowing the intercell flows $[Q]$, the concentrations $\{C(t)\}$ at time t can in theory be determined by solving the set of equations, Eqn (4).

Although this is not normally needed in ventilation measurements, the solution to Eqn (4) has some relevance as it allows strategies to be developed for dispersing the tracer in a multi-celled building and subsequently collecting various samples for analysis.

At the moment, analytical solutions to Eqn (4) can be derived only for the simplest multi-cellular flows. Work is, however, underway to extend the solutions to problems involving more complex, but realistic flows.

2.3 Inverse problem of determining inter-cell air flows from concentration measurements.

The inverse problem of determining inter-cellular flows Q_{ij} from observations of gas concentrations $C_i(t)$ is encountered in ventilation studies. .

The quantities $\{C\}$, $\{\dot{C}\}$, $[V]$ and $\{f\}$ of Eqn (4) are either known or can be measured. There are, however, $(N^2 - N)$ unknown values of Q_{ij} and Q_{ji} (for $i \neq j$) together with 2N unknown values of Q_{io} and Q_{oi}. To satisfy this total of $(N^2 + N)$ unknown values, we have at our disposal N equations from Eqn (2) and N equations from Eqn (4).

By satisfying N unknowns by the N equations of Eqn (2), there are only N equations from Eqn (4) left to satisfy the remaining N^2 unknowns. The only way these unknown values can be found is by generating (N-1) independent sets similar to Eqn (4).

The gth generated set of Eqn (4) can be written as,

$$\{K\}^{(g)} = [Q].\ \{C\}^{(g)} \tag{6}$$

where g = 1, 2, ..., N. The elements of the column vector $\{K\}^{(g)}$ are given by

$$k_i^{(g)} = V_i.\dot{c}_i^{(g)} - f_i^{(g)}$$

for i = 1, 2, ..., N.

By algebraicaly manipulating the N^2 equations given by Eqn (6), we can write the unknown airflows Q_{ji} in column form and the concentration matrix $\{C\}$ in square form so as to form a system of equations given by

$$^{(l)}\{Y\} = [A] \cdot {}^{(l)}\{X\} \tag{7}$$

Here, equation relation to a cell l is denoted by the raised prefix (l) where l = 1, 2, ..., N. The ith element of the column matrix $^{(l)}\{Y\}$ and j^{th} element of $^{(l)}\{X\}$ are given by

$$^{(l)}y_i = V_l \dot{C}_l(i) - f_l(i) \tag{8}$$

$$^{(l)}x_j = q_l(t) = - S_l^{\bullet}\delta_{lj} + Q_{jl}\ (1 - \delta_{lj})$$

and the a_{ij} element of square matrix $[A]$ by,

$$a_{ij} = C_j(i) \tag{9}$$

Note that the raised prefix (l) refers to the set of equations relating to l^{th} cell whilst the lowered suffice l refers to a quantity relating to the lth cell.

Given $^{(l)}\{Y\} \neq 0$, Eqn (7) may be regarded as a non-homogeneous system of N linear equations in the N unknowns, $^{(l)}\{X\}$. The system has a unique solution, if and only if, the determinant of $[A]$ is not equal to zero, ie

$$|A| \neq 0 \tag{10}$$

Once the flows S_l and Q_{il} (excluding i = l) are solved for all l, the exchange of air between any cell l and the outside, Q_{ol} and Q_{lo}, can be calculated from Eqn (2).

2.4 Possible strategies in constructing N sets of conservation of tracer equations

N sets of Eqn (6) can be generated by any one of three possible strategies or by a hybrid of these. The sets can be formed by either

(a) observing $\{K\}$ and $\{C\}$ at N different time points, or

(b) varying the injection rates $\{f\}$ a total of N times, or

(c) using N different tracer gases.

As Sherman et al (4) point out, it is simpler to use N different gases to give the required N^2 equations for Eqn (7) than to increase the length of the experiment. This reduces operator effort and ensures that changes in external weather conditions do not violate too strongly the time invariancy of the inter-cellular flows, $\{X\}$. The main disadvantage of this procedure is that it requires sophisticated multi-injection and sampling hardware as well as analytical equipment. The system may also be fully automated and under on-line computer control.

The concentration matrix $[A]$ will be non-singular if each gas is injected into one and only one cell. In this case, element $^{(l)}y_i$ of Eqn (8) is modified to

$$^{(l)}y_i = V_l\ \dot{C}_l(i) - f_l(i)\ \delta_{li} \tag{11}$$

2.5 Commonly used experimental techniques

Three techniques utilising tracer gas are commonly used to determine ventilation rates and airflows within single- or multi-celled buildings:

decay; constant concentration; and constant flow. A fourth technique of recent origin and still not in widespread use is that of long term averaging.

i. Tracer decay

In this technique, quantities of the tracer gas are released within each test space, mixed as thoroughly as possible with the internal air and then allowed to decay by interchange with the other spaces and the outside air. The element $f_1^{(i)}$ can then be dropped from Eqn (8) which then reduces to,

$$^{(1)}y_i = V_1 . \dot{C}_1^{(i)} \tag{12}$$

ii. Constant concentration

In this case the concentration of a tracer gas is held constant in each cell. If N different tracer gases are used and only one gas is injected to any one cell, the vector matrix $\{\dot{C}\}$ drops out of Eqn (7) reducing the term $^{(1)}y_i$ in Eqn (8) to

$$^{(1)}y_i = - f_1^{(i)} \delta_{1i} \tag{13}$$

The direct infiltration rates, Q_{01} (1=1,2,....,N), from the outside to the 1 cells can be found only from one set of Eqn (7) by carrying out a special experimental procedure in which one gas is used as the tracer and the concentration of this gas is kept constant at $\bar{C}$ in all cells. Eqn (7) then reduces to the set of equations,

$$-f_1 = \bar{C} \sum_{f=i}^{N} \left[-S_1 \delta_{1j} + Q_{j1}(1-\delta_{1j}) \right] \tag{14}$$

for cells 1 = 1,2,....,N. The raised suffix (i) is dropped since only one tracer is used.

Substituting Eqns (2) and (3) in Eqn (14), we have

$$Q_{o1} = f_1/\bar{c} \tag{15}$$

for 1=1,2,....,N.

The overall infiltration rate into the building is then simply,

$$\sum_{1=1}^{N} Q_{o1} \tag{16}$$

iii. Constant flow

The essence of this technique is to inject, at a constant flow rate, a tracer into each cell. This is a simpler technique than constant concentration but still requires the solution of N sets of Eqn (6). This means that N different tracer gases must be used, each being injected into only one cell.

If injections are continued until equilibrium concentrations are reached, the elements (given by Eqn 8) of matrix $^{(1)}\{Y\}$ in Eqn (7) are simplified to

$$^{(1)}y_i = -F_1^{(i)} . \delta_{1i} \tag{17}$$

where F is the constant injection rate. A disadvantage is the long time needed before equilibrium concentrations are reached as no allowance is made in the conservation equations for changes in the airflow rates $[Q]$ during the test period.

iv. Long term averaging

In some circumstances, it may be necessary to determine the long-term averages of airflows in a building where the averaging period is about a month. Sherman et al (4) point out that, in principle, the three techniques considered above could be used over that period but this would tie-up equipment which are both expensive and complex. A simpler system, which could determine the averaged terms of the corresponding equations, in a straightforward manner, would be preferred.

If the long term averages are denoted by an overbar, Eqn (4) can be simplified to

$$- \{\bar{f}\} = [\bar{Q}] \{\bar{C}\} \tag{18}$$

2.6 Error analysis

The total error in any measurement can be made-up from errors of accuracy and precision. Precision errors are defined as the standard deviation of results from repeated trials under identical conditions whilst accuracy relates to the deviation of the expected value of the measurement from the true value.

Standard techniques of error analysis can be used to estimate the errors involved in determining airflow rates in multi-cell ventilation measurements.

If Δy_i and Δa_{ij} are the absolute standard errors of y_i and a_{ij} in Eqn (7), then

$$\Delta y_i \simeq \sum_{f=1}^{N} a_{ij}\, \Delta x_j + \sum_{j=1}^{N} \Delta a_{ij} . x_j \tag{19}$$

where Δa_{ij} are the corresponding errors in a_{ij}.

Eqn (19) can be put into matrix notation, from which:

$$\{\Delta X\} = [A]^{-1} \{\Delta Y\} - [A]^{-1}[\Delta A] \{X\} \tag{20}$$

The elements in the error matrices $\{\Delta Y\}$ and $[\Delta A]$ are given by,

$$\Delta Y_i = \sqrt{(\dot{C}_1{}^{(i)} \Delta V_1)^2 + (V_1 . \Delta \dot{C}_1{}^{(i)})^2 + (\Delta f)^2} \tag{21}$$

where $$\Delta \dot{C} = \left[\sqrt{(\Delta C_2^2 + \Delta C_1^2)} / \delta t\right] + \left[\delta C \sqrt{\Delta t_1^2 + \Delta t_2^2} / (\delta t)^2\right] \tag{22}$$

in which $\delta t = t_2 - t_1$

$\delta C = C_2 - C_1$

and $$\Delta a_{ij} = \Delta C_j{}^{(i)} \tag{23}$$

For large values of δt, Eqn (22) reduces to

$$\Delta \dot{C} \simeq \sqrt{C_2^2 + C_1^2} / \delta t \qquad (24)$$

2.7 Problems of imperfect mixing of tracer gas with internal air

A major problem which may arise during airchange measurements is imperfect mixing between tracer and internal air. Even with perfect initial mixing with the use of fans, there is no guarantee that this will persist during measurement. In all cases of imperfect mixing, the location of the sampling points will determine the airchange rates that are measured.

Hunt (4) describes three extreme cases in which the tracer does not mix perfectly with the room air. In the first case, some of the air in the test space can stagnate even after initial perfect mixing of the tracer. Secondly, the space behaves as two separate parts with different airchange rates. Finally, the third case of imperfect mixing occurs when entering air displaces the air already present without mixing with it.

3 <u>Experimental procedures</u>

There are essentially four different techniques which span the complete spectrum of ventilation measurements using tracer gas;

(a) The 'grab' technique, whereby the airchange rate data can be obtained by non-specialist personnel at low cost;

(b) The highly sophisticated 'automated' systems which can monitor changes in ventilation and infiltration rates in complex buildings over any defined period;

(c) Techniques for measuring averaged airchange rates over long and extended periods;

(d) Techniques to determine flows between spaces within complex buildings.

3.1 Tracer gases

Gases as tracers in ventilation measurements are preferred to particulate and liquid aerosols because of the relative ease of dispersion, collection and analysis (5). Many authors (5,6,7) have set out the properties required of an ideal tracer, but no gas can satisfy all these. A compromise has to be made by choosing the tracer most suitable for any particular test.

For single-tracer measurements, sulphur hexafluoride (SF_6) satisfies most of the necessary and desirable properties required of a tracer gas. This gas can be detected continuously in the ppm range by using an infrared analyser. If detection is however required at lower concentrations, as for instance in large buildings, then an electron capture analyser may be used. With the electron capture detectors presently available in the commercial market, it is not possible to make a continuous measurement of concentration.

The two families of the Freon and some of the perfluorocarbon gases, (those known commercially as the Flutec gases), appear to be suitable for multi-gas measurements. Either infrared or electron-capture analysers could be used to detect them in real-time measurements.

3.2 Grab sampling

The essential feature of this procedure is the intermittant nature of the sampling and subsequent off-site analysis. On site, experimenters distribute the tracer gas following a simple practical routine, and then after a suitable period, take air samples at various times and locations. These samples are then sent to a central laboratory for analysis.

This procedure gives a broad indication of the magnitude of air infiltration in a simple building or of ventilation rates in different zones of a complex building. It is suitable for use by relatively unskilled personnel and is particularly appropriate for remote locations in energy-audit or retrofit programmes when a large number of buildings need to be monitored.

In order to carry-out this procedure, a known quantity of tracer gas is withdrawn from a supply container and injected into containers for dispatch to a site. These containers could either be graduated syringes (8) or flexible, polyethylene bottles (9). The volume of tracer in each container should be approximately set to a level determined by,

- the volume of the space
- the expected airchange rate;
- the range of the instruments that will analyse the collected samples.

The concentrations in each container do not need to be known with great accuracy nor do they need to be precisely equal to one another because the aim is to measure only the relative concentrations which occur during collection of samples.

The tracer can be distributed by walking slowly around the building and injecting into each separate space an amount of tracer proportional to that volume. Every effort should be made to even out tracer distribution, and to this end the gas should be injected slowly. Mechanical ventilation systems, when available, should either be switched on for a short while to provide initial mixing or run at full recirculation throughout the test. Otherwise, floor fans are useful. Grot (8) suggests that 30 minutes to 1 hour is required for adequate mixing to take place.

Once mixing is achieved, the sample containers can be filled. These containers can be purpose-built sample bags made from FEP Teflon, Saran or Mylar (11,12). They can be filled with the sampled air either by a hand pump or a small battery-powered pump. As an inexpensive alternative, polyethylene bottles could be used. Ten successive squeezes, each yielding 50 percent contraction in volume, of 500 ml polyethylene bottles provide a valid sample of the tracer concentration in a sampled space (3,11).

A large number of samples can be collected if decay curves are to be plotted, but if this is not possible and reduced accuracy is acceptable, an initial and final sample container may be enough. In both cases, time intervals <u>must</u> be known. The number of samples taken will affect the standard error of estimate of any 'best-fit' to the exponential decay.

Harrje et al (9) notes that in houses with good internal air circulation, sampling at one central location is adequate. In larger buildings, or buildings where internal mixing is limited, multi point sampling should be carried out as airchange rate will vary from one zone to another (10).

Containers carrying the sampled air are then sent back to the laboratory for analysis. Drivas et al (13) believe samples, say of SF_6 could, if necessary, remain in polyethylene bottles for up to two weeks.

3.3 Automated systems

Any part or all of the injection, sampling and analysis procedures carried out in ventilation measurements can be automated. This has become easier with recent developments in micro processor technology. Automated systems can carry out complicated tasks which are beyond the scope of manual control and allow additional information to be obtained. Two such techniques which require automated systems are those of constant concentration so as to determine infiltration and, multiple tracer measurements so as to determine internal air movements as well as the infiltration rates. Systems, so far built and reviewed below, have been designed mainly for single tracer gas use.

(i) Infiltration rate

As discussed in Section 2.5.ii., the single-tracer gas constant concentration technique depends upon the fact that the tracer gas in all cells in a building are maintained at the same concentration so that the effect of flows between the cells is eliminated. This allows the fresh air infiltration into any cell as well as the total infiltration into the whole building to be determined.

The British Gas 'Autovent' system (14) uses this technique. Two sets of solenoid valves are controlled by a microcomputer. One set controls sample lines whilst the other is used to control the injection of the tracer gas. Each line is sampled in turn and the amount of tracer needed to maintain concentration to a preset value is calculated. This is then converted into a time period in which the relevant preset injection valve is opened to deliver the required amount of tracer. Mixing between the tracer and the room air is allowed for by,

- ceasing to inject tracer into that room for six seconds before it is scheduled to be sampled,
- using a purge pump on the sample line, and
- terminating the injection lines in areas of high speed air, either near radiators or small mixing fans.

Collet (15) describes a similar system. This differs in the use of fixed jets in the injection lines rather than the stable pressure regulators used in the Autovent system.

Condon et al (16) have developed a system which uses two calibrated dump volumes to deliver a known amount of tracer gas to the test space. The filling of these volumes and the consequent injection is carried-out by energising and de-energising various solenoids under computer control.

A continuous injection tracer gas system is described by Sherman et al (4) in which a micro-computer in the system controls the injection of the tracer and a mass flow controller holds the flow rate constant. The gas is injected in half hour cycles and the tracer flow rate for any cycle is set equal to the target concentration times the calculated ventilation flow rate from the previous cycle. Within a cycle, the computer selects the sampling port used during that cycle and concentration data is collected at regular intervals from the gas analyser, and retained by the computer so that, at the end of a cycle, the airchange rate can be calculated. These calculated values are then stored on computer disc.

Hartmann and Muhlebach (17) describe a simpler automated system for measuring ventilation rates using the tracer decay technique. They used a purpose-built controller to inject and to sample their N_2O tracer which was then analysed with an infrared detector.

The systems so far reviewed have used IR detectors to measure concentration levels. Kumar et al (18) describe an automated system for use in either the decay or constant concentration mode. Their system uses a Hewlett Packard 9815A programmable desk-top calculator. The calculator can handle one injection and one sample line through a specially designed interface unit. Under control of the calculator, the tracer gas (in this case, SF_6) is discharged by trapping a volume of the tracer between two solenoid-operated gas valves and then releasing one of the valves. An interval timer in the interface unit can then be triggered to sample the gas concentration. The electron capture detector, modified by the replacement of the manually operated sampling valve with an electro-pneumatically operated valve, is triggered from the calculator. The signals from the analyser are then digitised and the SF_6 peak is isolated and retained for subsequent analysis.

The systems described so far have been mainly used in small, residential buildings. However, three studies in larger buildings have been carried out in the USA on the 7-storey (37 000 m^3 volume) Norris Cotton Building in New Hampshire (19), the 11-storey (28 000 m^3) Administration Building at the National Bureau of Standards, Washington (20) and the 7-storey Administration Building at Princeton University (10). In each of these cases, the systems used were semi-automated and the tracer had to be injected manually into the repsective ventilation system. The ventilation systems were then used to mix the tracer gas with the internal air. To determine tracer concentrations, samples from various locations were automatically accessed in sequence and at regular intervals, so that the corresponding decay curves could be determined. The samples were analysed with electron-capture detectors and the resulting data stored on magnetic tapes for subsequent off-line computation.

Grot et al (21) has described a fully automated and sophisticated system for measuring airchange rates using the tracer decay technique in two large buildings; the 4-storey (68 000 m^3) Collins Building near Glasgow, UK and the 26-storey (450 000 m^3) Park Plaza Building in Newark, USA. The key components of the system include a micro-computer, an electron-capture gas chromotograph, a ten-port sampling manifold and five injection units.

(ii) Infiltration rate and internal air movement

Of the few systems built so far for measuring inter-cell airflows, that of I'Anson et al (22) can measure airflows between three tracer gases; Freon 12, Freon 114 and BCF released in three zones and measured using an electron-capture gas chromotograph [following earlier work of Foord and Lidwell (23)]. The whole process is labour intensive and, at the moment, each test takes about two to three hours to complete. The variability of weather conditions over this long time period adds uncertainty to the determination of these inter-cell air exchange rates. The authors quote an approximate accuracy of 20%.

Littler (24) describes a technique designed to study movement within rooms and between rooms. Five of the Flutec gases are dispersed at various points using a method proposed by Dietz et al (25) in which the gas is emitted at a steady rate by placing a small elastomer sponge of Fluorel (Buna-N) in contact with the Flutec liquid. At each sampling point, the room air will then be drawn in turn through eight sample tubes packed with chromosorb 102. The sample tubes are subsequently connected to a gas chromotograph employing flame ionisation detection to obtain concentrations of the components and the time histories of the change in gas composition.

3.4 Long-term average techniques

Sherman et al (4) describes an inexpensive system, designed to determine long-term airchange rates in dwellings. It consists of two small suitcases, each containing a solenoid pump pulsed by an interval timer. One system is set-up to empty a bag containing tracer gas and the other is used to inflate an initially empty bag with room air. The equipment can be left unattended on site for the duration of the experiment and then brought back to the laboratory for analysis. The average airchange rate then can be calculated by measuring the concentration of the tracer in the sample bag, the volume of gas injected from the injection bag and the time period over which the experiment was carried out.

The perfluorocarbon tracer technique developed by Dietz et al (25) and used by Littler (Section 3.3.ii) is also suitable for long-term measurements of the average airchange in a building. The tracer source is a diffusion tube consisting of the fluoroelastomer plug impregnated with a known mass of the perfluorocarbon and crimped within a metal shell. The perfluorocarbon tracers then diffuse from the tube at a known rate.

According to Dietz et al, these sources have useful lifetimes of about five to ten years. A battery operated pump is then used to draw air through an adsorption tube, placed at any location, over the required time period. If an electron-capture detector is used, Dietz et al claim that the wide dynamic range of the detector would allow an effective sample period from one to hundreds of days.

4 Discussion

This report presents the underlying theoretical basis for measuring airflows in complex, multi-cellular buildings. It also reviews the experimental procedures which may be used to carry-out these measurements. The following progress has been made:

(a) Theoretical work has begun to find solutions to the problem of determining tracer concentrations in multi-celled buildings when inter-cell flow rates are assumed. From this it will be possible to determine the best strategy for distributing tracer gas.

(b) Equipment to carry-out grab sampling has been purchased as well as a infrared analyser to simultaneously measure the concentrations of several tracer gases.

(c) A micro-computer controlled automated system has been built to inject and sample tracer gases. Both the injection and sample modes can use up to 12 separate channels.

(d) Preliminary design work is under way to build a simple and portable long-term averaging system.

Preliminary tests will be carried out in office buildings at the Building Research Establishment (BRE) to validate this work before making measurements on suitable buildings outside BRE.

References

1 Department of Energy, Digest of UK Energy Statistics 1979, HMSO, London, UK, (1979)

2 B A Taylor, Natural ventilation and the PSA estate, Proceedings of the CIBS Symposium on Natural Ventilation by Design held at the Building Research Establishment, Garston, Watford, UK (1980)

3 F W Sinden, Multi-chamber theory of air infiltration. Building and Enviroment, Vol 13, pp 21-28 (1978)

4 M Sherman et al, Air infiltration measurement techniques. 1st Air Infiltration Conference, on Air Infiltration Instrumentation and Measuring Techniques, Windsor, England (1980)

5 G F Collins et al, A preliminary evaluation of gas air tracers. J of the Air Pollution Control Association, Vol 15(3), pp 109-12, (1965)

6 C M Hunt, Air infiltration: a review of some existing measurement techniques and data. ASTM Special Publication No 719, Philadelphia, pp 3-23, (1980)

7 E R Hitchin and C B Wilson, A review of experimental techniques for the investigation of natural ventilation in buildings. Building Science, Vol 2, pp 59-82, (1967)

8 R A Grot, A low-cost method for measuring air infiltration rates in a large sample of dwellings. ASTM Special Publication No 719, pp 50-59 (1980)

9 D T Harrje et al, Sampling for air exchange: rates in a variety of buildings. ASHRAE Trans., Vol 88(1), (1982)

10 D T Harrje et al, Air infiltration site measurement techniques. 2nd AIC Conference on Building Design for Minimum Air Infiltration, Stockholm (1981)

11 B E Saltzman et al, Halogenated compounds as gaseous meteorological tracers. Analytical Chemistry, Vol 38(6), pp 753-58 (1966)

12 C A Clemons and A P Altshuller, Responses of electron-capture dector to halogenated substances. Analytical Chemistry, Vol 38(1), pp 133-36, (1966)

13 P J Drivas et al, Experimental characteristics of ventilation systems in buildings. Env. Sci. and Tech., Vol 6(7), pp 609-14 (1972)

14 D K Alexander et al, Experimental techniques for ventilation research. 1st Air Infiltration Conference, on Air Infiltration Instrumentation and Measuring Techniques, Windsor, England (1980)

15 P F Colet, Continuous measurements of air infiltration in occupied dwellings. 2nd AIC Conference on Building Design for Minimum Air Infiltration, Stockholm (1981)

16 P E Condon et al, An automated controlled-flow air infiltration measurement system. ASTM Special Publication No 719, Philadelphia, pp 3-23, (1980)

17 P Hartmann and H Muehlbach, Automatic measurements of air change rates (decay method) in a small residential building without any forced air-heating system. 1st Air Infiltration Conference, on Air Infiltration Instrumentation and Measuring Techniques, Windsor,

18 R Kumar, An automated air infiltration measuring system using SF_6 tracer gas in constant concentration and decay methods. 1st Air Infiltration Conference, on Air Infiltration Instrumentation and Measuring Techniques, Windsor, England (1980)

19 C M Hunt, Ventilation measurements in the Norris Cotton Federal Office Building in Manchester, NH. ASHRAE trans., Vol 85(1), pp 828-39 (1979)

20 C M Hunt and S J Treado, Air exchange measurements in a high-rise office building. Proc. ASHRAE/DOE Conference on Thermal Performance of the Exterior Envelopes of Buildings, Florida (1979)

21 R A Grot et al, Automated infiltration measurements in large buildings. 1st Air Infiltration Conference on Air Infiltration Instrumentation and Measuring Techniques, Windsor, England (1980)

22 S J I'Anson et al, A multiple tracer gas technique for measuring air-flow in houses. Proc. of CIB W67 3rd Int. Symp. on Energy Conservation in the Built Environment, Dublin (1982)

23 N Foord and O M Lidwell, A method for studying air movement in complex occupied buildings such as hospitals: Halocarbons as gas tracers using gas chromotography. Building Services Engineer, (1973)

24 J Littler, Detailed study of air movement in buildings (Unpublished)

25 R N Dietz et al, An inexpensive perfluorocarbon tracer technique for wide-scale infiltration measurements in homes. Proc. Int. Symp. on Indoor Air Pollution, Health and Energy Conservation, University of Massachusetts (1981)

QUESTIONS AND DISCUSSION

Pilavachi (DG XII) : I understand that you have represented a room by one cell. Would it not be useful to represent a larger room by two or more cells ?

Answer : Yes, if different parts of the room have very different ventilation rates.

Comment - Collet (BTI) : as long as there is a reasonable mixing in the big cell I do not think that it matters very much if you divide the big cell to smaller cells because the geometry of the internal mixing is a steady state situation which means that the concentrations will be for example $C^1 \simeq C^2 \times 0{,}8 \simeq C^3 \times 0{,}5 \simeq C^4 \times 0{,}75$ at any given time which means

$$n_1 = \ln \frac{c_1^1}{c_2^1} \qquad n_2 \simeq \frac{\ln \frac{c_1^2}{c_2^2}}{t_1 - t_2} \simeq \ln \frac{c_1^1}{c_2^1} \frac{\times 0{,}8}{\times 0{,}8}$$

Collet (BTI) : You said that your equations could take absorbtion into account, by your measurement method. But how can you in every room detect if a decrease is caused by air flow between rooms and between rooms and outside. Your equations do they take a storage into account ?

Answer : Absorbtion is only and can possibly only be taken into account if it is measured by other measurements.

Dewsbury (BSRIA) : 1. What is the lowest concentration of the Flutec gases that can be measured by electron capture ?
2. Is air drawn through all the sampling lines continuously ?

Answer : 1. About 10^{-12} to 10^{-15}. They are used for meteorological studies, because such low concentrations can be detected.

2. Yes. There is a purge pump which operates continuously.

Eastwell (BSRIA) : 1. What is the time required to analyse a fuel set, say 12 samples. i.e. what is the cycle time of the combined system ?
2. In "samplevalve" is gas drawn continuously ?

Answer : 1. $\simeq$ 10 min ($\simeq$ 1.5 min per gas).

2. Yes.

MEASUREMENTS OF AIR INFILTRATIONS RATES IN DWELLINGS WITH ONLY EXHAUST VENTILATION

Author	:	P. F. Collet
Contract number	:	EE-A-5-052-DK
Duration	:	26-X-82 to 31-X11-83.
Head of Project	:	P. F. Collet
Contractor	:	Teknologisk Institut
Address	:	Gregersensvej DK-2630 Taastrup Denmark

Summary

The aims of this part of the project is to see, if it is possible to allocate the air change to the room wanted by exhaust ventilation and ventilation openings correctly placed and operated.

Measurements of Air Change Rate in Prototype Dwellings at Galgebakken and Brøndby Strand.

Galgebakken

General Remarks. The natural fresh air change rate was measured continuously in 2 atrium houses and in 6 terrace houses. During the periods of measuring the inmates used their dwellings as usual.

The measurements show that the basic fresh air change rate is comparatively small, viz. 0.25 h^{-1} for atrium houses and 0.20 h^{-1} for terrace houses, - in outdoor conditions generally corresponding to the mean winter climate.

The fresh air change rate has turned out to vary very considerably when the dwelling is being used; as an average it is 1-5 times higher than the basic fresh air change rate when the dwelling is not used.

Prototype Houses

The Atrium House, Field 4-4. (Measurements cf. Supplement 1)

The natural fresh air change rate was very low, viz. about 0.1 h^{-1} at $t_u \cong 4^oC$ and v = 0-2 m/s.

When the mechanical exhaust is on, it is possible to a certain extent to lead the fresh air into different rooms according to one's wishes as long as the free passage between the inner room partitions is generally unhindered.

The measurements also show that although the natural fresh air change rate is very low, the house is not air tight to an extent which influences the total consumption of mechanical exhaust of the house in case of closed ventilation openings.

The measurements performed by the Danish Building Research Institute of draught at the measured exhaust air volumes showed a pressure difference of about 5 Pa via the house envelope.

If the 5 Pa is representative of the under pressure within the house compared to the outdoor pressure, it will be possible, using the equation

$$Q = C \times \Delta P^{2/3},$$

to calculate the air change rate, viz. $n \cong 3.1\ h^{-1}$ at an under pressure of 50 Pa.

These values come close to the ones demanded for tight houses in Sweden; compared to Danish demands the above tightness is extreme.

The measurements also show that up to 30 per cent of the fresh air enters the halls No. 6 and No. 7 plus the bathroom No. 8. If the doors and windows of these rooms are sealed better, it will improve the possibility of directing the air flow through the house as the under pressure will be increased by about 50 per cent.

Conclusion:

The use of mechanical exhaust in the kitchen and in the bathrooms will render it possible to direct the fresh air into different rooms according the one's wishes - at least to a reasonable extent.

It is a condition that the doors within the house are not tight or that another reasonably unhindered air passage from room to room exists. Furthermore the front doors must be sealed and the total exhaust area must exceed 300 cm^2.

Judged from the measurements of the natural fresh air change rate with open and closed ventilation openings, respectively, and with no mechanical exhaust it is not possible immediately to demonstrate any improvement. However, it is quite possible that the fresh air change is improved at higher wind velocities than the ones measured. Nevertheless, a comparison with Square 5-5 shows clearly that the increase of the ventilation area obtained by the ventilation openings is far too small.

The Terrace House, Square 5-5. (Supplement 2.)

Here, too, the natural fresh air change rate is rather low, 0.2-0.3 h^{-1}, compared with ordinary dwellings.

The measurements in rooms Nos. 4, 5, 7, and 8 show that opening of the ventilation openings in these rooms can cause an increase of the air change rate of 50-100 per cent and besides cause some increase of the air change rate of the bedrooms.

However, a really satisfactory improvement of the natural fresh air change rate is not obtained until all ventilation openings are open, - equivalent to a total opening area of about 260 cm^2.

Conclusion:

The house is tight compared with ordinary houses.

The natural fresh air change rate is too low when many persons are present at the same time.

The ventilation area - with the small metal grids of a total area of 110 cm^2 - is insufficient as regards an increase of the fresh air change rate. The ventilation area must be increased to at least about 250 cm^2, equivalent to the total area of all the ventilation openings.

Brøndby Strand, Hallingparken 52, I. left. (Supplement 3.)

The natural air change rate is very low, 0.1-0.2 h^{-1}.

The measurements show that the consumption of mechanical exhaust varies according to open or closed ventilation openings.

On the other hand, the distribution of the fresh air change rate is altered significantly in accordance with the use of the ventilation openings and also in accordance with open or closed doors within the house.

When the ventilation openings are open and the doors within the house are ajar, 50 per cent of the fresh air will travel into the bedrooms, whereas only 20 per cent will do so when the ventilation openings are closed and the doors within the house are closed, too.

Because of the wind direction the fresh air will preferably travel into the two sitting-rooms when the exhaust is off.

Conclusions:

The flat is very tight compared with ordinary flats.

By means of mechanical exhaust and open ventilation openings in the bedrooms it is possible to direct about 50 per cent of the fresh air change rate straight through these rooms when the doors within the house do not shut tight.

If the location of the ventilation openings is wrong or if they are open while the bedroom doors are closed, the fresh air change rate may mainly work in the other rooms instead of the bedrooms.

Brøndby Strand, Tranumparken 1.3.4. (Supplement 4.)

The natural fresh air change rate is rather low, 0.2-0.3 h^{-1}.

The measurements show that the fresh air change rate by mechanical exhaust depends very much on open or closed ventilation openings.

This is probably more due to the changeable weather conditions - plus the fact that a great many flats are linked in parallel on the ventilation shaft - rather than to the flat being tighter than the dwellings mentioned above.

In this case it does not immediately appear that is is feasible to direct the fresh air into different rooms as requested. This fact is mainly due to the changeable weather conditions during the measuring.

The measurements show that some air enters the horizontal division of the house, probably from the flat below.

Conclusion:

The flat is tight compared with ordinary flats.

With the given pressure in the ventilation shafts the fresh air change rate will probably mainly depend on the total area of the ventilation openings.

Conclusion of the Measurements Undertaken at Galgebakken and Brøndby Strand

1. It appears that in dwellings with mechanical exhaust the fresh air change rate depends little on the ventilation openings being open or closed.

   This is due to the fact that even at very small pressure differences the required amount of air will travel through the natural leaks.

2. In dwellings with mechanical exhaust it will be possible (by means of relevant ventilation openings) to direct the fresh air into different rooms as requested if the doors within the house (or corresponding measures) are not tight.

   It is important that major leaks are sealed, e.g. by front doors, in order to improve the distribution of fresh air to the different rooms as requested.

3. The ventilation openings applied at Galgebakken will procure a modest increase of the air change rate, but they are too small to achieve the necessary increase of the fresh air change rate by natural air change and also too small to provide the required distribution of the fresh air change rate.

   Even with the addition of mechanical exhaust these ventilation openings are too small to ensure a satisfactory distribution of the fresh air change rate to the different rooms.

4. Dwellings on the same ventilation shaft and leaks between the flats will reduce the possibilities of establishing the proper distribution of the fresh air change rate in the individual rooms, cf. Tranumparken.

5. In all likelihood the applied ventilation areas in the house envelope will prove far too small.

6. In order to ensure a reasonable distribution of the fresh air change rate mechanical exhaust must be applied, the ventilation openings of the house envelope must be more than 250-300 cm^2, and they must have a relevant room distribution with a view to the rooms into which the fresh air is to be directed.

7. In order to ensure that the natural ventilation is reasonably satisfactory under normal conditions the total area of the ventilation openings must exceed 250-300 cm^2 and they must have a reasonable distribution, too.

   In order to ensure a reasonably good natural ventilation even during transition periods and at low wind velocity, the total ventilation areas will probably have to exceed 4-500 cm^2 - possibly a little lower in terrace houses and a little higher in atrium houses.

Galgebakken Field 4-4

Ventilation Conditions during the Measurements

During all measurements the doors within the house were ajar, windows were closed and front doors were closed as well. Specific conditions for the individual measurements were:

Measurement No. 1:	Metal ventilation openings open. Fresh- and Vento valves closed. Mechanical exhaust on.
Measurement No. 2:	All ventilation openings closed. Mechanical exhaust on.
Measurement No. 3:	Metal ventilation openings open in rooms 2, 3, 4, and 5. All other ventilation openings closed. Mechanical exhaust on.
Measurement No. 4:	All ventilation openings closed. Mechanical exhaust off and closed.
Measurement No. 5:	Metal ventilation openings open in rooms 3, 4, and 5. Fresh valve open in room 1. Vento valve open in room 2. Mechanical exhaust off and closed.
Measurement No. 6:	Adjustment of ventilation openings as in No. 5. Mechanical exhaust on.

The Climate during the Measurements

30.3 Measurements Nos. 1, 2, 3.

8 a.m.	$1.0^{o}C$	3	Beaufort	east
2 p.m.	$5.8^{o}C$	3	"	"
9 p.m.	$3.2^{o}C$	1	"	north

31.3 Measurements Nos. 4, 5, 6.

8 a.m.	$0.5^{o}C$	0	Beaufort	
2 p.m.	$8.8^{o}C$	1	"	north west
9 p.m.	$3.0^{o}C$	1	"	west

The climatic data obtained from weather station Torsbo.

Results

The results of the 6 measurements performed 30.3 and 31.3 are shown in the diagrams below. The diagrams etc. show the fresh air change when the dwelling is not used and by mechanical exhaust during the period 82.04.02 - 82.04.06.

Air change (number of times per hour).

Measurement \ Room	1 Living room	2 Small bed-room	3 Bed-room	4 Small bed-room	5 Small bed-room	6 Hall	7 Hall	8 Bath-room	9 Bath-room	House
1	0,54	0.90	0,29	0,29	0,55	0,71	1,52	0,77	0,03	0,67
2	0,59	0,63	0,45	0,30	0,39	1,11	2,00	1,40	0,02	0,67
3	0,41	1,11	0,94	0,46	0,79	0,90	1,43	1,31	0,19	0,72
4	0,11	0,02	0,08	0,09	0,08	0,16	0,18	0,12	0,01	0,10
5	0,11	0,11	0,13	0,07	0,10	0,20	0,25	0,13	0,02	0,12
6	0,56	1,03	1,01	0,41	0,74	0,56	1,35	0,85	0,20	0,74

Air Change (m^3 per hour)

Measurement \ Room	1	2	3	4	5	6	7	8	9	House
1	57	33	25	7	10	7	29	7	0	176
2	62	23	14	7	7	11	38	13	0	176
3	43	41	29	11	14	9	27	12	2	189
4	12	1	2	2	1	2	3	1	0	25
5	12	4	4	2	2	2	5	1	0	31
6	60	40	31	10	13	6	26	8	2	193

Air Change (percentage)

Room / Measurement	1	2	3	4	5	6	7	8	9	House
1	32	19	14	4	6	4	17	4	0	100
2	35	13	8	4	4	7	22	7	0	100
3	23	22	15	6	8	5	14	6	1	100
4	48	3	10	8	6	6	14	4	1	100
5	37	13	13	5	6	6	15	4	1	100
6	31	20	16	5	7	3	13	4	1	100

Air change (number of times per hour corrected for standard condition).

Standard condition: temperature difference $\Delta t = 17^{o}C$

wind velocity $V = 4.6$ m/s

Room / Measurement	House
4	0,16
5	0,21

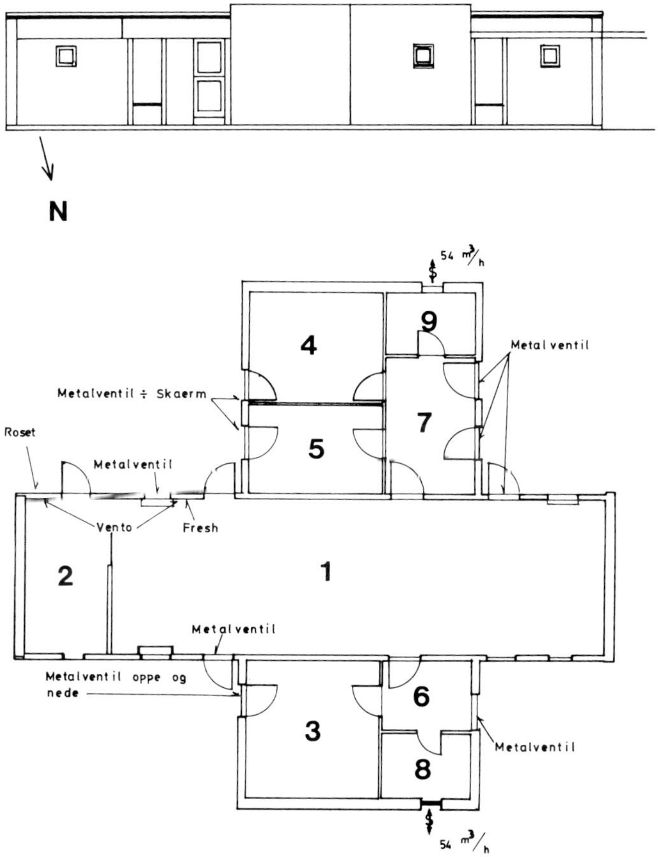

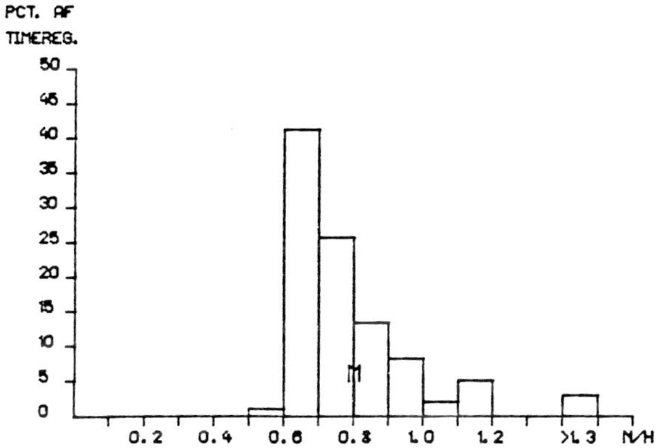

GALGEBAKKEN MARK 4-4 : APRIL 82

D. 2 KL. 14 TIL D. 6 KL. 15

FIG. 1.2

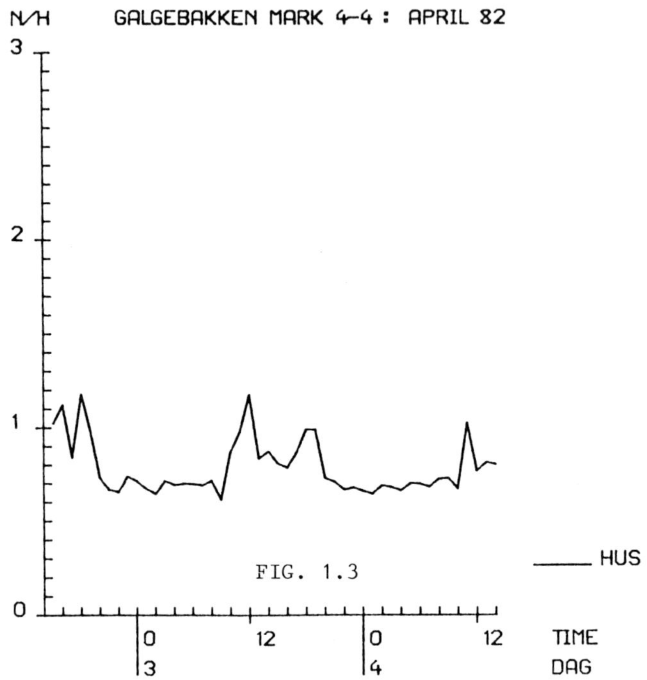
N/H
GALGEBAKKEN MARK 4-4 : APRIL 82
3
2
1
0
0
12
0
12
TIME
3
4
DAG
HUS

FIG. 1.3

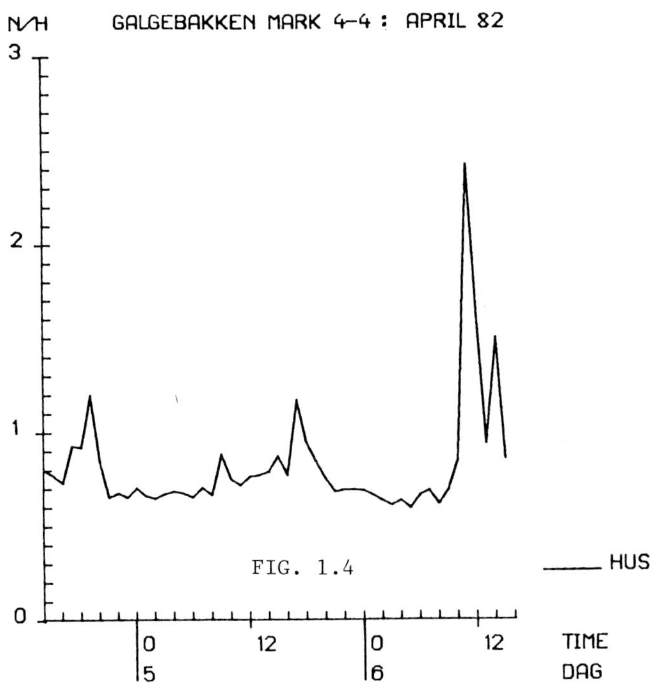
N/H
GALGEBAKKEN MARK 4-4 : APRIL 82
3
2
1
0
0
12
0
12
TIME
5
6
DAG
HUS

FIG. 1.4

Galgebakken Square 5-5

Ventilation Conditions during the Measurements

During all measurements the doors within the house were ajar, windows were closed and outer doors were closed as well. Specific conditions for the individual measurements were:

Measurement No. 1: All Ventilation openings open.

Measurement No. 2: All ventilation openings closed.

Measurement No. 3: Metal ventilation openings open. Fresh- and Vento valves in living room open.

Measurement No. 4: Metal ventilation openings open except for ventilation openings at the bottom of the doors.

Measurement No. 5: All ventilation openings closed.

Measurement No. 6: Ventilation opening at 1. floor open. Fresh valve in living room open. Metal ventilation openings in kitchen and hall open.

Measurement No. 7: All metal ventilation openings open.

Measurement No. 8: All ventilation openings closed.

Measurement No. 9: Metal ventilation openings in hall, kitchen, and small bedrooms open. Vento valves open.

The Climate during the Measurements

14.4 Measurements Nos. 1, 2, 3.

8 a.m.	2.9°C	1	Beaufort	west
2 p.m.	9.1°C	3	"	"
9 p.m.	6.0°C	2	"	"

15.4 Measurements Nos. 4, 5, 6.

8. a.m.	5.9°C	3	Beaufort	north	west
2 p.m.	10.8°C	3	"	"	"
9 p.m.	6.5°C	0	"		

16.4 Measurements Nos. 7, 8, 9.

8. a.m.	5.1°C	2 Beaufort	west
2 p.m.	11.6°C	4 "	"
9 p.m.	7.4°C	3 "	north west

The climatic data obtained form weather station Torsbo.

Results

The results of the 9 measurements performed 14.4, 15.4, and 16.4 are shown in the diagrams below. The diagrams and histograms show the fresh air change when the dwelling is not used during the period 82.04.15 - 82.04.19.

Air Change (number of times per hour)

Room / Measurement	1 Bath-room	2 Hall	3 Bed-room	4 Living-room	5 Bed-room	HOUSE
1	0,06	0,86	0,48	0,77	0,55	0,65
2	0,00	0,57	0,26	0,46	0,17	0,36
3	0,02	0,51	0,50	0,74	0,60	0,63
4	0,00	0,63	0,36	0,48	0,25	0,40
5	0,01	0,64	0,13	0,30	0,15	0,24
6	0,00	0,46	0,27	0,31	0,51	0,32
7	0,01	0,19	0,22	0,44	0,45	0,37
8	0,00	0,37	0,22	0,34	0,11	0,26
9	0,04	0,27	0,79	0,61	0,56	0,58

Air Change (m^3 per hour)

Measurement \ Room	1	2	3	4	5	HOUSE
1	1	5	13	85	15	120
2	0	3	7	51	5	66
3	0	3	14	81	17	116
4	0	4	10	53	7	74
5	0	4	4	33	4	44
6	0	3	8	34	14	59
7	0	1	6	50	13	68
8	0	2	6	38	3	48
9	1	2	22	67	16	107

Air Change (percentage)

Measurement \ Room	1	2	3	4	5	HOUSE
1	1	4	11	71	13	100
2	0	5	11	77	7	100
3	0	3	12	70	15	100
4	0	5	14	72	9	100
5	0	8	8	75	9	100
6	0	5	13	58	24	100
7	0	2	9	71	18	100
8	0	4	13	77	6	100
9	0	1	21	63	15	100

Air Change (number of times per hour corrected for standard conditon).

Standard Condition: temperature difference $\Delta t = 17^{o}C$
wind velocity $V = 4.6$ m/s.

Room / Measurement	House
1	0,82
2	0,39
3	0,69
4	0,41
5	0,28
6	0,34
7	0,39
8	0,23
9	0,50

FIG. 2.2

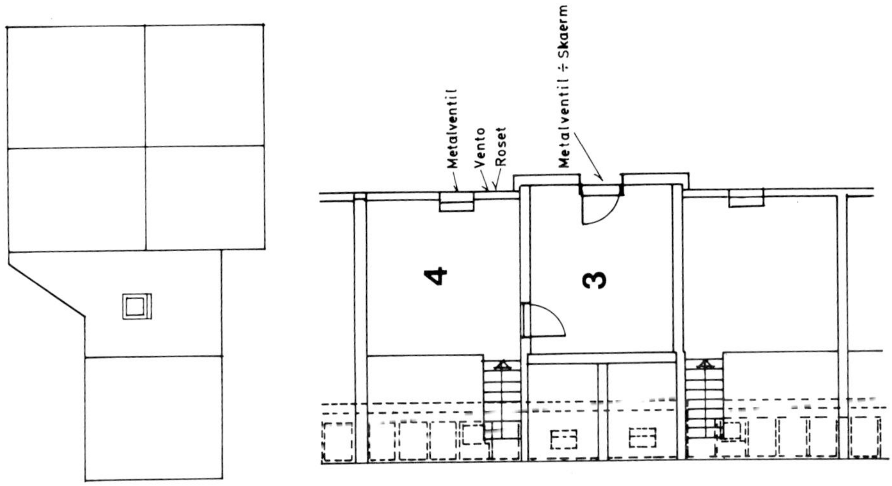
Metalventil
Vento
Roset
Metalventil ÷ Skaerm
4
3

FIG. 2.1

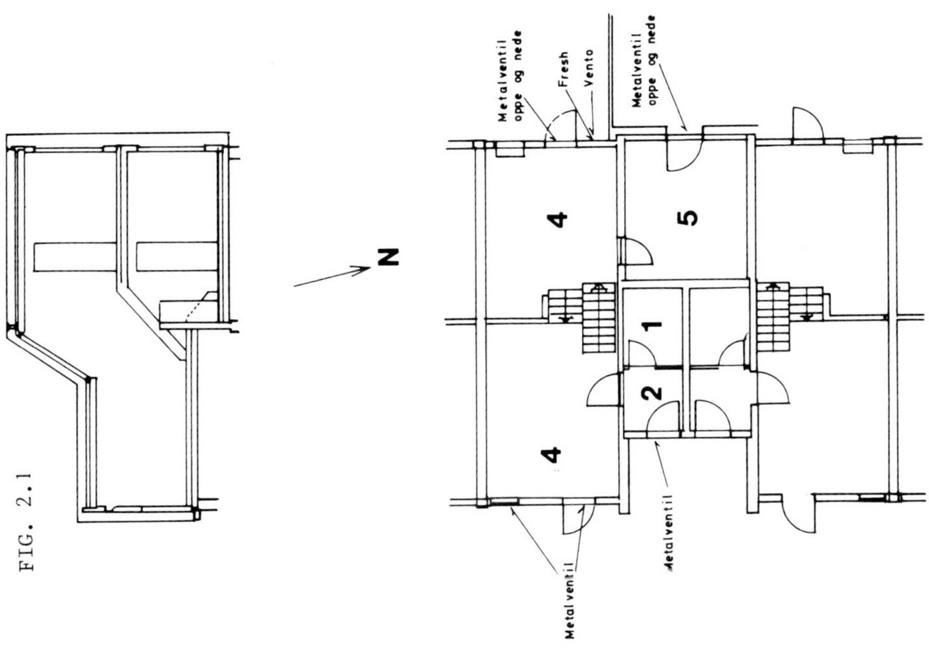
Metalventil
oppe og nede
Fresh
Vento
Metalventil
oppe og nede
N
4
5
1
2
4
Metalventil
Metalventil

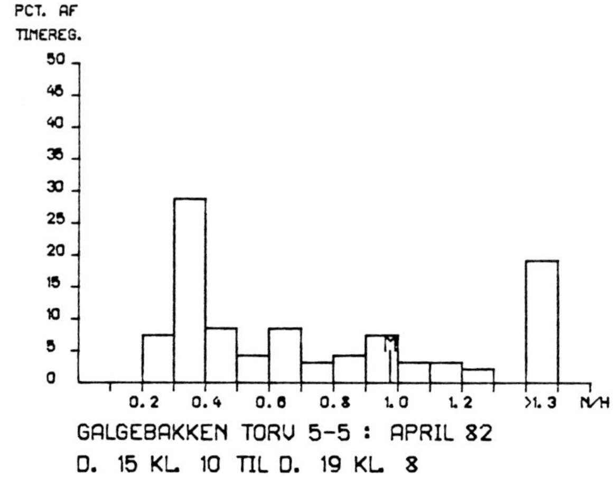
PCT. AF
TIMEREG.
50
45
40
35
30
25
20
15
10
5
0
0.2
0.4
0.6
0.8
1.0
1.2
>1.3
N/H
M
GALGEBAKKEN TORV 5-5 : APRIL 82
D. 15 KL. 10 TIL D. 19 KL. 8

FIG. 2.3

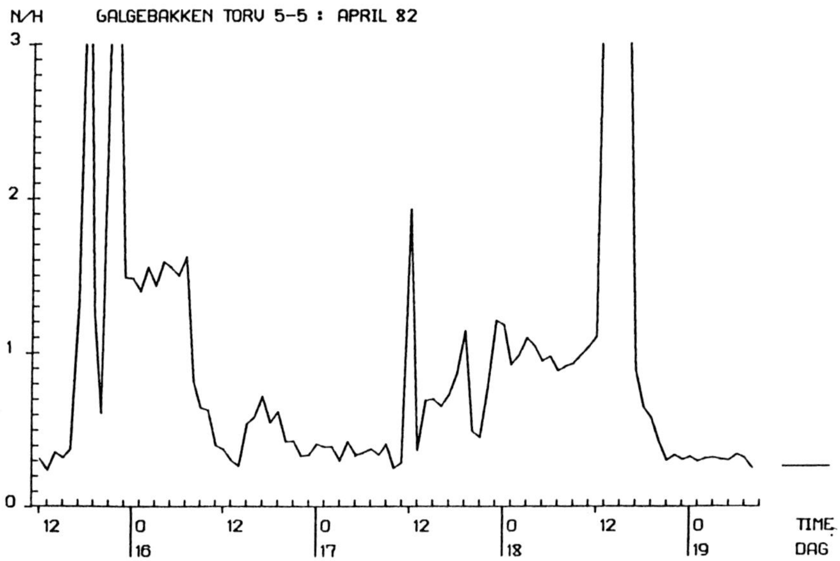
N/H
GALGEBAKKEN TORV 5-5 : APRIL 82
3
2
1
0
12
0
16
12
0
17
12
0
18
12
0
19
TIME
DAG

FIG. 2.4

Brøndby Strand, Hallingparken 52, 1.left

Ventilation Conditions during the Measurements

During all measurements windows and outer doors were closed.

Air quantities exhausted mechanically were put at the following values:

Room 7 (small bathroom)	approx.	40 m^3/h
Room 6 (big bathroom)	"	40 m^3/h
Room 1 (kitchen)	"	70 m^3/h

Specific conditions for the individual measurements were:

Measurement No. 1:	Doors within the house ajar. Mechanical exhaust on. All ventilation openings open 10 mm.
Measurement No. 2:	Doors within the house ajar. Mechanical exhaust on. Ventilation openings closed.
Measurement No. 3:	Doors within the house closed. Mechanical exhaust on. Ventilation openings closed.
Measurement No. 4:	Doors within the house ajar. Mechanical exhaust on. Ventilation openings in bedrooms and living room open.
Measurement No. 5:	Doors within the house closed. Mechanical exhaust off. Ventilation openings closed.

The Climate during the Measurements

24.4 Measurement No. 1

8 a.m.	6.0°C	0	Beaufort	north
2 p.m.	11.1°C	0	"	"
9 p.m.	7.7°C	2	"	"

25.4 Measurement No. 2

8 a.m.	6.4°C	0	Beaufort	north
2 p.m.	12.2°C	2	"	north west
9 p.m.	10.0°C	4	"	west

26.4 Measurement No. 3

8 a.m.	8.4°C	5 Beaufort	north west
2 p.m.	12.5°C	7 "	" "
9 p.m.	8.4°C	5 "	" "

27.4 Measurement No. 4

8 a.m.	6.6°C	4 Beaufort	north
2 p.m.	11.8°C	3 "	"
9 p.m.	11.2°C	5 "	"

28.4 Measurement No. 5

8 a.m.	9.4°C	6 Beaufort	west
2 p.m.	11.1°C	6 "	"
9 p.m.	7.9°C	5 "	"

Results

The results of the 5 measurements performed during the period 82.04.23 - 82.04.28 are shown in the diagrams below. The air change (number of times per hour) is also shown in the diagrams.

Air Change (number of times per hours)

Room / Measurement	1 Kitchen	2 Living room	3 Hall	4 Bedroom	5 Bedroom	6 Bathroom	7 Toilet	HOUSE
1 (23-24)	0,72	0,63	0,38	1,70	0,88	0,01	0,01	0,73
2 (24-25)	0,89	0,85	0,63	1,25	0,49	0,10	0,16	0,75
3 (25-26)	0,91	0,90	0,71	0,83	0,44	0,14	0,02	0,72
4 (26-27)	0,54	0,83	0,31	1,89	0,98	0,03	0,02	0,76
5 (28-28)	0,28	0,30	0,02	0,00	0,00	0,00	0,00	0,14

Supplement 3 (cont.)

Air Change (m^3 per hour)

Room / Measurement	1	2	3	4	5	6	7	HOUSE
1 (23-24)	33	29	10	39	26	0	0	138
2 (24-25)	41	39	17	29	14	1	1	142
3 (25-26)	42	41	19	19	13	2	0	136
4 (26-27)	25	38	8	44	28	0	0	144
5 (27-28)	13	14	1	0	0	0	0	27

Air Change (percentage)

Room / Measurement	1	2	3	4	5	6	7	HOUSE
1 (23-24)	24	21	8	28	19	0	0	100
2 (24-25)	29	27	12	20	10	1	1	100
3 (25-26)	31	31	14	14	9	1	0	100
4 (26-27)	17	27	6	30	20	0	0	100
5 (27-28)	47	51	2	0	0	0	0	100

Air Change (number of times per hour corrected for standard conditions).

Standard conditions: temperature difference $\Delta t = 17^{\circ}C$
wind velocity $V = 4.6$ m/s

Room / Measurement	House
5	0,15

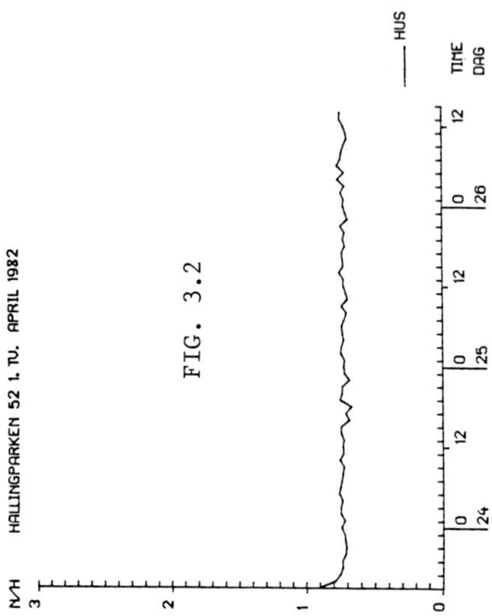

FIG. 3.2

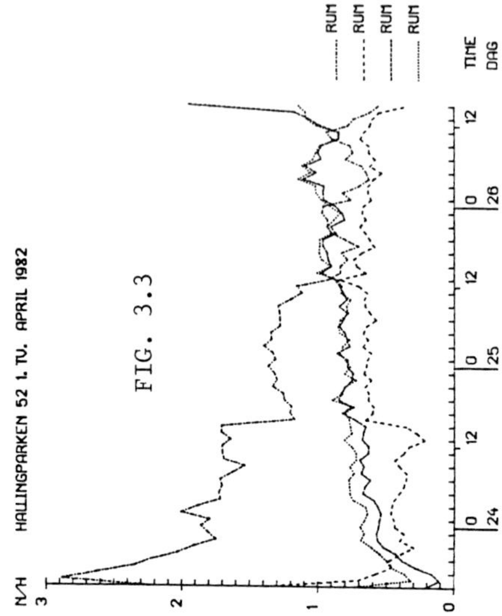

FIG. 3.3

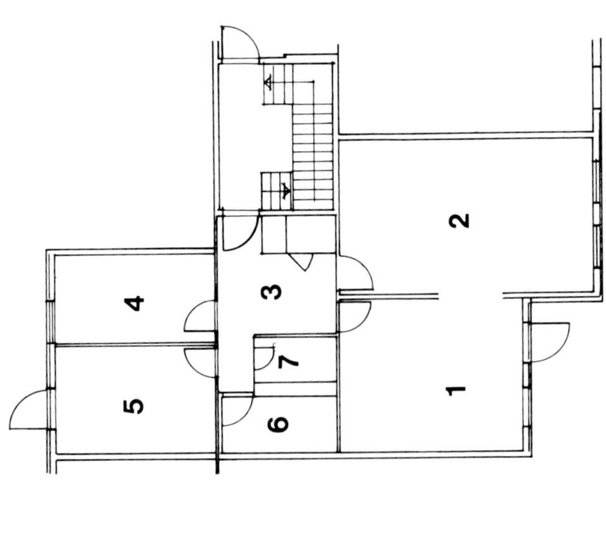

FIG. 3.1

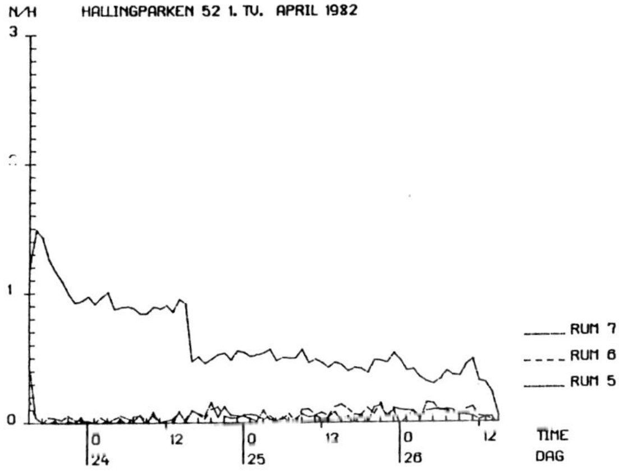

FIG. 3.4

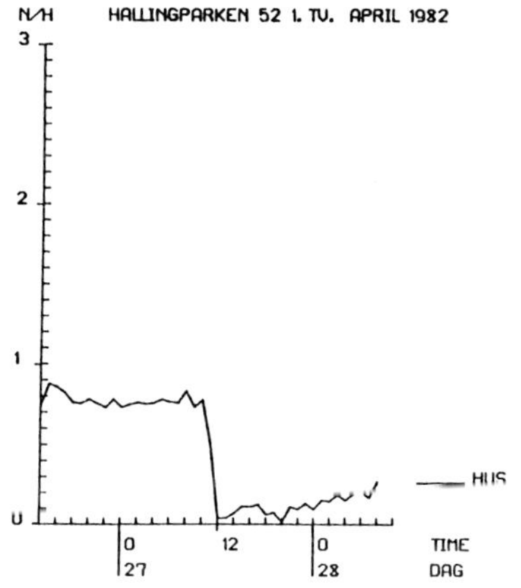

FIG. 3.5

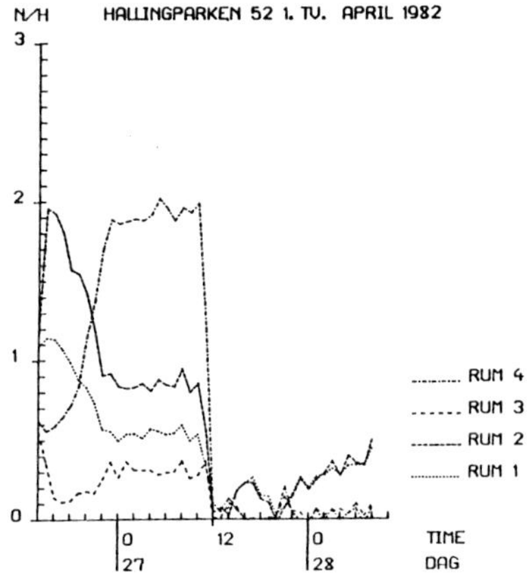

FIG. 3.6

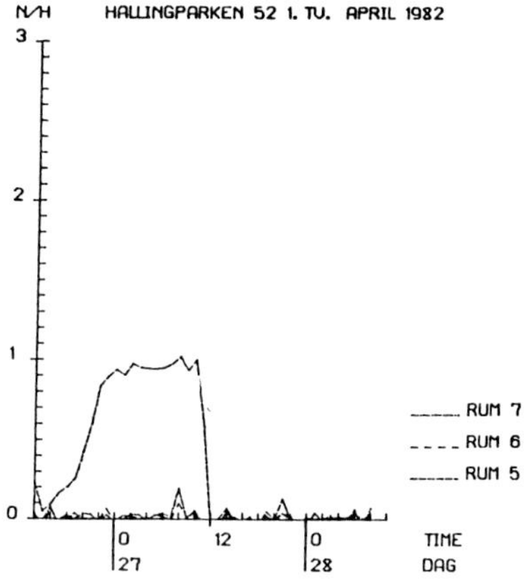

FIG. 3.7

Brøndby Strand, Tranumparken 1, 3, 4

Ventilation Conditions during the Measurements

During all measurements the doors within the house were ajar, windows were closed and outer doors were closed as well.

Air quantities exhausted mechanically were measured and the following values were registered:

Room 3 (big bathroom)	approx.	30 m^3/h
Room 4 (small bathroom)	"	40 m^3/h
Room 9 (kitchen)	"	70 m^3/h

Specific conditions for the individual measurements were:

Measurement No. 1: Mechanical exhaust on. Ventilation openings closed.

Measurement No. 2: Mechanical exhaust on. Ventilation openings open.

Measurement No. 3: Mechanical exhaust off. Ventilation openings closed.

The Climate during the Measurements

29.4 — Measurement No. 1

8 a.m.	6.9 °C	4 Beaufort	west
2 p.m.	6.9 °C	2 "	"
9 p.m.	4.5 °C	1 "	"

30.4 — Measurement No. 2

8 a.m.	3.2 °C	2 Beaufort	west
2 p.m.	9.4 °C	5 "	"
9 p.m.	5.8 °C	4 "	"

1.5

8 a.m.	7.8 °C	3 Beaufort	west
2 p.m.	10.1 °C	5 "	"
9 p.m.	4.2 °C	1 "	"

2.5 Measurement No. 3

8 a.m.	4.1°C	4	Beaufort	west
2 p.m.	7.0°C	4	"	"
9 p.m.	5.0°C	2	"	"

Results

The results of the 3 measurements are shown in the diagrams below. The air change (number of times per hour) is also shown in diagrams for measurements Nos. 2 and 3. A curve showing the concentrations at a cycle with 3 hours' dosage and 3 hours without dosage is enclosed. Rooms Nos. 3 and 4 are exceptional. The curve for Room 3 shows the tracer gas concentration between the flooring and the concrete deck. The curve for Room 4 shows the concentration in a wardrobe in the hall.

Air Change (number of times per hour)

Room / Measurement	1 Living room	2 Bedroom	3 Big bathroom	4 Small bathroom	5 Small bathroom	6 Bedroom	7 Small bedroom	8 Hall	9 Kitchen	10 Hall	HOUSE
1	0,22	0,47	0	0,35	0,60	0,38	0,66	2,94	0,54	0,35	0,47
2	0,85	1,19	0	0,04	0,63	0,45	0,95	1,24	0,18	0,03	0,63
3	0,26	0,49	0	0	0	0,02	0,21	0,51	0,11	0,10	0,21

Air Change (m^3 per hour)

Room / Measurement	1	2	3	4	5	6	7	8	9	10	HOUSE
1	14	18	0	3	11	10	13	24	15	10	117
2	54	45	0	0	12	12	18	10	5	1	157
3	17	19	0	0	3	1	4	4	3	3	52

Air Change (percentage)

Room / Measurement	1	2	3	4	5	6	7	8	9	10	HOUSE
1	12	15	0	2	10	9	11	20	13	8	100
2	35	29	0	0	8	8	11	6	3	0	100
3	32	35	0	0	5	1	8	8	6	5	100

Air change (number of times per hour corrected for standard condition).

Standard condition: temperature difference $\Delta t = 17^{o}C$
wind velocity $V = 4.6$ m/s

Room / Measurement	House
3	0,19

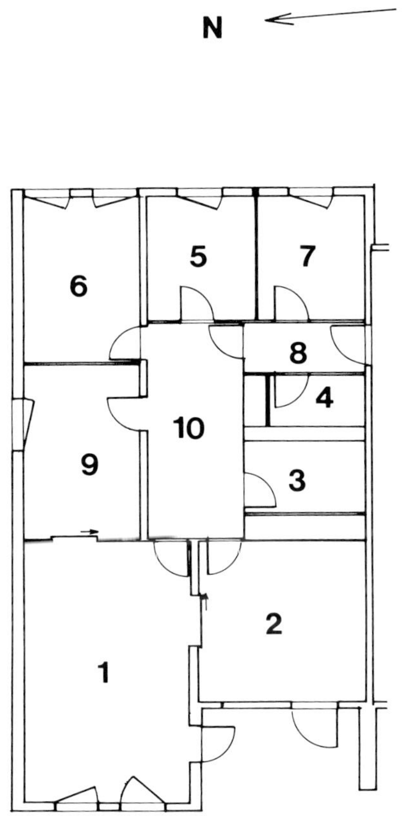

FIG. 4.1

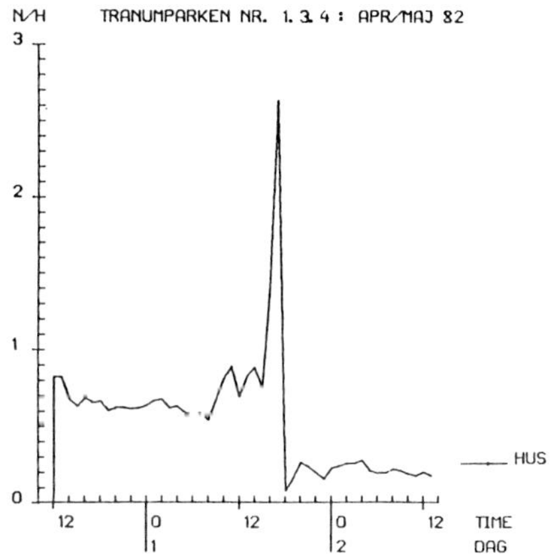

FIG. 4.2

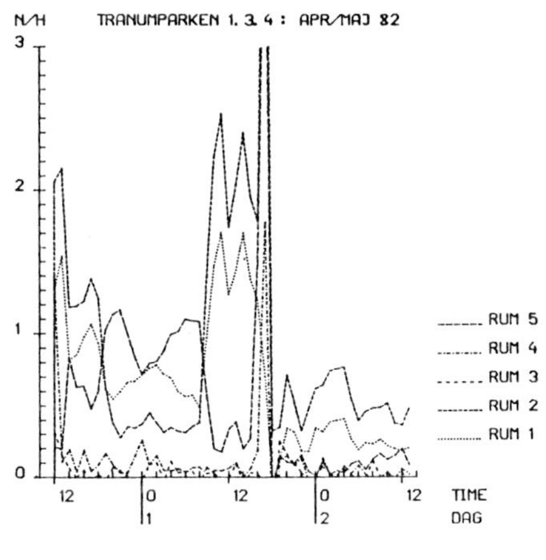

FIG. 4.3

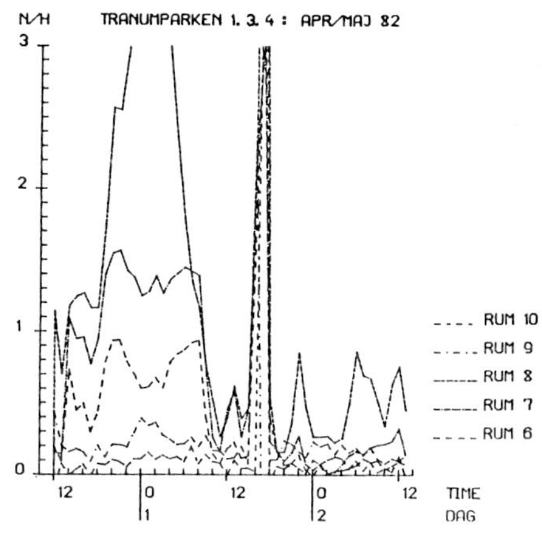

FIG. 4.4

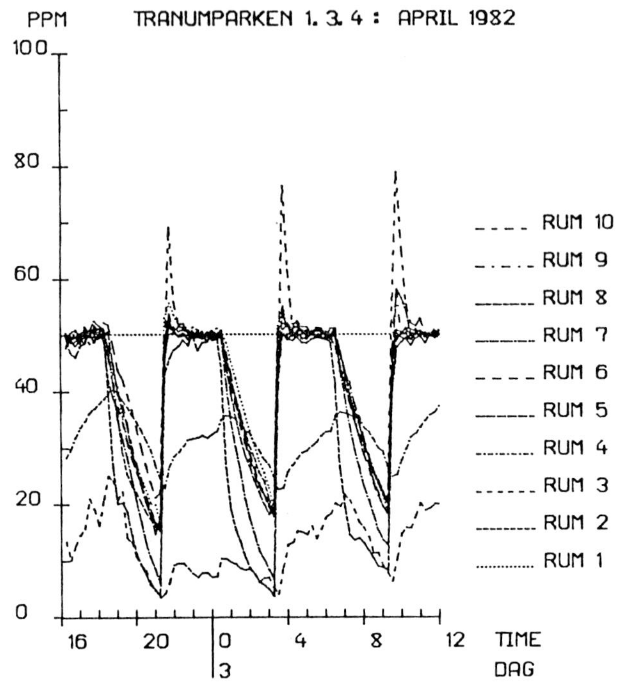
PPM
TRANUMPARKEN 1. 3. 4 : APRIL 1982
100
80
60
40
20
0
16
20
0
3
4
8
12
TIME
DAG
RUM 10
RUM 9
RUM 8
RUM 7
RUM 6
RUM 5
RUM 4
RUM 3
RUM 2
RUM 1

FIG. 4.5

QUESTIONS AND DISCUSSION

Eastwell (BSRIA) : 1. Actual vs. "ideal" are different problems. In order to legislate it is necessary to define conditions carefully and this implies no occupancy. Personal choice is an addition to this basic requirement.
2. UK experience suggests that possession of controls does not result in their use. Thus openable area for minimum ventilation is leaky to result in under ventilation.

Comment - Dewsbury (BSRIA) : It should be recalled that UK buildings have, until quite recently, been intentionally "leaky" to provide adequate ($\sim$ 0.5 ac/h) fresh air.

Uyttenbroeck (Chairman) : The behaviour of occupants is much more influencing the a.c.r. than the basic infiltration in the house. Should we then ask for very tight houses ?

Answer : With natural a.c.r. it is not necessary to build very tight houses. It is a reasonable guess that the a.c.r. should be around 0,2 - 0,4 h^{-1} which can be improved by slots operated by the occupants.

Dewsbury (BSRIA) : Have you found that wind speed and direction affect the distrubution of fresh air between the fresh air inlets ?

Answer : Yes, but we do not believe that this matters very much. This is because the air in the house is quite well mixed and it does not matter very much if the air change per example in a bedroom comes from the outside or from the corridor.

Uyttenbroeck (Chairman) : Do you have in Denmark a standard for air infiltration ?

Answer : No, but we have a standard for obtaining the necessary ventilation.

CO_2 BASED VENTILATION IN BUILDINGS

Authors : F. THELLIER, R. GROSSIN

Contract number : EEA-5-053-F(D)

Duration : 12 months 1 December 1980 - 1 December 1981

Head of project : J.L. BOY-MARCOTTE, Head of Heat Transfer Department

Contractor : Société BERTIN & Cie

Address : Société BERTIN & Cie
BP 3
78370 PLAISIR, France

Summary

A new energy conservation technique based on the control of the "fresh air intake" through measurement of CO_2 content in extracted air was examined for different kind of buildings (houses, offices, stores, show-rooms). Analysis of air contaminants showed that CO_2 was the most satisfactory air quality parameter for ventilation control. Various techniques for measuring CO_2 concentration were examined in view to design a low cost sensor (around 200 US $); this led to the conclusion that only infrared absorption technique was able to meet the requirements of a use in habitat.
Computational simulation was performed in order to determine annual energy savings with regard to standard ventilation rates ; a technico-economic analysis showed that a CO_2 based ventilation system could provide a twice better pay-back period than an air-in/air-out heat exchanger recuperator. Future development of CO_2-based ventilation system is promising more especially as an enquiry made within European CO_2 sensor manufacturers has shown they could design and produce CO_2 appliances at a market price around 200 US $. Before developing such a device, it is necessary to experimentally ascertain the figures in energy saving yielded by the numerical simulation ; carrying out an experiment in actual conditions will also give the manufacturers the level of performance required for the sensor to achieve a fair energy saving/comfort compromise.

1. Aim of the study

The aim of the study is to assess the possibilities of conserving heat energy in a building through control of the fresh air intake. Fresh air flow rate allowed in will be varied according to the carbon dioxide content of the air inside the building which depends on how many people are present. Contemplated buildings are large business premises (show-rooms, stores, offices) as well as family residences.

The scope of the study is as follows :

- standards in force in various countries,
- analysis of real situations,
- daily numerical simulation allowing to compute the heat energy conserved,
- selection of the CO_2 sensing technique,
- conclusion.

2. Standards applying in various countries for air renewal

Standards are extremely various; the present situation is evolving as the heat insulation scheme is spreading and induces substantial heat losses due to air renewal.

The following ranked instances summarize the situation (data refer to a 100 m^2 area apartment (250 m^3) :

i) USA, GERMANY, UNITED KINGDOM

Minimum flowrate is stated in m^3/hr per person depending on the activity and the type of premise.

For a 250 m3 apartment : minimum flowrate 110 m^3/hr.

ii) FRANCE, DENMARK

Minimum flowrate per room (Denmark) or per dwelling (France).

For a 250 m^3 apartment : minimum flowrate 170 m^3/hr for France, to be reduced to 120 m^3/hr in 1982 ; in Denmark only 80 m^3/hr are required.

iii) THE NETHERLANDS

Minimum and maximum flowrate depending on the type of room (bedroom, kitchen,...).

The standard implies 250 m^3/hr for a dwelling of 250 m^3.

ITALY and SPAIN : no standards applicable.

In JAPAN the standard refers to a maximum CO_2 content not to be exceeded : 1000 ppm.

3. Analysis of real situations

Experiments in foreign countries are scarce.

In JAPAN, Nakahara and his team have experimented such a system in offices ; in the USA,experiments have been carried out in a school by a Berkeley team ; data relevant to family dwelling have not been published so far.

On the basis of these few bibliographic elements, the possibility to install flowrate controllers governed by the CO_2 content has been examined for existing buildings (Centre Pompidou (Paris), hospital, department store). Such analyses have shown that the prime condition for operating a flow control is to have separate fresh air blowers per volume controlled ; when heating is through warm air, air renewal rate can be controlled with a CO_2 sensor but if the volume to be heated can be considered as having a uniform CO_2 content : such a system can, for example, be installed in a cinema with a single hall ; but, if several halls are connected to a single heating system, the air renewal common to all halls cannot be controlled to assure comfort in all of them.

4. Daily numerical simulation of a dwelling equipped with a CO_2 content-control

To calculate the reduction in heat loss due to air renewal, brought in by the CO_2 content-governed flow control, a daily simulation allowing to check at every moment the CO_2 content has been worked out.

First, an assumption on rooms occupancy over 24 hours was established; from this assumption various strategies on start up and stopping of blowers (one or two speeds) were run. The result of the simulation is figured out by an average air flow to be compared with the steady flowrate that would result from applying the standard ; the comparison allows to assess the heat energy saved.

Examination of this simulation shows that the mean flow over 24 hours can be reduced by 60 % relatively to the French standard, which means an annual saving of 1500 kWhr for a 100 m^2 area apartment (250 m^3). Comfort is improved with respect to an air exhauster running at a constant rate of 1 vol/hr : when the rooms are over-occupied for several hours, automatic switch on the high speed of the blower, at 2 vol/hr for instance, allows maintaining the comfort conditions by holding the CO_2 content below 1000 ppm and the odors at a low level. Such a simulation is presented in Fig. 1.

Moreover, the system is low sensitive to poor airtightness of the house of about 0.1 to 0.2 vol/hr : CO_2 sensing acounts for natural aeration by reducing the forced ventilation.

5. Selection of CO_2 measuring technique

Various techniques for measuring the CO_2 content have been examined with a view to design a sensor costing a few thousand French Francs. Only the infrared absorption technique meets the whole of the following specifications :

- detection of a threshold between 500 and 5000 ppm
- accuracy on threshold detection : $\pm$ (100 ppm allowed)
- shift : <10 ppm per year
- yearly calibration allowed
- power supply : mains
- size : about 5 dm^3
- durability : MTBF : 10 years
- selling price : 1000 to 2000 FF

Call for bids over major European makers showed that the target price could be reached for a mass production of about 10000 units per month.

This tentative price, which needs of course to be confirmed, will allow the CO_2 content-fresh air control to be used in family residences, which means a potential energy conservation of many M TOE/year.

6. Conclusion

The computational simulation carried out in this study have shown that the heat energy savings achievable on heat lost through ventilation are of same magnitude whether the fresh air flow is controlled from CO_2 content or an exhaust air-intake air recuperator is installed.

The enquiry made of European instrumentation makers suggests that an infrared absorption-based CO_2 sensor could be produced at a selling price of 1000 to 2000 French Francs for a production of 10 000 units/month. This would permit to extend the use of this system with a good pay-back to the family habitat, providing potential for energy conservation of several M TOE/year.

Cost comparison with the air-in/air-out recuperator is shown below for a 250 m^3 volume house :

	Installed price French Francs	Annual saving kWhr	kWhr price FF/kWhr/year
Recuperator	9000	2000	4.5
CO_2 control	2500 to 5000	1500	1.5 to 3

The pay-out time of the CO_2-control system is thus twice shorter than for the recuperator : this comes essentially from the fact that there is no air ducting to install.

As no air ducting is required, the system can be fitted to existing buildings having an air exhauster running permanently at the standard's rate.

Development of this system is dependent on the development of a low-cost CO_2 sensor : before such a development is launched, it is necessary to experimentally ascertain the figures in energy saving yielded by the numerical simulation. This experimental approach will also give the instrument makers the level of performance required for the sensor to achieve an acceptable saving/comfort trade-off.

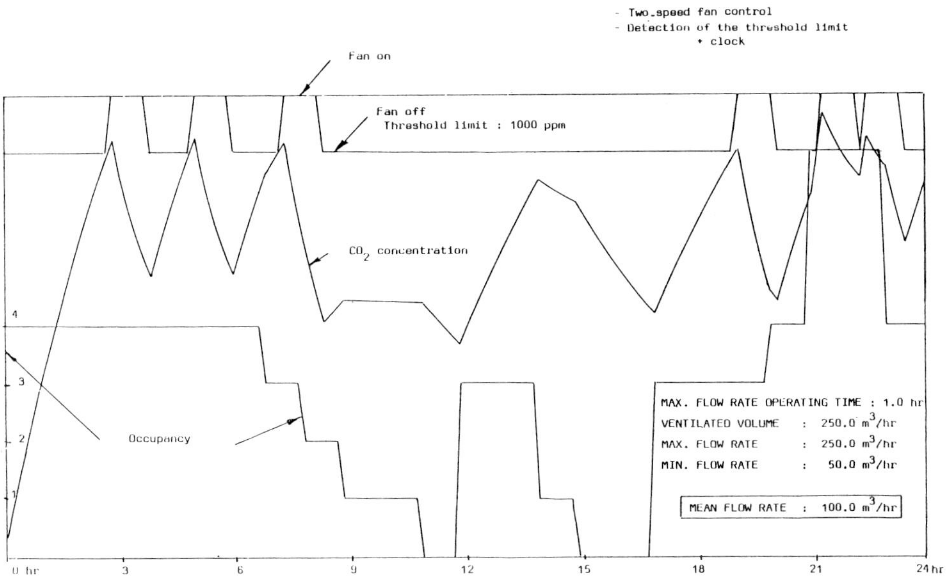

Fig. 1

QUESTIONS AND DISCUSSION

Pilavachi (DG XII) : CO_2 is a good key component to use to control ventilation. However in some cases other components may also be important e.g. vapours in the kitchen or bathroom, CO produced from ovens or heaters, etc.

Answer : Yes. So, in parallel with automatic control loops, we need a manual swich to run the fan during a given time (one hour for example).

Eastwell (BSRIA) : In the simulation was the dwelling assumed to be fully heated throughout and what weather conditions were assumed ?

Answer : The simulation is very simple and operates only on a daily profile. The savings demonstrated are based on this daily figure.

Uyttenbroeck (Chairman) : An acceptable price for the CO_2 devices probably can be obtained if the manufacturers can expect a large market (houses or dwellings). But first it has to be established whether the control of the CO_2 content in the exhaust air of a building alone, can result in an adequate air distrubution in the different rooms.

SESSION VII - CHARACTERISTICS OF HOUSES

Chairman: C. BOFFA

INTEGRATION OF ENERGY SAVING TECHNIQUES FOR DWELLINGS

Author : M. GUILLAUME

Contract Number : EEA-5-056-B (G)

Duration : 36 months 1 July 1980 - 20 June 1983

Head of project : J. UYTTENBROECK, Centre Scientifique et Technique de la Construction

Contractor : Centre Scientifique et Technique de la Construction (C.S.T.C.)

Address : C.S.T.C.
Avenue Pierre Holoffe 21
B - 1342 LIMELETTE

Summary

The aim of the research is the integration of the different energy saving possibilities in one-family houses, and the research work already completed has resulted in the design and construction of two identical buildings, covering a ground area of 107.6 m^2 in each case and with heat losses under extreme external temperature conditions (namely -10 °C) of approximately 7 kW and with an estimated energy requirement of 7000 kWh.
In order to satisfy these requirements one of the buildings will be fitted with a glycol water/water heat pump linked in one direction to 25 m^2 of solar panels producing a free heat gain of about 320 W/m^2 and in the other direction to a radiant floor heating system; in this case the operating coefficient is estimated to be 2.3 and corresponds to an energy consumption of 3038 kWh.
The other building will be heated with a more traditional installation consisting of, among other elements, an FOD boiler with an effective output of 14 kW and an efficiency at reduced load (average seasonal load factor = 35 %) of 77 %; the energy consumption characteristic of this system is estimated, on the basis of an overall seasonal efficiency of 60 %, to be 11,508 kWh or approximately 1150 l of FOD.

1. Introduction

The research has the following aims :
- improvement of the seasonal efficiency of heating installations
- integration of the results obtained previously on isolated themes
- the obtaining of more accurate estimations of the performances of installations
- the definition of a methodology of IN SITU measurement of the performances of installations
- the drafting of a code providing the basis for an integrated design of new and traditional thermal equipment.

The C.S.T.C. is carrying out this research with the assistance of the two following external laboratories :
- Laborelec
- the Building Physics Department of Liège University.

The main phases of this research are :
- the design, by Professor Dupagne of Liège University and based on Guidelines for Architects, of two identical buildings
- the calculation of classical heating plant by the C.S.T.C.
- the calculation and monitoring of alternative heat sources by Laborelec
- the testing, monitoring and interpretation by the C.S.T.C.

2. Description of the Installations

2.1 The buildings

The main geometrical characteristics of these two buildings are the following :
- overall ground-plan area : 107.6 m^2
- cellar space : 212 m3
- habitable space : 235 m3
- glazed area : 25.8 m2, i.e. 24 % of the plan area.

They are illustrated by the diagrams in figures 1 and 2.

2.2 The heating installations

The two buildings are fitted with radiant floor heating, but only that in the house heated with an alternative heat source (a single heat pump) will be used.

2.2.1 The "alternative" heating installation

The installation is illustrated in figure 3 and consists of the following main elements
- 25 m^2 of air-water solar panels on the roof oriented southwards (see also figure 1)
- a cold buffer store
- a single glycol water/water heat pump
- a warm buffer store
- heat emitters, which will be either a floor slab (radiant heating), or oversized radiators for heating with low-temperature water (45 °C)

2.2.2 The "classical" heating installation

This is illustrated in figure 4 and consists of the following elements :
- a boiler with a 14 kW FOD burner
- a one-pipe distribution circuit
- radiators fitted with 4-way one-pipe valves and sized for a water temperature range of 90 °/70 °C.

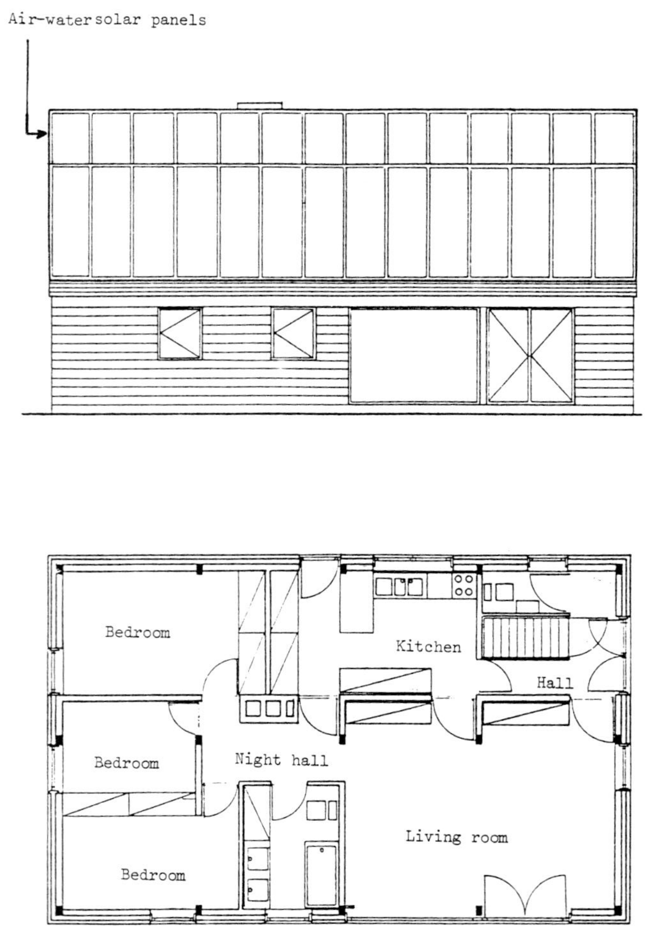

Fig. 1 - South facade and ground plan

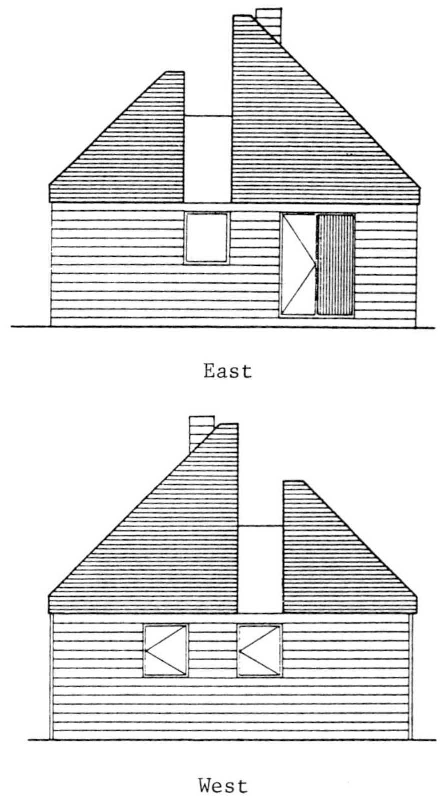

Fig. 2 - East and west facades

2.3 Measuring equipement

2.3.1 Number of measurement points

The readings, distributed over a total of 150 different measurement points in the two buildings, will be provided by :

- 5 integrator heat meters on the hydraulic circuits
- 7 integrator electricity meters to measure the consumption of the heat pumps, the boilers and the auxiliary equipment
- 6 integrator electricity meters for the consumption of the auxiliary electrical convector heaters
- 2 volumetric pulse meters for the FOD consumption
- 10 measurement points for the climatic conditions
- 60 measurement points for the dry and resultant internal air temperatures
- 60 measurement points for the water temperature in the hydraulic circuits.

2.3.2 Data acquisition and processing

A Motorola data acquisition system will be used to regularly log the 150 measurement points described above at intervals which vary from one channel to another.

The structure of this system and of the processing system for the data obtained are outlined in figure 5.

2.3.3 Hydraulic meters

The main component of the integrator heat meters is a volumetric hydraulic meter which generates electrical pulses.

We were unable to find transmission meters on the market with sufficient precision for this purpose, i.e. meters transmitting more than 10 electrical pulses per litre of water measured.

We therefore used meters of a widely available model (1 revolution per 22.2 cm^3) and linked them to Hewlett Packard encoders giving 500 pulses per revolution, i.e. 500 pulses per 33.3 cm^3. Under these conditions and with a water flow of 180 l/h, which is a typical value in our installations, one obtains an output signal (voltage) at a frequency of 750 Hz - a frequency which is perfectly acceptable for the conversion of frequency into voltage at the input to the heat integrators.

An example of a calibration curve is given in figure 6.

3. Results and State of Progress of the Research

Up till present we have not yet had the opportunity to make any measurements in the actual buildings.

Thus only the sizing calculations and the predictive calculations of performance can be examined here.

3.1 Heat losses from the buildings

These losses are calculated in accordance with the German standard DIN 4701 and the results obtained are given in figure 7. It can be seen that for a reference external temperature of - 10 °C the maximum losses amount to 6.96 kW, taking into account night set-back and with an (volumetric) average daytime internal temperature of 19.9 °C.

3.2 Solar gain factor

The theoretical (clear sky) solar gain factor has been calculated for the twelve months of the year (see figure 8).

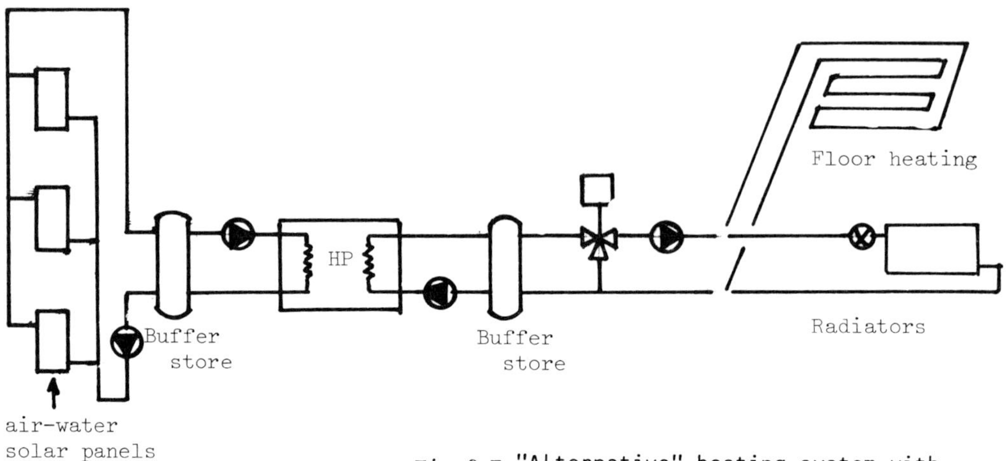

Fig.3 - "Alternative" heating system with a heat pump and solar collectors and floorheating or radiators.

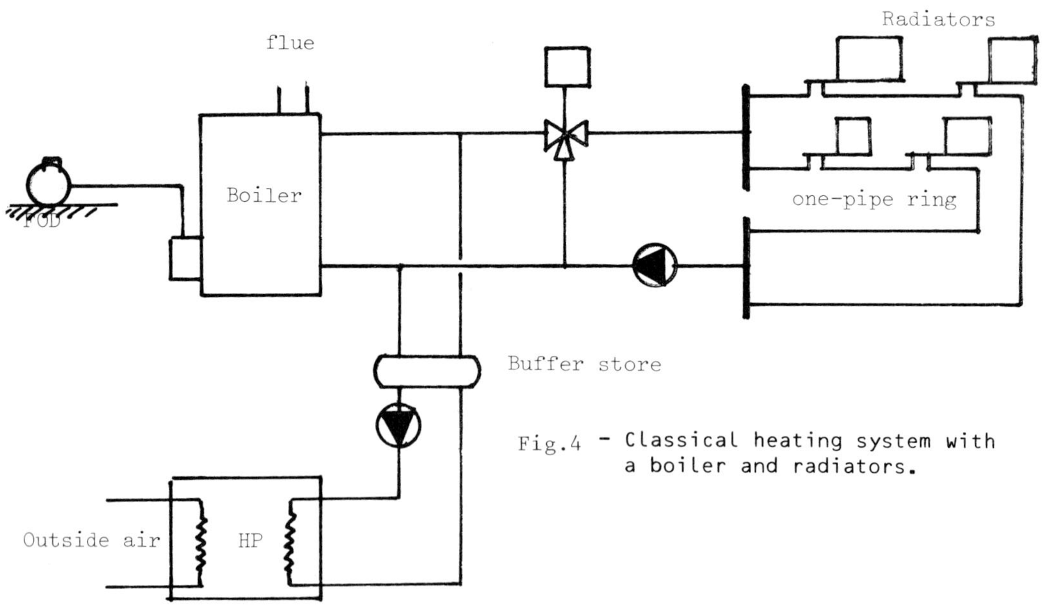

Fig.4 - Classical heating system with a boiler and radiators.

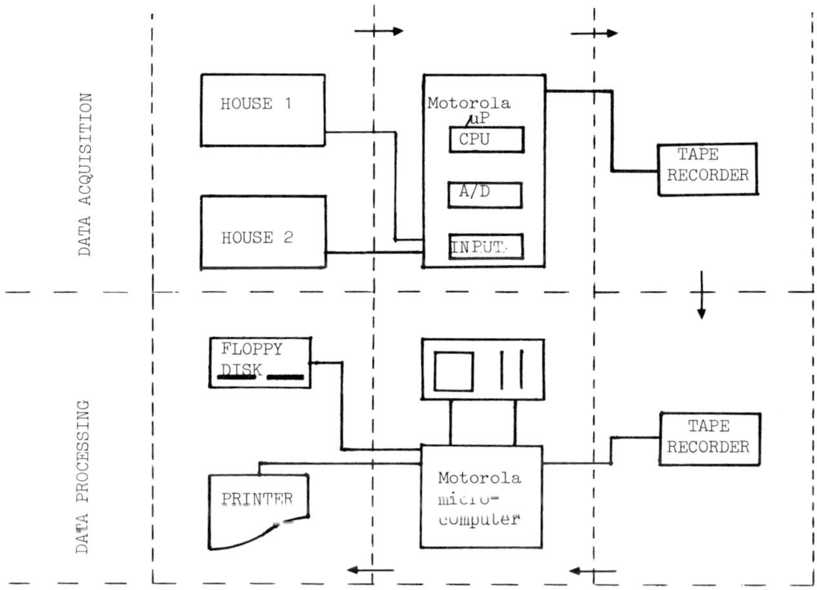

Fig. 5 - The Motorola data acquisition system which regularly logs 150 measuring points

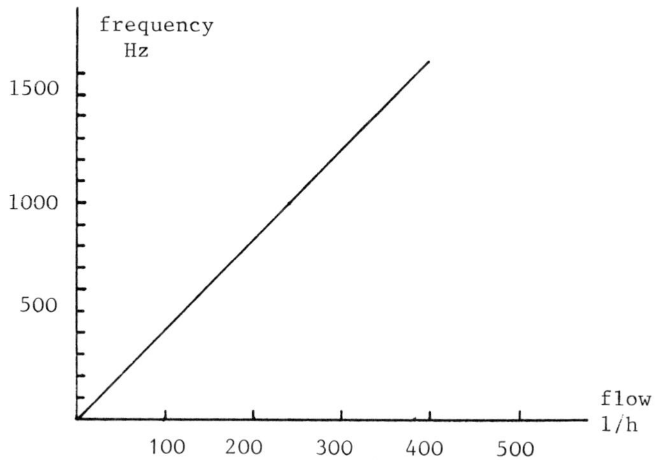

Fig. 6 - Calibration curve for a hydraulic meter, which is used as a heat meter

ROOM NUMB	TEMP.	TRANSM. LOSS	R	D	VENTIL. LOSS	TOT.HEAT LOSS
	°C	W	%	%	W	W
1.	22.	1958.	7.	-5.	910.	2907.
2.	24.	570.	7.	-5.	196.	777.
3.	18.	193.	7.	-5.	243.	440.
4.	18.	259.	7.	0.	217.	494.
5.	18.	373.	7.	5.	349.	767.
6.	20.	708.	7.	5.	382.	1175.
7.	15.	119.	7.	5.	43.	176.
8.	15.	110.	7.	5.	106.	229.
		4289.				6964.

VENTILATED VOLUME : 227.1 M3/H
HEATED VOLUME : 227.1 M3

MEAN INDOOR TEMPERATURE : 19.9 °C
OUTDOOR TEMPERATURE :-10.0 °C

INSULATION LEVEL-NBN B62301

KS (W/M2K)	V/S (M)	KSREF (W/M2K)	KA
0.49	0.8	1.00	K 49.

Fig. 7 - Heat losses of the building calculated according to the German standard DIN 4701

The average value for this factor, namely 12.637 °C, would produce an average temperature without heating of 13.21 °C in the buildings, assuming :

- a ratio of actual insolation to theoretical clear sky insolation of 0.588
- an average external temperature during the heating season of 5.78 °C.

3.3 Energy consumption

3.3.1 "Traditional" heating

The overall efficiency of this installation will be calculated using the following values :

- production efficiency, as indicated in figure 9	77 %
- emission efficiency of the radiators	91 %
- regulation system efficiency	90 %
- distribution efficiency	95 %

Under these conditions the overall seasonal efficiency of the installation is 60 %.

The energy consumption can be calculated from the following relationship

$$E.C. = \frac{24 \times d \times (U_{av}A + 0.36\ \tau V)\ (t_{ai} - t_{wh})}{1000\ \eta_{os}}$$

in which the following values are used :

d	= duration of the heating season : 210 days
$U_{av}A + 0.36\ \tau V$	= building heat loss for 1 degree of temperature difference = 225.3 W/°K
t_{ai}	= average internal temperature = 19.9 °C
t_{wh}	= temperature without heating = 13.21 °C
η_{os}	= overall seasonal efficiency of the installation = 60 %.

Thus one obtains an energy consumption of 11,508 kWh, or in other terms 1151 l FOD.

3.3.2 "Alternative" heating

In this case the glycol-water/water heat pump is linked to solar panels with an estimated heat output of 32 W/m^2 °C.

Under these conditions, and based on the average external temperature during the heating season, the average coefficient of performance (COP) of the heat pump can be estimated to be 2.8 as indicated in the graph in figure 10.

Using the following values to calculate the overall operating coefficient for the alternative installation

- emission efficiency of the floor slab	95 %
- regulation system efficiency	90 %
- distribution efficiency	95 %

one obtains a value of 2.3.

The energy consumption for this system is then estimated to be 3038 kWh.

3.4 Laboratory measurements

3.4.1 Measurements made on boilers

The 14 kW boiler which will be used for the traditional heating installation was tested in the laboratory in order to determine its efficiency characteristics

- with a constant water output temperature (about 70 °C)
- for different load factors.

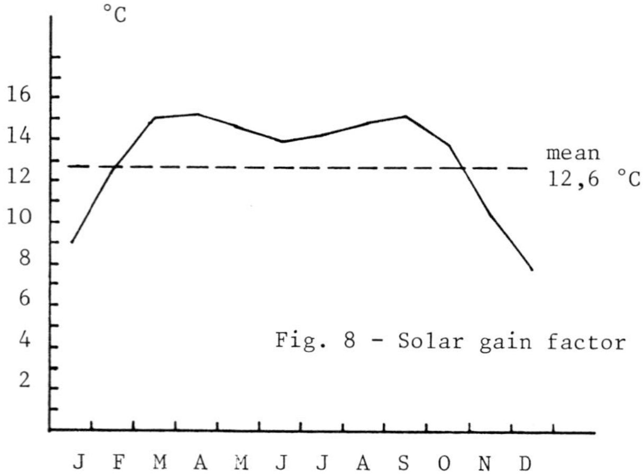

Fig. 8 - Solar gain factor

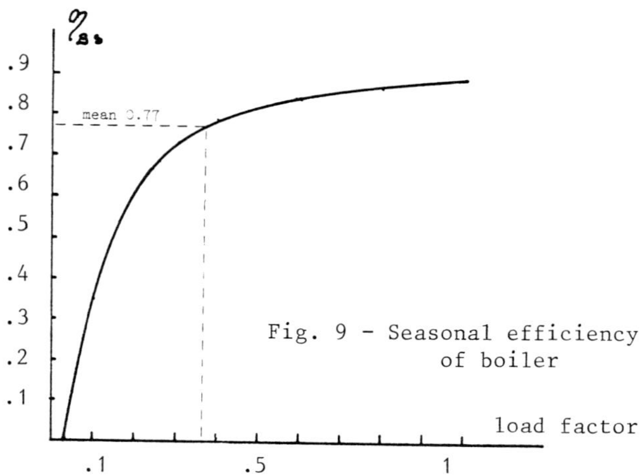

Fig. 9 - Seasonal efficiency of boiler

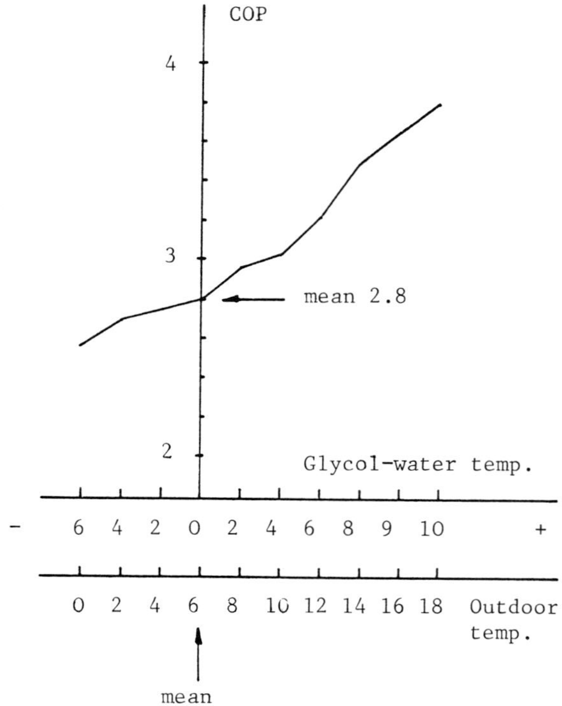

Fig. 10 - C.O.P. value of the heat pump as a function of the outdoor temperatures (with heat produced at 45 °C)

The results obtained are given in figure 9.

3.4.2 Measurements made on the heat pump

The heat pump which replaces the boiler was tested in the laboratory by Laborelec.

The main results obtained are represented on the graph in figure 11.

4. Conclusions

The five principal phases which constitute the research programme are outlined below.

- PHASE 1 - Bibliography

This is at present in preparation but has already allowed :

. Professor Dupagne to design the two buildings following the Guidelines for Architects
. the bringing together of disparate information on the functioning of boilers and, among other things, the publication of an article on this subject and the laboratory testing of a mathematical model which will be used for the analysis and interpretation of the actual measurements.

- PHASE 2 - Design and Construction of the Buildings

This important phase has been completed, as can be seen from figures 1 and 2.

- PHASE 3 - Calculation of the Installations

This has been completed and certain of these calculations have resulted in :

- the placing of the floor slabs (radiant heating)
- the purchase of the heat pumps and the boiler
- the predictive calculation of the overall performances of the different systems
- the checking in laboratory of their power, efficiency and COP characteristics.

- PHASE 4 - Thermal Balancing - Monitoring

The thermal balancing will begin during the month of June and will continue until October this year; it will probably be repeated as a check after the first heating season, i.e. from May next year. The monitoring has been made possible by the purchase of data acquisition and processing systems, which we are at present programming, and by the development of specific equipment such as the heat meters described previously.

- PHASE 5 - Interpretation of the Results

Certain models, based on well-known classical methods, for the interpretation of results have been developed; the deviation found between the actual measurements and the predicted values will enable us to :

- refine the methods used
- highlight those of our measurements which are of importance and representative for the IN SITU testing of the performances of new and traditional heating installations.

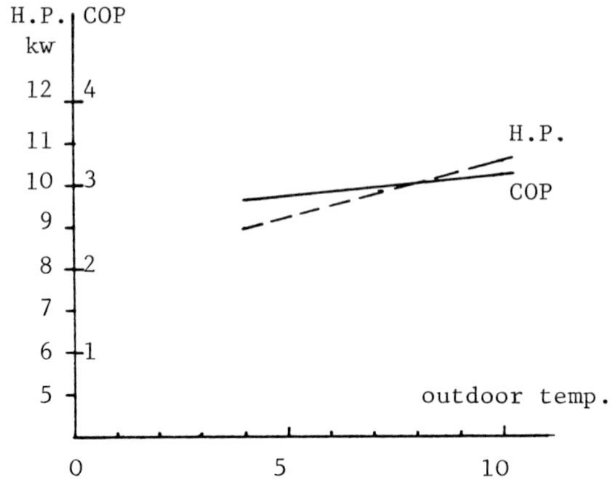

Fig. 11 - Heat capacity and C.O.P. of a heat pump as a function of the outdoor temperatures (with heat produced at 45°C).

QUESTIONS AND DISCUSSION

Clarke (University of Strathclyde) : It took the I.E.A. two years to monitor a building. What did you do in this field ?

Answer : This will be described in the guidelines for architects which we are preparing. I agree with you that it is difficult to develop a measuring system but it is often even more difficult to interpret the measurements.

Clarke (University of Strathclyde) : Often we don't know whether the data are correct or meaningful.

Guillaume (CSTC) : You have to start somewhere. We started to use the nominal boiler efficiency that was wrong and now we use the boiler efficiency with at 35 % heatload.

Prof. Boffa (Chairman) : Which computer programme did you use for the simulation of the thermal behaviour of houses which are not heated ?

Answer : The programme was developed by the University of Liège.

Prof. Boffa (Chairman) : Do you expect at the end of the project to have clear data on how much energy can be saved by for example replacing oversized heating systems.

Answer : Yes.

MAIN CHARACTERISTICS FOR ENERGY SAVING IN LOW COST HOUSING

Authors : H. HEIKHAUS, L. LARET, J. LEBRUN

Contract number : EEA-5-057 B

Duration : 01.01.80 - 31.12.81

Head of project : H. HEIKHAUS

Contractor : Centre Scientifique et Technique de la Construction

Address : C.S.T.C., Rue du Lombard, 41, 1000 BRUXELLES

Summary

The main results of the research presented here consist in :

- an enormous amount of experimental data collected on the reduced scale models and on the corresponding micro-climate during nearly three years in Limelette;
- a detailed statistical analysis of the energy consumed by the (continuously) heated models as function of the corresponding temperatures of the unheated ones.
- Some more limited analysis of the dynamic behaviour of the heated (continuously and intermittently) and unheated models;
- a detailed comparison of different solar radiation calculation methods among themselves and with the numereous measurements realized on site;
- a proposal of practical estimation of the ground losses;
- a practical method to estimate the heating energy demand through tables of "equivalent" degree-days,
- a second order dynamic approach which could easily be used in many practical problems;
- a statistical analysis of Belgian meteorological data which allows a rational choice of the heating design temperature. As the calculation of equivalent degree-days, their analysis could easily be transposed to any other climate.

Introduction

The results here presented are those of the third and last two-year period of research program initially devoted to the "real fuel consumption for the heating of houses". The starting point of this research has consisted in evaluating the real effect of different energy saving measures (mainly improvements of the enveloppe) in existing but unoccupied houses.

The work done during the first period was mostly a provocation. |1| It showed among other that :

1) the energy saving measures considered were giving the expected results, at least separately, but their superposition was rather disappointing;
2) the results were significantly affected by the solar free gains;
3) these solar free gains could be taken into account through the simple concept of "temperature without heating" (t_{WH});
4) the temperature without heating was measurable and should also a priori be calculable.

In order to overcome the numereous uncertainties that occur in the monitoring of real houses, an "intermediate step" of experimental analysis has been introduced into the second research period : it consists in measuring the "floating" temperature and heating energy consumption of a dozen of very simplified (and therefore physically well-defined) reduced scale models.

This new experimental tool has been used parallelly with other attempts of measurements and calculations on real houses and with the development of the theoretical methods |2| |3| |4| |5| |6|.

It has been demonstrated that :

1) a steady state analysis could effectively provide a satisfactory estimation of the (average) heating energy demand of a house;
2) the prediction of the temperature without heating t_{WH} could still be improved by a better choice of solar and infra-red radiation hypotheses;
3) a low order dynamic model (including no more than one or two time constants) was "rich" enough to represent the oscillations of the building temperature when its heating control was turned off or overloaded.

The third "period", which has just been achieved includes the latter experimental and theoretical developments as well as a synthesis and a practical valorization of the previous results. Specific tasks results and conclusions of this work are briefly presented hereafter.

2. Experimental and theoretical tools

2.1. Real houses

One more limited, but rather satisfactory verification of the prediction of t_{WH} has been realized on two research houses monitored by the Building Research Establishment (B.R.E. Garston, U.K.) |7|.

2.2. Reduced scale models

Twelve cubic models of 1 m^3 of internal volume were built, installed and monitored by the C.S.T.C. in Limelette during the preceeding research period. They are of 6 different "types" differing from each other by insulation, thermal capacity, tightness, orientation, fenestration (with or without window) and orientation (South or East). One standard model was tested under different configurations in a wind tunnel of the University of Liège.

The monitoring of the twelve reduced scale models has been continued by the C.S.T.C. in Limelette. A new "pair" of "high inertia" models was installed on the same site for the last year of measurements (1980-1981). One more "standard" model was also installed in the vicinity of Liège in 1980.

2.3. Theoretical tools

2.3.1. Detailed dynamic behaviour of the reduced scale models (as well as of real buildings) have been and are still studied by our reference computer programs LPB-1 and LPB-2 : these two programs are calculating the "environmental" temperature and/or the corresponding thermal loads in one or several separate zones |8| |9|.

2.3.2. The simplest steady-state and one-zone approach which has been developed on the bases of the first results of this research is the program LPB-3.

It uses the concepts of recuperating coefficient and "equivalent degree-days" :

1) By establishing a very simple steady state heat balance of the building, at different seasons (practically on the 15th of each month) one can estimate its temperature without heating (t_{WH}) in clear sky conditions (CS). The difference between this term $(t_{WH})_{cs}$ and the outside air temperature t_o is called "recuperating coefficient" :

$$R = (t_{WH})_{cs} - t_o \qquad (1)$$

2) The very simple hypothesis which has been used up to now in LPB-3 to allow a heat balance in any other weather conditions, is to assume that the building overheating is proportional to the solar intensity measured on a horizontal surface (I_{tH}) :

$$t_{WH} = t_o + R \ J \qquad (2)$$

with $J = I_{tH}/(I_{tH})_{cs} \leqslant 1$

It will appear later how this method can be improved (cf § 4.1.).

3) As was theoretically and experimentally established, the steady-state heating energy demand (Q) of the building can be expressed as follows :

$Q = K\ (t_i - t_{WH}) - Q_{occ}$

with K = global heat transfer coefficient

Q_{occ} = occupancy free heat

$$\text{or } Q = K\ (t_{NH} - t_{WH}) \qquad (3)$$

with $t_{NH} = t_i - Q_{occ}/K$

t_{NH} = "temperature of no heating"

Thus, by integrating this term on a certain period :

$$\int Q\, d\tau = \int K\ (t_{NH} - t_{WH})\, d\tau$$

$$\simeq K(\Delta\tau)\ \Sigma\ \overline{(t_{NH} - t_{WH})}$$

with $\Delta\tau$ = duration of one day = 86400 seconds

$\Sigma\,\overline{(t_{NH} - t_{WH})}$ = "equivalent" degree-days of the period.

This last term is pre-calculated day by day on the basis of the two meteorological variables, t_o and $\underline{J}$, and for different values of the two parameters, t_{NH} and R.

2.3.3. The extension of LPB-3 to a multi-zone problem is very efficiently realized by the LPB-4 program. |10|

2.3.4. Finally, and also within the frame of the present research program, simplified dynamic approaches have also been developed and compared to the experimental results. LPB-5 is an example of solution, based on a second order schema |3| |4|

3. Analysis of the main results

3.1. Steady state analysis of the behaviour of the reduced scale models

More details on this analysis can be found in a previous paper |11|. All transient effects were eliminated from the present analysis by treating them in 10-days averages |2| |3|.

3.2.1. Heating energy consumptions

In order to check the validity of equation (3), the measured values of Q, t_i (=t_{NH}) and t_{WH} have been analyzed through the following linear regression :

$$Q = C_o + C_1(t_i - t_{WH}) \tag{4}$$

Main results and a graphical example are given in Table 1 and Figure 1 respectively.

Table 1 : Regressions Q : $C_o + C_1\ (t_i - t_{WH})$ Period : 1979-81

Model	Normal	Insulated	Inert	East Orient.	Without Ventil.	Without Window	New Inert
Co|W|	-4.87	-3.81	-1.73	-4.14	+0.02	-1.74	-0.54
C_1|WK^{-1}|	7.08	6.05	6.96	7.36	6.21	5.87	7.08
Correlation factor	0.996	0.996	0.996	0.995	0.999	0.996	0.993

Figure 1
Correlation Q : f(t_i,t_{WH})
"normal" model

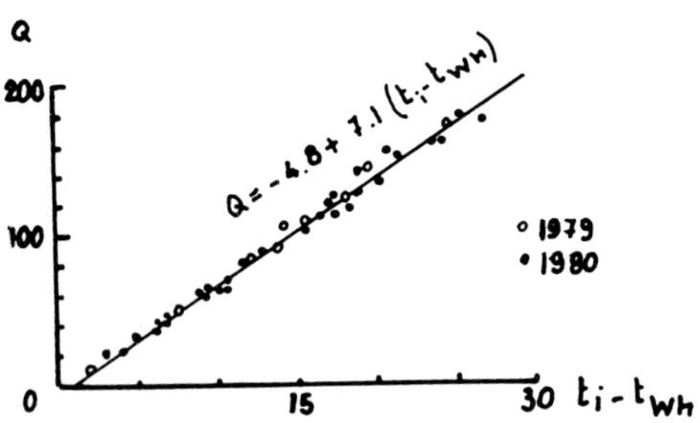

According to equation (3), we should have :

$C_o = 0$
$C_1 = K$ = global heat loss coefficient of the model.

In fact K is also depending on the wind speed (4) and slightly also on the temperature difference (buoyancy effect). These two effects explain respectively :

1) a certain scattering of the results, i.e. an uncertainty of C_1;
2) a slight non linearity : $Q \div (t_i - t_{WH})^{1+n}$

i.e. a negative value of C_o in certain cases.
The wind effect has effectively been taken into account by using a double regression :

$$Q = C_3 (t_i - t_{WH}) + C_4 (t_i - t_{WH}) . u \qquad (5)$$

The results are given in Table 2

Table 2 : Regressions $Q : C_3(t_1 - t_{WH}) + C_4(t_1 - t_{WH}) . u$ Period 1979-81

Model	Normal	Insulated	Inert	East Orient.	Without Ventil.	Without Window	New Inert
$C_3 \|WK^{-1}\|$	5.93	5.04	6.08	6.23	5.96	5.12	6.35
C_4 $\|WK^{-1}m^{-1}s\|$	0.29	0.26	0.25	0.29	0.086	0.22	0.23
Correlation factor	0.999	0.998	0.999	0.998	0.999	0.999	0.998

These results are very satisfactory; they are in agreement with the theoretical characteristics of the model (partially validated in Laboratory). Figure 2 is another illustration of the significance of equation (5) : the "calculated" values of Q are here based on the theoretical characteristics of the model considered...

Figure 2
correlation
$Q_{measured} = f(Q_{calculated})$
model "normal"

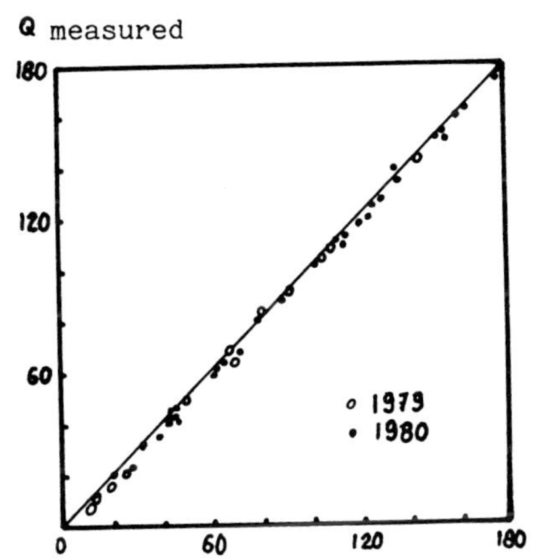

3.2.2. Temperatures without heating (t_{WH})

Measurements of $(t_{WH} - t_o)$ were also compared with theoretical calculations.

The results presented in Table 3 correspond to two bases of calculations :
1) equation (2);
2) measured solar intensities (I_t) on the four wall orientations.

Table 3 : mean differencies Δ and corresponding standard deviations δ (in degrees C) between calculated and measured t_{WH} (over 56 ten-day periods)

Model		Normal	Insulated	Inert	East Orient.	Without Ventil.	Without Window
equation (2)	Δ	0.01	0.47	- 0.22	- 0.40	- 0.01	0.01
	δ	0.64	0.89	0.74	0.62	0.77	0.29
measured solar intensities	Δ	0.05	0.49	- 0.19	- 0.04	0.02	0
	δ	0.29	0.62	0.38	0.37	0.27	0.27

The analysis of these results demonstrates that our definition of R should be improved : R cannot be considered as remaining constant whatever the weather may be. This is particularly critical for a South orientated wall in winter

(See, for example, the results obtained with the standard, South-orientated model in Figure 3)...

Figure 3
Prediction of t_{WH}

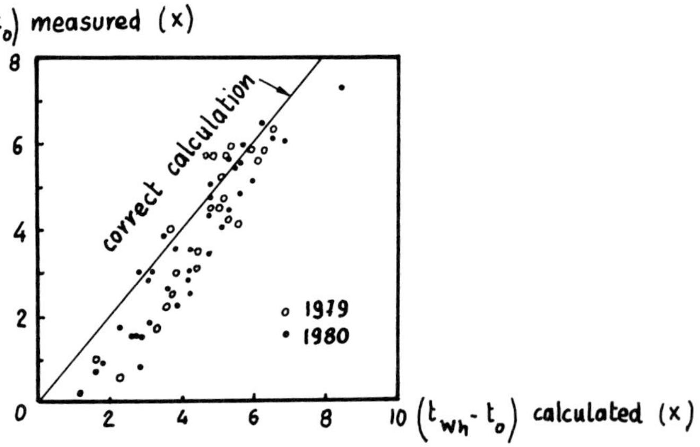

Obviously, this effect will also appear with the "solar aperture" approach; the largest discrepancy will be also recorded for rather equivalent radiative gain in South orientation and other orientations in average sky conditions.

3.3. Dynamic behaviour of the models

3.3.1. Floating temperatures in unheated models

The floating temperatures of unheated models have been continuously recorded as all other variables during the whole measuring period.

A small part only of these observations have been used until now in order to validate our dynamic calculation models.

As in other tests realized in laboratory conditions |3|, it was found here that a higher model than a second order cannot be obtained significantly.

A second order dynamic model is "complex" enough to approximate the natural temperature variations in all the laboratory experiments as well as in the field.

One example of this sort of comparison is given in Figure 4.

Figure 4 : temperature of the normal unheated model

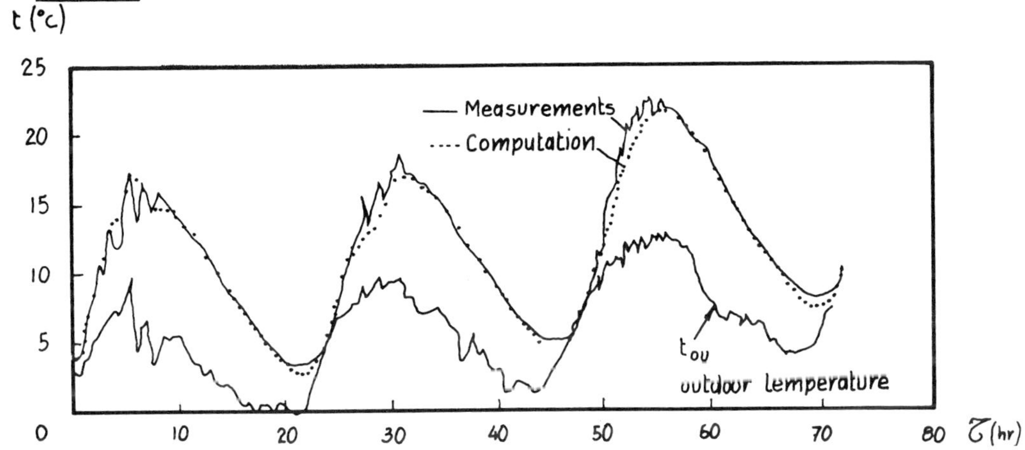

Of course, due to the scaling effect, the actual inertia of the models remains much lower than in full-scale buildings which exhibit much higher filtering effects periods.

3.3.2. Floating temperatures in heated models

Two causes of floating have been observed separately and also combined in certain conditions :

- the inside temperature may float above its set point occasionnally, when the solar free gains are overcompensating the heat losses (during some of the sunny hours);
- the inside temperature may float below its set point during the "OFF" periods in intermittent heating.

3.3.2.1. Overfloating :

Due to the relatively low inertia of the model, this occurs during most of the heating season as shown in Figure 5.

An example of this case is given in Figure 6. To this overfloating corresponds obviously an overconsumption (increase in heating degree-days) |11|...

3.3.2.2. Underfloating :

This phenomena could only be observed without interference with other causes of oscillations during enough cold periods and with a relatively high set point of the thermostat (25 to 26°C). Example : Figure 7.

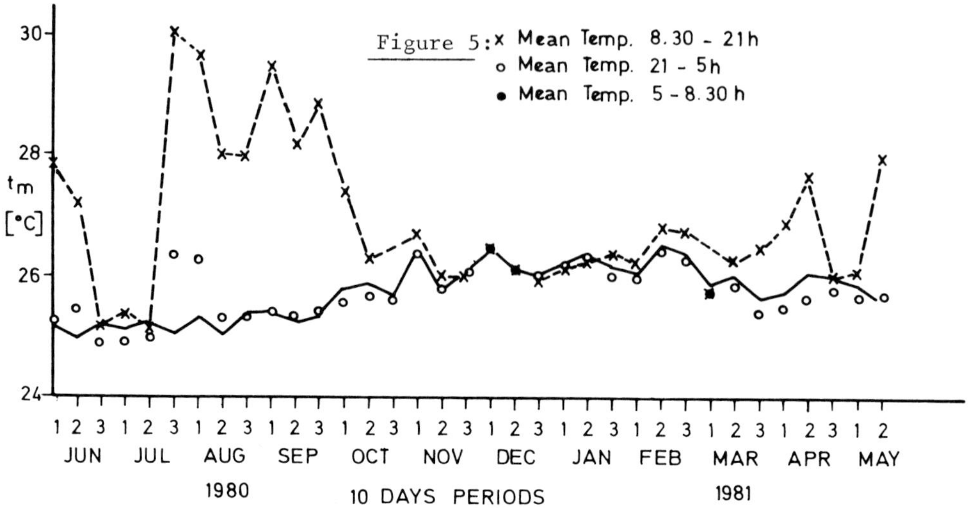

Figure 6 : example of "overfloating" in a continuously heated model

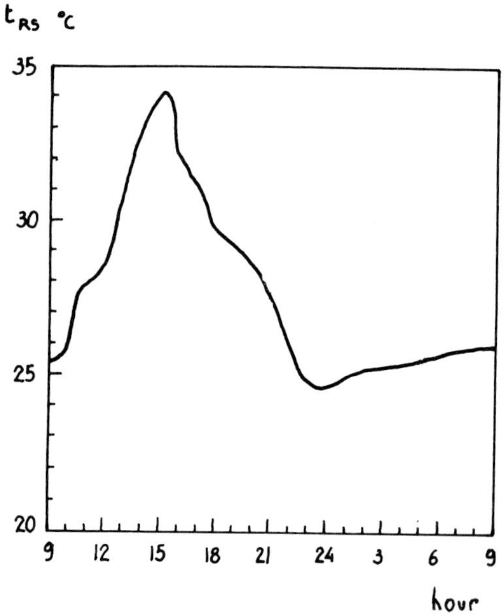

Figure 7 : Example of "underfloating" due to the heating intermittency

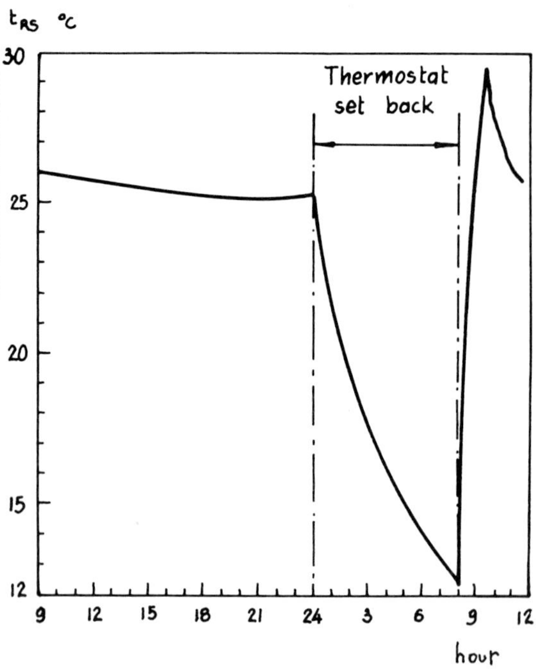

3.3.2.3. Combined oscillations :

Figure 8 : example of "combined" oscillation of temperature"

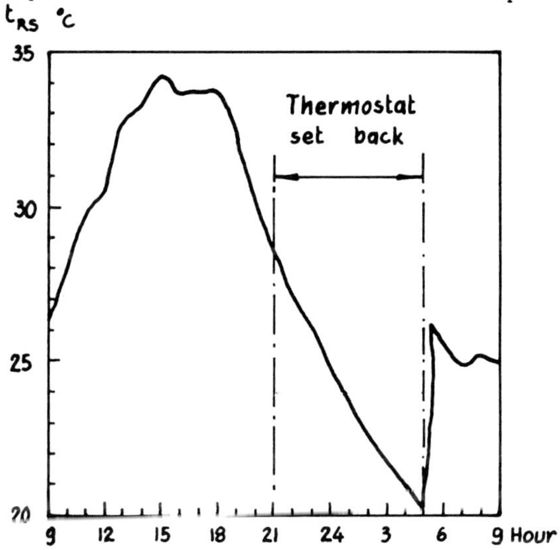

The intermittency underheating doesn't only appear combined with occasionnal overheating as shown in the example of Figure 8. It is also perturbated by the control unefficiency : in comparison with the central globe thermometer which has been taken as reference, the thermostatic probe is more influenced by the temperature of the envelope, against which it is positioned and less influenced by any (direct and indirect) radiant effect of the heating source. This is clearly illustrated by the diagram of Figure 9: the actual set point is a decreasing function of the outside temperature.

In this regime of combined oscillation, an accurate calculation of heating consumption can hardly be achieved. Detailed comparisons between the measurements and the results of full dynamic calculations (LPB-1, LPB-2) are still on the way.

Figure 9 : control unefficiency

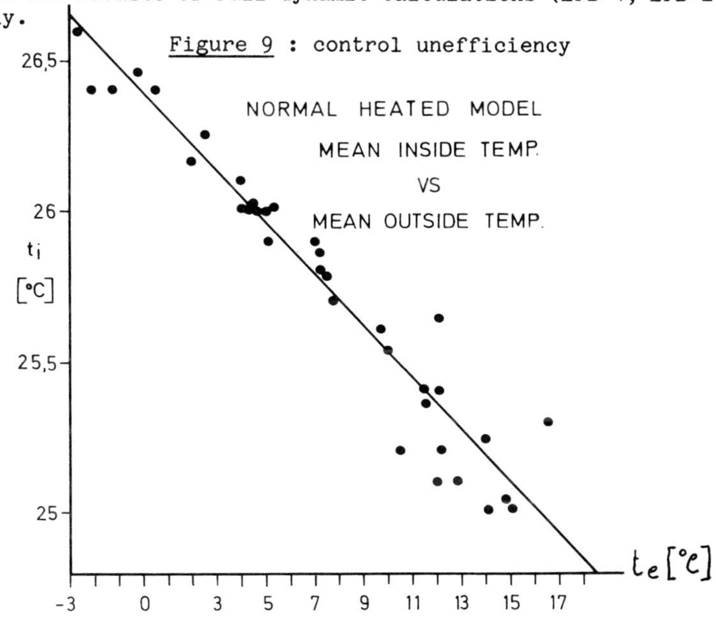

3.4. Calculations and measurments of radiant exchanges

The experimental results show clearly that our equation (2) :

$t_{WH} = t_o + R.\mathfrak{J}$

should be improved at least when the building has a significant proportion of glazing on South orientation. This is illustrated by the curve of Figure 3, the present hypothesis gives an overestimation of t_{WH} which should be corrected in order to avoid a corresponding underestimation of heating energy requirements ...

In fact the equation (2) implies that all the radiant effects on the building are always proportionnal to the solar intensity on a horizontal surface (I_{tH}). We should have in particular :

$I_t / I_{tH} = (I_t / I_{tH})_{cs}$ for the total solar radiation

and $I_R / I_{tH} = (I_R / I_{tH})_{cs}$ for the long wave "sky" radiation (< 0) for any orientation of the receiving surface.

In fact this similarity is only realized for the direct component I_b ("beam") of solar radiation and with no significant noise only for a short time of integration (about one hour) : $I_b / I_{bH} = (I_b / I_{bH})_{cs}$ (6)

The ASHRAE hypothesis consists in applying the similarity also to the diffuse component (I_d) :

$I_d / I_{dH} = (I_d / I_{dH})_{cs}$

There is a third hypothesis, as simple, which simply considers that the diffuse radiation is fully isotropic : then

$I_d / I_{dH} = (1 + \cos p)/2$ (7)

with p = slope of the surface.

These different hypotheses and a linear combination of the last two have been tested by comparison with measured values on different orientations. The results show clearly that the ASHRAE hypothesis drives us to significant errors which systematically depend on the season and on the weather .These systematic errors almost vanish when applying the hypothesis of isotropic diffuse radiation, which appears as the best practical compromise.

In the practical evaluation of I_t ,it appears that the ALBEDO of the ground is a cause of uncertainty. Almost all the measurements made within the frame of this research show the same tendancy : they correspond to quasi-no reflectance from the ground area when a ALBEDO of about 0.2 would be normally considered (grass cover). This fact could justify further studies on real (diffuse and non diffuse) reflections ...

3.5. Calculations of ground losses

Too little attention has been paid up to now to the ground losses : the relative contribution of this term is much higher in the yearly energy demand than in the heating power. The hypotheses used for classical heating calculations are therefore to be improved.

Among the existing methods which have been investigated up to now, the two which are the most interesting are proposed by the C.S.T.B. (France) and by the A.S.H.R.A.E. (U.S.). They define a rational equivalent heat transfer coefficient taking into account a simple model of the multidimensionnal conduction heat flows in the soil. The last one contains moreover an attempt of rational definition of the reference temperature which should be improved because :

- the methodology is only defined for heat load computation;
- the extrapolation for different soils and for other countries is not possible;
- the definition of the equivalent yearly mean temperature doesn't depend on the buildings.

A sensivity analysis has indicated the need for a better definition of outdoor temperature to be chosen : simply the yearly as well as the daily mean outdoor air temperature, or a sinusoïdal outdoor air temperature with a given amplitude. This amplitude depends on a first order dynamic approach of soil behaviour. It is defined from the equivalent heat loss scheme of Figure 10.

Figure 10 : equivalent scheme of the soil

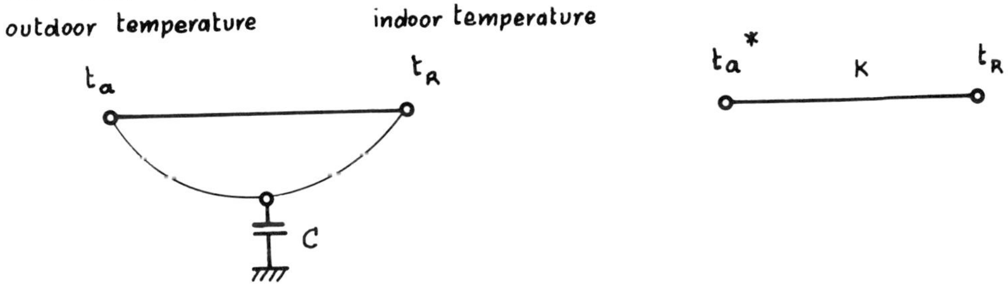

Equivalent first order dynamic approach of soil heat losses

Equivalent results obtained with t_a^*, identical yearly mean value as t_a, and a sinusoïdal one year period where amplitude depends on the first order dynamic model of soil heat losses.

The results appear simple enough for practice and can easily be extrapolated for many conditions.

4. Exploitation of the steady state and dynamic models.

4.1. Equivalent degree-days

Previous tables of equivalent degree-days were based on two parameters : t_{NH} and R |3|.

The recuperating coefficient was defined in clear-sky conditions and assumed transposable to other weather conditions.

The curves of Figure 11 show the variation on the terms I_t/I_{tH} which have been effectively identified. A good compromise (accuracy-simplicity) seems to consist in defining a new R by linear interpolation between the two external reference conditions :

$$R = R_{cs} \left[1 - (1-R_{cc}/R_{cs})\ (1-J)/(1-J_{cc})\right] \qquad (8)$$

with R_{cs} = R previously defined in clear sky conditions

R_{cc} = R in covered sky conditions

$J_{cc} \simeq 0.35$

New tables of equivalent degree-days have been computed on this basis : the third parameter R_{cc} is introduced under the form of a correction which can still be neglected or very roughly estimated in first approximation.

The following practical procedure is recommanded for determining R_{cs} and R_{cc} :

$R_{cs} = R^s_{cs} + R^r_{cs}$

$R_{cc} = R^s_{cc} + R^r_{cc}$

$R^r =$ sky long wave radiation contribution in R

with $R^r_{cs} = I_{RHcs}/K \sum_i A_i \varepsilon_i U_i (1 + \cos p_i) F^r_i / 2h_{oi}$ (< 0) (9)

$I_{RHcs} = -100 \ W \ m^{-2}$ sky long wave radiation (assumed here as isotropic)

K= global heat loss coeffiient of the zone (room or building) considered.

$A_i =$ area of wall i

$F^r_i =$ shading factor of longwave radiation for wall i at the outside surface of the wall i

$\varepsilon_i =$ emissivity of the wall i

$h_{oi} =$ global heat transfer coefficient of the wall i

$p_i =$ slope of the wall i

$\sum_i$ = extended to the whole envelope of the zone

$R^r_{cc} = 0.45 \ R^r_{cs}$ (< 0)
= best estimation presently available for (average) covered sky conditions.

$R^s =$ solar radiation contribution in R

$R^s_{cs} = I_{tHcs}/K \sum A_i S_i (I_{ti}/I_{tH})_{cs}$ (S_i = solar factor of wall i)

$R^s_{cc} = I_{tHcc}/K \sum A_i S_i (I_{ti}/I_{tH})_{cc}$

with I_{tHcs} depending mostly on the latitude and on the season. A set of 12 montlhly values is generally sufficient in the current practice : see table 4

$(I_{ti}/I_{tH})_{cs} = [(I_{bi}/I_{bH})_{cs} - (1+\cos p_i)/2] \ 0.75 + (I_{ti}/I_{tH})_{cc}$ (10)

with the hypothesis of isotropic diffuse and reflected radiations.

A set of 12 values of I_{bi}/I_{bH} (direct intensities) can easily be predetermined for different wall orientations : see for example the values proposed in table 4.

Table 4 : Brussels climate (proposed values for the 15th of each month).

	ItH_{cs} (W/m^2)	South	$(I_b/I_{bH})_{cs}$ East/West	North
JAN	50.7	3.665	.781	0
FEB	94.9	2.284	.672	0
MAR	158.2	1.379	.672	0
APR	235.3	.770	.517	.012
MAY	293.9	.466	.478	.057
JUN	317.5	.348	.465	.096
JUL	308.0	.387	.470	.083
AUG	263.3	.596	.498	.031
SEP	191.8	1.042	.552	0
OCT	120.8	1.804	.628	0
NOV	64.9	3.096	.730	0
DEC	41.3	4.224	.872	0

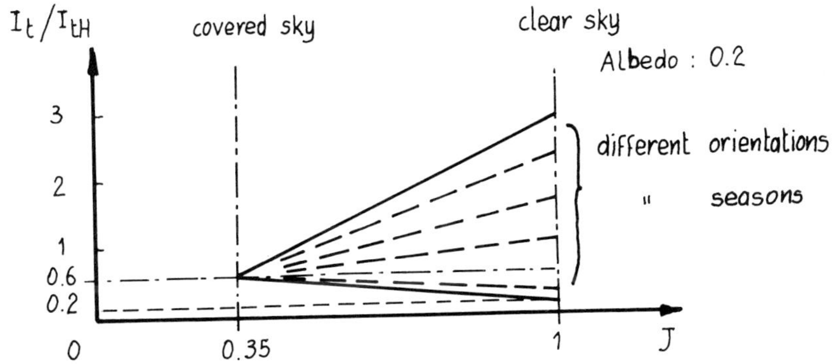

Fig.11

4.2. Simplified dynamic model

The second order model is also used as a computer model in LPB-5 program. This model is schematized in Figure 12. It contains 5 parameters. $1/R$ is the steady state heat loss coefficient and the 4 others are dynamic parameters ξ , θ , φ ,and C. For many purposes only 3 parameters can be used ($\varphi \ll 1$ and often one can use $\varphi = 0$).

Figure 12 : the second order model

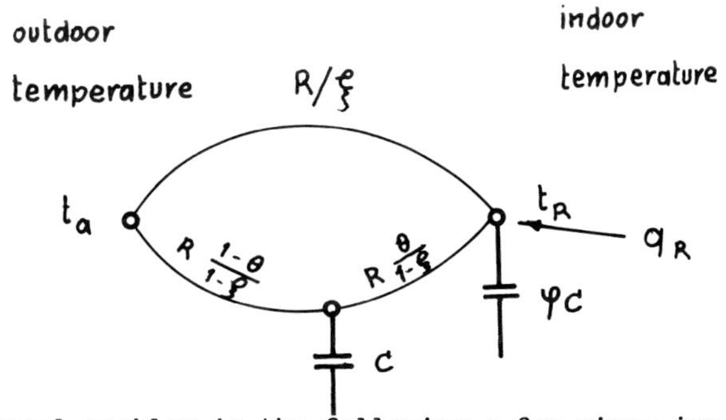

The general problem is the following : for given input-output analyses, limited range of frequency and some criterion of accuracy, what is the lowest order model to be used ?

We proposed an original practical solution in this field and we show that previous first order approaches do not allow enough accuracy |4| |5|.

The basic elements of this approach are the following |4| |5| :

- the output considered is the zone resultant temperature;
- there are two inputs of interest : the outdoor temperature and the heat input (all other inputs are to be expressed by an equivalent linear combination of these two effects);
- the zone is defined by the range of frequency which gives the "depth of penetration" of significant effects.

4.3. Example for the use of the recuperation coefficient and the equivalent degree-days

We want to know the benefice in energy consumption by installing a heat exchanger (recuperator) in the ventilation system of a one-family-house with "good" insulation and a"good" orientation.

The principal characteristics of the house are :

Volume 225 m^3

Ventilation $0.7h^{-1}$, i.e. a capacity flow $\dot{C}$: 57.5 WK^{-1}

Transmission heat loss ΣAU : 181.5 WK^{-1}

Global heat loss coefficient K : $AU + \dot{C} = 234\ WK^{-1}$

Mean inside temperature : $t_i = 18°C$

Occupation free heat Q_{occ} : 468 W

Temperature of no heating : $t_{NH} = t_i - Q_{occ}/K = 16°C$

Recuperation coefficient R (assuming here that R_{cc} R_{cs}) and corresponding equivalent degree-days for Brussels (Uccle) :

Month	1	2	3	4	5	6	7	8	9	10	11	12	TOTAL
R	3	5	7	7	7	6	6	7	7	6	4	3	-
degree-days	331	245	147	124	17	3	1	0	2	42	242	366	1520

The annual heating energy needs (not the consumption, which depends on the efficiency of the heating system) will then be :

$E = K \sum (t_{NH} - t_{WH}) \Delta \tau = 234 \times 1520 \times 24 \times 3600 = 30.7 \times 10^9$ J (8536 kWh).

We suppose an over all efficiency of 70 % for the heat exchanger that is, the ventilation loss will be reduced to $\dot{C}_1 = \dot{C} \times (1 - 0,7)$ and K will consequently reduce to $K_1 : K - 0.7\ \dot{C} = 197,3\ WK^{-1}$ or, in relative value, = - 15.7%.

If we neglect that the number of degree-days will also be smaller the energy consumption E will be reduced by 15,7% too. But this would be a very bad estimation as we shall see hereafter.

the smaller number of degree-days comes from the combined effects ot t_{NH} and R. We have : $(t_{NH})_1 = t_i - Q_{occ}/K_1 = 15,63°C$ (instead of 16°C) and $R_1 = R\ K/K_1$ for every month, because the solar gains are unchanged. We obtain :

Month	1	2	3	4	5	6	7	8	9	10	11	12	TOTAL
R	3,56	5,93	8.3	8.3	8.3	7.12	8.3	8.3	7.12	4.74	3.56	-	
degree-days	310	220	115	105	13	1	0	0	1	30	221	345	1363

and $E1 = 197.3 \times 1363 \times 24 \times 3600 = 23.2 \times 10^9$ J (=6.452 kWh)

So, for a reduction of K by 15.7%, we obtain a reduction of 24.4% in annual energy needs.

In the following table are indicated the errors which would have been committed on the heat recovery estimation by applying various hypotheses :

Hypothesis	Heat recovery (10^9/year)	Relative error
present method	7.5	0
conventional 20/20 degree-days	10.9	+ 45 %
" 18/18 " "	8.7	+ 16 %
" 15/15 " "	6.8	- 9 %
equivalent degree-days without taking into account the changes of t_{NH} and R	4.8	- 36 %

The errors are impressive; they illustrate clearly the interest of the present method. Most paradoxal could appear the error of 16%, corresponding to the conventionnal 18/18 degree-days :

It is true that the ventilation heat loss depends effectively on air temperatures and the inside average has been taken at 18°C. But, during a certain part of the mid-season, the ventilation heat loss is yet compensated by the solar and/or occupancy gains. Heat recovery is then useless. Of course, it had been more profitable with a worse architectural design of the house.

The combination of several energetic improvements would still be more critical : a particular number of degree-days corresponds to each of these and to any combination between these. That's why the energy savings cannot be directly added to each other...

5. Practical conclusions and prospectives

5.1. Conclusions

- the steady state analysis of these results confirms us the validity of the t_{WH} approach in calculating any building heating energy demand;
- the calculation of the t_{WH} itself requires a careful definition of solar and infrared radiations;
- a building which contains a large south glazing area ("solar passive"), cannot be characterized only through one curve of solar recuperating coefficient (i.e. its natural overheating in clear sky conditions) as was initally expected;
- the "equivalent degree-days" method can easily be improved by introducing, in the preceeding case, a curve of covered sky recuperating coefficient and by the using linear interpolation for all the "intermediate" sky conditions;
- a second order allows a very satisfactory representation of almost the whole dynamic behaviour of the reduced scale models;
- the very high accuracy of the steady state analysis of the results doesn't eliminate the necessity for some dynamic calculation in the current practice ("design" and "audit") : even in continuous heating the time average of inside temperature t_i is generally a priori unknown and can differ significantly from its set point, due to excessive solar and occupancy free heat.

5.2. Prospectives

The equivalent degree-days method is now fully operationnal and should be used in any comparison of heating energy saving techniques.

The steady-state analysis of experimental results is almost complete and usable for calculation validation by any interested research team.

The dynamic analysis could be continued among others to validate the calculation of overheating and/or heating intermittency effects. These experimental results are also available for any international use.

The experiment on reduced scale models could be extended in testing "solar passive" devices. It has proved to be a very valuable way of analysis. A limited set of very well standarized models could also be used as reference in order to define experimentally the climate in various sites of Europe. A model could also be installed in the vicinity of any building to be monitored as a very useful checking point for the measurements.

A new (and hopefully even more accurate) model is to be installed in the vicinity of Liège for a detailed dynamic analysis...

6. References

|1| H. HEIKHAUS, J. UYTTENBROECK
"Mesure de la consommation d'énergie dans dix maisons neuves".
Revue n°1 du C.S.T.C., mars 1978

|2| H. HEIKHAUS
"Etude de l'influence des paramètres d'isolation des bâtiments sur la consommation réelle de combustible".
Rapport final de la convention C.S.T.C. - I.R.S.I.A. N° 2649. Novembre 1977.

|3| J. BADOT, L. LARET, J. LEBRUN, P. NUSGENS
"Influence de certains paramètres sur la consommation réelle de combustible"
Rapport final de la recherche "CSTC-IRSIA Seneffe II, convention 2649". Liège, novembre 1979.

|4| L. LARET
"Contribution au développement de modèles mathématiques du comportement thermique transitoire de structures d'habitation".
Thèse présentée pour l'obtention du grade de Docteur en Sciences Appliquées de l'Université de Liège, mai 1981.

|5| J. LEBRUN, J. UYTTENBROECK, H. HEIKHAUS
"Temperature without heating : a fundamental building characteristic for calculation of heating energy consumption. Experimental results".
Symposium "CLIMA 2000", Budapest, September 1980.

|6| A. DUPAGNE, H. HEIKHAUS, J. LEBRUN, J. UYTTENBROECK
"Practical method to evaluate the effect of building characteristics in the energy needs for heating".
International Congress Building Energy Management, Povoa de Varzim, Portugal, May 1980.

|7| H. HEIKHAUS
"Caractéristiques principales des habitations sociales permettant de réduire la consommation d'énergie pour le chauffage. Complément de recherche sur la température sans chauffage".
Rapport de 1ère biennale de la Convention C.S.T.C. - I.R.S.I.A. n° 3279, 1980.

|8| Ph. TROKAY, Y. DELORME
"Application des facteurs de réponse à l'étude dynamique d'un ensemble de locaux. Description du programme LPB-1".
Collection des Publications de la Faculté des Sciences Appliquées. Université de Liège, 1979.

|9| Ph. CONSTANT, H. SOHET
"Evaluation du comportement dynamique d'un local par la méthode des différences finies".
Collection des Publications de la Faculté des Sciences Appliquées. Université de Liège, 1979.

|10| F. LORENZ
"A static computation programme for architectural design".
Paper submitted to "Building & Environment", published by Pergamon Press Ltd.

|11| H. HEIKHAUS, J. LEBRUN
"Thermal monitoring of a dozen of reduced scale models of houses".
Symposium organized by the C.E.C., DGXII-C "Energy research program", in collaboration with the Building Physics Laboratory : " Comparative experimentation of low-energy houses", Liège, May 1981.

|12| R. SONDEREGGER, D.T. GRIMSRUD, D.L. KRINKEL
"An instrumental microprocessor-assisted residential audit".
International colloqium CEE, Comparative experimenation of low-energy houses, Liège, May 1981.

PLANNING DESIGN, EXPERIMENTAL REALISATION AND ENVIRONMENTAL CONTROL OF A RESIDENTIAL UNIT

Author : S. LOS, G. MANCUSO, N. PULITZER

Contract number : EEA-5-058-I

Duration : 17 months

Head of project : Ing. G. ZACCARIA

Contractor : AGIPNUCLEARE S.p.A.

Address : Viale Brenta, 29
I - 20139 MILANO

Summary

The aim of this research work is the design of a flexiable prefabricated building system for mass production. Such a system

(a) may be adapted to different climatic areas;

(b) is developed according to the "open" system philosophy, so to grant the interchangeability of some of its components; and

(c) is aimed at building multi-family housings with low capital and operating costs.

In order to pursue this aim a programme has been developed consisting of a series of steps.

1 Introduction

A building may be considered as a couple of twin or "conjugate" systems:

- a space system consisting of a set of spaces and their interrelations;
- a technological system consisting of a set of components and equipments which define and enclose such spaces.

The design of a building is completed when both space and technological systems are determined. It is clear that the space system has to be determined first.

In the design process two different levels of definition may be distinguished:

- at performance level the varios functions performed by the building are determined and the corresponding performance requirements are translated into performance specifications;
- at detailed plan level one specific solution is chosen from a set of possible solutions meeting the performance specifications.

Both space and technological systems must be designed at both performance and detailed plan levels.

2 Design process steps

2.1 Context definition

2.1.1 Climatic environment

Reference to four European countries - France, Spain, Italy, and Greece, generally establishes,as a common climatic feature, the typical climate of temperate areas presenting four seasons - cold or cool winter, warm or hot summer, and short mid-seasons.

The case studies are referred to three homogeneous climatic areas:

- a mainly continental area, typical of inland or alpine regions, exemplified by Bolzano and Lyon (46° N Lat.);
- a coastal area, characterized by a windy marine climate, with cool winters and warm summers, exemplified by Genoa, Nice, and Marseille (44° N. Lat.);
- a typical mediterranean area, mostly hot, exemplified by Messina, Athens, and Lisbon (38° N Lat.).

2.1.2 Users

The users taken into consideration have: (a) an income range similar to that provided of by public housing projects; (b) a job away from home; (c) live in well serviced urban areas of middle or large towns; (d) are 3 persons families for the first two areas, 4 persons families for the third climatic area.

Average floor area of each dwelling unit is 70-80 sq.m, in accordance with european standards.

2.2 Building types selection

Professional and research experiences done in the last years on the subject of residential housing allow to select from usual residential building types the ones most relevant and most suited to energy requirements; they are: clustered single-family houses (raw- and courtyard-houses); unilateral or bilateral multistory buildings, with the main axis oriented east-west, north-south, or according to the heliothermic axis; and tower houses, indifferent as to orientation.

2.3 Definition of building and energy functional units of the technological system

A bioclimatic approach to architectural design cannot be restricted to building envelope or other building components. The whole building, and its subdivision in building and energy functional units, has to be involved in the design process in order to take into account all energy interactions.

2.3.1 Building functional units

S - Support
OE - Opaque enclosure
TE - Transparent enclosure
PF - Partitions and finishing

2.3.2 Energy functional units

C - Collector - for collecting solar and converting it into thermal energy
TS - Thermal storage - for heat (or cold) storing
D - Diffusion - for distributing heat (or cold) to spaces;
E - Dissipator - for dissipating the excess heat.

Energy flow between two units may be direct and/or indirect, i.e. through a "transportation" duct or pipe (T).

Finally, the different operation modes of a system may be determined by introducing the control meta-function (R).

More building and/or energy functions may be performed by the same physical component (e.g. a massive wall may function as collector, storage and partition, as in the case of direct gain). Inversely, more physical components may perform the same building or energy function (e.g. the floor and a wall may function as storage).

2.4 Definition of "dwelling" modules common to the varios building types

In order to analyse the energy functions of the various building components, they should be assembled in such a way as to form an elementary inner space, which is then defined by external and internal walls, floor and ceiling forming a dwelling unit that we call "dwelling" module. The "dwelling" module is assumed as base module in order to provide a significant description of residential building types at technological system performace level. This module consists of a parallelepiped, corresponding to a dwelling unit, that generates the different building types by horizontal and vertical combination.

2.5 <u>Performance requirements definition</u>

Performance requirements of "dwelling" modules are here defined for space and technological systems, taking into account the necessary interaction of building and energy aspects.

2.6 <u>Performance design of the "dwelling" module</u>

The performance design of the "dwelling" module defines the space system of the "dwelling" module, but it also defines the performance specifications for its technological system, as to allow the use of computer programs to simulate and test the energy efficiency of the proposed system.

2.7 <u>Building system product-components definition</u>

Designs of components of the "dwelling" modules are developed, following the above defined subdivision into functional units, on the base of the performance specifications established in the preceding step.

2.8 <u>Energy behaviour testing</u>

In reference to the three climatic areas, analyses are performed to test by computer simulations the energy behaviour of specific "dwelling" modules located in critical positions in relation to the building types.

2.9 <u>Rules definition</u>

Our reference to an "open" building system assumes a set of variable components connected one to the other by an invariant relation structure, which defines a set of rules determining the restraints that have to be respected in order to grant the interchangeability of the system (defined by the components catalogue). These rules relate to several conventions concerning:

- the modular co-ordination of components
- the quality control of components
- the interface between components
- the joints.

2.10 <u>Selection and detailed design of components</u>

A catalogue of components is developed in accordance with the set of rules previously established.

2.11 <u>Massing at the performance system level</u>

Examples of the building system applied to the 4 building types previously selected are provided.

2.12 Detailed plan of a "dwelling" module

Detailed palns of a "dwelling" module are developed for each climatic area.

QUESTIONS AND DISCUSSION

Dupagne (University of Liège) : Is the design methodology for North, Central and South Italy the same or do you change the design strategy to take into account the rather big differences in temperature ?

Answer : The design methodology is the same, it states the correlation between house form (structure) and local climate so it must be the same. The design process is the same but the strategy is quite different as the problems are different : for the northern areas we invest more money in conservation strategies and less in solar collection ones, in the south areas instead we can save more energy by collecting solar radiation than by conserving it. The relative importance of parameters is varying following the variation of climatic factors, the design method has been developed to take in account this relationship.

Dupagne (University of Liège) : What about the occuracy of the computer programme when applied to different climatic conditions ?

Answer : The computer programs have been studied to cope with different climatic conditions, and applied to analyze buildings in different areas of the South and the North; but up till now they have not been tested experimentally for houses or cells.
We have proposed and built a real scale test cell, by which it is possible to check the computer programs for different climatic areas, but since now we have not performed these checks.
Computer programs and test cells have been studied within the context of the CNR Progetto Finalizzato Energetica 1° research work.

Prof. Boffa (Chairman) : This model should be convalidated. I wonder whether that type of work could not be done by the European Community.

Ehringer (DG XII) : We don't like to put too much emphasis on computer simulation.

Prof. Boffa (Chairman) : It is not necessary to start a measuring programme to provide the data for such a convalidation. It is possible to use pre-existing data.

Clarke (University of Strathclyde) : The IEA tried to convalidate simulation models but ran into severe technical problems.

Los (TNSO) : In the Italian solar energy programme we have built a device which can be used for the verification of computer models.

BASIC CHARACTERISTICS OF LOW-COST HOUSES IN ORDER TO REDUCE THE ENERGY CONSUMPTION FOR HEATING. LINK PROGRAMME - COORDINATION AND CONSEQUENCES FOR THE ARCHITECTURAL DESIGN (SYNTHESIS)

Authors : A. DUPAGNE, J.M. HAUGLUSTAINE, L. LARET

Contract number : EEA-5-060-B (G)

Duration : 36 months 1 January 1981 - 31 December 1983

Total budget : B.F. 18.320.000 CEC Contribution : 50 %

Head of project : A. DUPAGNE

Organization : Université de Liège
Laboratoire de Physique du Bâtiment

Address : Avenue des Tilleuls, 15, Bât. D1
B-4000 LIEGE (Belgium)

SUMMARY

The first research undertaken for the Commission of the European Communities has led to a first manual version of a design methodology, mainly orientated to the conception of low-cost houses, integrating the fundamental thermal concepts into the design process without any use of sophisticated technology (1).

The work to be carried out in the present project will concern :

- the verification of the efficiency of the design methodology by its practical application on the design of a few low-energy houses and the building of these houses;
- an experimental study, which will check the first evaluations given by the application of the design methodology;
- the development of the lacunas so discovered in the design methodology,
- the achievement of a conversational computerized version of the design methodology;
- in the monitoring of the houses, the analysis of the internal partitioning and of the usable fraction of the building thermal inertia.

The final product of this research project will contain a revised version of the manual design guide and its computerized version; the energy savings actually obtained will be described, with a complete analysis of the resulting global economical cost; the perspectives of energy savings in the low-cost houses sector will be assumed.

1. INTRODUCTION

The first research undertaken for the Commission of the European Communities has led to a first manual version of a design methodology, mainly orientated to the conception of low-cost houses, integrating the fundamental thermal concepts into the design process without any use of sophisticated technology.

The work to be carried out in the present project will concern :

- the verification of the efficiency of the design methodology by its practical application on the design of a few low-energy houses and the building of these houses;
- an experimental study, which will check the first evaluations of operating and maintenance costs given in the application of the design methodology, and the comfort conditions ; it will look for the reasons of the dispersion of the results and evaluate it;
- the development of the lacunas so discovered in the design methodology, such as the dynamic approach of the buildings, the data management and the costs evaluations (investment, operating and maintenance costs);
- the achievement of a conversational computerized version of the design methodology;
- in the monitoring of the houses, the analysis of the internal partitioning and of the usable fraction of the building thermal inertia.

The final product of this research project will contain a revised version of the manual design guide and its computerized version; the energy savings actually obtained will be described, with a complete analysis of the resulting global economical cost; the perspectives of energy savings in the low-cost houses sector will be assumed.

An interesting application of the design methodology has been done within the frame of the IDEE research work (CEC Contract n° EEA-5-056-B (G)) under the leading of Jos Uyttenbroeck : the architectural handbook has been used for the design of 2 experimental low-energy houses, where the CSTC intend to experiment different heating systems.

A second collaboration with other scientific teams is planned with the Construction Division of An Foras Forbartha (Dublin), for the contract n° EEA-1-007-EIR (N). In the frame of this collaboration, we will use our static building load computation program LPB-4 for the houses they monitor. A calibration of the program will be possible owing to the comparison of the obtained simulation results, with the monitoring measurements. Moreover, the theoretical results will enable the Irish team to orientate their measurements to parameters which would seem to be very important.

2. PROGRESS OF WORK

2.1. Architectural design

2.1.1. General situation of the site

Located at about 1,5 km from the centre of Marchin and 10 km from the city of Huy, the proposed site is distributed among 3 parts, bordered by a small valley of an affluent of the Hoyoux, and a road to the centre of Marchin. The site is crossed by two roads: an accommodation road leading to a couple of farms, and a more important one leading to Huy, but with little traffic. A filtering station lies at the lowest point of the site. The part of the site nearest to the small valley rises rapidly and is timbered; it rises afterwards with a mean slope of 16%, is orientated North-North-West. This site seems to be actual and the application of the design methodology lets hope to obtain energetic results better than those obtained with plans usually proposed to the Belgian National Housing Society for such a type of site : the living area is usually orientated to both North and South, without any difference between the glazing areas on North or South façade.

2.1.2. Settlement layout

A general houses laying out as parallel as possible with the contour-lines of the ground is the cheapest, due to the reduction of the excavation works for the implantation of the houses clinging to the ground. Furthermore the contour-lines are mainly orientated so as to allow the houses a good bipolar North-South orientation, the best one to give a good solar recuperation in the winter, without any risk of too high summertime temperatures. This disposition gives moreover the occupants a nice view on the small valley. The breadth of plots, usually of 8 m, has been raised up to 10 m, in order to increase the South façade area.

Concerning the accesses to the houses, the West part of the land, included between the top of the small valley and the main access road, offers indeed an existing accommodation road, which can be kept and fitted for the traffic. The East part of the site is presently accessible with a road already covered with asphalt and which will only be extended punctually to give access to the groups of houses and manoeuvring areas. This settlement of access roads is sufficient, as simple as possible and very cheap. Figure 1 shows the resulting settlement layout.

Given the present access roads, the general site is then distributed into two sorts of houses : the houses numbered 1 to 12 and 26 to 30 are accessible by their South-East side, whereas the others by their North-West part. The architectural solution for the first type of houses will be more difficult than for the second one, because the South part must incorporate both the general entrance (not heated) and the living-room, with a window area as large as possible, in order to give the house a good solar recuperation. The garden facing the living-room is directly connected with the road ; this requires special architectural precautions to avoid any loss of intimity in the living-room.

In the architectural program of the houses, the National Housing Society has required one garage for each house. In order to meet this wish, we have retained two solutions to be applied according to the configuration of the terrain. The houses on the South side of the ground (numbered 13 to 25 in Figure 1), have their garage incorporated into the house volume ; it easily takes place in the excavated part of the ground, without any overcost.

For the other houses (numbered 1 to 12 and 26 to 30), the incorporation of the garage into the house volume was rejected, for the following reasons :

- Placed in the garden site of the houses, it requires an additional road and then a resulting additional cost.
- If it is placed in the road site of the houses, i.e. South-East, its large door surface cannot participate in the solar recuperation of the house.
- Used as a party volume, it annihilates any grouping interest, because of the intercalation of not heated rooms between the heated houses.

Then we proposed the alternative of garage elements separated from the houses and laid out along the access road. This solution offers a couple of advantages :

- The relatively low volume, moved away from the house, doesn't give too prejudicial a mask for the solar recuperation of the house.
- The volume intercaled between the public road and the private South-West garden places a visual screen which preserves a relative intimity

and privacy to the garden.-The proximity of the garages to the road avoids a visual dispersion of cars in the landscape, maintained in a limited space.
The mass of the garages will act as an interesting screen against the noise coming from the street.

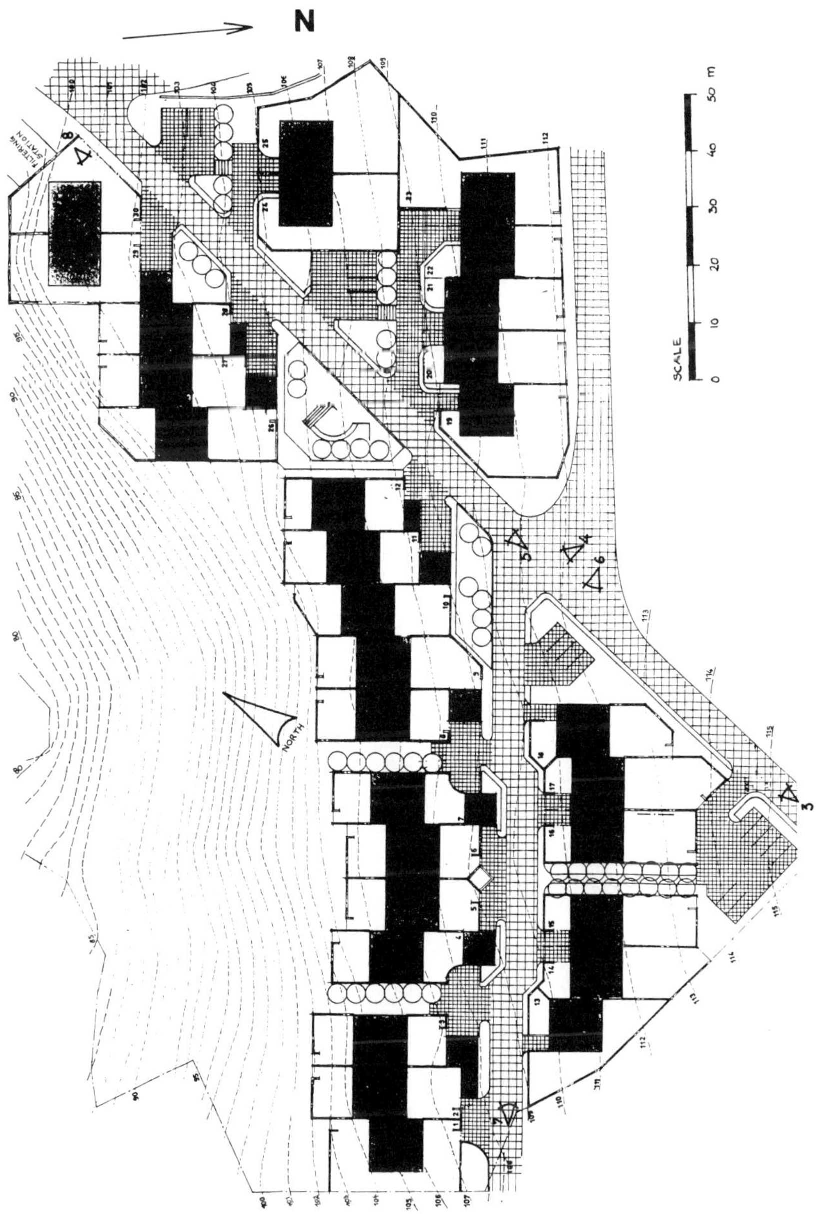

Figure 1 : Settlement layout of the 30 houses in Marchin

Concerning the mechanical resistance of the ground to compression efforts, seismic soundings have been carried out by the Geology Laboratory of the University of Liège, which specify that the mechanical characteristics of the substratum would be in a favourable range for the construction of light buildings (such as houses), except for the zone of houses numbered 26 to 30, where the present embanking is not sufficiently compact. This is the main reason why we came to the decision, with the National Housing Society, not to build the group of houses numbered 26 to 30.

2.1.3. Wind influence on settlement layout

The most important wind parameters to take into account for thermal considerations are the cold wind prevailing directions, and the mean speed of wind, both in the local microclimate. In the first stages of the design, it is not possible to install in the virgin site an expensive meteorological station, without any risk of vandalism or theft. It would have been simpler to extrapolate the observations made in the nearest meteorological station, or the surrounding ones. The extrapolation must consider the modification of the ground roughness, which depends on both the relief and the average height of the terrain roughness elements (e.g. houses, trees, etc). But our site gathers both the influence of a small well-timbered valley and an overhanging pasture-land plateau : this combination increases the uncertainty by a hypothetic choice of the terrain roughness; consequently, the accuracy of the extrapolation for the wind speed or its direction is compromised.

The last possibility for a good wind analysis of the microclimate is to build a scale model of the site to be placed in a wind tunnel, which is the only tool available for the prediction of the wind conditions at pedestrian level for new building developments. Our terrain presents a very specific configuration, with the adjacent valleys and the important proportion of vegetation. This increases the difficulty for the right choice of the limits of the scale model in the surrounding country. Furthermore the modelization of the existing vegetation is another source of error and requires very specialized techniques and sophisticated materials; therefore it costs a lot of money and takes a lot of time, without any certainty of accuracy and out of the budget and time of our project. It will be more realistic to take in a first approach the usual wind speeds (already considered in the meteorological data in the design methodology), the average value of which is 4 m/s.

A second wind parameter is the cold winds prevailing directions, with consideration of wind temperatures having an influence on the heating energy needs, i.e. below 18°C (which represents the internal average temperature of buildings) and of the wind speeds having an influence on the infiltration losses, i.e. exceeding 0,5 m/s (2). The wind graphs calculated for all the Belgian meteorological stations during the heating season allow to determine two large orientation sectors corresponding to the most unfavourable winds to the infiltration losses. These sectors are respectively North-East and South-West (figure 2). A more accurate wind graph would be calculated on the basis of in situ measurements with a local meteorological station : it represents the ideal solution, but expensive and tainted with the dangerous abovementioned risks).

MELSBROEK
Moyenne (Saison de Chauffe)

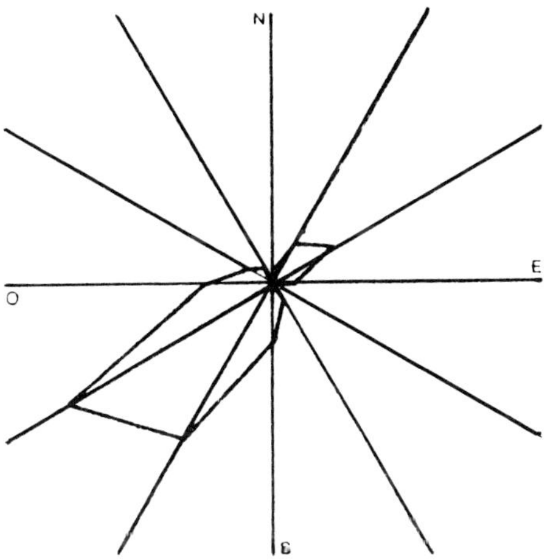

SAINT-HUBERT
Moyenne (Saison de Chauffe)

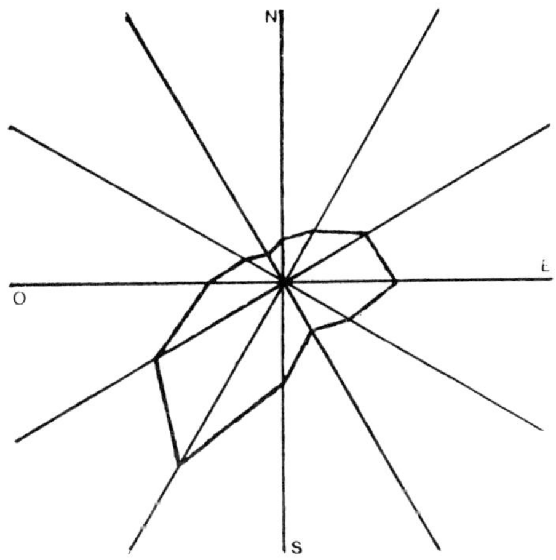

Figure 2 : wind graphs for speeds > 0,5 m/s, weighted by the winds temperature during the whole heating season (3).

The settlement layout has taken these wind graphs into accoutn by reducing the exposition of the houses to these winds and so to offer them a microclimate.

2.1.4. Use of stereographic sun-path diagrams during the design of the settlement layout

The sun-path diagrams are usually constructed and used for the design of shading devices. They can, however, also be used for the prediction of insolation (sun penetration) and the determination of overshadowing by external obstructions, i.e. the ground, the garages or other houses. It is important not to priviledge a few houses, better orientated, to the detriment of other houses, which are downwards. An interesting step can be taken from an energetical point of view, if we consider that each spherical quadrilateral of the sun-path diagram defined by the lines of two successive hours and between two successive dates, corresponds to a period of insolation and coïncides with a certain amount of energy received by the point where the overshadowing period is calculated. For example, one square meter of a vertical façade orientated South-South-East has received, for the 8th hour only, an amount of 26,1 kWh from the 20th of February until the 21st of March and from the 21st of September until the 21st of October. With meteorological data and corresponding calculations, each spherical quadrilateral can be filled in by the right amount of energy (Figure 4). This figure directly gives the energetical value of the overshadowing period and can be used as a design tool for the comparison of a few settlement solutions.

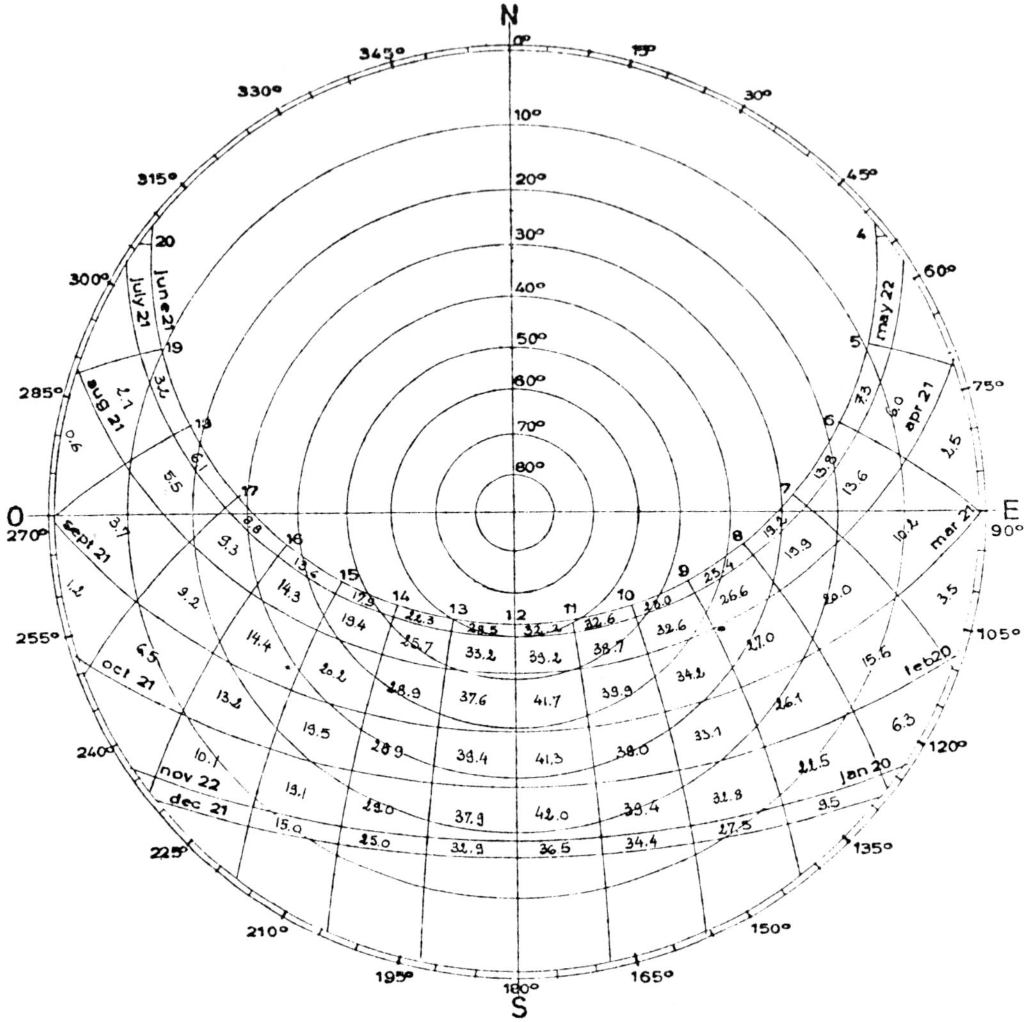

Figure 3 : Energetical sun-path diagram for a vertical surface orientated South-South-East and for the geographical latitude of Marchin.

In the groups themselves, it was important to move sometimes the houses, the contour lines of the ground are not strictly straight lines, and the houses must be moved to a few places in order to follow as best as possible the existing ground. This disposition of houses has the second aesthetic advantage of avoiding rectilinear and monotonous rows of houses. It was nevertheless interesting to use this method manually in order to assess energetically the positions of a few houses in their own group and the interrelative positions of groups.

We have calculated the periods where the lowest point of a South-South-East façade is overshadowed by the party-wall of the next house (with one or two floor levels), for houses movements from 1 m to 5m. With a resulting solar incident lost energy of 5% maximum, between next houses. We have chosen to avoid-wherever it was possible - a movement longer than 3 m.

The same procedures has been used for the mutual positioning of houses 5 and 6 (and their garages) and houses 14 to 17. The garages (low volumes) have higher incidence than the houses (limited to 2 storeys). The diagram was the reason why we tried to give the houses 14 to 17 as few storeys as possible, or to prevent a too important overshadow mask by a tilted roof, which removes the highest point of the house farther than for an horizontal roof. The resulting settlement layout is given in Figure 1.

2.1.5. Application of the methodology to the design of the houses

As already mentioned previousely, the wind and shading masks effects on the settlement layout are a first preliminar, but not negligible step of the application of the design methodology. The next actions concern :

- the space allocation, related to the architectural program given by the National Housing Society, with the required area for each room, depending on the number of occupants of the house (5 to 6) and the number of bedrooms (3 to 4) ;
- the study of the shape factor F_S, which is the ratio between the envelope area (i.e. the total gross area of all the fabric elements surrounding the heated space) and the net area of the heated space ;
- the calculation of the heating energy needs for the designs suggested.

The space allocation has at first taken into account standard room areas required by the Belgian National Housing Society and has distributed these areas into storeys by considering the slope of the ground, the disposition of access roads and the possible orientations of the rooms themselves. The configuration of the terrain allows us to imagine split levels, the influence of which on the shape factor has been assumed : the splitting of levels has no significant importance on the calculated values of F_S.

On the other hand, a few considerations decrease the importance of the shape factor, which doesn't take reductions of ground losses into account and is penalized by large South façades more recuperating than the differently orientated façades. The only more accurate way is the calculation of heating energy needs by the use of graphs proposed in annex MET 3 of the design methodology (1).

This calculation is then realized in a couple of steps, for each design :

1. calculation of the internal comfort temperature, which is inspired by the human data and weighed according to the occupancy profile ;
2. five progressive hypotheses were made on the insulation degree of all the fabric elements (opaque or glazed) and on the glazing factor ;
3. calculation of the global heat loss coefficient K" ;
4. use of the method proposed in annex MET 3 of the design methodology.

Table 4 gathers the calculated values of the global heat loss coefficient K" ($W/m^2{}^\circ C$) and the annual heating energy consumption E" (kWh/m^2) for the different drafted designs and the chosen hypotheses.The coefficient K" is very sensitive to the compacity of the house, and we have a new example of this general remark : hypothesis n° 5 (the worst insulated houses) reveals the energetical characteristic of the designed envelope, and shows that the designs B and F to H appear very poor, from an energetical point of view. The differences decrease with better insulation levels.

As a general conclusion to the study of the global heat loss coefficient, we must be very attentive to badly compact designs, except if the excrescence of the envelope is used in order to increase the solar recuperation, for example by a larger South facing windows area. The global heat loss coefficient doesn't take any solar recuperation into account. It gives a tendency of the design without any consideration on its good or bad orientation.

But this tendency is confirmed by the values of E" : the designs B and F to G consume more than 300 kWh/m^2 of heated area (for the most pessimistic hypothesis of insulation). The difference decreases with an increasing degree of insulation. The split levels are not always interesting from an energetical point of view : if the design D is the most efficient, the designs F to H offer less interest. On the other hand, the use of split level increases both the risk of overcosts (due to the complication of the building) and the risk of a partitioning badly realized in the house. We have then chosen to look for more simple designs, with storeys developed on the whole surface of the house.

According to these results of the design methodology application, an outline sketch for each type of house (differenciated by the position on the site and the possible access to the house) has been proposed to the Belgian National Housing Society, who have approved it.

2.1.7. Detailed design

The detailed design of the houses includes the detailed fenestration of the windows. The North ones should be as small as possible, with considerations for visual comfort (i.e. the view on the valley) and the artificial lighting needs. A special effort will be devoted to the search for the optimal surface, according to energy consumption for both electric lighting and heating needs. The South windows will be as large as possible, in order to give the houses a good recuperating factor ; but the maxiumum area will be limited by the risk of too high summertime temperatures. Then special precautions will be taken to find a compromise between very large areas and comfort temperatures, with the help of overshadings or other solar protections.

The detailed design of the houses will also include the right dimensioning of the insulating materials thickness and the choice of their positioning in the fabric. This last stage of the design assumes the use of our LPB-4 program, which will give us the sensitivity of the energy consumption to the insulating materials thickness. The optimum thickness and positioning will be found according to the energy consumption, the cost of the insulation and the respect of the Belgian regulations related to the global heat loss coefficient.

2.2. Design of experiments

2.2.1. Aims of the research

We can synthesize the problem as : the acquisition of information on dwelling energy performances on the field. It includes the correlated analysis of building structure characteristics in function of locate climate, climatying system and occupant intervention.

Two kinds of analyses are possible : a large size sample with a low number of measurements in each dwelling or a low size sample analyses with detailed measurements. The large size sample approach allows a significant statistical interpretation on a class of dwellings and of the occupant's behavior.

The low size sample approach allows detailed comparisons with numerical computations and the progress in the knowledge of the phenomena involved.

SUGGESTED DESIGN	VALUES OF K" Hypothesis n° 1	2	3	4	5	VALUES OF E" Hypothesis n° 1	2	3	4	5
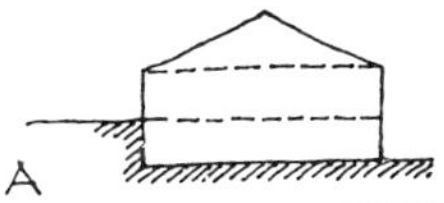 A	2.32	2.76	2.80	3.23	6.44	85.6	101.7	108.6	124.1	293.9
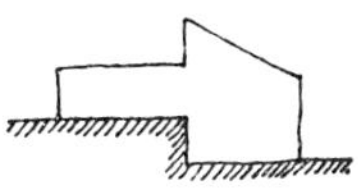 B	2.48	2.96	3.18	3.67	7.28	91.7	103.8	119.6	127.3	314.7
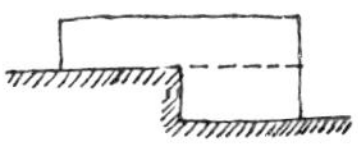 C	2.21	2.60	2.71	3.11	6.30	90.0	103.0	114.2	129.3	289.1
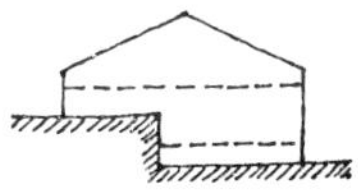 D	2.17	2.54	2.57	2.94	6.28	82.1	97.5	100.6	116.6	288.4
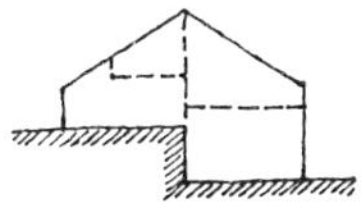 E	2.40	2.86	2.96	3.42	6.45	89.4	107.3	113.6	135.2	286.8
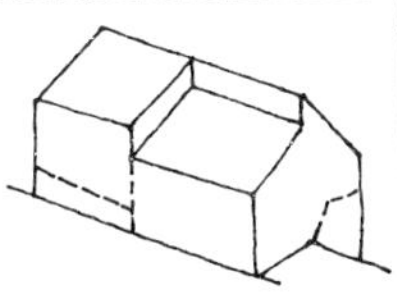 F	2.67	3.29	3.25	3.84	7.31	98.7	117.7	124.2	142.1	310.1
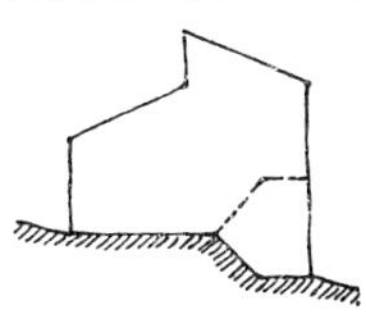 G	2.67	3.13	3.38	3.84	7.67	103.8	121.9	139.1	149.5	321.4
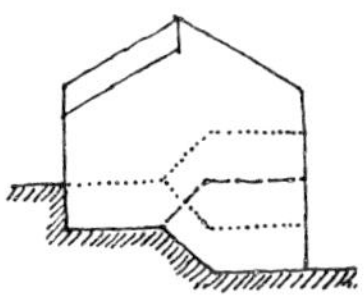 H	2.69	3.26	3.32	3.89	7.33	98.2	114.5	125.7	146.9	316.9

Table 4 : Calculated values of K" and E" for the different drafted designs and the chosen hypotheses.

The main aim of this research is the comparison with the results of the proposed design methodology. This comparison involves on the one hand the estimation of the level of accuracy between the measurements on the field and the proposed techniques, and on the other hand possible implementations if necessary in practice.

2.2.2. Principles

The measurements will be recorded in five houses, two of them will be unoccupied, the local climate will be recorded : air temperature, solar radiation and wind speed. The estimation of the net energy produced by the equipement is required as accurately as possible, and also the air infiltration rate.

The audit by the CSTC will help us to check the basic performances : steady-state behavior (coheating) and air permeability (pressurization test and tracer gaz technique). All measurements will be taken at high sampling rate with real time logical analysis. Past experience has shown that the reliability of data mainly depends on immediate detection of anomalies by intercorrelated tests to be developed during the experiment itself.

The greatest difficulties in interpretation will appear especially for these points :

- the measured temperature in each zone and its relation to the characteristic temperature (air or resultant temperature) depending on the anisothermy; many a resultant measurement of temperature in each zone will help us and also a more detailed recording in the unoccupied houses for typical rooms as the living-room;
- all the multizone problems as natural air convection between rooms and air infiltration orientation; this latter problem would provoke uncertainties on results because these parameters (important in high insulated dwellings) cannot be defined properly and direct determination is not possible.

Once more, the audit by the CSTC can give some information on these problems. For example, the comparisons between the estimated heat loss coefficient for each room and the measured value is a fundamental information in this interpretation. It is interesting to consider the ground losses which can have higher effects in the chosen structure than in classical dwellings. The high ability for the houses to recuperate solar radiation will produce internal convective behavior, difficult to model with high accuracy : experiments will give us interesting and probably unexpected results. We must foresee a methodology to be able to provide enough information to interprete these phenomena.

2.2.3. Methodology

The basic principle is to provide the highest flexibility on data acquisition and treatment on the site, about 40 km away from our laboratory and computers. We retain a technique based on a microcomputer data processing on the field with phone connection with our laboratory. This method gives us all the flexibility of a powerful multiuser software in the data treatment in real time; in parallel with new developements.

An intelligent logger allows a high level relation between probes and the microcomputer package. We consider we need about 120 measurements of temperatures and about 30 counters of energy or flows, with the capacity to install easily a new measure during the experiments.

The following scheme (figure 5) roughly describes the installed system of data processing on the field.

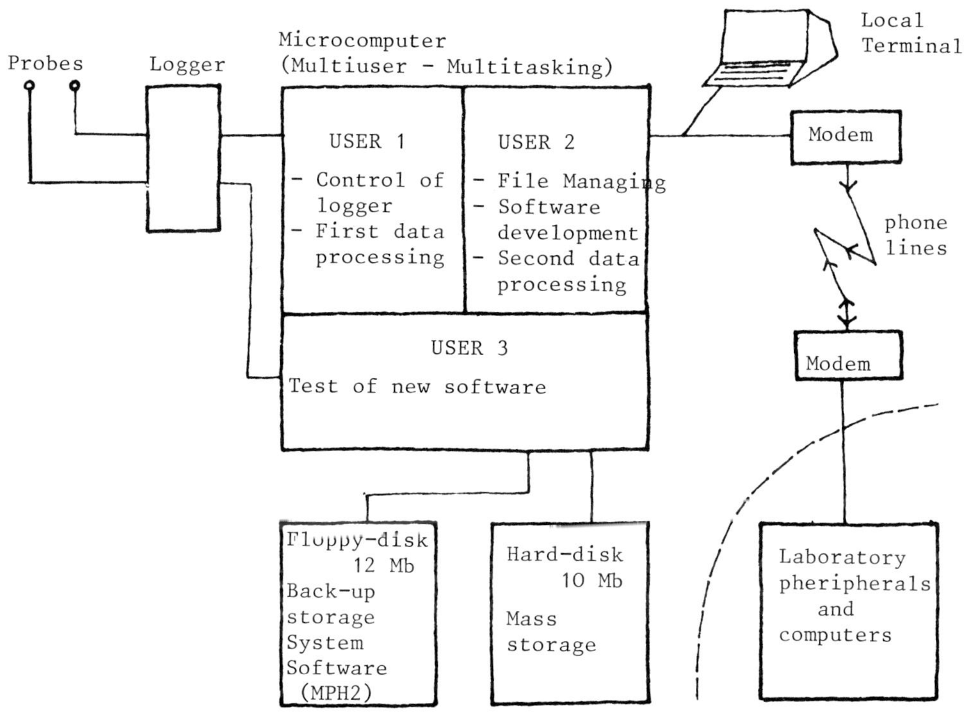

Figure 5 : Installed system of data processing.

The data processing will be a three-level-technique.

- first level on the field : the multiuser-multitasking microcomputer will store filtered data on high mass memory (hard disk system) ; a floppy disk will be used for back-up and transportation of information. The phone connection with the laboratory will allow direct control of data processing, new software developement and data transmission if immediate analyses are necessary. The first level analysis will concern the logical treatment of data with the detection of anomalies in real time. Some actions would also be possible by automatic procedure;
- the second level will concern the treatment of data in laboratory, especially for efficient presentation and parametric estimation. A wide variety of computers, softwares and terminals can be used at this level;
- the third level relates to theoretical computations of existing methods as ESP (Glasgow), LPB-1, LPB-4, etc...Detailed analyses will only be possible with powerful computers.

The following scheme (figure 6) resumes these three stages of data analysis.

2.3. Coordination work

Some coordinated actions have been included in the research program. All are concerned with a more or less partial application of the design methodology produced by the first research work (contract n° 615-78-1EEB).

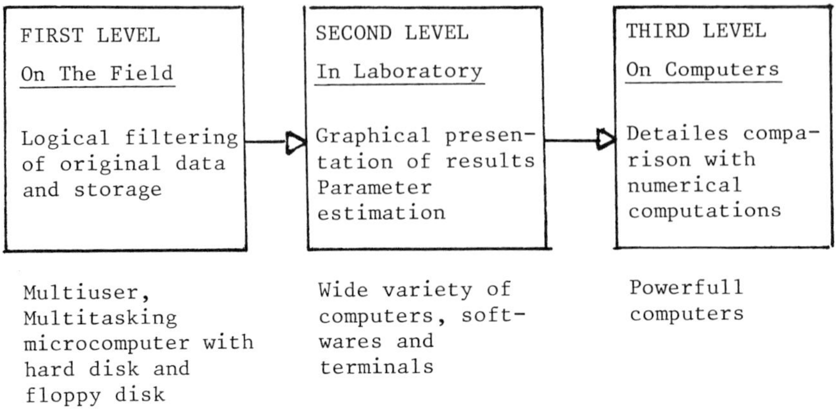

Figure 6: Stage of data analysis

2.3.1. Participation in contract n° EEA-5-056-B(G)

An interesting application of the design methodology has been done in the research work called IDEE under the leading of Jos. Uyttenbroeck : the architectural handbook has been used for the design of two experimental low-energy houses, where the CSTC intend to experiment different heating systems (see figures 7 and 8).

Due to their specific design the two houses can reach three major targets that were imposed to the design team :

- the energy consumption is minimum as well in winter as in summer (no cooling is required);
- thermal bridges and vapor condensation are avoided;
- the interchangeable building structure allows the easy modification of internal and external walls (this allows to analyse the influence of inertia on heating energy consumption).

An estimated quantity of energy used during the heating period has been assumed by the LPB-4 program in steady-state conditions.

Thermal bridges have been evaluated and reduced by the design team using the "Koudebrug" program developed by the KUL University.

2.3.2. Participation in contract n° EEA-1-007-EIR (N)

A very fruitful contact has been taken with Tom Fuller from An Foras Forbartha, during the CIB meeting held in Dublin on March 82. Tom Fuller organized a complete visit of the two houses, the monitoring of which was started by the thermal calibration during winter 1981-82. The organization of the computer program LPB-4 has been studied.

As soon as climatic and building data will be received in Liège, the program will be run for the different intermittence cases required by the Irish team.

3. RESULTS, COMMENTS AND FURTHER DEVELOPEMENTS

3.1. Architectural design

The site analysis represents a first preliminar step in the whole design, but very important because it decides -mainly in groups of houses- on the prevailing orientations to give the houses. Since the design fees and the construction costs decrease with the number of houses to be built, the housing societies always prefer to plan the construction of large groups of houses.

We have then taken the opportunity of the settlement layout in order to add an interesting module to the existing methodology, with the intention to give the designer practical methods to take energetic considerations into account already in the site analysis, such as with the calculation of the shading masks and their resulting insolation losses for overshaded houses. But the proposed method is relatively heavy : the whole façade considered must be divided into a great number of elementary surfaces as close as possible to points, the sun-path diagrams of which have been drafted and -finally- converted into energetical losses.

A second gap in the design methodology was the difficult consideration of the dynamic behavior of the building, with a resulting difficulty to estimate the summertime air temperatures or the intermittence efficiency. We have then introduced a simplified method which allows to calculate, by the use of diagrams, the damping and the time lag of variations of external temperature and internal solar flux and therefrom the maximum amplitude of the internal temperature variaiton (t_i) max. The maximum efficiency of the thermostat set-back utilization is also found by a graphic way, as a funtion of the jump, the time constant, the total period (in most cases 24 h or one week) and the period of set-back (8h, 16h, one week-end). (4)

The application itself of the design methodology seems to be too expensive because of the time spent by an energy conscious architect, if no additional fee to be paid by the client has been foreseen for such an energetical study in the architect fees scales of architects' associations.

The computerization is then a solution which will give the architect the computation tool he needs in order to reduce the time necessary for the application of the design methodology. The computerization will give the possibility of a more accurate site analysis than the manual method, which divides the façades into a small number of surfaces. The computerization will be a simplification of the inputs of the static LPB-4 computation program, which will become a real building design language between the designer and the computer.

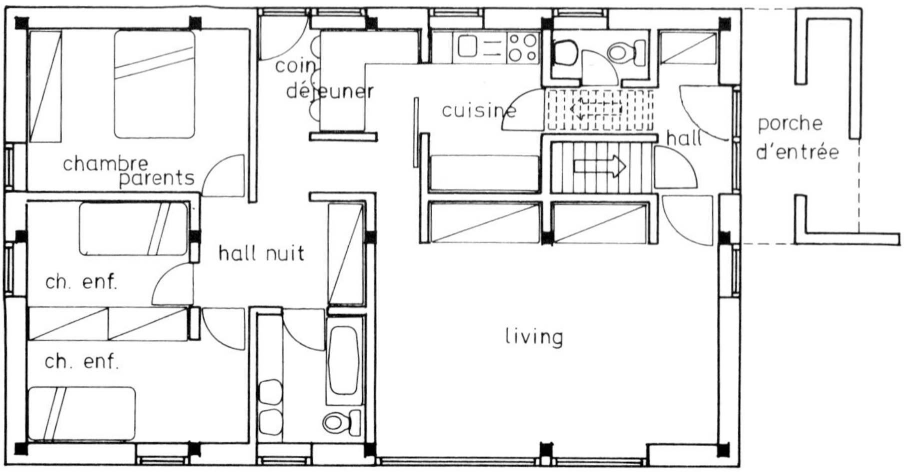

Figure 7: Plan

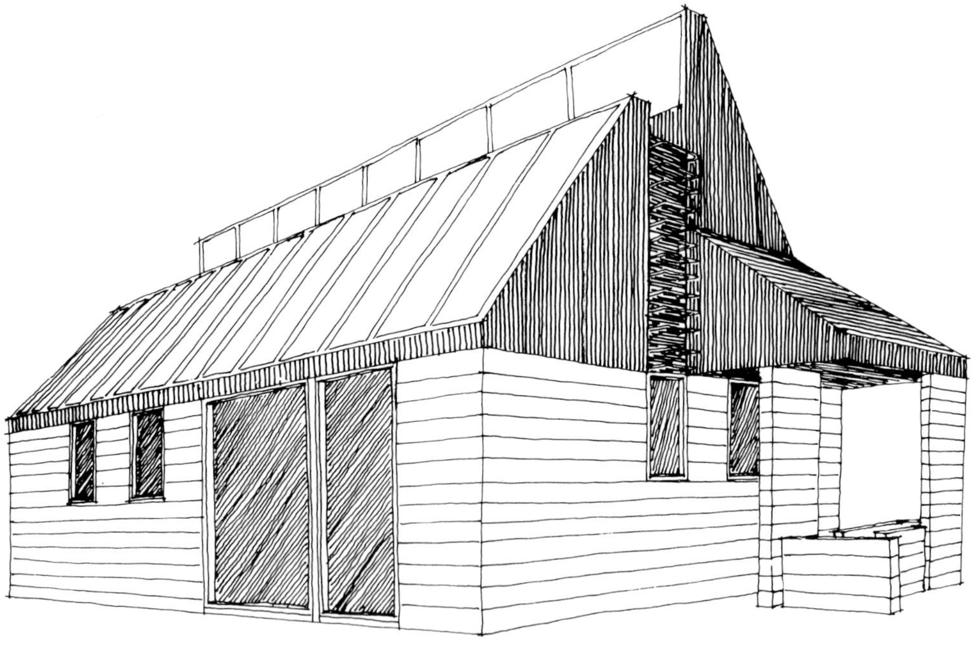

Figure 8: Perspective

3.2. Design of experiments

The main point is to get enough information (such as many "well chosen" points as possible) in order to understand unexpected experimental results which is the basic interest of such a research.

Some of these facts will only appear during experimentation. Thus a maximum flexibility of the whole data acquisition system will be preserved.

The quality of information lies in the smallness of confidence interval size. It is necessary, if we want to be able to estimate significant building parameters and discriminate measures significantly with several theoretical computations.This quality would be improved by accurate measure of air infiltration (i.e. continuous tracer gaz measure) and check of building envelope properties (i.e. thermography). The repetition of the audit measurements at some time intervals during the experience would guarantee the time evolution of the main parameters. We must be conscious that it will be impossible to measure actual air infiltration, especially the orientation of air flow in the multizone structures. This uncertainty will probably interfere with the results due to the magnitude of this term in highly insulated dwellings.

It is also interesting to foresee now the possible extension of the monitoring ot the 25 houses. The chosen microcomputer will bear this extension.

4. REFERENCES :

(1) A. DUPAGNE, M. CORNET, Basic characteristics for low-cost houses in order to reduce the energy consumption for heating. Link programme - Coordination and consequences for the architectural design (synthesis). Architectural design handbook. CEC contrat n° 615-78-1EEB, February 1981.

(2) H. RIETSCHEL, W. RAISS, Traité de chauffage et de climatisation, Tome 1, Dunod, 1973.

(3) L. MUYLDERMANS, Préoccuation énergétique en urbanisme, Programme National Belge R & D Energie, Services de Programmation de la Politique Scientifique, Belgique, 1981.

(4) M. CORNET, Méthode simplifiée pour la détermination des températures d'été et de l'efficacité de l'intermittence, in Final report of E/VI/1 research "Développement d'un modèle opérationnel de référence pour l'évaluation de la consommation énergétique des bâtiments, SPPS 1982 -Programme National Belge R & D Energie.

QUESTIONS AND DISCUSSION

Los (INSO) : We are going to a specification and certification for building materials in Europe. Also an energy saving design methodology is useful provided that it is not complicated and can be used by architects.

Ehringer (DG XII) : Is it possible to introduce the cost in the model ?

Answer : Yes this will be done.

SYSTEM SIMULATION FOR BUILDINGS

A new project in the IEA Implementing Agreement on Buildings and Community Systems

Jean LEBRUN

Laboratoire de Physique du Bâtiment
Université de Liège
Faculté des Sciences Appliquées
15 avenue des Tilleuls - Bât. D1

B - 4000 LIEGE (Belgium)

The International Energy Agency is an organization which tries to bring about a better cooperation in the field of energy use, production and R, D and D. Most of the European countries, the USA, Canada, Japan, Australia and New Zeeland are members of this organization. One of the ways to achieve the objective is to bring about collaborative R, D and D in the form of Implementing Agreements which often are subdivided in specific projects described in a series of annexes, which may be jointly funded or tasksharing. In the Implementing Agreement on Buildings and Community Systems a new Annex X is now proposed on "System Simulation for Buildings". This work is closely related to work which has been described in the previous paper and it was judged appropriate to include a short description of this project in the proceedings of this contractors meeting.

1. INTRODUCTION : SOME FACTS ...

1.1. Considerable efforts have been made until now in developing more and more efficient models for building thermal simulations. Far less has been made for the simulation of the H.V.A.C. equipments and there are very few solutions enabling realistic energy balance of the whole system, which is constituted by the building, its occupants and its H.V.A.C. equipment.

1.2. Because of this lack of good simulation tools researchers have to resort to many artifices in the theoretical and experimental works :

- Highly simplified building space heating demands are considered as conventional inputs in the experimental tests and in the simulation of the equipments;
- Highly "idealized" equipments (or no equipment at all !) are included in many building energy analyses.

1.3. The preceeding artifices avoid a serious contradiction in building energy management : more "efficient" is the building and generally worse are the conditions for its equipment, which has to run most of the time in very reduced load, to fullfill a very variable and very selective space heating demand.
Combined economical results of a "good" building and a "good" equipment are also desappointing ...

1.4. But that is not the whole of it : various unrealistic hypotheses are dissimulated inside the simulation of each sub-system : the building and its equipment.

1.4.1. The building space heating demands are often (experimentally or theoretically) defined on the basis of very crude references : strong differences can occur between the (space and time) temperature profiles which are required by the occupants and the profiles which are effectively measured or computed in the building (taking into account the building space and time "irreversibilities" : coupling effect between zones and thermal inertia). One important cause of occasionnal overheating is the space and time distribution of solar and occupancy "free heat" (another one is the losses of the H.V.A.C. distribution circuits where considered ...).

1.4.2. The performances of H.V.A.C. equipments are affected by many underestimated (or even ignored) unefficiencies : control, storage (if any) and distribution losses and all the "auxiliary" consumptions (pumps, fans, control devices, ...).
These contributions explain why the true global energy balance of a heating system can be so desappointing. For example, the global efficiency of a fuel oil system can very well decrease from 85 to 50 % even below, when passing from nominal to yearly average conditions. Another typical example is the global C.O.P. of a heat-pump system which hardly overpass 2 (mainly due to the auxiliary consumptions required to supply the evaporator and condensor circuits) when most of the theoretical considerations would allow to a minimum of 3 (50 % higher !) ...

2. ENERGY MANAGEMENT : THE TOWER OF BABEL.

The preceeding facts are only very few among many other. Researches and developments in Energy Managements have already accumulated many experimental and theoretical results, which could allow the establishment of significant strategies and tactics ... or to the building of a new Babel which will never be achieved because of misunderstanding among the researches; each specialist is using his own highly specialized terminology which seems to have been invented in order to forbid any comprehensive comparison among the different solutions. "Solar ratios, C.O.P." and all more conventional "efficiencies" are some of the many concepts which should be reduced to strictly equivalent terms ...

Even more dangerous is the fact that each specialist has generally to "re-invent the wheel" when trying to express the results of his research in terms of global energy and economy balance : distribution and control unefficiencies have to be "rediscovered" at various levels; fans, pumps and heat exchanger characteristics must also be reintroduced in many parts of "classical" as well as of "non-conventional" systems with a wide range of hypotheses (which are not always "confessed" by the authors) ...

In short, the present confusion is such that almost anything could yet be demonstrated by system simulation in the field of building energy management. This confusion is likely to increase with the richness of results, which are supplied by the various research and development teams. Therefore, we are challenged in the same way at the different national and international levels : how each researcher should present his (very partial) results so that the others could combine them with their own and vice-versa ? More generally, how could we effectively use all the results which are available ? Is it possible to integrate them into a true global optimalization when looking for the most attractive energy saving measures ? ...

3. TOWARDS A NEW "ESPERANTO"

A common language is urgently needed to enable this integration and to garantee the effectiveness of further researches and developments : all the "elements" (hypotheses, data and characteristics of the components) should be expressed in very well defined formats. As many of these elements should be made available at the widest international level through a data-bank.

Two main typical uses of the data-bank can be considered from now on :

1. systems simulation in order to estimate the contribution of a particular component. This component has to be combined with other standarized parts of the system to make possible any objective comparison with other technical solutions. By using the data-bank the user will not only save time of research and avoid useless explanations but also give his simulation results the maximum of transparancy and therefore the maximum of reliability (possible cross comparisons with other results available) ...

2. Audit procedure : establishment of the global sketch ("portrait robot") of an existing system. In this case, the system simulation as no other goal than to estimate the effectiveness of some energy saving measures (retrofits, financial incentives, etc, ...). Many terms of the energy balance will then remain unidentified and a strong standardization of many data and hypotheses is therefore required in order to garantee the objectivity of the comparisons : the intensive use of the data-bank appears as the only way of rationnally filling the many lacks which are inherent to such a study ...

4. A "TECHNICAL FORMAT"

An agreed methodology has to be developped for describing the performance of a range of standard H.V.A.C. components. This could be done through a "technical format" containing, among others, the following informations :

4.1. Device definition :

4.1.1."Description" of the function, identification of the parameters, variables, hypotheses, dimensionless grouping of variables and parameters, fundamental and "simplifying" hypotheses, etc, ...

4.1.2. Function laws and corresponding errors : the laws can be expressed in an analytical way, in a graphical way and/or in a graphical form.
The errors are due to :
- simplifying hypotheses;
- simplified analytical expression of the laws;
- imprecise values of parameters.

4.2. Consumptions :

With distinction between "direct" and "indirect" consumptions : the last ones are caused in other devices. They are generally of two types : mechanical consumptions (for instance the consumption required to move the fluids through a heat exchanger) and thermal or EXERGETIC consumptions (for instance the thermal losses and the thermal resistance imposed by a heat exchanger).

4.3. Technical and economical data

At last a rough information should be given on investment, maintenance and running costs, related to the life time and the possible ageing process (for example, the fouling of a heat exchanger). A classification could be proposed among several level of prices and qualities.

4.4. References

Each technical form is a summary (and possibly a compromise) of theoretical and experimental researches. These researches would have to be clearly identified to allow any further improvement of the technical data.

5. THE NEW TASK PROPOSED BY THE INTERNATIONAL ENERGY AGENCY

This project is actually started in the frame of the I.E.A.'s Implementing Agreement on "Research and Development in Building and Community Systems". The main objection of this project consists in establishing the new system simulation ESPERANTO, i.e. the data-bank.

The work is organized on a task sharing basis; it includes the development of the technical format as well as the practical organization of the data-bank itself. Furthermore, to avoid the risk of producing inadequate simulation tools (as a new informatistic "monument" which could never be used !) some practical "exercises" are proposed from the beginning : the participants are asked to simulate various H.V.A.C. systems for two basic types of building, domestic and commercial. The overall energy consumption of the building will be calculated in each case on the basis of detailed specifications. This work is expected to allow us :

1. to check the reliability of the simulation methods;
2. to get some examples of very detailed energy saving analyses which could be considered as references;
3. to start effectively the feeding and the using of the system simulation data-bank.

ENERGY CONSERVATION ON EXPOSED DOMESTIC SITES

Authors : F. O'FARRELL, G. LYONS, G. LYNSKEY.

Contract number : EEA-05-054-EIR (H).

Duration : 24 months 1 September 1981 - 31 August 1983

Head of project : Mr. F. O'Farrell, An Foras Taluntais.

Contractor : An Foras Taluntais

Address : An Foras Taluntais, Headquarters,
19 Sandymount Avenue, Dublin 4.

Summary

The internal heat gains of an isolated dwelling resulting from external climatic factors are being recorded in two separate dwellings A and B. The internal ambient temperature of house A was monitored at the site at Oakpark Research Centre in Carlow. External climatic data was obtained from the Kilkenny Meteorological Station and from the Oakpark weather station. The resultant data is being processed to produce a theoretical simulation model for predicting internal ambient temperatures for different weather conditions. A similar size house B but with a different fabric construction and different surrounding ground profile has also been monitored, but the data has yet to be processed. The fabric of the second house is being altered for further monitoring.

1.1 Introduction

This project is concerned with the conservation of energy in isolated dwellings, particularly farm dwellings on exposed sites. While the majority of the population of Europe lives in towns and cities, consequently the greatest amount of energy required for domestic purposes is consumed in these conurbations, a sizable number live in isolated houses scattered throughout the countryside. In Ireland in 1971 there were 225,445 permanent housing units with agricultural land out of a total of 701,180 houses in the State. All those housing units with agricultural land may not be classified as farm dwellings (the number of farmers in the country were approximately 180,000). However the larger number of dwellings all have the similar siting attributes of having land around them, in other words, "open aspect" and therefore similar advantages and disadvantages with regard to energy demands and conservation.

While the complete results of more recent census returns in Ireland are not available estimates would indicate that the proportion of isolated houses has not greatly altered, 287, 883 out of 902, 408 in 1979. This would indicate an annual net energy demand of 39 GJ x 10^6 for such dwellings. If the gross annual energy consumption per farm dwellings was up to the average for the country, 142 GJ., then this figure would be even higher. As it is, it represents well over half the total energy input into Agriculture. This compares with the European scene generally, when it has been estimated that 70% of the conventional energy used in agriculture goes into space heating of rural houses and domestic water heating. Again the figures for farm holdings throughout Europe gives an indication of the amount of dwellings involved, the approximate number of holdings are: Ireland - 260,000, UK - 270,000, Denmark - 127,000, Luxembourg - 5,400, Belgium - 102,000, Netherlands - 141,000, France - 400,000 and Germany - 300,000.

1.01 Extent of energy used

Even among isolated houses and farm dwellings the amount of energy used can vary enormously. Energy usage is very closely related to the standard of household management practiced and the life style of the owners. At the commencement of the 'oil crisis' the mean net energy consumption per house in Ireland was 98 GJ, in Denmark 118 GJ and in America 137 GJ, illustrating the tremendous energy requirement of the dwelling house in Western Society. Consequently it becomes apparent that the way a household lives will be one of the largest determinants on the amount of energy it requires or demands.

Similarly while it has long been acknowledged that the greatest energy savings in the domestic sphere can be realised from adopting conservation measures, the adoption of many of these may require a change of attitude or life style, e.g. in a fairly recent survey in Ireland, it was found that " 14% of the farm households kept the back door open all day even during the heating season ". A first and obvious change of life style is called for if ventilation rates and consequently, heat losses are to be reduced.

1.02 Alternate life style

The large expansive Western way of life has been severely questioned over the last few years, probably spearheaded by Schumacher, but certainly followed by many other individuals and groups, the general tenor being that it cannot, and should not, continue.

However, while much has been written about low energy stretegies, and even acted upon by different small groups, in practice, the general trend is still concerned with large developments and towards large ranch farms and of course, large energy consuming houses. These two approaches, or philosophies perhaps, are poles apart and seen ireconcilable. Yet, some of the implications of the alternate approaches and the philosophies behind them are worth examining and indicate that isolated houses may become more important. Many feel that the balanced use of energy and the development of a meaningful way of life are inextricably bound up with the land question. It is one thing to suppose that all new isolated houses will be large sophisticated and highly mechanised farm dwellings complete with all 'modern services and amenities'. But at the same time to hope that the majority of farmers may live in such houses and that they will be highly energy efficient is frankly contradictory. As Illich (1973) argues 'better housing then can be occupied only be those who are well off or by those on whom the law bestows direct rent subsidies'.

1.03 Household type

Historically and geographically households are in continual change. These range from groups of families, through large extended families, nuclear families, non families to single person households. In Ireland, as elsewhere, it is obvious that the conservative traditional family structure is changing. Yet in 1971, in Ireland, two thirds of the farm households were family households. It is expected that this proportion will increase in the 1981 census. Thus, in Ireland the family remains the basic farm household and it is for this the farm dwelling must be designed and it is this that is worth examining. The energy opportunities and uniqueness of the isolated farm dwelling as a housing unit must be appreciated and exploited.

1.04 Uniqueness of isolated dwellings

The isolated dwelling is unique as a housing form. Because it is in open countryside, it can make optimum use of orientation, siting and shelter with none of the constraints imposed by tight urban situations. Unique by virtue of its demographic structure, it is also unique by virtue of the skills available or by the attitude of the farming community to undertake such skills. Much building at farm level is undertaken by the farmer himself, at least at some stage of their careers. The extent of self build can vary enormously from architect designed houses to houses that received little thought in advance. One of the most successful well designed do-it-yourself vernacular in these parts must be that of Walter Segal, particularly his housing in Lewisham, United Kingdom.

1.05 Appearance in countryside

In addition to the energy conservation aspects of exposed dwellings there is the element of the conservation of the rural countryside. Thus all building in rural areas particularly exposed buildings such as dwellings must be designed so as to be in harmony with the countryside. This applies not only to the building itself, but also to its immediate surroundings including planting, screening or alternations of levels. Thus the objective of this project is to examine the effects of external climatic factors on the heat demand of exposed domestic sites. Also to examine the microclimate obtainable from different site conditions and thus access the saving that optimum siting and internal retro-fitting can have on the energy consumption and to do this within the context of buildings which compliment the countryside and rest in harmony with it.

2.00 Description

The sites selected for this project are situated at the Oakpark Research Centre of the Agricultural Institute. This is situated about 4 km. almost due North of Carlow town, lat. 52° 15' 12" N., and long, 6° 54' 15" W. This centre is responsible for research on tillage, agricultural engineering and biomass. The estate comprises 292 hectares which include a wild life sanctuary consisting of a 16 hectare artificial lake and several wooded islands.

The sites themselves are exposed to South and South West. The wind being from these directions for 31% of the year.

The mean daily duration of bright sunshine can vary from 2.14 hours in January to 6.62 in May and the annual horizontal global solar radiation is 3,657 Mj/m^2. The average maximum temperatures in summer is 19.6°C and average minimum in January is 1.3°C. There are an average 2284 degree days below 15°C and the average annual rainfall is 818 mm.

The land generally is flat open land with the odd tree. However as even in exposed buildings some shelter is usually provided it was considered desirable that the houses to be monitored should be protected from the North and East. This decision also meant that monitoring could examine the effect from total exposure on one face and protection from another.

The two sites chosen are within 100 m. of each other. Both sites have the same general characteristics, being exposed to the South and protected from the North. Both houses have been built over three years so that in each case construction has completely dried out and therefore stable thermal performances can be obtained.

2.01 House 'A'

House 'A' is a 112 m^2 bungalow of fairly recent construction, but poor conservation standards. The accommodation consists of five bedrooms, kitchen/living room, sitting room, bathroom and utility room. It is orientated to face S.S.E. The calculated u values are given in Table 1.

TABLE 1: Fabric Conductance - Test house 'A'*

Building Element	Area (m^2)	U Value	U x A (W/°C)
Roof	124.965	0.60	74.919
Floor	124.865	0.68	84.908
Ext. Wall (280 cavity)	21.883	1.7	37.201
(225 hol. block)	70.628	2.0	141.256
Windows N.	20.63	4.48	92.422
E. & W.	2.14	4.48	9.587
S.	9.56	5.67	54.205

* Utility has been ignored

This house was considered to have two zones. Zone 1, the southerly one, consisting of the sitting room and two front bedrooms - The back section of the house being zone 2.

2.02 Test House 'B'

This house is also a bungalow, of a similar floor area. The accommodation consists of kitchen/living room, sitting room, four bedrooms, utility room and two bathrooms. This house has three heating zones. The zone I consists of kitchen/living room, sitting room, and a bedroom. This southernly zone receives direct insolution through South windows. Zone 2 consists of most of the northern half of the house including three bedrooms and zone 3 consists of the utility room and second bathroom.

The house includes some energy conservation and passive solar elements in its construction, thus part of the south wall is a Trombe wall of high density concrete block. The other external walls are of cavity wall construction, the two leaves being 200 mm. hollow concrete block and the 200 mm. cavity containing 150 mm. fibre glass. The calculated u values are given in Table 2.

TABLE 2: Fabric Conductance - Test house 'B'

Building Element	Area (m^2)	U-Value ($W/^oCm^2$)	UxA ($W/^oC$)
Roof	121.5	0.11	13.365
Windows N.	2.18	5.00	10.90
Windows S.	11.63	3.97	46.17
Windows E.	0.86	5.00	4.30
Windows W.	3.60	4.48	16.13
Doors N.	1.98	4.00	7.92
Doors E.	1.98	3.50	3.93
Insulated Wall	66.44	0.30	19.07
Trombe Wall	32.11	1.03	33.07
Floor	121.5	0.48	58.32

3.00 Methodology

One of the problems that occur in studying the effect of external macro and micro climate on internal ambient temperatures is the difficulty of obtaining data on which to make comparative analysis. One of the more usual approaches is to take two houses, very often a pair of semi-detached houses, incorporate different energy conservation standards in each house and monitor them for a fixed period. However, this tells very little about the effect of the surrounding site conditions on internal gains and in fact can only examine the effects of different fabric conductance levels. Even in this area one cannot rely on calculated u values as buildings may not achieve such values in practice. Similarly in the case of two semi-detached houses the orientation of each house in practice is not identical and therefore the solar gain must vary to a greater or lesser extent. In other words, it is difficult to achieve identical configurations which would receive identical external climatic data inputs. Not withstanding, this form of monitoring has very obvious advantages and a limited form of such monitoring is incorporated in this project. The two test houses 'A' and 'B' as described are on very similar sites and therefore climatic configurations will both be monitored.

The other analytical approach is to use the same house but with altered fabric conductances and site configurations. This removes all difficulties or discrepancies which would have arisen from identical houses that are not quite identical, but of course, raises its own problems. The obvious one being that if a house is monitored for a fixed period and subsequently altered and remonitored, it will be done under completely different climatic conditions. Nevertheless, from the resultant figures, recorded over a long period of time, predictions could be made for other weather conditions. Thus though different recording times are involved meaningful comparisons may be made. Again the longer the monitoring periods, say even twelve months, the more accurate the predictions.

However, time is not always available on such a generous scale, nor indeed are facilities. The present project seeks to investigate the relationships of external climatic to internal heat gain in considerable detail but over fairly long periods of time. Thus one of the main trusts of the project will involve the development of the methodology. This will involve the establishment of formulae of a steady-state nature by which internal ambient temperature may be generated from external meteorological data. This will be set up by the application of multiple regression techniques to internal and external data recorded over a period of months. From this a theoretical computer simulation model for predicting internal ambient temperatures on an hourly basis will be constructed. By the application of these techniques, the potential for conservation for exposed dwellings will be examined.

3.01 Heat Gains

Casual heat gains in any building originate from three main sources:

1. Human emission.
2. Activity of domestic appliance.
3. External factors.

By examining unoccupied and unheated buildings, the first two factors can be eliminated. It is also hoped to keep monitoring as simple as possible. No matter how many sensors or how complicated a regime may be, it is still the internal ambient temperatures which will reflect any and all gains due to climatic factors. Thus the analysis will follow the general pattern of internal temperatures without heating being dependant on external temperatures without heating on the previous day. Hourly based regression models will be developed so that the steady state assumption applies only between estimated hourly values. In this way the continuously varying climatic and solar radiation influences are accurately reflected by internal temperatures.

At present the Irish Meteorological Service are preparing, along with their European partners, a standard meteorological year. By adopting such a base for the predictions from the different monitoring regimes, it should be possible to provide an even more meaningful and useful study.

Passive solar heat gains and the auxiliary heating requirements will be evaluated through the application of an hour-by-hour heating load analysis to the different zones of the test houses. A computer programme (SOLHET) is being developed for this analysis using the general heating demand design approach.

3.02 Monitoring Regimes

Several different monitoring regimes will be undertaken. After each regime has been completed, some aspect of the test house or site will be altered and a subsequent regime commenced.

Regime 1. Monitoring internal ambient temperatures in test house 'A' for a five month period.

Regime 2. Altering the ground profile to test house 'B' and monitoring internal ambient temperatures for a four month period.

Regime 3. Altering the fabric conductance of test house 'B' by double glazing and monitoring for three months period.

Regime 4. Erecting shelter belt to the South of test house 'B' and monitoring for three monts period.

Regime 5. Altering the shelter belt and monitoring to continue for further three months.

3.01 Sub regimes

One of the constraints of this project is the time factor. The longer the period of time for each regime that is monitored, the more accurate will be the regression analysis equation. Obviously the length of time available for the overall project places considerable restrictions on the alterations and variations that can be incorporated into it. Not withstanding there are many additional variations that are worth examining. So in addition to the main regimes, it is hoped to examine several sub regimes. These will be examined with a view to improving the methodology for such detailed alterations and also to procuring results which while maybe not statistically accurate, would never the less be obviously sound. The sub-regimes would run for days rather than months and would include the following alterations:-

Regime 5. The addition of triple glazing and the sealing of airleaks.

Regime 6. The addition of thermal shutters to windows which would be opened by day and closed by night.

Regime 7. The performance of the house resulting from snow cover conditions not usual in Ireland but relevant to Continental conditions.

3.03 External micro-climate

In addition to monitoring the general external climatic conditions, more detailed monitoring of the micro-climate will also be undertaken. Wind direction, wind speed and ambient temperature will be monitored in different positions.

3.04 Final assessment

Study of the different monitoring regimes and their predicted casual heat gains will be used to generate an economic assessment of the measures. All monitoring, site alterations and house alterations will be made available to the technical University of Lyngby, Copenhagen for subsequent wind tunnel studies. Finally recommendations will be made for further research and study.

4.00 Progress

Two of the monitoring regimes have been completed and part of the analysis of regime 1.

4.01 Regime 1. As it was considered essential that every effort should be made that the project could be completed in time set for it, it was decided to commence monitoring as early as possible rather than to wait around for every item of equipment, provided that that would not interfere with the monitoring itself. Internal ambient air temperature was adopted as the main thermal indicator within the dwelling and measured in both zones of the house. The data thus collected consisted of hourly internal ambient temperatures, hourly external ambient temperatures, hourly wind-speed and hourly global solar radiation.

The external parameters were monitored at the Oakpark weather station (approx. 300 m. from house 'A' and 200 m. from house 'B'). However, as this station provided only daily averages it was necessary to use the more detailed hourly records from Kilkenny meteorological station situated 35km. distance. To determine if this data accurately represented climatic conditions in the neighbourhood of the test house, a series of statistical comaprisons were applied to daily average meteorological records from both the Oakpark and Kilkenny stations. This was conducted by applying statistical F - test to the mean daily values from both locations to establish the significance of the difference between the two records. The results of these tests showed a non significant difference for external air temperature, wind speed or global solar radiation.

Internal air temperatures were measured by means of copper-constantan thermocouples and recorded on a Phillips PM 8235 Multipoint chart recorder which incorporated antomatic cold-junction temperature compensation. Supplementary measurements were taken on bi-metallic strip thermograph pen recorders, to maintain continuity of data during equipment calibration or failure. All probes were checked weekly against a standard thermocouple temperature meter.

4.02 Regime 2. Before monitoring was commenced, the erection of wind deflecting mounds around house 'B' were designed. Models of the house were made and these mounds were examined in a water table to simulate wind flow across the house. In the three dimensional model flow from six directions were examined, E. and S.E., were omitted as the cover from surrounding trees provided considerable protection from this direction. In the two dimensional model flow was examined only from North or South.

The final design and formation of the mound as shown on the drawings also sought to incorporate aspects of design suitable for the countryside. Thus on the south side it is kept fairly near the face of the house, but such that the trust of any flow from the south would be deflected to the eaves level of the house. (The mound could not be higher than 1 m. above f.f.l. of the house itself to allow the occupants a clear view over the mound while seated). The mound was not continued around the house as this would be visually depressing. The planting was also selected to enhance the environs of the house and to assist wind protection.

The house itself, as mentioned, contained several passive solar elements and ventilation louvres and grills. However, it was decided that if all ventilation louvres were continually adjusted to gain optimum solar benefit, it would yield very biased results. Consequently, the position of all louvres were kept open during the monitoring and will remain the same for all subsequent regimes. The data monitoring in house 'B' has been undertaken by means of a Microdata logger M 1600 L and p.r.t. probes over a four month period. This data logger has given considerable trouble. It was in operation being tested before the starting date and so far three of the cards have been faulty and have had to be returned for repair. The latest fault which occurred was a malfunction of the integrated circuit on the C 1139 buffer and scan control card. This caused several errors on the recorded data for the first recording regime on house 'B'. However the extent will not be statistically relevant and a programme has been prepared to clean up the tape. At present, some difficulty has arisen with interfacing the tape from the data logger with the PDP 11 computer at Oakpark, this should be solved shortly. The data will then be transferred to floppy disc and moved to the Vax computer in H.Q. for further analysis.

4.03 Regime 3. The fabric conductance of experimental house 'B' has now been altered by means of additional glazing and a new monitoring regime, similar to regime 2, is at present on going.

5.00 Equipment

Further external monitoring equipment, windvanes, anemometers and solarimeters are under order and will be commented on in subsequent reports.

5.00 Analysis

The data generated from regime 1 are now being used to develop analogue functional relationships to equate mean zonal internal temperatures with the selected variables. The range of variables included in the analysis are:

1. Internal temperature at time i ($t_{int._i}$), the dependent variable. (°C)
2. Internal temperature at time i - 1 ($t_{int._{i-1}}$) (°C)
3. External air temperature at time i ($t_{ext._i}$) (°C)
4. Wind speed at time i (W_i). (knots)
5. Global solar radiation received at a horizontal surface, at time i (R_i). (J/cm^2).

The two main hourly based regression models are presented in Tables 3 and 4.

TABLE 3 : Regime 1 - Zone 1 - hourly based multiple regression model

	Constant Term			Variables		
		$t_{int.1_{i-1}}$	$t_{ext._1}$	W_1	R_i	R^2
Regression Coefficient	0.656	0.909	0.042	-0.025	0.003	0.962
T-value		161.100	8.520	-8.355	15.745	
β Coefficient		0.909	0.049	0.029	0.058	

TABLE 4: Regime 1 - Zone 2-hourly based multiple regression model

	Constant Term			Variables		
		$t_{int.2_{i-1}}$	$t_{ext._i}$	W_i	R_i	R^2
Regression Coefficient	0.022	0.960	0.039	-0.005	0.008	0.996
T-value		509.571	20.739	-4.473	11.215	
β Coefficient		0.960	0.044	0.006	0.015	

Further analysis has yet to be undertaken including analysis of data from subsequent regimes.

6.00 Conclusions

There are no firm conclusions at this stage of the project, save perhaps that things often take longer than one thinks.

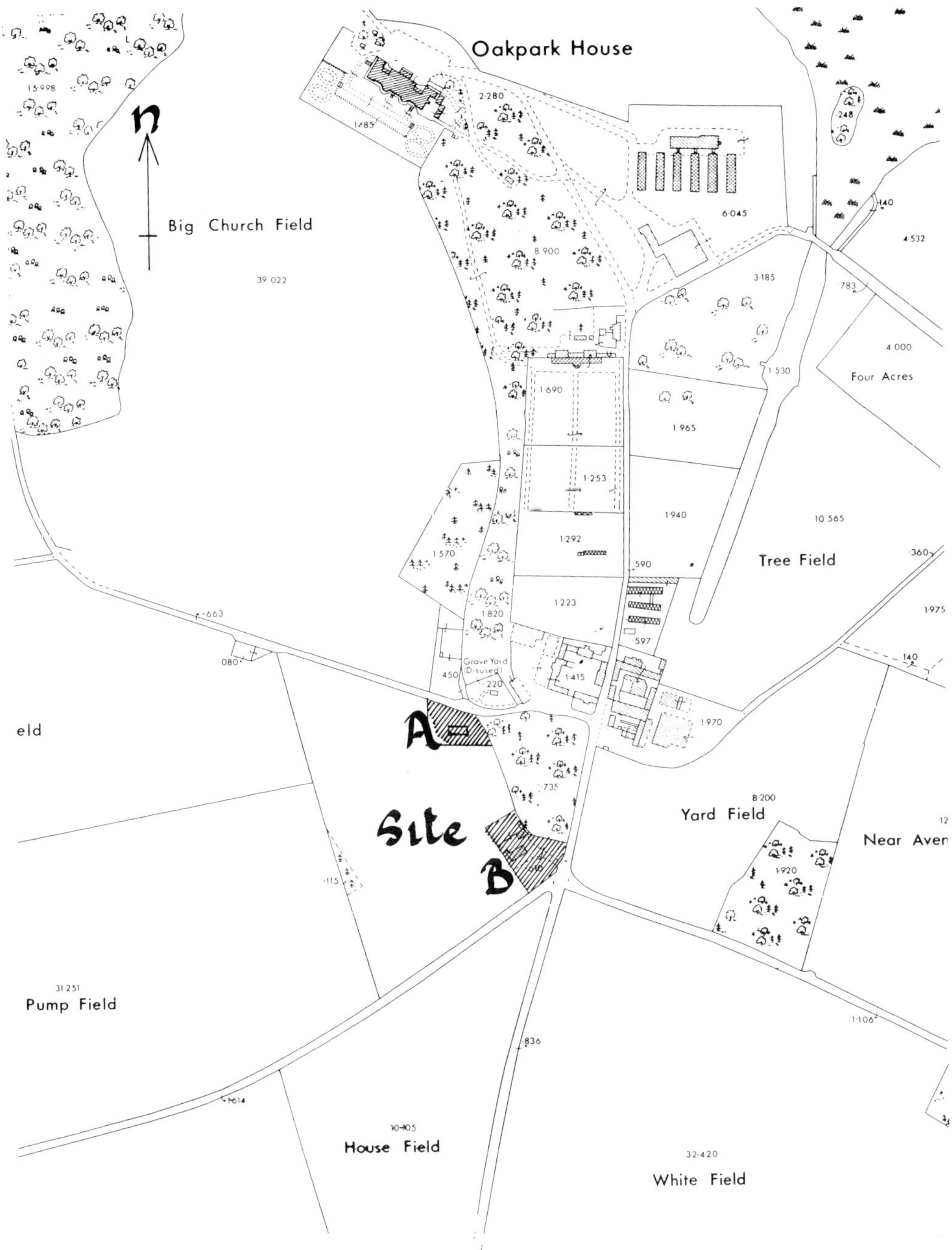

Oak Park Research Centre
Carlow

Elevation

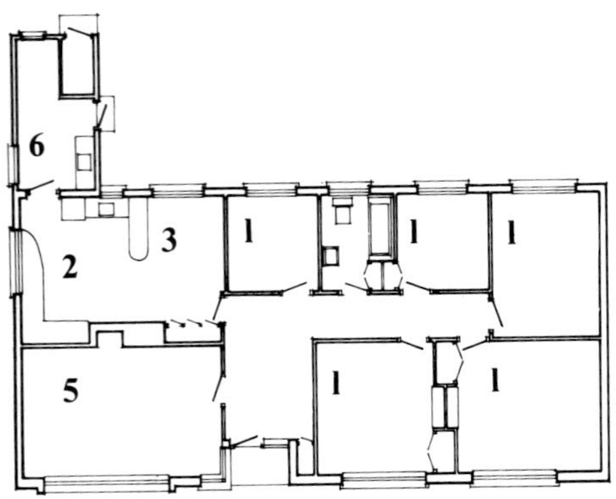

Plan

legend

1. Bedroom
2. Cooking
3. Dining
5. Sitting
6. Utility

Test House 'A'

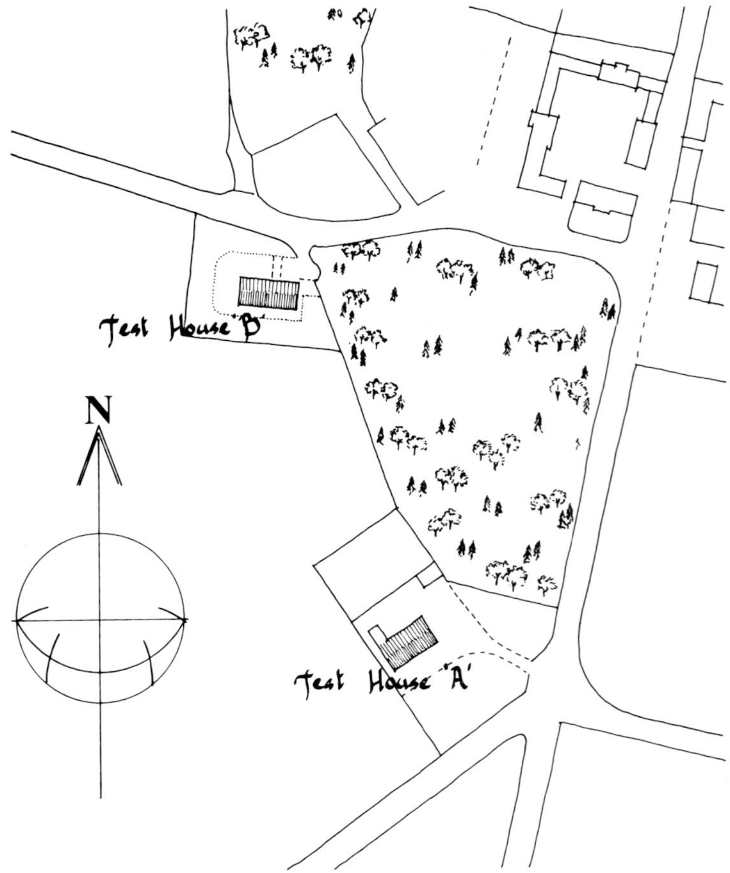

Site Location

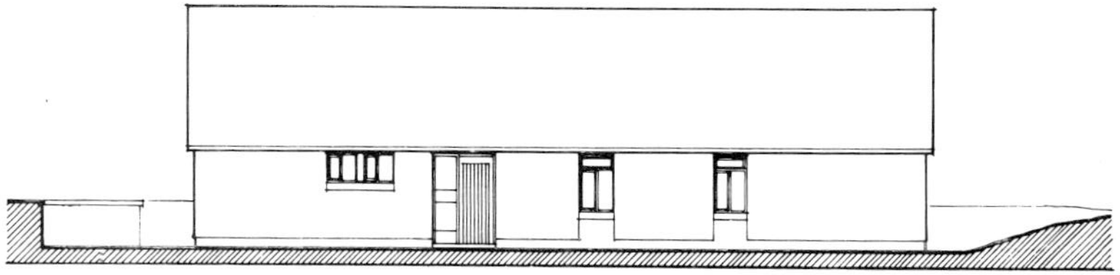

N. Elevation

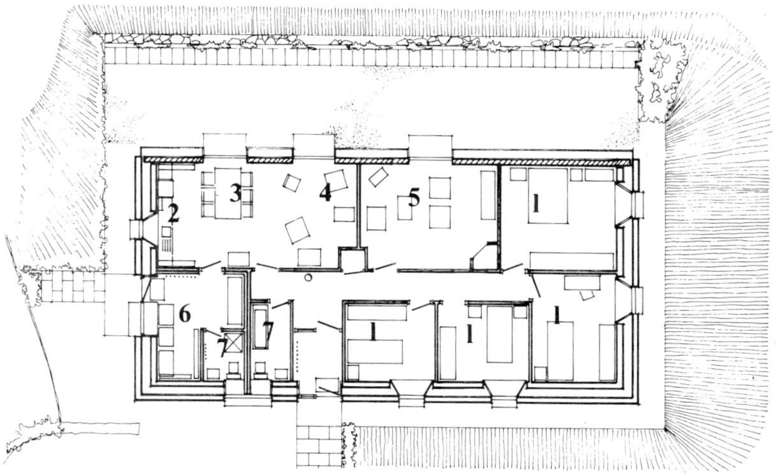

Plan

Legend

1. Bedroom
2. Cooking
3. Dining
4. Living
5. Sitting
6. Utility

Test House B

Test House 'B' ~ South Aspect

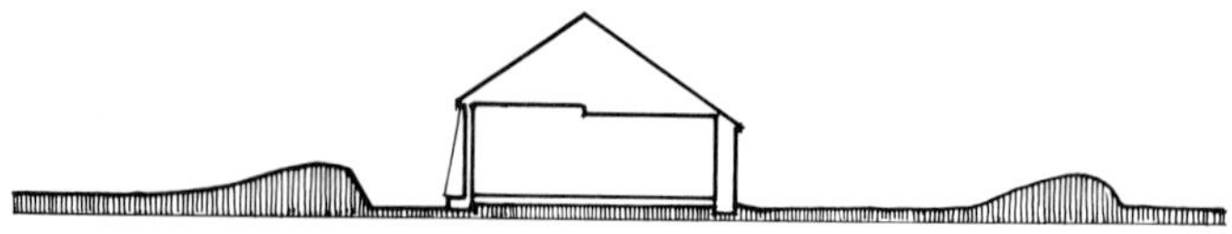

Cross Section

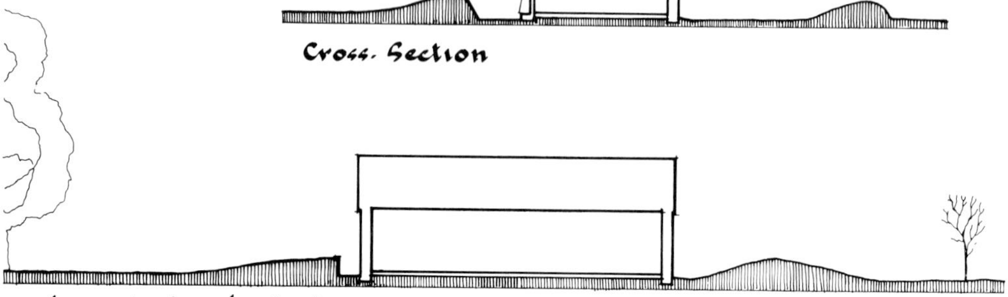

Longitudinal Section

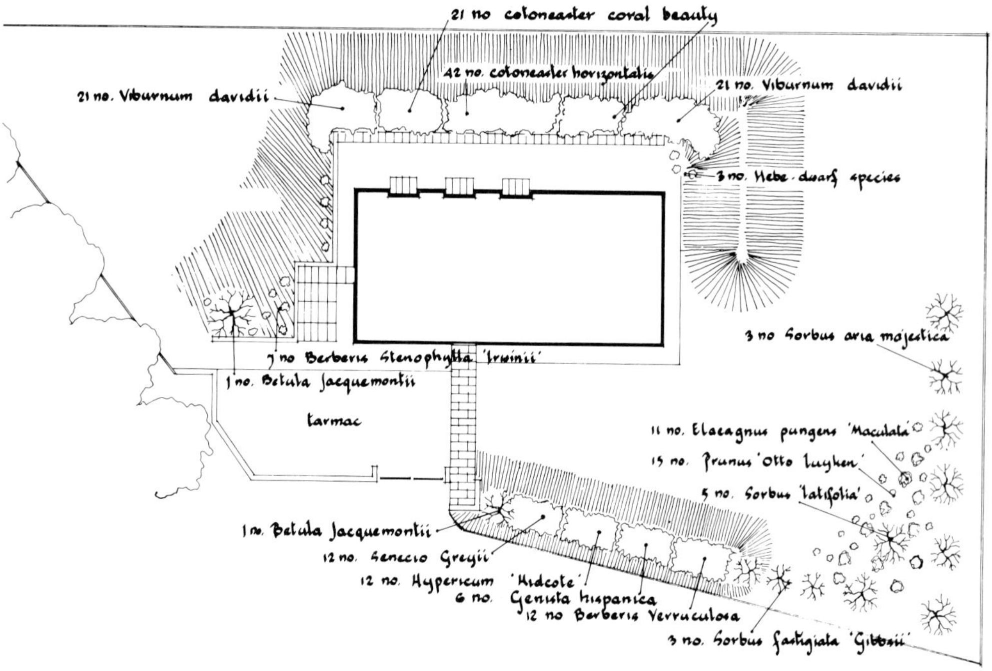

Site Plan showing planting

Test House 'B'

THE ESP SYSTEM : PROGRESS REPORT ON EXTENSIONS AND APPLICATION TO HOUSE DESIGN

Authors	:	J.A. CLARKE, D. MacRANDAL
Contract number	:	EEA-5-059-GB
Duration	:	18 months October 1981 - April 1983
Head of Project	:	Professor T.W. Maver
Contractor	:	Architecture and Building Aids Computer Unit, Strathclyde (ABACUS)
Address	:	ABACUS Department of Architecture University of Strathclyde 131 Rottenrow GLASGOW G4 0NG U K

Summary

Over the period 1975-80 ABACUS developed the prototype of a computer-based simulation model known as ESP (Environmental Systems Performance). The ESP system deals by explicit means with the dynamic energy transfers within multi-zone problems ranging from large commercial developments to housing applications. Since then a grant under the EC Energy Conservation R & D Programme has allowed further developments on three complementary fronts:

1. a range of extensions have been implemented which undoubtedly enhance the model's range of application, ease of use and relevance to design problem solving.
2. ESP has been used as the technical 'engine' in a pilot study to develop a simple technique for assessing the dynamic energy implications of housing location and design.
3. the ESP system is being implemented on an inexpensive mini computer operating under UNIX.

This paper summarises the work to date and future intentions in each of these areas.

1.1 The ESP System and Recent Extensions

ESP dynamically models the energy transfers within, outwith and between a number of zones linked together to form a building - where each zone can be regarded as a general polyhedral space bounded by any combination of opaque and transparent surfaces. This is achieved by the use of an advanced implicit finite difference technique linked to a proprietory matrix handling scheme which allows the detailed modelling of the zone flowpaths shown in figure 1.

The objective of the ESP system is to make possible a rigorous appraisal of any proposed or existing building either as an aid to an economic selection of installed plant and associated control schedules or to give an indication of appropriate energy conserving design strategies.

The system can be operated either in 'interactive graphics' or 'batch' mode and has sophisticated input management and output facilities which enable a user to answer such questions as:

- what and when are the peak loads and what are the various causal contributions?
- what are the environmental conditions during extreme climatic influences or during restricted or no plant conditions (passive solar)?
- what are the energy requirements over any period?
- what will be the effect of some design change such as cavity infill, change of glazing type or distribution, re-zoning etc?
- what is the most energy conserving zoning strategy?
- against which control schedule should installed plant be operated?
- what is the best mechanism to determine the optimum start-time?
- what is the most appropriate set-back temperature?
- how should the elements of the design best be arranged to effectively utilise solar energy?
- conversely how can undesirable solar gains be eliminated?
- how effective is phase change material and what is the most efficient distribution strategy?
- what is the optimum arrangement of constructional elements to encourage load levelling and hence efficient plant operation?
- what are the energy consequencies of non-compliance with current energy regulations?

and so on. This allows the designer to identify potential problem areas, appropriate design modifications, best-fit plant operational strategies, comfort levels, fabric performance, condensation risks and, most importantly, the interrelationship between design and performance parameters which can become the seed solution of future design projects.

Figure 2 shows the connections between the various program and database elements comprising ESP. These include:

- ESPIMP a program which allows the interactive construction of the data files describing some or every zone within the building to be subjected to some weather influence and simulated over time. All items, as input, are processed through legality and 'within acceptable range' checks before being located in an output zone file for transfer to the simulation engine ESPSIM. Other facilities include a range of interactive editting commands and automatic (and optional) access to construction and casual gain profile databases.

- ESPCON a program to manage (create, modify, delete, list) a primitive constructions database containing the thermophysical properties (conductivity, density, specific heat, solar absorptivity and emissivity) of a number of miscellaneous homogeneous elements. This program will also allow the creation and editting of a second database containing whole composite multilayered constructions formed from elements extracted from the primitives database.

- ESPPRO a program to manage a casual gain profiles database containing any number of standard or project-specific profiles defining the time varying heat gain from occupants, lights etc.

- ESPSHD a program to predict the time-series shading of external opaque and transparent surfaces as caused by surrounding site and facade obstructions. This program need only be invoked if a detailed treatment of shading is required: a simplified treatment of shading is possible via ESPIMP.

- ESPINS a program to predict the time-series insolation of internal opaque and transparent surfaces as caused by solar penetration through windows. This program need only be invoked if a detailed treatment of insolation is required: simple insolation treatment is possible via ESPIMP.

- ESPVWF a program to compute view factors between the surfaces bounding any multi-faceted body. Again need not be invoked if, for example, final geometry is not certain. In this case ESPSIM will default to time-averaged values for radiation coefficients rather than computation, from first principles, based on the computed view factors.

- ESPAIR a program which simulates the flow of air throughout any multi-zone system (where each zone is connected by flow resistances representing windows, doors, cracks, etc) subjected to boundary wind loading. If not invoked a simplified treatment of airflow is possible via ESPIMP in which infiltration and inter-zone air flow can be specified as time varying design air change rates.

- ESPCLM a program to manage the climatological database allowing creation (by prediction), modification and analysis of the hourly time-series values of the various climatic parameters retained.

- ESPSIM is the central simulation 'engine' which simulates energy flow by a heat balance matrix inversion technique of implicit enumeration. The technique is unconditionally stable for all computational time-increments and operates in conjunction with hourly values of various climatological parameters recovered from the climatological database. As the model steps through the user-defined time-increments it continuously predicts the environmental and energy status of the simulated building. This is done by placing 'nodes' at many points of interest, for each node performing a heat balance between all interacting regions (over time and space dimensions), and then simultaneously solving the resulting set of equations at each time-step.

- ESPOUT is the output program which operates on the simulation results located in the results database by ESPSIM. A variety of output facilities are available: visualisation; result interrogation; graphical display; statistical analysis and tabulations.

Much of the foregoing owes its existence to the EC contract: many of the facilities of the input management program; the profiles database management program; the insolation patch prediction program; the integration of the air flow module; the extension of the climatological database (to 50 years of 1 hour values of the main climatological parameters); the devleopment of a generalised view factor computation technique; the development of a control language for the simulation 'engine' ESPSIM; and the incorporation of sophisticated output features (variable time-step plotting, colour etc) in the module ESPOUT.

The contract has also allowed a number of other aspects to be investigated and specified. For example, currently awaiting integration are a range of 'cost-in-use' routines and a number of systems representative of those commonly found in buildings (VAV airconditioning, wet central heating, electrical storage, solar collection systems, etc).

Extensive documentation is now available for the ESP system (1) which, to date, has (or is being) implemented on 17 mainframes throughout Europe and has been adopted as the reference model by the EC passive solar modelling group.

1.2 Dynamic Energy Implications of Housing Location and Design

Emerging dynamic models (such as ESP) are substantially different, in a number of respects, from the previous generation they seek to replace: they are dynamic rather than static; highly predictive in use; and, at the same time, design-oriented, research-oriented and rigorous.

Their increasing flexibility in use has given designers a tool to explore a number of interesting design problems in an attempt to increase understanding of the complex energy subsystem. They also provide the means to develop complementary techniques which, while preserving the dynamic properties of building hehaviour, are simple to apply and so overcome many of the problems associated with large system implementation, management and operation.

This section describes attempts to utilise ESP as the technical basis of a project to develop an Index of Climatic Severity. This Climatic Severity Index (CSI) is building type specific and indicates, by a single number on a dimensionless scale, the stress placed on a buildings's energy system by an given environment.

Such an index would have many uses - ranging from energy related decisions on building location to the determination of design climatic collections.

The CSI project was a pilot study, now completed, funded by the Scottish Development Department and with substantial input from the EC contract in the form of the cost associated with the long term simulation of a number of distinct house types.

The technique has a number of potential applications in the domestic sector:

Design Location Decisions

Many national, regional and urban agencies have a range of alternative sites which can be developed for housing. Of course many factors will affect their decision but one should be the energy consumption consequences. For this purpose the sites must be compared on a standard energy-related index.

Building Control

Energy legislation is concerned primarily with achieving a defined level of thermal comfort for a reasonable expenditure of fuel. Since labour mobility, wage rates, tax legislation, social security benefits and so on are normally factors operating as uniform systems within national boundaries, it is reasonable (is it not?) to aim for fuel expenditures which are similarly uniform, irrespective of the location of the house. Therefore, thermal standards have to be modulated, to become proportionately more stringent as the climate becomes more severe. To achieve this modulation an index of climatic severity is needed which will accommodate careful design based on non-mandatory codes rather than complex legislation.

Optimum Distribution of Scarce Resources

In many EC countries a substantial portion of the resources being invested in buildings, especially housing, is in the form of improvement to existing stock. An important element of this investment is the upgrading of inadequate insulation, heating systems, ventilation etc. To determine this, existing houses will have to be assessed in terms of the climate in which they are located and their existing thermal properties. One approach is to work by a CSI which takes into account the range of house types actually existing in a particular region by matching them in various proportions against a range of standard or model houses.

Justice in Distribution of Social Security Assistance

Generally social security payments - supplementary benefit, unemployment benefit, fuel subsidies etc - attempt to take into account both social factors (family size, the presence of the sick or elderly) and some crude physical factors (size of house, type of heating etc). But they generally fail to make allowance for the climate or the real thermal condition of the dwelling - even to such an elementary extent that a house in mid-block is rated as being in similar need to one at the extremity of a block; even though this single variation could cause fuel consumption differences of the order of 20 - 40%. It would be possible for relatively unskilled personnel to assess, from a range of standard house types, which one approximates most closely to an applicant's, and then to select from a standard CSI map a value upon which benefit payments would be based.

Determination of Starting Design Hypothesis

Because the CSI technique allows the rapid appraisal of the interplay between insulation level, capacity level, wind permeability and solar utilisation or exlcusion - it can be used to establish the approximate level of each parameter best suited to the house type, design location, financial limits etc. This then forms the basis of a proposal which can be carried forward for refinement by, for example, a simulation model such as ESP.

Determination of Design Climatic Data

The index - by allowing the simple assessment of the stress placed on a building by any declared climate collection - readily lends itself to the classification of climatological data and hence to the generation of design reference collections which are building type specific and thus overcome the problems inherent in a 'standard year' for all building types.

Of course CSI's already exist - for example based on the degree day technique - but these are inadequate because they exclude the effects of important climatic parameters and only include building properties in the most rudimentary way. As a consequence different heating seasons with near equal degree day totals will result in different energy consumption figures depending on the building's ability to effectively harness solar radiation or the effects of wind or insulation amount and position.

1.2.1 CSI Derivation

Figure 3 details the major tasks comprising the study. Considering each task in turn:

- the ESP system was used throughout the study to predict the dynamic energy behaviour of selected house types when each were subjected to a range of climates.

- the 10 year climate collections issued by the UK Meteorological Office relating to KEW (Thames Valley), ABERPROTH (Wales) and ESKDALEMUIR (Scottish Borders) were processed by ESPCLM to extract monthly mean values of each of the main climatic parameters for use later in the statistical part of the project.

- with regard to the range of standard house types included in the pilot study, the assumption was made that house heating energy characteristics can be adequately summarised by the three properties 'mass and insulation', 'solar admittance (or exclusion)' and 'wind permeability'. If 3 levels of each property is sufficiently representative, this gives a combinatorial possiblity of 27 house types and for the purpose of the pilot study 3 house types were selected : a 6 apartment detached chalet bungalow (H1); a Scottish Special Housing Association 3 apartment flat (H2); and a 4 apartment timber framed bungalow (H3). Including the effects of rainfall as three levels of fabric humidity would have increased the range of standard house types to 81. This was not done however since preliminary calculations suggested that with contemporary insulation levels, the energy effect (of the related thermal property fluctuation) is small.

- at this stage a number of long term energy simulations were conducted effectively processing each house type against each 10 year climate collection. In these simulations living and bedroom areas were treated as different zones with different comfort levels imposed. Thus, assuming an 8 month heating season, this amounted to 3 climate collections by 10 years each by 8 months per year by 3 house types and by 2 zones per house <u>or</u> 1440 zone months of ESP simulation at a 1 hour time-step.

- then for each zone month a complete energy breakdown was extracted (ESPOUT) and passed, along with the synoptic climate data, to the statistic section of the project where the data was fed to a linear regression model allowing stepwise regressions in the search for the best fit relationship. The final regression equations for the three house types where:

$$\text{H1} \quad E = 1682.5 - 83.3\bar{t} - 1.54\overline{Rd} - 2.97\overline{Rf} + 75.4\bar{V}$$
$$\text{H2} \quad E = 729.3 - 48.5\bar{t} - 1.12\overline{Rd} - 1.83\overline{Rf} + 54.6\bar{V}$$
$$\text{H3} \quad E = 1213.9 - 66.1\bar{t} - 0.78\overline{Rd} - 2.30\overline{Rf} + 68.8\bar{V}$$

where,

E = required energy (KWh)
$\bar{t}$ = average monthly temperature (deg.C)
$\overline{Rd}$ = average monthly direct normal radiation (w/m^2)
$\overline{Rf}$ = average monthly diffuse horizontal radiation (w/m^2)

V = average monthly wind speed (m/s)

- at this stage, the effect on the foregoing equations, of a variety of micro-climatic modifiers, can be determined by directly modifying the climatic data (to simulate shelter belts for example) or by superimposing such factors as shading on the ESP simulations.

- the CSI technique can now be generated. This is a 4 stage process:

  1. by entering the maximum and minimum climatic conditions (table 1) - to which the house types were subjected - into the appropriate regression equation, table 2 can be constructed which shows the energy difference between the maximum and minimum loading for each of the climatic variables and for each of the house types.

  2. by then computing the ratio of these energy differences to the overall totals, the contribution of each particular climatic variable and be ascertained (table 3).

  3. if the best combination of climatic variables (lowest energy case) is made, arbitrarily, to correspond to a CSI of 2 and the worst combination a CSI of say 10, the contribution of each individual climatic parameter to the CSI can be determined by multiplying these upper and lower CSI values by the previously determined ratios. Figure 4 gives the CSI contributions for the 6 apartment detached chalet bungalow case. It is important to realise that although individual contributions will vary for different house types, the previously outlined technique ensures that the total CSI must necessarily be the same. Therefore it is possible to develop a general solution in which the CSI can be determined purely as a function of climate. This means that CSI maps can be constructed for all locations with available climatic data.

  4. a family of curves (figure 5) can now be produced relating climatic severity to house energy consumption for the 27 standard house types.

1.2.2 Use of the CSI

For any country for which a CSI map has been produced, the CSI for any locality can be read off and used, in conjunction with figure 5, to determine the relative energy consequences of adopting alternative housing mixes or locations.

Alternatively the curves of figure 5 can be redrawn for the case of constant energy as shown in figure 6. This allows, for any CSI established from a CSI map, the determination of a CSI adjustment to preserve energy equity between house types. That is, whilst preserving energy equity, one house type can be relocated (relative to another house type) anywhere in a region defined by relating the adjusted value to the CSI map.

Future work will see the computer processing of the remaining house types to complete the set of 27, validation of the technique and implementation on a micro capable of supporting simple graphical interaction.

1.3 Small Machine Implementation

It is increaingly acknowledged that the building design profession will, over the next decade rely more and more on design models which predict, easily and accurately, the cost and performance attributes of buildings. As planning and design problems become more complex - in response to the

Variable	Max.	Min.
$\bar{t}$	16.3	-1.0
$\overline{Rd}$	119.7	13.1
$\overline{Rf}$	85.0	10.4
$\bar{V}$	5.9	2.5

Table 1

	Energy Difference		
Variable	H1	H2	H3
$\bar{t}$	1441.1	839.1	1143.5
$\overline{Rd}$	162.2	119.4	83.2
$\overline{Rf}$	221.6	136.6	171.6
$\bar{V}$	256.4	185.6	233.9
	2083.3	1280.7	1632.2

Table 2

	Energy Contribution (pro. of 1)		
Variable	H1	H2	H3
$\bar{t}$	.692	.655	.700
$\overline{Rd}$	.079	.093	.051
$\overline{Rf}$	.106	.107	.105
$\bar{V}$	.123	.145	.143

Table 3

need to utilise natural resources more efficiently - the importance of computer based design models will undoubtedly increase.

Effective proliferation of advanced energy simulation software is not an easy task. Currently three scenarios are being pursued at ABACUS: the direct provision of system source coding for implementation on a mainframe where available; the establishment of a network of Energy Advisory Centres offering energy advice to the design profession based on simulation modelling; and the installation of the ESP system on a powerful but relatively inexpensive mini computer. The latter task has been funded entirely by the EC contract and what follows in this section is an account of the current status and future plans for the packaging of ESP on a PDP11/24.

Dynamic energy modelling of buildings is proving to be an important tool for designers and managers of buildings. Unfortunately, by the very nature of these systems, they are large, complex programs processing very large quantities of information. This means that nearly all such systems have been developed on large mainframe computers. This has the side effect that the vast majority of potential users of such systems do not have access to them, mainly due to the cost of time on such computers. For example, the ESP system consists of 70,000 lines of Fortran code, uses a total of 200K words of memory and generates up to 8M words of output. To analyse an average sized building costs, on a DEC10, 50p/zone day on top of the time required to prepare the necessary data.

The current project is an attempt to remove this problem by implementing ESP on a small minicomputer, the PDP11/24. This was chosen because it is typical of the sort of computing power that will be available for less than £10,000 when the next generation of super-micro arrives, hopefully within 1-2 years. The PDP11/24 has a moderate amount of memory, 256K, a large disk, 150Mb, and a magnetic tape unit. Interaction with the computer, and with ESP, is via a Tek 4014 and a small plotter and printer. It was also decided to use the UNIX* operating system as this seems likely to be the most popular operating system for small computers over the next few years. Therefore, besides providing a good environment for the ESP implementation and development, UNIX will also ensure that the system will be easily transported to other machines.

In transferring ESP from its original DEC10 mainframe to the PDP11, several problems have to be solved:

1. Firstly, and most easily solved, the difference in the dialects of Fortran used in the different machines had to be reconciled. It was decided that, at the same time, all non-standard Fortran would be removed and the source code would be brought into line with the ANSII Fortran 77 standard and all DEC10 dependencies removed. This should further improve portability of the system.

2. Secondly, the low level utilities used by ESP had to be rewritten for the PDP11. Some of these were trivial, e.g. the different format of reporting the date, but some, such a file handling, had to be completely rewritten to deal with the different file handling philosophies of the DEC10 and UNIX.

3. Thirdly, and the biggest problem, is to 'shrink' ESP onto the PDP11/24. Since the PDP 11/24 is a 16 bit machine, program instructions can only access 64K bytes of data. To overcome this limitation some method of

---

*UNIX is a trademark of Bell Laboratories

switching between different blocks of 64K bytes is needed. The system has hardware memory management to allow operations of this kind, theoretically allowing a program to access all the available memory (256K in this case). But, although the system allows several different programs to be run in different areas of memory at the same time, it does not permit user programs to access the memory management hardware and so restricts program and data size to 64K bytes.

Overcoming this problem requires circumventing the UNIX system. The problem has been split into two parts:

Program
The program can be split into small (8K) segments, each containing one or more subroutines, and these segments can be overlaid, i.e. when a subroutine is called, the overlay handler reads the relevant segment into memory replacing the calling segment and starts execution of the subroutine; when the subroutine returns, the overlay handler replaces the calling segment and resumes execution at the appropriate position.

Data
By only using 16K bytes, 8K for the overlay handler and library routines and 8K for the program, 48K bytes are left for the data. Overlaying data is not as straightforward as for program instructions because data is used more or less randomly by the program and is not accessed in a linear manner. The solution here is to segment the data but make the individual routines read in the data they require, in one block if it is less than 48K, otherwise in several pieces. This will require changes to the source code of ESP and, unfortunately, will severely restrict the ease with which the system can be modified and expanded.

Obviously this program and data subdivision scheme will increase the number of disk accesses and hence increase the execution time of the program. However since the program would already be quite slow on such a small machine, it is felt that the reduction in speed will be acceptable, especially since the extra computer time does not cost anything. It is possible that a very big, detailed simulation, e.g. 10 zone/year, may have to be run overnight.

1.3.1 Current Status

Before starting to implement ESP, we had to set up the PDP11/24 and set UNIX running on it. An appreciable amount of time was spent gaining experience of both the PDP11/24 and UNIX, since both of them are new to this group. To date the ESP source has been transferred to the PDP11 and converted to Fortran 77 format. The necessary sections of the utility library have been rewritten and are operational. Some of the smaller ESP sub-systems, those not requiring overlaying, are working. We are currently trying to obtain the overlay handler described above from the European UNIX Users Group, and are examining the ESP source to decide how it should be chopped up into segments. At the same time, we are investigating various strategies for segmenting the data into manageable pieces.

We are hopeful however that a virtual memory version of UNIX, or at least a method of keeping data segments in memory rather than on disk, will become available in the future. This would simplify later extension/modification of ESP.

2. <u>Conclusions</u>

The ESP System is now at an advanced stage of development with the current EC Contract contributing on a number of fronts: a number of important extensions have been researched, implemented and tested; the system has been used to develop a simple to apply, but inherently dynamic, technique relating to decisions on housing location and design; and the purchase of a PDP 11/24 for implementation of ESP operating under UNIX.

These measures have undoubtedly improved the model's ease of use as well as predictive validity and will, due to the small machine implementation, contribute significantly to the future dissemination of large energy systems throughout the design community: certainly the need for such techniques has never been so great.

<u>References</u>

1. The ESP System: Documentation Set, ABACUS Publication, University of Strathclyde, February 1982.

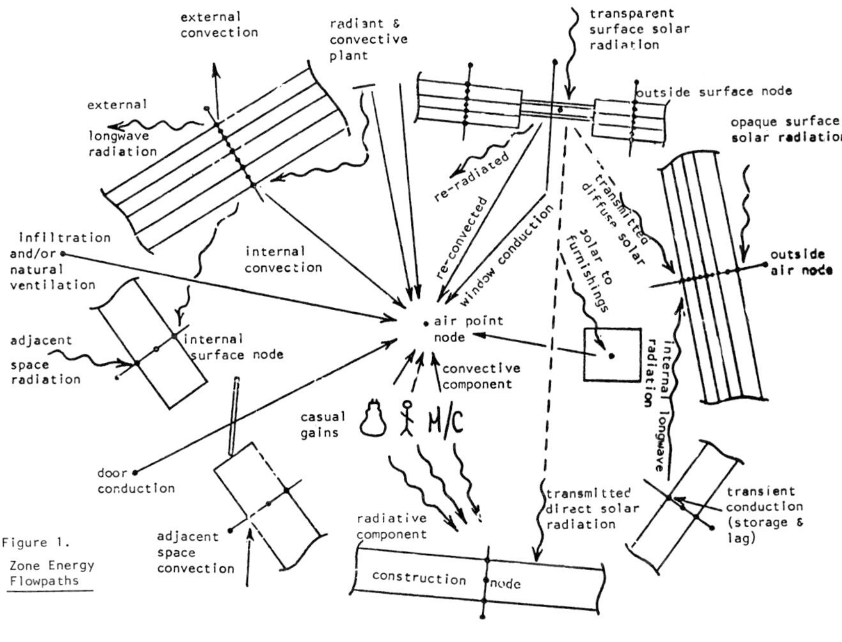

Figure 1.
<u>Zone Energy Flowpaths</u>

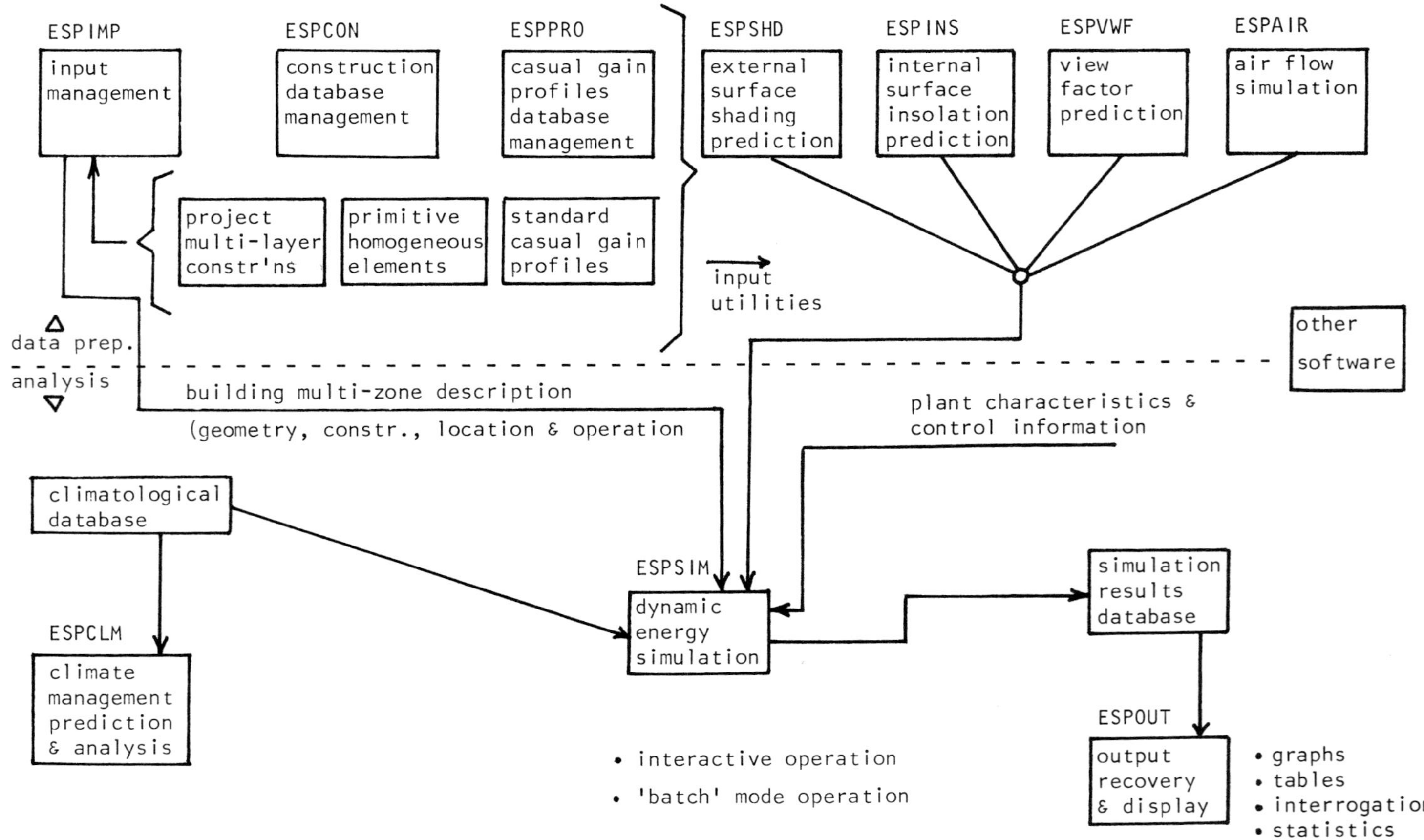

Figure 2. <u>The ESP System</u>

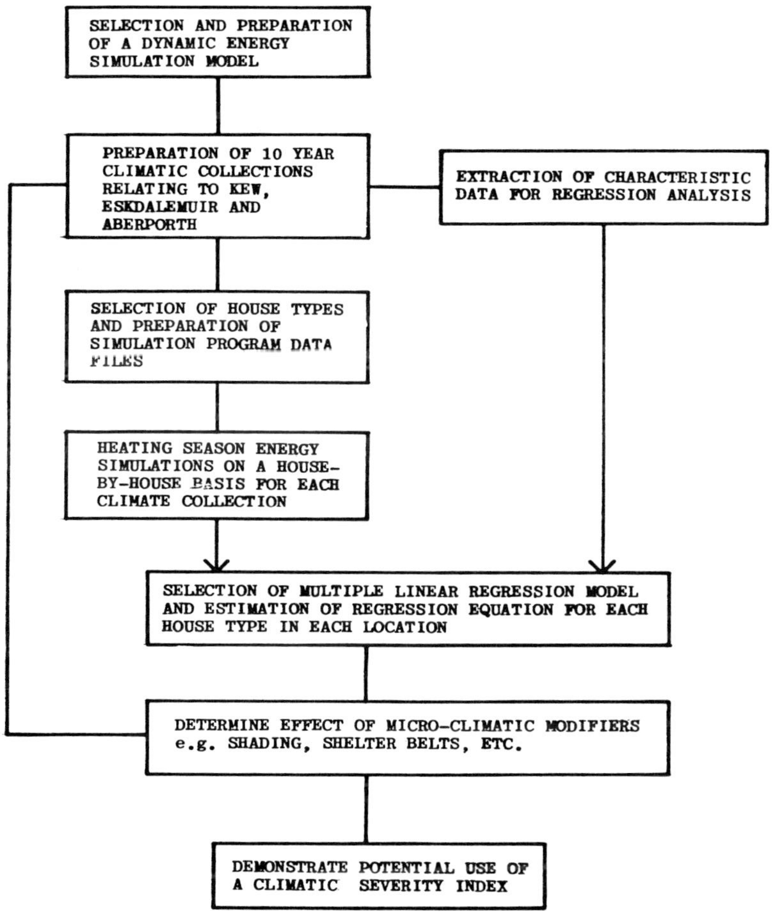

Figure 3. Climatic Severity Index; Project Tasks

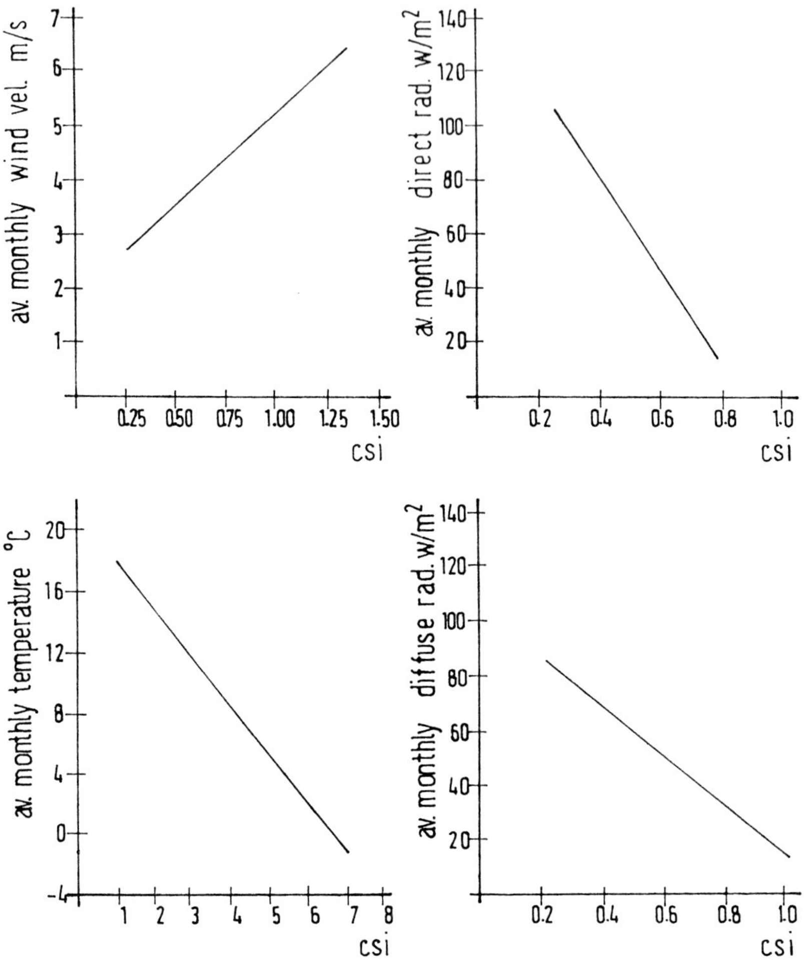

Figure 4. <u>Contribution of the Climate Parameters to the CSI for H1</u>

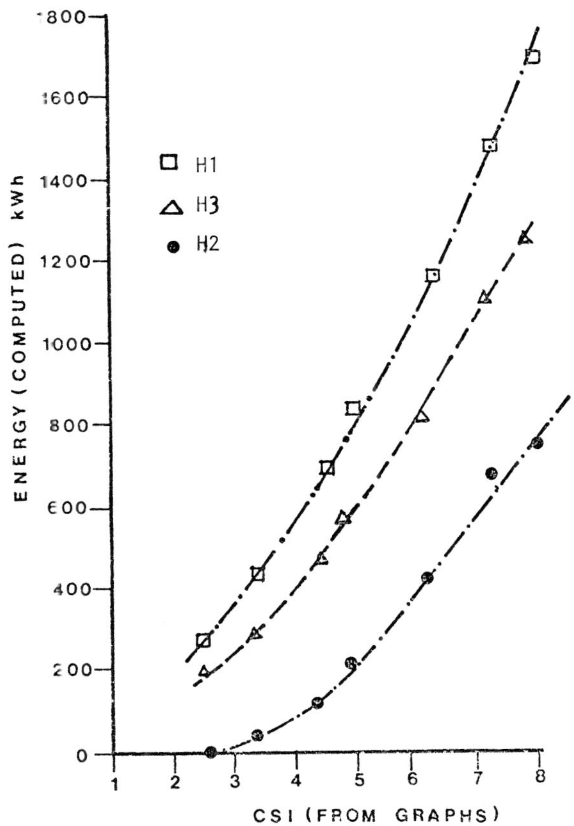

Figure 5 The SCI/Energy Relationship for Different House Types

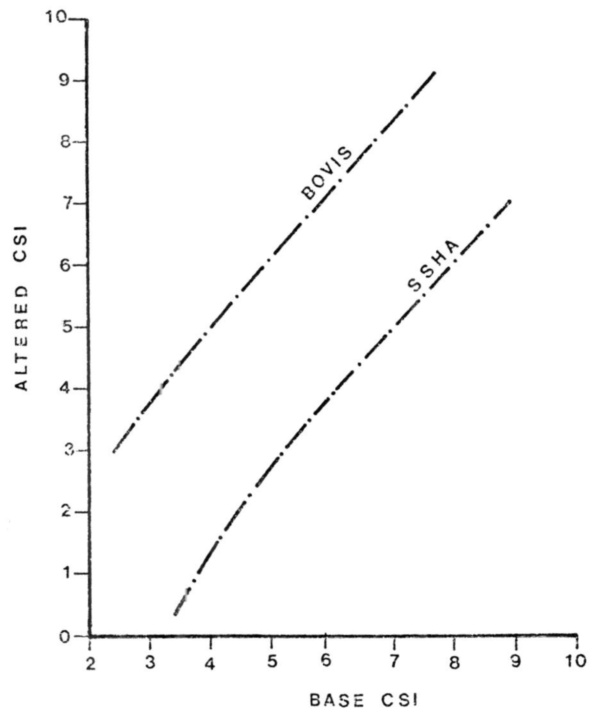

Figure 6 Adjustment to the CSI to Preserve Energy Equity

QUESTIONS AND DISCUSSION

Lebrun (University of Liège) : I have a question on the regression analysis. Which assumption did you introduce concerning the heating system efficiency ? What is variable or constant ?

Answer : For the moment we have assumed a heating system efficiency of 1. This is of course totally unrealistic but does not alter the CSI methodology. We think that the heating efficiency can be incorporated into the general regression equations by applying it as an external coefficient k such that :

$$CSI = k\ (a\bar{t} + b\bar{R}d + c\bar{h}_1 + d\bar{V} + e)$$

Lebrun (University of Liège) : The climate variables are inter correlated. Does this not affect the significance of your results ? In other words : does it make sense to introduce the effect of the temperature t°, wind and solar radiation etc. separetely ?

Answer : There is evidence that the different pairs of the four climatic variables are correlated. Clearly this is the case for direct and diffuse radiation; it is also the case for both radiation terms and air temperature. The correlation between air velocity and the other three variables is very low. None of the variables are perfectly correlated (that is the coefficient R is never equal to 1). The question of the correlation is important in so far as it determines the likelihood of extremes of various climatic variables occurring coincidentally. For instance, the most extreme heating stress will be placed on a house on days of low temperature, low direct and diffuse radiation and high wind velocity. If we take the distribution curves for each of these four climatic variables and look at the values that occur in the, say, 5 percent of the distribution curve, then the combined probability of these four values occurring simultaneously would equal the product of the four individual probabilities but only of they were totally independent. If the four variables were totally interdependent then the combined probability would be the same as a probability of any individual factor - that is, 0.05. For the independent case the product of the four probabilities obviously lies somewhere between 0.05 and 0.00000625 with the actual position depending on the degree of correlation between the variables. This correlation will not be constant but will vary with latitude and climatic type even along the same latitude. In the CSI study the generated CSI equations had associated R values in excess of 0.99 !

The study set out explicitly to generate a technique which allowed an assessment of the separate effects of the four main climatic parameters. This is important to allow the range of applications envisaged.

Zegers (DG XII) : What is the cost of a computer design study ?

Answer : A simple question on the danger for overheating in a building, may cost £ 50. A complete energy design of a building will cost somewhere between £ 5000 and £ 50.000. Such a £ 50.000 study is very detailed and may last as long as 18 months.

Los (INSO) : Is it possible to apply your model to large buildings in different climates ?

Answer : Yes and you will obtain the results much faster than with measurements.

SESSION VIII - DOMESTIC APPLIANCES

Chairman: C. BOFFA

WASHER WITH ELECTROLYTIC WATER DISSOCIATION

Author : Riccardo Bertolino
Contract number : EEA.3.024.I
Duration : 12 months 1st July 1981 - 30th June 1982
Head of project : Mario Morello
Contractor : Riber S.p.A.
Address : Riber S.p.A.
via Manzoni, 4
I - 10092 BEINASCO (Torino)

Abstract (Summary)

The goal of the present paper is to illustrate the research that has been carried out up to now by Riber reaserchers to evaluate the actual chances of constructing a washing machine adopting the process of electrolysis too. As it will be later shown in detail, this process could allow saving energy and improving washing results, especially for medium dirty clothes and white materials. We must point out since now that the results we achieved are to be considered as e preliminary step. The experimental tests have been carried out, at present, in a laboratory electrochemical cell. These tests allowed the choice of the electrode material, the process efficiency determination and the industrial cell design. Therefore we have not yet tried to perform any industrial washing test. The difference between adding to water chemical oxidizers (like commercial ipochlorite) and producing them "in situ" will be illustrated. We can conclude this abstract in the following way: what we have done is already enough to allow us to say that electrolysis has good chances to succeed in reducing energy washing, especially that arising from warming water.

1. Introduction

The aim of our work is to construct a washing machine with minimal energy consumption. Thus it is necessary to perform all washing operations at the lowest possible temperature for a two-fold reason:
on one hand most of the usual washings do not require high temperature to be cleaned; on the other hand energy is mostly wasted in warming water and clothes from 20° to 60° - 90° centigrades.

We can add to these two remarks another one; most clothes need only a bleaching action that is usually achieved by adding some oxidizers to water.

Another possible method is that of producing the same compounds "in situ"; this can be easily obtained by means of an electrochemical cell. As it is well known the bleaching action is due to the chemical reaction of the oxygen (derived from the oxidizers decomposition) with the insoluble organic compounds, which constitute the dirt. Together they make soluble organic compounds which leave from the clothes.

Therefore the action of an oxidative agent consists of a reaction between the agent itself and the oxidable compounds. Its efficiency will depend on both its theoretical oxidative power (that is on the oxygen pressure which balances the decomposition) and its reaction rate.

Thus we are not sure "a priori" that compounds containing more oxygen are better oxidative agents, since they could be very stable (e.g. molecular oxygen is of no use here since it does not act at temperatures lower than 100° and high pressure).

Some chemical oxidative compounds easily release atomic oxygen. Among them hypochlorites and perborates are widely used as bleaching agents, while not very much is known about persulphates, chlorates and perchlorates.

Since we are mainly interested in cleaning fabric clothes we restrict ourselves to considering three different classes of electrolytical oxydations of salts, yelding to:

a) Chlor-Oxygen compounds (hypochlorites, chlorates, perchlorates);

b) perborates;

c) persulphates.

Colouring, expensive and too corrosive compounds have been ruled out for the previous reasons.

We began our work by considering the Sodium hypochlorite first. According to the list given above, we are now going to briefly discuss and classify the different compounds.

Chlorides can be oxidated at the anode of a cell giving rise to chlor oxygenated compounds ClO^- (ipochlorites) ClO_3^- (chlorates) ClO_4^-(perchlorates) At a temperature higher than 40°C the first easily turns into the second.

If we start with a solution of common table salt we cannot turn Cl^- into ClO^- completely because the latter turns into ClO_3^- which oxidates to ClO_4^-.

As far as the other peroxides are concerned the task is simpler due to the absence of chain reactions. Electrolytical conditions must however allow oxidation without production of molecular oxygen. This cuts down the available anodes, which can be built up by Pt, C, PbO_2, Fe_3O_4.

If we perform the electrolysis of a Sodium chloride solution we obtain Hydrogen at the cathode and Chlorine at the anode. The complete reaction is:

$$NaCl + H_2O = NaOCl + H_2$$

In order to obtain complete reaction of Cl and NaOH, the electrodes should be as close as possible to one another. With this arrangement, however, the resulting hypochlorite could reduce itself at the cathode. Since we cannot use oils and chromates (as is usually done) to prevent these reactions between cathode and products, the only way out we are left with is to provide an electrode that easily produces molecular Hydrogen rather than atomic one.

Fe or Ni cathodes are suitable to this purpose.

Besides the temperature these processes depend on the pH too.

If temperature raises ipochlorites decompose into chlorates. If pH exceeds 10, the chlorates are favourite and vice-versa. We obtain perchlorates from oxidation of chlorates at the anode. Concentration and temperature must be higher than 20 g/l and 25° - 40° centigr. respectively at current density less or equal to 2500 A/m^2.

One must only care that Sodium salts have to be used to prevent salts from depositing on the anode. One can also think of using electrolysis of carbonates (temper. about 20°, steel cathodes, current density 1000 - 2000 A/m^2) the solution being composed by up to:

1) 40 g/l Sodium metadrate;
2) 2 g/l caustic soda;
3) 140 g/l Sodium carbonate.

The last oxidates to peroxycarbonate which decomposes into NaOH and H_2O_2. This latter turns the metadrate into perborate. The last compound we think it is worth thinking over as possible electrolyte is ammonium sulphate.

One has to adopt steel cathodes and PbO_2 anodes and almost sature solutions

The SO_4^{2-} ion can be electrolytically oxydated into $S_4O_8^{2-}$ ion which is an intermediate step of H_2O_2 preparation.

If the temperature exceeds 30°C and we keep a current density of about 1000 A/m^2 all the persulphates decompose into corresponding sulphates and oxygenate water. The efficiency of perborate and percarbonate production has still to be tested.

Now we have exausted the classification of the possible useful electrolytes we draw our attention over a second main question, namely that of building up suitable electrodes.

In this field, as in the previous one, a general survey allowed us to restrict our choice between two cases.

In fact even if platinum is the most suitable metal to be used in electrolytical cells, it is too expensive. Different compounds have been selected to overcome this drawback, namely oxides such as PbO_2, magnetite TiO_2, RuO_2 or, on the other hand, insoluble elements such as graphite.

To our purpose however, we have to fulfill some essential requirements; an anode should have the following patterns:

a) it should not pollute the solution;

b) it should not corrode;

c) it should have good conducibility.

Thus we can rule out magnetite anodes, since they do not conduct enough. Moreover PbO_2 is easier to be fixed over compact graphite. Graphite itself has a great drawback; it is attacked by the oxidizers produced during the electrolysis and it crumbles easily. Besides PbO_2 anodes we are left with Ruteniun doped TiO_2 anodes, whose behaviour we still have to test. Our choice fell upon a special kind of Lead dioxide anode, the GSLD (Graphite Substrate Lead Dioxide).

It would be conceivable to use anodes made out of PbO_2 only, but this cannot be done since it is very difficult to obtain a good contact between PbO_2 and the current carrier. On the contrary this is easily done, using the graphite to which the PbO_2 has previously been fixed.

GSLD anodes are used to production chlorates and perchlorates in the industries.

2. Description of the testing equipment and tecnique

The hypochlorite production was carried out in a glass cell of 500cmc. We used electrolytic solutions constituted by NaCl, chemical reagents and commercial dish-washing machines salt, dissolved into distilled water. We adopted Platinum foils as electrodes; their surface was 5 square cm. and the distance between them was 10 cm.

The tests were carried out at a constant 25°C temperature and under galvanostatic conditions, namely 0.75 A current and 0.15 A/cm^2 current density.

During the test the solution has been continously stirred to keep it homogeneous. It should be noticed that these are the normal conditions in a washing machine. The electrolysis lasted from half an hour through 1, 1½ 2,3, up to five hours. We collected all our results in two plots, figures I and II, we will comment a bit later.

After testing the solutions with platinum electrodes, the tests have been repeated with GSLD anodes because they are easy to prepare in an electrochemical laboratory.

The anode preparation is carried out in an electrochemical cell using a lead sheet as cathode and graphite anode. The electrolytic solution contained 320-330 g/l Lead nitrate and 25-30g/l Copper nitrate.The pH of the solutions was initially 4. During the electrolysis it was maintained at 1.5 by the addition of Lead carbonate or Copper carbonate.

Deposition was carried out under galvanostatic condition at a current density of 3-5A/dm^2 and at a temperature of 58°C-65°C. The graphite anodes have been precleaned by electrolysing the anodes in a 10% Sodium hydroxide solution, using an Fe sheet as cathode.

The thickness of the lead dioxide deposits was about 3 mm. According to literature to avoid pinholes and pores in the deposits, quaternary ammonium surfactants can be used.

Although GSLD anodes have given very good results, no life-tests have been carried out up to now. As a further step towards the actual washing conditions, we looked for a cell to be used directly in the washing machine

We are now testing the cell whose longitudinal section is shown in fig. III.

It is placed along a pipe going from the bottom of the tub to the top of the front side of the drum of the machine (see fig. IV).

In the PVC pipe of the industrial cell a stainless steel cathode (-) and a GSLD anode (+) are placed, the GSLD being downstream. Two plugs connect them with the electric circuit of the machine. We want to point out that this is just the first arrangement, a temptative one.

To conclude this section we give a brief description of the procedure we used to analize the solutions. To evaluate the amount of produced oxidative agent we adopted a volumetrical method, which is based on the iodometric titrating of the oxidizer.

To a known volume of a given solution a 20% KI solution is added and then the produced I_2 is titrated by means of N/10 tiosulphate.

3.4. State of the work, analysis and comments of results

We can now supplement the general discussion we showed before by adding it the first specific results of our research.

Our point of view was slightly modified by the possibility of producing bleaching agents rather than oxygen directly inside the washing machine.

Even if the concentration of hypochlorite in the commercial solutions varies depending on how much time has elapsed since the solution were prepared, it is possible to see from drawings I and II, that under our experimental conditions, after a period ranging from ½ to 1½ hour it is possible to obtain a solution of Sodium hypochlorite as good as the commercial one.

The difference between the two lines in each drawing is in the concentration of the two solutions. The steeper line corresponds to higher concentration. We see that both common table salt and Hidrella (dish washing machine salt) give eventually the same amount of hypochlorite. Also at low concentrations we see that within a reasonable time a good amount of hypochlorite can be obtained.

We would like to point out, as a final remark, that in the practical arrangement the cell is placed on the washing circuit (see drawing IV). Therefore almost all of the produced hypochlorite is immediately used without requiring high concentrations in the solution.

5. Conclusion

We would like to resume the work we have done up to now in the following way.

First of all we carried out a general inspection of the possible choices of solutions and electrodes relying on literature and commercial experience. This preliminary study was very useful to rule out some kinds of electrodes and to draw our attention on the possible classes of hypochlorites Our main idea is to produce the bleaching agent inside the washing machine since in this case small concentrations are required and widely available compounds can be used.

We also point out that this "electrolytical programme" integrates the usual washing programmes rather than substituting them.

Our goal is to replace by the electrolytical production of bleaching agents the warming processes responsible for energy wasting. As we have discovered through our experiments, this is not pure fancy but can rather be obtained choosing suitable materials both as electrolytes and as electrodes constituents. The same hypochlorite concentration which is obtained by adding bleaching agents to water can be achieved, with low current densities and normal temperatures, by producing them directly inside the machine. This can be considered a well promising success.

The results obtained represent the experimental background to continue our work. The future programme includes some laboratory and industrial tests.

The laboratory tests will allow:

i) to establish the GSLD anode life;

ii) to compare the behaviour of GSLD anodes with anodes constituted by doped TiO_2;

iii) to determine the efficiency of perborate and percarbonate production.

The industrial tests will allow:

i) to determine the actual efficiency of the electrochemical process under washing conditions;

ii) to measure the efficiency of cloth washing at low temperature;

iii) to determine the real energy saving.

When all this programme is exhausted we plan to come to a more systematic use of our cells in order to discover if any other of the cleansing agents contained in common detersives can be obtained by electrolysis.

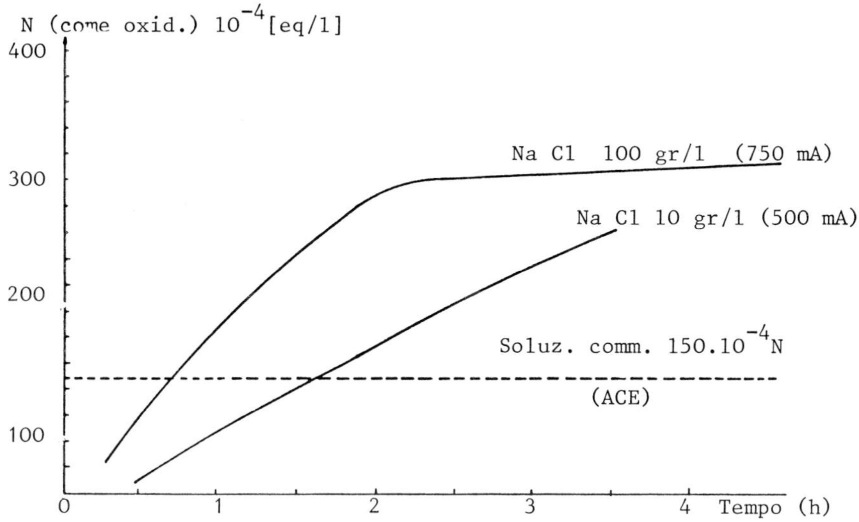

Fig. I - Diagramma

Tolieranze per dimensioni senza indicazione a dis. Foro H 13 albero h 11
Angoli di fusione $\pm$ 2° Angoli di lavorazione $\pm$ 0'30"
Raggi non quotati $\overline{R}$ = 0,5 Raggi non quotati per lavorazione lamerie
R = spessore lamiere

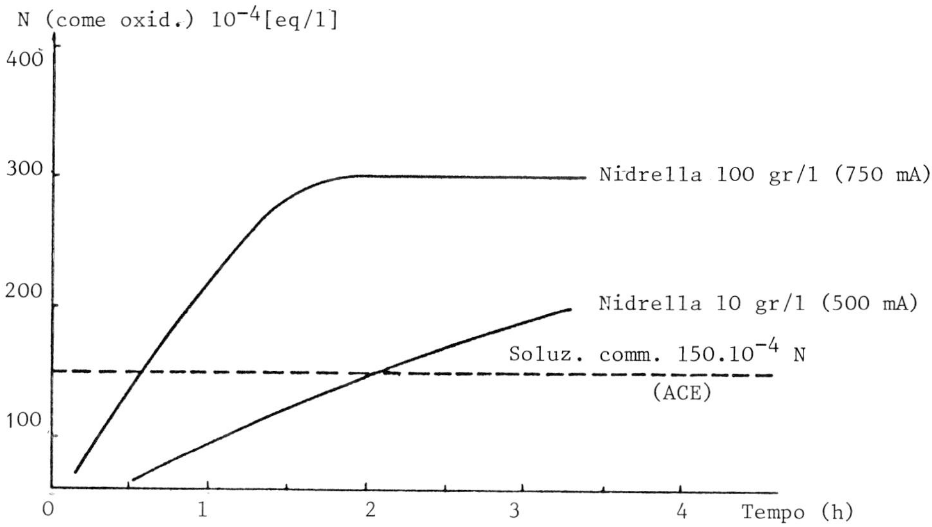

FIG. II - Diagramma

Tolieranze per dimensioni senza indicazione a dis. Foro H 13 albero h 11
Angoli di fusione $\pm$ 2° Angoli di lavorazione $\pm$ 0'30"
Raggi non quotati $\overline{R}$ = 0,5 Raggi non quotati per lavorazione lamiere
R = spessore lamiere

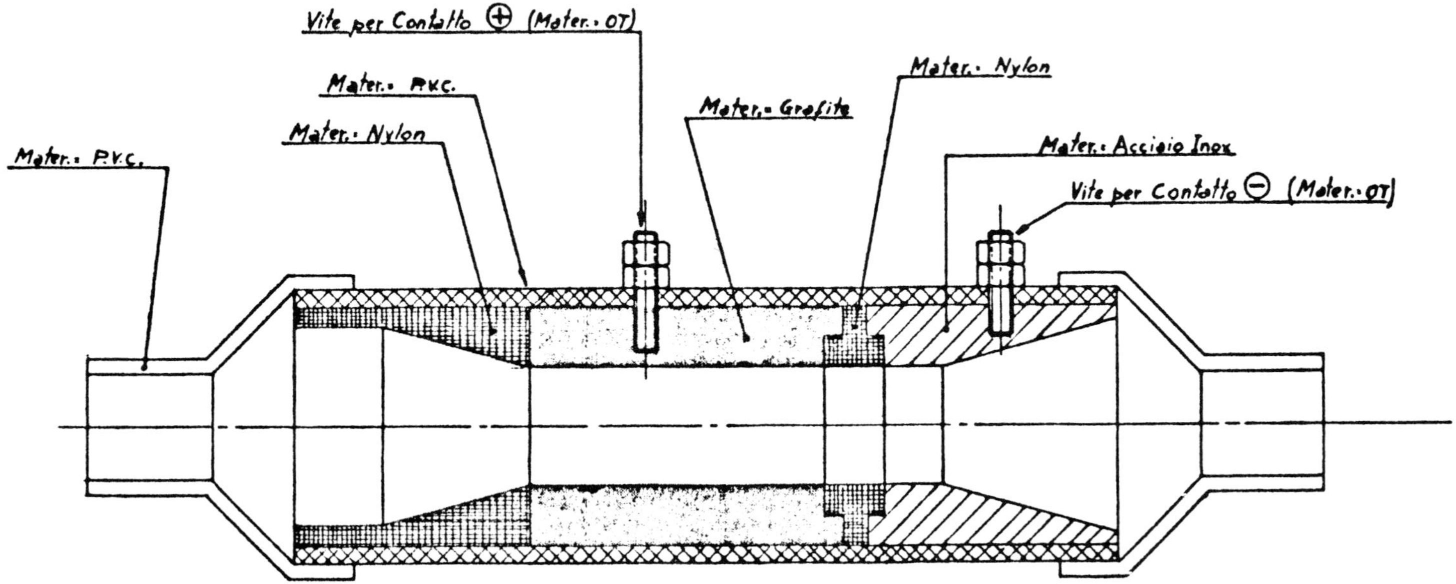

Fig. III - Cella per elettrolisi

Tolieranze per dimensioni senza indicazione a dis. Foro H 13 albero h 11
Angoli di fusione $\pm$ 2° Angoli di lavorazione $\pm$ 0'30"
Raggi non quotati $\overline{R}$ = 0,5 Raggi non quotati per lavorazione lamiere R = spessore lamiere

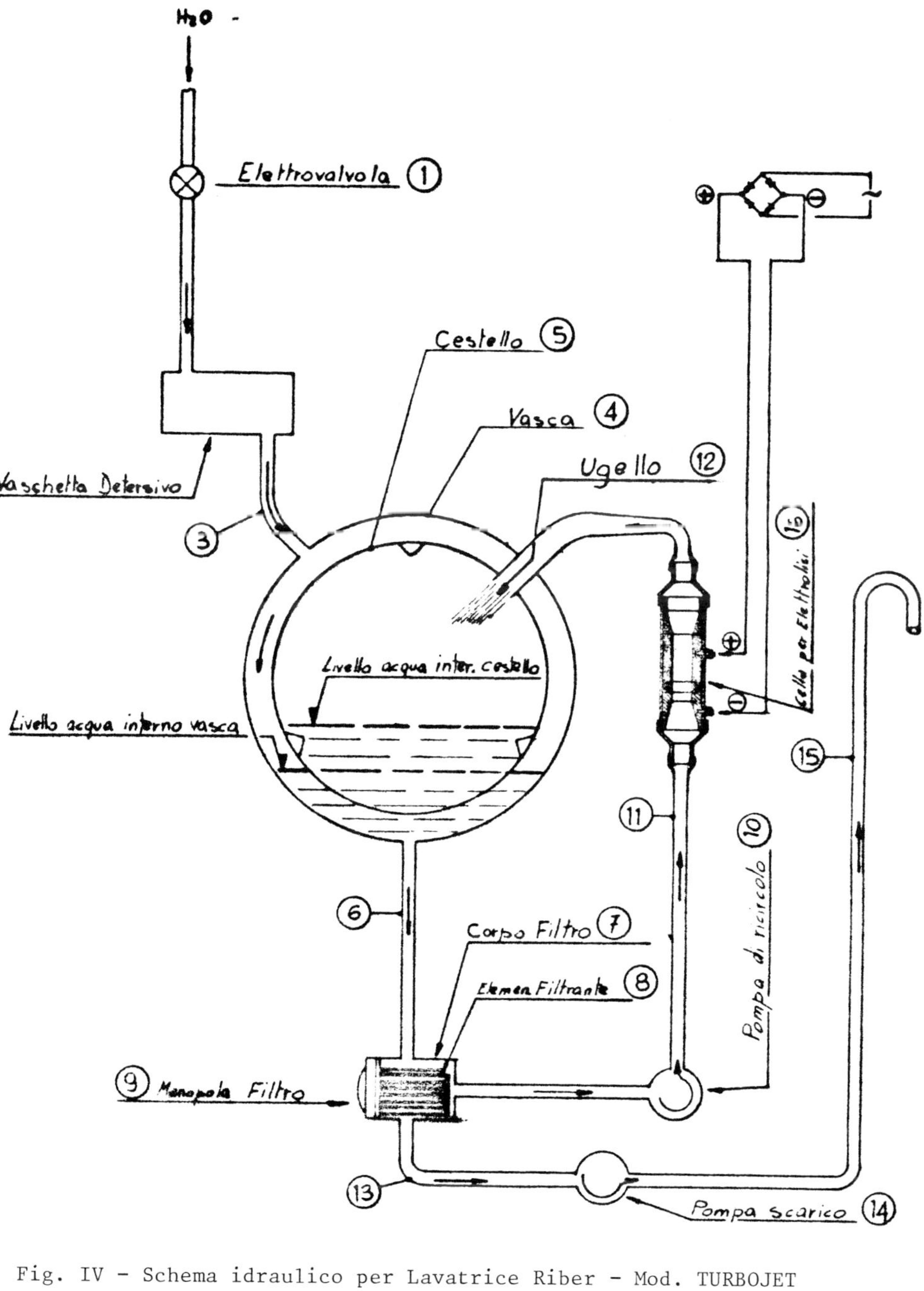

Fig. IV - Schema idraulico per Lavatrice Riber - Mod. TURBOJET

QUESTIONS AND DISCUSSION

Zegers (DG XII) : 1. On expiration of the contract will a washing machine, working with detergents at low temperature, be available ?
2. How do you expect to obtain such a low temperature ?

Answer : 1. A washing machine with an appropriate low temperature cycle will be available at the end of the contract.

2. The low temperature will be obtained by means of an electrolytic washing, cutting out the heating element.

Nørgard (DTH) : 1. Your method of generating bleaching agents inside the washing machine is energetically profitable if compared to the wash at high temperatures, but do you consider it more economic than producing such bleaching agents in a factory ?
2. Can you produce an energetic balance between your process and the normal one ?
3. Both for instantaneous and total input.
4. Thermodynamic yield of the process : to what extent is the absorbed electric energy is really used ?

Answer : 1. It depends on the bleaching agent. Using hypochlorite does not mean it is more economic because on the industrial point of view we have unemployed chlorine, but keeping into consideration transport, packing and waste costs we deem it is more convenient to produce it in the necessary quantity "in loco".

2. 3. 4. The data we have reached up-to-now are not the final ones, we will give an answer on the subject as soon as they are conclusive.

Boffa (Chairmain) : What is the :
1. Size of the cell.
2. Pollution of solutions.
3. Position of detergent manufacturers : will special electrolysis detergent be produced ?
4. Energy content of 1 kg. of detergent. How much energy is necessary to produce it ?
5. Electrolytic cell required input.
6. Total energy consumption for the complete cycle. Comparison with the traditional one.

Answer : 1. This is shown in Fig. V of the supplement to the report.

2. The drained solutions are not more polluted than for conventional working machines.

3. Up till now it is not necessary the use of special electrolysis detergents as the common commercial salt (NaCl) is sufficient.

4. We are not able to answer this question as research is still going on.

5. Input is $\leq$ 200 W/h.

6. A full and detailed answer will be given in the final report. Meanwhile we can say that the energy savings with the new washing machine will be 30-40 % as compared to the traditional ones.

Pilavachi (DG XII) : From Figg. I and II it is shown that the NaCl gives the commercial solution concentration after about 45 mins. Do you not consider such a long time a disadvantage ?

Answer : The hypochlorite production is constant and continuous during the electrolysis process. Consequently only a small quantity is needed. To obtain the same effect the concentration aiming electrolysis can be much smaller than the concentration resulting from commercial products.

WASHER WITH ELECTROLYTIC WATER DISSOCIATION

(Supplement to previous report)

Riccardo BERTOLINO (Riber S.p.A.)

In the period from 1st January to 1st March 1981 the laboratory and the industrial tests, as provided in the programme we pointed out in due time, were done.

1. Laboratory tests

The tests foreseen in the programme were comprehensive of:

a) measurements in order to establish the GSLD anodes life;

b) comparing measurements between GSLD and doped TiO_2 anodes;

c) measurements concerning the production efficiency of perborate and per carbonate.

1.1. Tests (a)

The first GSLD anodes have been fitted on the test washing machines and have been working for more than three months (320 working hours) without giving rise to any inconvenient.

Being the anodes till working and washing tests periodically performed the sole provision we made was the cell rinsing with tap water at the end of each electrolysis cycle.

1.2. Tests (b)

It has not been possible to compare the two types of anodes since the doped anode TiO_2 is not yet available even if the manufacturer has been required two months ago. As soon as the bureaucratic problems regarding the product privacy are overcome we will be given the first prototype to assemble on the electrolityc cell.

1.3. Tests (c)

We have studied the electrolytic production of perborates in our laboratory, the tests have just begun and, even if at present we have not final data, it seems that the perborate production occurs regularly but with major difficulties for hypochlorite.

The produced perborate quantity corresponds to 1/5 of the relative quantity of hypochlorite approximately. (see Fig. IV)

2. Industrial tests

The tests foreseen in the programme were comprehensive of:

a) measurement of the electrochemical process efficiency in washing conditions;

b) low temperature linen washing tests;

c) determination of the energetic saving obtained.

2.1. Tests (a)

The tests we carried out on the washing machine equipped with the electrolytic cell having the configuration as indicated in the first report showed that the limit electrolysis conditions, namely the conditions below which the bleaching efficiency rapidily decrease, are the following:

- NaCl concentration 20 g/l
- Intensity of current 2 A
- Cell voltage 20/25 V
- Water temperature 35°C
- Time for electrolysis 60/50 minutes
- Absorbed energy $\leq$ 75Wh

The voltage applied to the cell is due, in the major part, to the ohmic drop in the graphite and in the lead dioxide layer, as consequence it could be reduced by means of a different cell configuration.

In order to increase the productiveness of the electrolytic cell and to decrease the salt concentration, we have planned a new electrolizer, drawing No. P005490, in which the electric field between anodic and cathodic surface is more uniform. (Fig. I).

The active area is now increased and the electrolytic conditions can be stressed. The improved conditions are the following:

- NaCl concentration 4 g/l
- Intensity of current 8-9 A
- Cell voltage 15/20 V
- Water temperature 15/20°C
- Time for electrolysis 45/60 minutes
- Absorbed energy $<$200Wh

2.2. Tests (b)

The linen washing tests showed satisfactory results giving the same results already obtained in the laboratory.

The spots of wine, coffee etc. completly disappear thanks to the action of the few hypochlorite produced.

Therefore tests showed that if the oxidizers are produced in solution containing already soap, the relative bleaching action is greatly reduced.

Such a result, already experimented for hypochlorites several times, makes us to suppose that the oxidizer agent can act more rapidily with soap than with dirt of clothes.

It is necessary to carry out some tests in order to verify such a hypothesis, which is based on the spectrophotometer IR/UV solution analysis of soap, electrolysed soap and electrolysed soap with oxidizers.

2.3. Tests (c)

2.3.1. Preliminar practical tests

With reference to the previous paragraph, we decided to divide the washing cycle into two stages.

The first one, pre-wash, performed with electrolytic action by means of NaCl 4 g/l at room temperature.

This stage action is obtained by the liquid recycling passing through the electrolytical cell (turbo action) (see drawing IV of the previous report).This stage takes 45/60 minutes during which it is produced a little quantity of sodium hypochlorite (NaCl) having the capacity of removing oxidable spots, red wine, tea, coffee, ect. and generally the commonest dirt of linen.

The second stage consists of the washing cycle with a commercial detergent at a temperature from 35° to 45°C, i.e. at a low temperature. Then usual rinses, spinnings and final drainage will follow.

Notes:

The first tests performed as described before gave favourable results both for the dirt removal and for the energy consumption which were between 1.3/1.5 kW instead of the usual 2.5/2.8 kW necessary for a normal linen load.

2.3.2. Comparing tests

We will begin the energetic consumption comparison tests on two washing machines: a traditional and an electrolytic type, after the fulfilment of preliminary tests we explained in paragraph 2.3.1.

Thanks to the obtained experience we have considered to carry out such a comparison on the basis of the two following washing programmes. (Fig. II Fig. III).

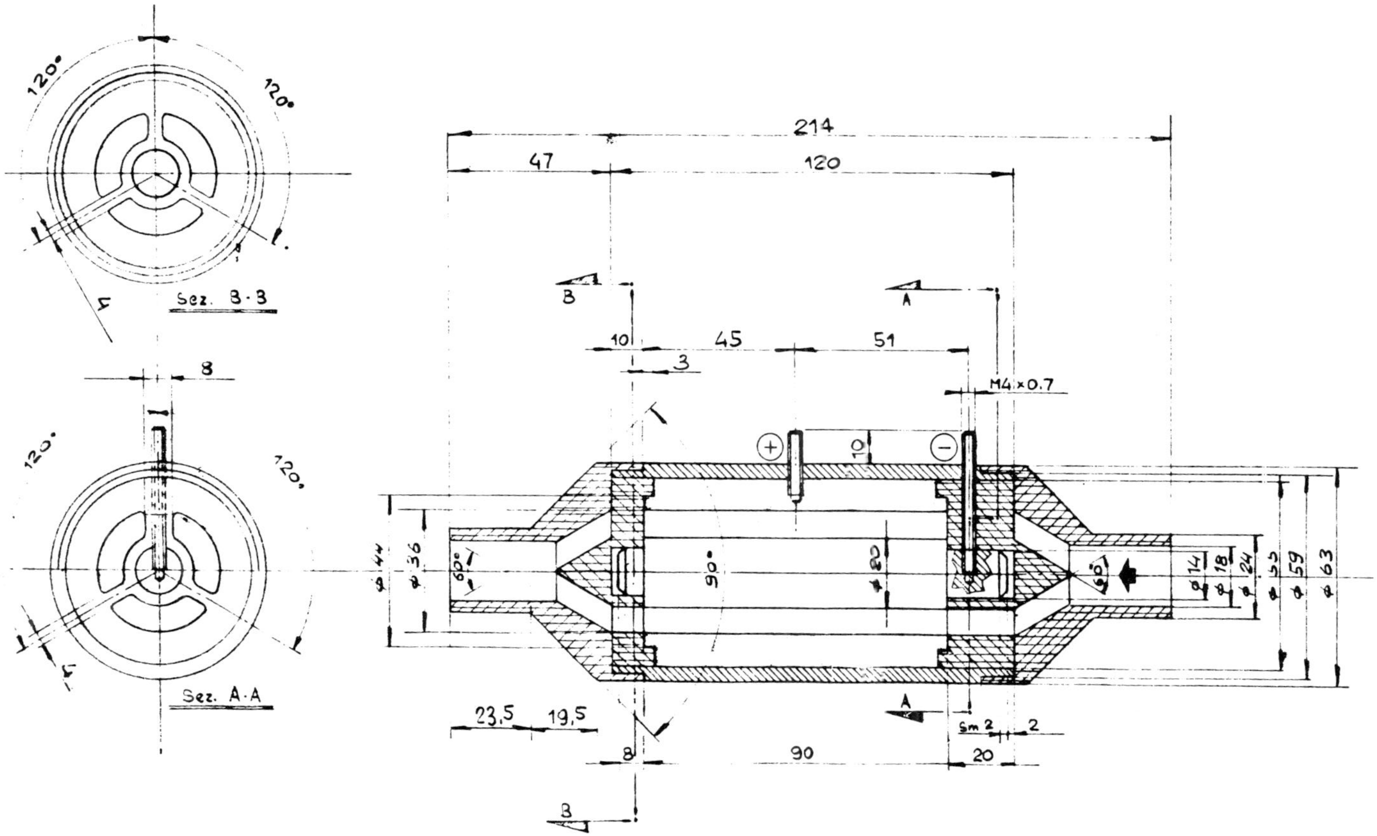

Fig. I - Cella per elettrolisi n° 2

Fig. II

Electrolytic Turbo washing machine			Traditional washing machine		
stage	parameters	cost	stage	parameters	cost
Electrolytic pre-wash	Liters of water :35 in total		Traditional pre-wash	Liters of water : 25	
	Water temperature : room temp.			Water temperature : 35°C	
	Commercial detergent:0 g.			Commercial detergent: 80 g.	
	NaCl :140 g.			NaCl : 0 g.	
	Time :60 min.			Timer time : 15-30 min.	
	Linen load :4 kgs.			Linen load : 4 kgs.	
	Standard soiled cloth			Standard soiled cloth	
	IEC No. 105 type			IEC No. 105 type	
	Cloth leucometer			Cloth leucometer	
	Reading			Reading	
	Consumption kW			Consumption kW	
TOTAL COST			TOTAL COST		

Fig. III

Electrolytic Turbo washing machine			Traditional washing machine		
stage	parameters	cost	stage	parameters	cost
Low temperature washing	Liters of water : 25 in total		Traditional washing	Liters of water : 25 in total	
	Water temperature			Water temperature	
	(1st test) 40°C			(1st test) 50°C	
	(2nd test) 50°C			(2nd test) 60°C	
	(3rd test) 60°C			(3nd test) 70°C	
	(4th test) 70°C			(4th test) 90°C	
	Commercial detergent: 120 g.			Commercial detergent: 120 g,	
	Real washing time : 50 min.			Timer real washing time:30-90min.	
	Linen load : 4 kgs.			Linen load : 4 kgs.	
	Standard soiled cloth IEC No. 105			Standard soiled cloth IEC No. 105	
	Cloth leucometer Reading			Cloth leucometer Reading	
	kW consumption			kW consumption	
	(1st test)			(1st test)	
	(2nd test)			(2nd test)	
	(3rd test)			(3rd test)	
	(4th test)			(4th test)	
TOTAL COST			TOTAL COST		

Fig IV

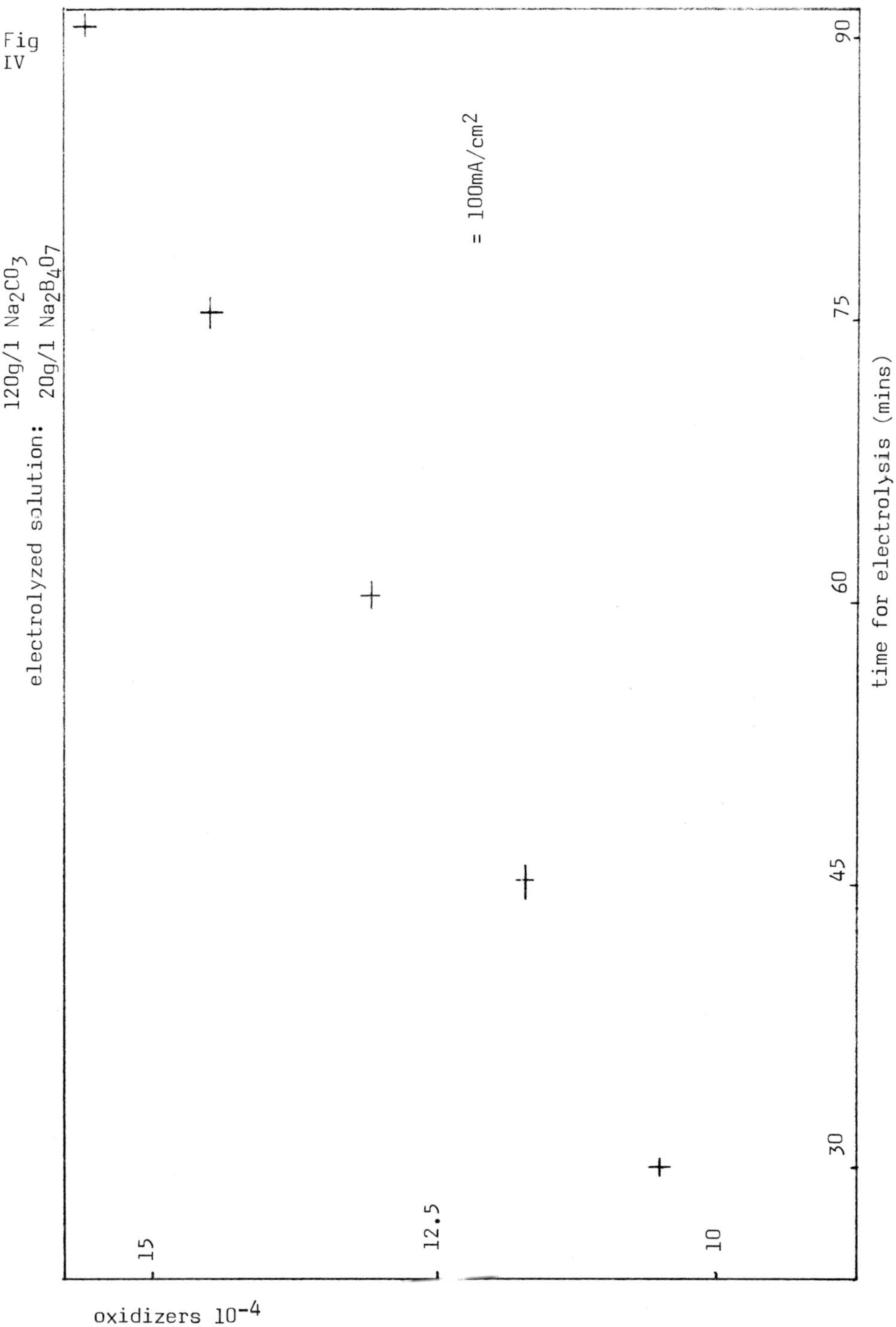
electrolyzed solution: 120g/1 Na_2CO_3
20g/1 $Na_2B_4O_7$
= 100mA/cm^2
time for electrolysis (mins)
30
45
60
75
90
oxidizers 10^{-4}
15
12.5
10

ENERGY CONSERVATION IN CATERING APPLIANCES

Author : J. ROBERTS, C.Eng., M.I.Mech.E., F.I.Gas E., M.Inst.E., M.B.I.M.

Contract number : EEA-3-026-UK

Duration : 36 months 1 July 1980 - 30 June 1983

Head of project : G. Percival, British Gas Corporation

Contractor : British Gas Corporation, Watson House

Address : British Gas Corporation
Research & Development Division
Watson House
Peterborough Road
GB - LONDON, SW6 3HN

Summary

Catering appliance design and cooking techniques and practices have little respect for operating efficiency - whilst Natural Gas resources dwindle and prices escalate. There is little information available on this market sector, but a modest saving of 10% would represent for the UK an annual saving of 56.3 x 10^6 therms (5.9 x 10^6 GJ) equivalent to approximately £12 x 10^6 at 1980 values.

The research programme reported here has primary objectives of investigating appliance design, control and usage, with a view to producing cost-effective reductions in energy consumption.

Activities are wide-ranging and include monitoring small commercial dish washers in operation; a survey of potential pan sensing techniques; efficiency studies of deep fat fryers, boiling tables, ovens, grills and griddles. These investigations are progressing at various speeds. At present the dishwasher monitoring exercise has been completed as also has that for the deep fat fryer design.

The main conclusions to date are that dishwashing machines operate at only a fraction of their full capacity and that they should be controlled automatically.

Conclusions for fryers are that "cool zones" are not effective and their elimination could increase efficiency and reduce operating costs. Other activities are proceeding.

1. INTRODUCTION

1.1. Background

It has long been recognised in the UK, and presumably in other countries, that catering appliances and cooking techniques and practices pay scant attention to efficiency.

Design, construction and materials are based on manufacturing costs; controls tend to be the minimum required for safety reasons; usage is based primarily on convenience to the user (who is probably not responsible for paying the fuel bills). Among the more wasteful practices is that of leaving appliances unattended and not in actual use for long periods whilst consuming fuel.

The fact that Natural Gas is thought of as a finite resource does not of itself sustain restraint in its use. However, the increasing cost of gas is a forcible argument for economising and avoiding waste, but only if the costs of such measures are recoverable from reduced operating costs in a short time. There is a dearth of detailed information on the UK catering market and its energy requirements, as there is also a lack of information for other member states of the European Economic Community. However, extrapolating from a previous British Gas estimate(1) suggests that in the UK Catering Market the Natural Gas requirement for 1980 would be 563 x 10^6 therms (59 x 10^6 GJ). Making a broad assumption that gas energy requirements are directly proportioned to population, then for the EEC as a whole the catering sector would account for something of the order of 2,815 x 10^6 therms (294 x 10^6 GJ). If a modest reduction of, say, 10% could be achieved then in the UK this would represent an annual saving of 56.3 x 10^6 therms (5.9 x 10^6 GJ), equivalent to approximately £12 x 10^6 at 1980 values.

1.2. Objectives

The objectives of this research programme are to investigate areas of appliance design, control and usage for the purpose of identifying potential cost-effective reductions in energy consumption. This data would then be used to stimulate appliance manufacturers and users into adopting the proposals.

1.3. Planned Programme

Broadly, the programme was planned to assess present levels of appliance efficiency and usage. This would be achieved by (a) laboratory examination and comparison of selected appliances and (b) monitoring the energy consumptions and, if possible, the waste heat of operational kitchens.

(a) Laboratory Examination. This aspect is intended to determine the range of efficiencies for each appliance category, consisting of ovens (natural and forced convection), hobs, grills and deep fat fryers. It will also include a feasibility study of methods of automatically controlling hob burners by sensing the presence (or absence) of a saucepan.

(b) Kitchen Monitoring. In this, suitable sites will be identified and appliances monitored to provide data on usage and energy consumptions.

1.4. Activity

Initial activity covered a broad front, consisting of (a) dishwasher monitoring, (b) a survey of means of sensing the presence of saucepans and (c) efficiency measurements on deep fat fryers and cooker hob burners. More recently, effort has concentrated on various aspects of the last item above.

This report covers the activity during the first eighteen months of this three year contract, which commenced on 1st July 1980.

2. TABLES AND FIGURES

For the dishwasher monitoring exercise, the machines were selected by rack capacity and included both the automatic and the manual type, shown in Figure I. Histograms have been produced to show the usage patterns for these machines, at each site monitored, of which Figure III is typical.

On appliance efficiency, Figure II compares gas and electric deep fat fryers and their "cool zones".

The indicated efficiencies obtained with both a flat-bottomed tank and a finned tank are shown in Figures IV and V respectively.

Typical appliances being studied are shown in Figures VI and VII, which are respectively deep fat fryers and a six-burner hob and oven.

3. PROGRESS

3.1. Hot Water Consumption of Dishwashers

In this exercise a total of six sites were monitored to provide data on the following: rinse water temperature, number of machine operations per day, water consumption for both the wash and rinse operations and the number of meals served. The sites monitored in this way consisted of three canteens, two hotels and one restaurant.

The information collected was analysed and a series of histograms produced relating to machine usage and time of day, as typified by Figure III.

This phase of investigation has now been concluded.

3.2. Pan Sensing Controls

A preliminary survey has been carried out to select suitable methods of pan sensing for this application.

It was based on earlier work[2] carried out in respect of electro-mechanical devices (which was unsuccessful).

Approaches to this problem, to be investigated, are:-

(a) Flame Rectification for sensing the spread of flame due to the presence of a pan. The main disadvantages are thought to be both technical and cost related.

(b) Micro-electronic, in which proximity switches would be used to sense the pan's presence. Here, also, the disadvantages may be related to the operating environment (heat and moisture) and the costs.

(c) Sonic detection. No effort has gone into this aspect since the initial survey, however it is expected to become active early in 1982.

3.3. Appliance Efficiency

3.3.1. Deep Fat Fryer Assessment

A study of appliance design suggested that "cool zones" (see Figure II) were ineffective in prolonging the life of cooking oil and in improving the quality of finish in cooked food.

Following a mathematical modelling exercise a shallow fryer having a flat-bottom and utilising high temperature radiant burners was constructed. This was based on a mathematical model, programmed into a Hewlett-Packard mini-computer. Unfortunately, the heat transfer performance did not come up to expectations (the actual results being indicated in Figure IV). A re-assessment of the mathematical model indicated that the heat transfer surfaces were inadequate and should be extended, which when applied to the laboratory model produced the results shown in Figure V.

This investigation has now come to an end and the theoretical and laboratory test results are currently being analysed and compared so that design recommendations mav be produced.

3.3.2. Cooking Techniques

A pilot study has been carried out for the purpose of providing a base line, for the comparison of cooking efficiencies of various appliances. The hamburger was selected as the test food because of its consistency of constituents, size and weight, combined with its low cost.

Cooking tests were carried out in which the cooking times and energy consumptions were monitored and the finished quality of the food assessed. Those appliances selected for this series of tests were: natural convection ovens, fan ovens, radiant grills, griddle plates, boiling burners and solid tops.

The data from this phase is currently being processed and will be included in a future progress report.

The next stage will be a comparison of the major design variants within each appliance type, for the purpose of assessing variations in both efficiency and general performance.

3.3.3. Boiling Table Study

This began with a study of the effect on efficiency of pan displacement. These tests were to quantify the relationship between efficiency and the movement of a saucepan from a central position. The results indicated that for displacements of up to 10% of the pan diameter, efficiency is unlikely to alter. Greater displacement produces a rapid decrease, so that a 50% movement will give rise to a 45% reduction in thermal efficiency.

The second phase was to quantify the effect of poorly adjusted flames in which the thermal losses due to flames licking the side of a saucepan were examined. The results indicated that for a fixed size of burner, efficiency will vary by 13% or more over a range of pan sizes from 190 mm diameter to 315 mm diameter. Furthermore, the results also indicated that badly adjusted aeration shutters can also contribute to inefficiency. This was found to produce variations over a range of 5% to 10% in thermal efficiency for a given burner.

Present activity covers four aspects, for which results are not available because the work is incomplete, which are as follows:-

- A theoretical and practical assessment of thermal losses due to the heat conductance of pan supports and the potential fuel saving for components of lower thermal mass.
- Verifying previous results and quantifying energy savings that might be expected from the use of saucepan lids. (This is being undertaken now because earlier work is untraceable).
- An initial study has been completed of the design and the materials used in the construction of solid top appliances. Traditionally these are made of cast iron which represent a comparatively large heat sink. A stress analysis has just started, the results of which will be used as a basis for re-designing in lighter materials.
- Lastly, an investigation is in progress to check the heat distribution patterns on radiant grills, flame grills and griddle plates as a starting point for a design assessment.

4. DISCUSSION

4.1. Hot Water Consumption of Dishwashers

The results of the monitoring exercise indicates that most dishwashing machines are under utilised or over-rated. Apparently most of their operational time is utilised at about 25% of their full capacity, which suggests that perhaps a smaller machine is necessary to handle the "off-peak" loads (i.e. two small machines to replace one larger machine).

An expected but nevertheless slightly disconcerting result was that showing the effects of human intervention in the dishwashing operation, in which hot water consumptions ranged from 25 L/week to 29550 L/week, dependent on the individual operating the machine. Although not originally part of this study, it was ascertained that chemical sterilisation was a growing technique in the USA. If this replaced the hot water rinse then it would reduce energy consumption and perhaps assist in overcoming the unhygenic conditions which may arise if low water flow rates are used. However, basically it is suggested that all dishwashing machines should be automatically controlled, for water flow rate, water temperature and washing time.

4.2. Pan Sensing Controls

At this stage there are no indicators by which predictions may be made regarding the likely outcome of this exercise.

4.3. Deep Fat Fryer Assessment

The results of laboratory tests indicate that the chief problems encountered were concerned with the physical sizes of both the heat exchanger and the combustion chamber. In that, the close proximity of burner and heat exchanger, necessary to achieve a high heat transfer efficiency was incompatible with the space required for good combustion. This prevented rapid clearance of the flue gases which in turn restricted the ingress of air for combustion. The result was unacceptable levels of carbon monoxide.

In providing adequate combustion space, the heat intensity from the radiant burner was dissipated over a larger area with consequent greater losses, leading to inefficiency. Improvements were only achieved by increasing the heat transfer area. However, the design proposals do indicate that the resultant appliance could be less costly to manufacture and to operate, because of its tank form and volume, whilst being easier to clean and having an efficiency some 10% higher than contemporary appliances.

4.4. Boiling Table Study

This exercise has many facets which are inter-connected, i.e. primary aeration of burners, gas rate and flame size, turn down ratio and saucepan size. The results confirm that generally inefficiencies are due to human foibles in operating the appliance, e.g. moving a pan to one side instead of turning down the flame or placing a small pan on too large a burner.

It is suggested that there are design features which could be "built-in" to overcome many of these problems, such as:-

- Fixed aeration burners
- Pan support design to prevent small pans being used on large burners
- Graduation marks around the gas tap.

It should, however, be remembered that these are limited in their effect and it does appear that education of the users of these appliances is probably of great significance in this area. This perhaps explains the need to verify the effects of using saucepan lids, which is an area under human control and yet makes a significant contribution to energy consumption.

5. CONCLUSIONS

5.1. Hot Water Consumption of Dishwashers

The chief conclusions drawn from this exercise to date are as follows:-

- An analysis of the operational histograms indicates that for most catering establishments the dishwashing machine operates at or near its maximum capacity for a brief period only. They mostly operate at about

25% of full load.

- Water consumptions monitored indicate that automatic machines maintain a consistent demand rate for hot water. Non-automatic machines show wide variations, due to human control (or lack of it).
- The extremes of minimum and maximum water consumptions represent respectively unhygenic conditions and waste with inefficiency.
- Automatic fill and timed machines reduce wastage of both water and energy.
- The use of chemical sterilisation should be investigated to determine its effectiveness as an alternative to a high temperature rinse.

5.2. Deep Fat Fryer

The conclusions relating to this study are:-

- It is doubtful that "cool zones" contribute either to the life of cooking oils or to the finished quality of food.
- Elimination of the "cool zone" would reduce the oil volume of the tank.
- A re-designed tank combined with radiant burners could increase efficiency by some 10% above present levels of contemporary appliances of this type.

5.3. Boiling Table

- The most significant benefits to be attained here are related to user techniques. This is heavily dependent on user education and personal discipline (perhaps allied to some inducement) to save energy.
- Certain aspects of design would assist in reducing energy consumption, e.g.
  - Fixed aeration burners.
  - Pan support design to limit the size of pan to that most suitable burner.
  - Clear wear-proof markings on control taps for ease of setting.
  - Design of burners to operate at low simmer rates.
  - Improved quality of manufacture of burners to provide a wider operating range.

REFERENCES

(1) Confidential British Gas Report

(2) Glew, G. Catering Equipment and Systems Design. Applied Science Publishers Ltd., 1977.

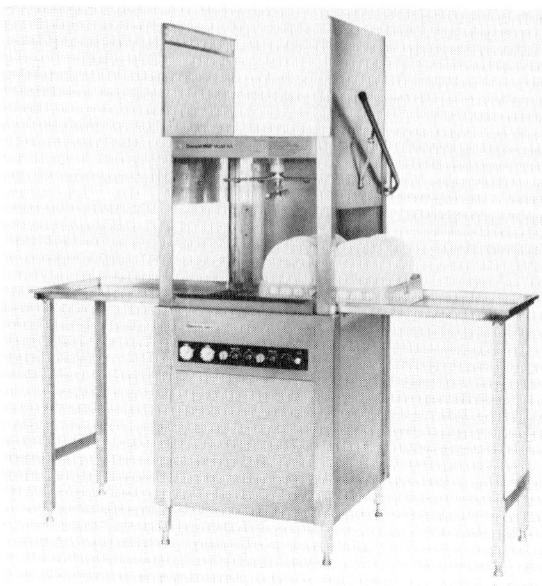

Fig. I Small (50 Trays/hr) Commercial Dishwashing Machine

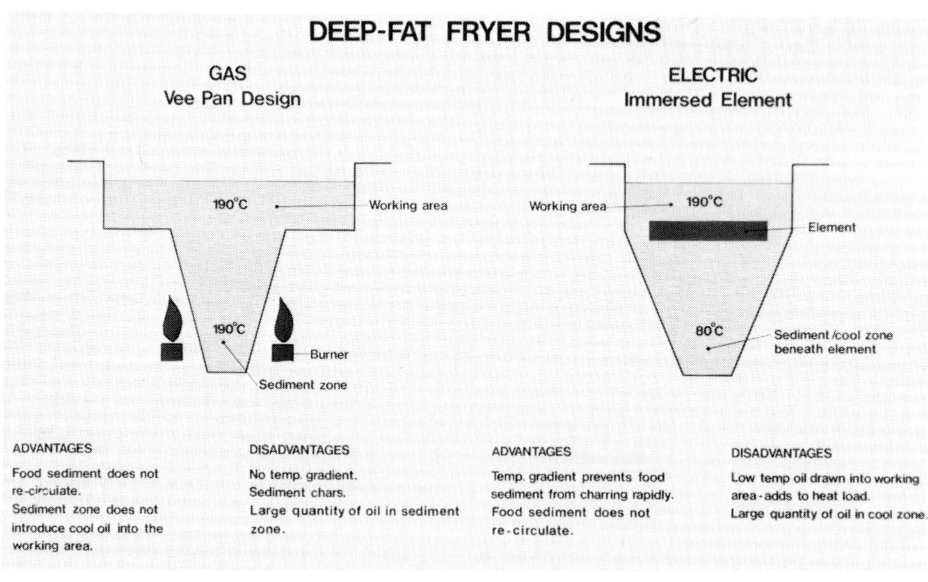

Fig. II Section of Fryer Tanks Showing "Cool Zones"

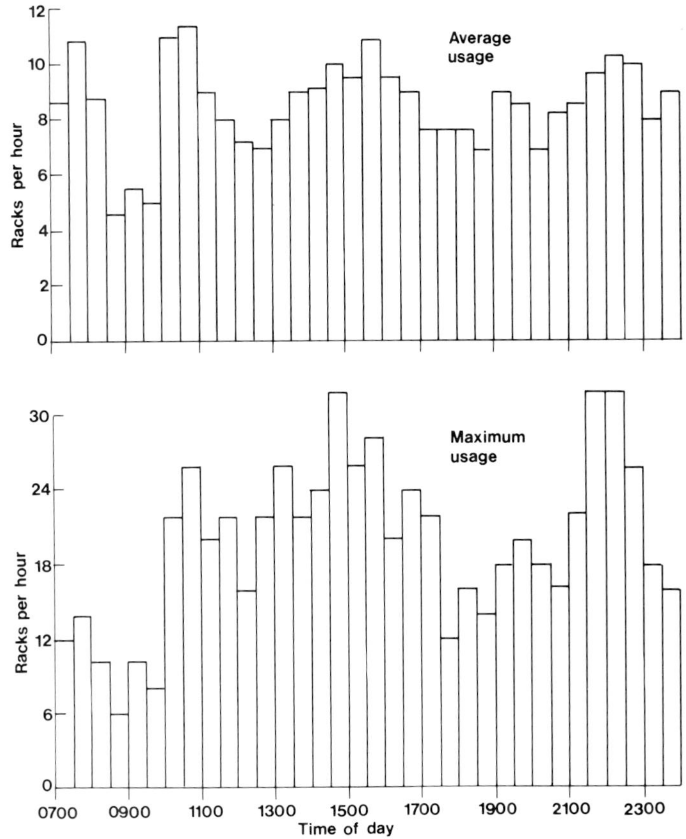

Fig. III Typical Dishwasher Average and Maximum Usage Pattern

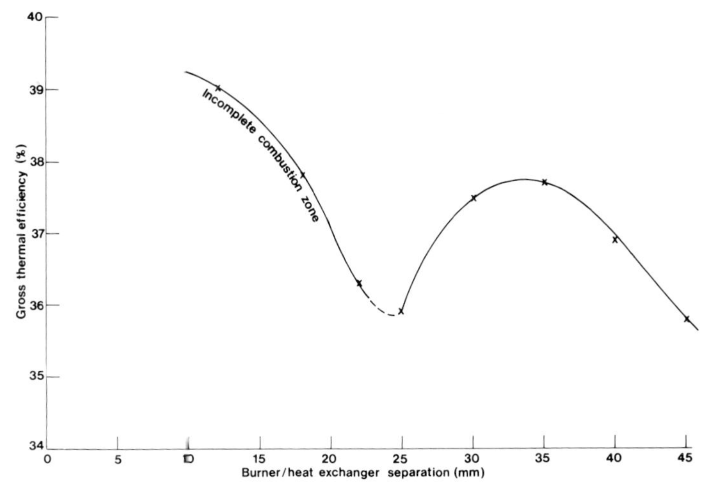

Fig. IV Heat Transfer Efficiency for Flat Bottom Tank

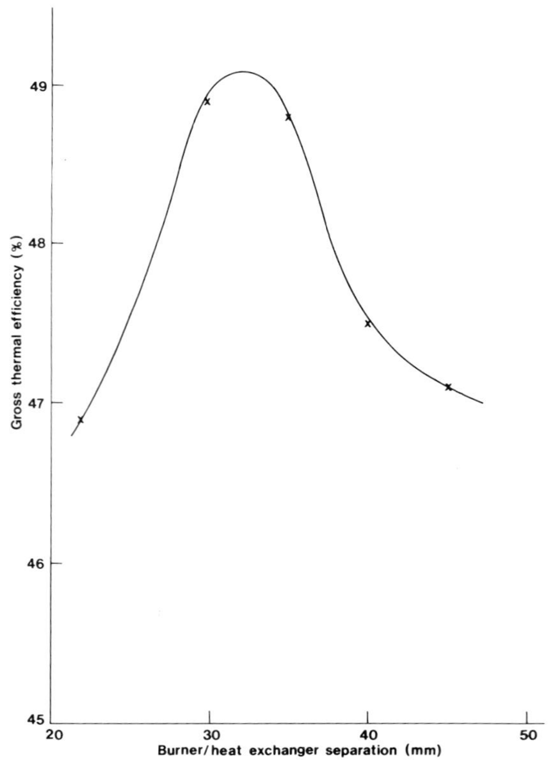

Fig. V Heat Transfer Efficiency for Finned Tank

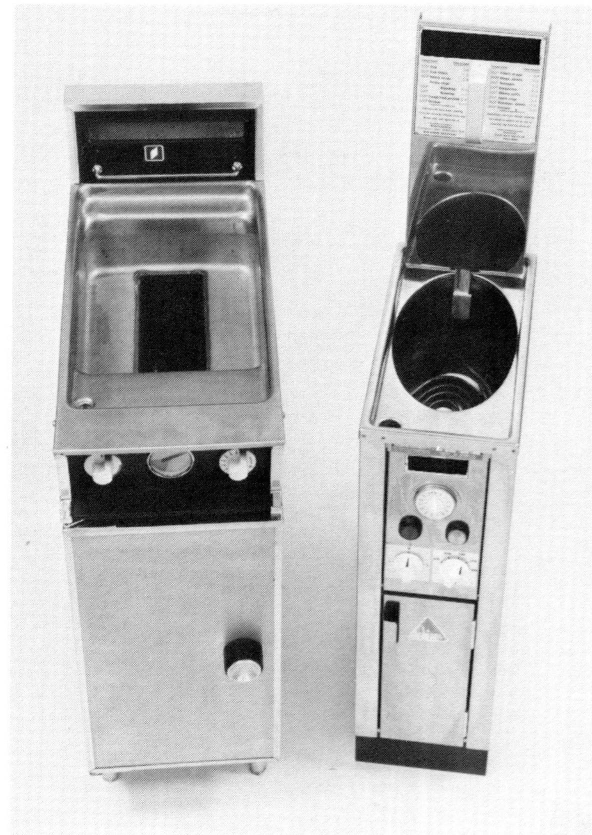

Fig. VI Typical Deep Fat Fryers Studied (Gas Heated on Left and Electric on Right)

Fig. VII Typical Six-Burner Hob and Oven

DEVELOPMENT OF ENERGY EFFICIENT ELECTRICAL HOUSEHOLD APPLIANCES

Authors : Jørgen S. NØRGARD, John HEEBØLL

Contract number : EE-A-3-025-DK

Duration := 24 months 15.IX.1981 - 15.IX.1983

Head of project : Jørgen S. NØRGARD

Contractor : TECHNICAL UNIVERSITY OF DENMARK

Address : DK - 2800 LYNGBY

SUMMARY

The background for this project is an energy analysis, which included investigation of the potentials for reducing electricity consumption of the household appliances to 1/3 of that of the 1975 stock.

Over the first months a cabinet for a 209 liter refrigerator with a 7 cm thick polyurethane foam insulation has been designed but not built. The refrigerator system is planned to have a COP of 1.5 or more, bringing the annual electricity consumption down to 100 kWh compared with 550 kWh for the 1975 stock and 250 kWh for the best models sold today. The economy in redesigning seems to be good.

For cooking we have designed a double walled pot and a small 12 liter efficient oven. A PID-control (proportional, integration, differentiation) system will be tried out in an experimental hot plate set up, which has been designed. Cooking equipment will be built in the coming 3 months.

INTRODUCTION

The background for the project is an energy analysis carried out in 1974 to 1979, investigating future energy options for Denmark, (1,2,3,4). This analysis included theoretical studies of how energy efficient the major household appliances could be designed. On the average it was found that their annual electricity consumption could be reduced through technical measures to about one third of what the 1975-stock of appliances consumed. For some appliances like refrigerators and freezers the potential savings were larger, and for others like cookers they were smaller.

Later studies confirm those findings and indicate a beginning redesigning by manufacturers towards energy efficient models (5,6,7).

It seemed natural to supplement these theoretical studies with some experimental investigations, including the building of prototypes of appliances. The purpose of the project is to construct and test such prototypes, designed for actual production.

Three types of appliances were chosen because of their large distribution and importance to many European households. The three are: Refrigerator, cooking devices and washing machine.

In the project we cooperate with four private companies, which manufacture 1) washing machines (Vølund), 2) cookers (Voss), 3) refrigerators (Gram), and 4) compressor units and control systems (Danfoss). Furthermore, we are in close contact with the Danish Government Home Economic Council, which is the official institute for testing appliances and other households consumer goods.

The targets for the project is to reduce electricity consumption for the three appliances as shown on figure 1,2 and 3 without significantly affecting the service and comfort we get from the appliances. The three graphs also show how much improvements have already been achieved by some manufacturers. The best models on the market today consume considerable less electricity than the 1975-stock. This has made it more difficult for us to reach our original targets in _relative_ terms. In _absolute_ terms, however, it is more likely that we can reach our targets with less radical design changes than was proposed in the first theoretical work.

For the washing machine the low _electricity_ consumption can be reached partly by using intake of hot water from the hot water supply system of the house instead of heating the water electrically. Cold water intake is maintained. The price for energy in the form of hot water is on the average slightly less than half the price for electricity in Denmark. For Europe the difference between energy as hot water and as electricity is even larger. The advantage of using intake of hot and cold water is especially significant when the hot water is supplied from combined heat and power plants. Our work on washing machines has not been started yet, partly because our target has in a way already been reached by our cooperating manufacturer, Vølund. The presumption behind achieving this goal is that people will be willing to adapt washing habits towards

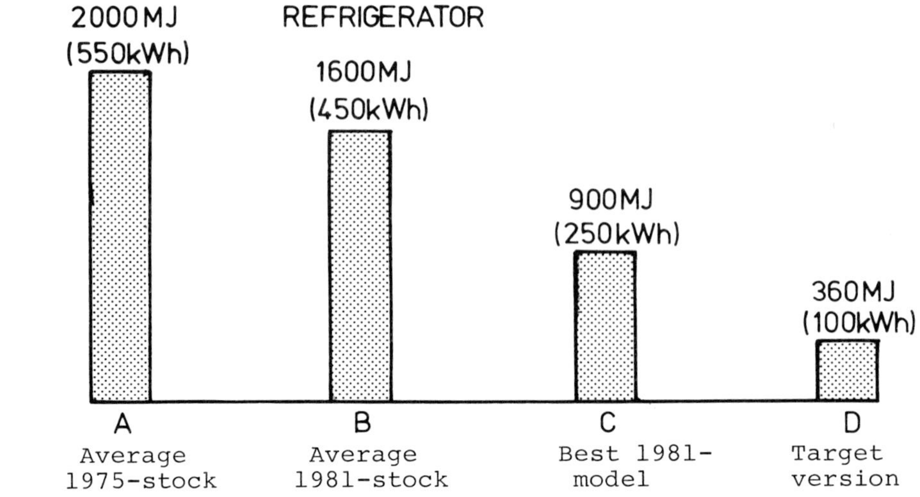

Figure 1. Annual electricity consumption for different versions of a 210 liter refrigerator, all in "normal" use.

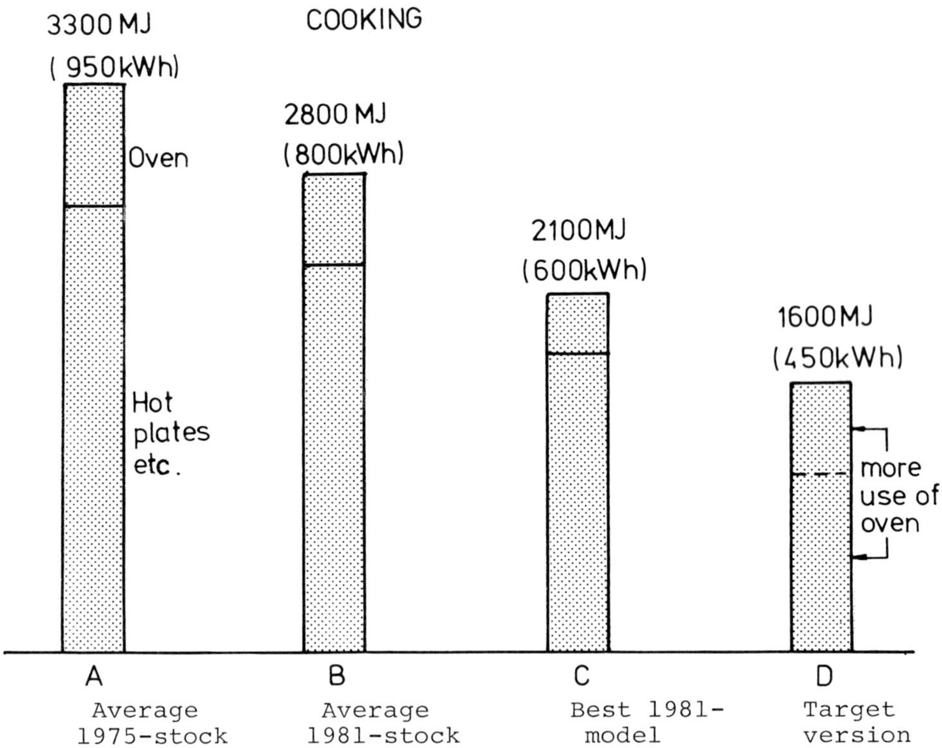

Figure 2. Annual electricity consumption for cooking equipment at "normal" use in a household with two to three persons.

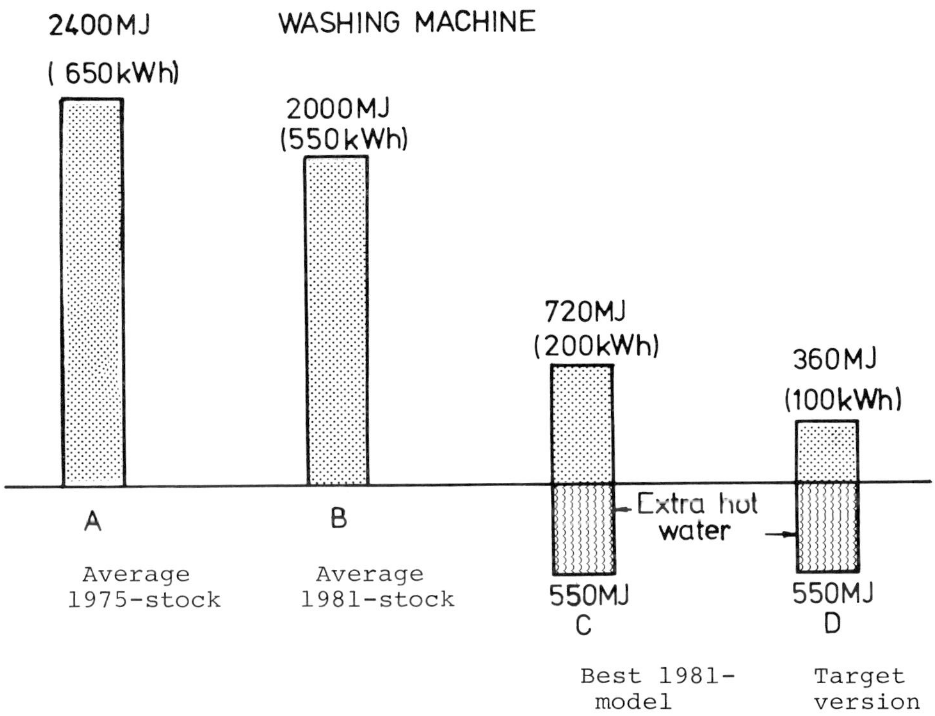

Figure 3. Annual electricity consumption and hot water consumption for different versions of washing mashines, all with todays Danish average of 5 uses per week. For "B" and "C" the fraction of hot washes (90°C) is the present Danish average of 1/3, while for the 1975-stock it was 1/2. "D" can be reached with the "C" version by changing washing habits towards essentially no hot washes.

lower temperatures, given that they can still achieve the same washing quality.

The first three months of the project have been used to establish contact with the companies and to make the following first designs of a refrigerator and some cooking devices.

EFFICIENT REFRIGERATOR

The most obvious measure for improving the energy efficiency of refrigerators and freezers is to increase their thermal insulation, given that their purpose is to maintain a temperature different from that of the surroundings. There are, however, other technical ways to reduce electricity consumption. In designing the improved 1981-models (Fig. 1) the manufacturers have focused on the refrigeration system and achieved a 55% reduction in electricity consumption compared to the 1975-stock.

The first version of an refrigerator we have designed is shown at Fig. 4. The inside volume is 209 liter. The outside dimensions are 59.5 cm (standard width) x 61 cm x 120 cm. Except for the thicker door, the outside dimensions correspond to those of a model sold today with 285 liter inside volume.

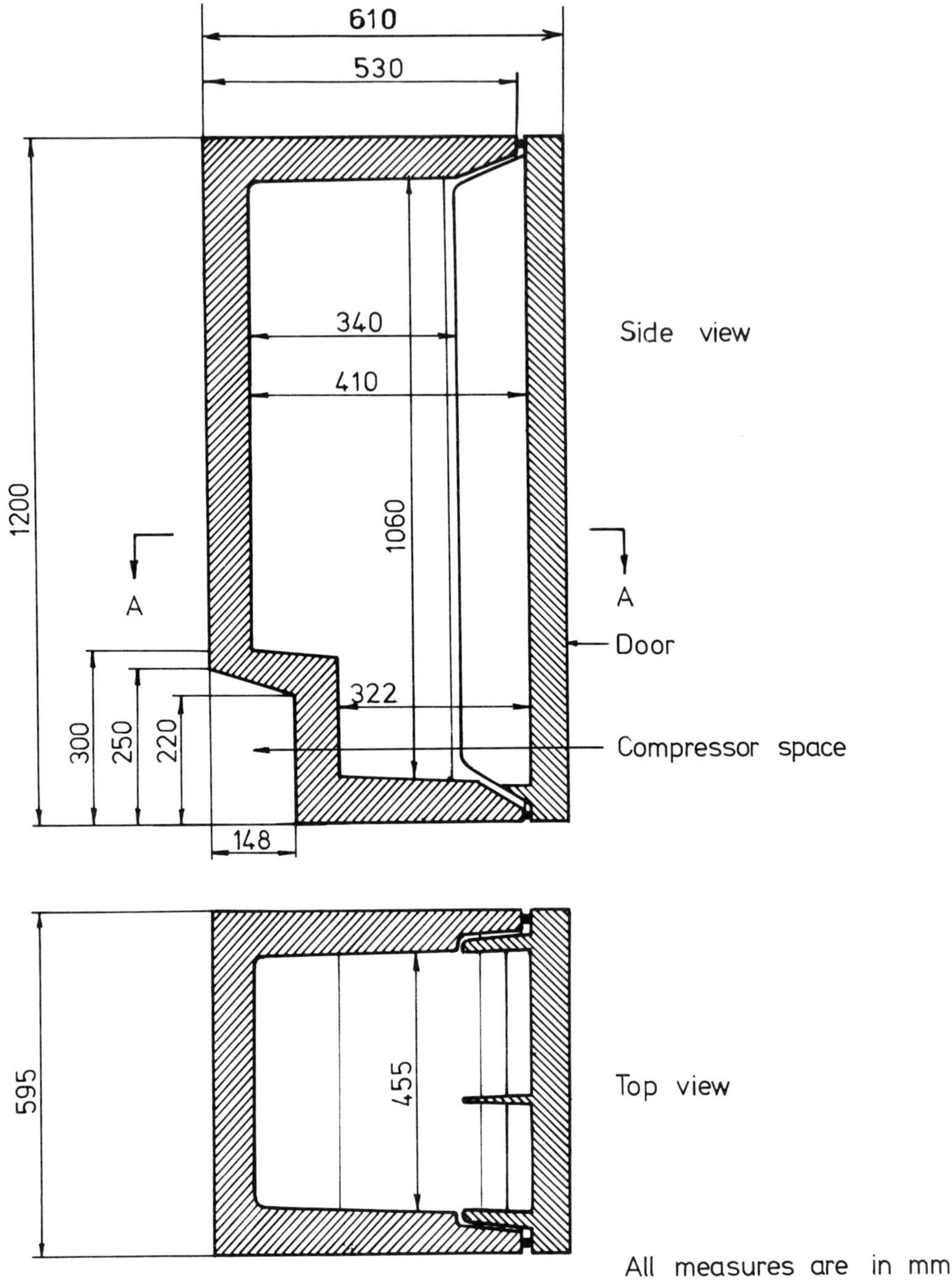

Figure 4. Sketch of the first design of an efficient 209 liter refrigerator cabinet with around 7 cm of polyurethane foam of insulation. The dotted lines indicate the original level of insulation.

Insulation material is polyurethane foam and its thickness is chosen to be around 7 cm compared to the normal 3 cm today.

For the refrigeration system required to keep the cabinet at Fig. 4 cooled we plan to increase the COP (coefficient of performance). COP for refrigeration system is here defined as the heat removed from the refrigerator divided by the electricity needed to do this. (If the system is used as a heat pump, the COP-value is 1.0 higher). Todays refrigerators - even the best - are often run by a system with a COP around 1.0. We aim at a COP in the range between 1.5 and 2. The measures include changes in the heat exchangers, that is the evaporator and the condensor.

If the surface area of the evaporator is increased, the refrigerator can be cooled with a higher evaporator temperature. Today an evaporation temperature of -20°C is not unusual in a refrigerator where the end use requirement is +5°C. If we exclude any frostbox an evaporator with a temperature around zero can manage the cooling. Considering the condensor, we plan to extend the surface area and probably also change the design for a better cooling. We have not yet decided the final design of the two heat exchangers. It is important to notice, however, that when the heat load on the refrigeration system is lowered through better insulation, a lower capacity compressor is needed. Thereby, the sizes of the heat exchangers relative to the compressor are automatically increased.

Usually, the lower capacity compressor units also have a lower efficiency. The compressor manufacturer Danfoss, however, will modify one of their standard units as to adapt it to the low capacity requirement of our prototype refrigerator.

We choose to compare with the 215 l refrigerator from the company Gram, which has an electricity consumption of 0.67 kWh per day, equal to 245 kWh per year, according to standard-tests with 25°C ambient temperature, 5°C inside, and no user-load. That is no door-openings and no exchange of contents. The insulation-layer of the prototype has been more than doubled, from some 3 cm to 7 cm. This will reduce the heat-load to approximately half, given that the gaskets and other coldbridges around the door are unchanged (8). If we assume a COP of 1.5 we should reach a test consumption of 0.22 kWh per day (82 kWh per year).

In practical use, however, both the electricity consumption of the present 215 l refrigerator from Gram and that of the prototype will differ from the test results. Kitchen temperature will be lower than 25°C, probably rather 18°, which reduces heat load and improves the COP. On the other hand, in actual use the door is opened, and food with room-temperature or even warmer is put into the refrigerator. The <u>relative</u> reduction in electricity-consumption is therefore expected to be less than according to tests. We aim at reaching 0.27 kWh per day (100 kWh per year) as shown at Fig. 1. In this case the electricity-savings will amount to <u>150 kWh per year</u>.

The economy of the energy conservation in refrigerators is evaluated by comparing the energy efficient 209 liter prototype with the above mentioned almost similar size refrigerator (215 liter).

The extra production cost includes a larger outer casing (corresponding to that of a 285 liter refrigerator), 110 liter extra insulation foam, larger evaporator and condensor, modification of a smaller compressor unit, and some other minor changes. Total extra production cost is found to be 150 Dkr. or 19 ECU in 1981 currency.

The retail price is in Denmark around 1.9 times the production cost for refrigerators, but 32% of the retail price for household appliances consist of government taxes. It is most likely that the marginal price/production cost ratio will be smaller than the average 1.9. It is also likely that government taxes in the future will be differentiated according to energy consumption with lower taxes on efficient models. It seems, therefore, that we will be on the safe side by assuming the retail price increase for our efficient model to be 1.9 times the extra production cost that is 280 Dkr. or 35 ECU.

It could be argued that the cost of the extra space, which the efficient refrigerator takes up in the house, should be included in the cost analysis. The extra space is 110 liter or 0.11 m^3, corresponding to 0.11/2.4 = 0.05 m^2 floor space. The marginal price for floor space is estimated to 1000 Dkr. or 125 ECU, and the cost of the 0.05 m^2 amounts to 50 Dkr. or 7 ECU. Total extra cost for the consumers to choose the efficient model will be around 330 Dkr. or 42 ECU.

The annual saving of 150 kWh corresponds, with Danish 1981 electricity prices of 0,70 Dkr. per kWh, to 105 Dkr. per year. The payback period will in that case be 3.1 year. Using European electricity prices of 0.10 ECU per kWh the payback period is 2.9 year. However, without the Danish taxes of 32% of retail price on the extra cost for the refrigerator we will with European electricity prices end up with a payback period of 2.0 year. The extra cost corresponds in Denmark to a 10% price increase, which is 1/4 of total taxes on the refrigerator.

When we assume the Danish average lifetime for appliances of 13 years, the rate of return corresponding to the 2.0 year payback period is more than 40% per year.

We plan to start building the refrigerator prototype in a couple of months after having discussed the design with the manufacturer and other experts.

EFFICIENT COOKING

Unlike refrigeration, cooking is not confined to one single appliance. Electrical cooking involves a cooker with hot plates, some pots, pans, and kettles, an oven for baking, etc. Energy consumption for cooking is also more dependent on behavioural patterns of the user, than in the case of a refrigerator. The electricity consumption for cooking indicated at Fig. 2 is estimated on the basis of cooking in average Danish household, which for instance implies 3 uses of the oven per week for the 1975-stock, and 2.5 uses per week for the 1981-stock and the best 1981-model. Our target includes the design of a small efficient oven, which will make it energy economic to substitute some of todays cooking on hot plates with cooking in the oven. Hence our target does include some changes in cooking habits, but not in a direction which can be considered a decline in comfort or in result. We assume 4 uses of oven per week where in 90% of the cases the new small oven is used.

It is found that today only about 15% of the electrical energy used for cooking actually ends up as heat in the food (1,2). The other 85% is lost to the surroundings or is used to heat hot plates, pans, water for boiling, etc. This points towards two principle measures for reducing energy consumption for cooking, namely

1) improved thermal insulation, and
2) reduced heat capacity of cooking equipment.

We have made some preliminary designs for insulated pots. One is an ordinary 2 liter pot, which for experimental purposes can be surrounded by a separate insulated cylinder and covered by an insulated lid as indicated at Fig. 6. Fireproof mineral wool fibres are used as insulation materials in a 10 mm thick layer sealed in between stainless steel plates.

A more likely final design is shown at Fig. 5. Here the walls of the pot and the lid are constructed of double stainless steel plates, and the 10 mm cavities are filled with fibres and sealed off. The bottom of the pot is made of copper or aluminium to ensure good heat conduction to the hot plate and an even heat distribution.

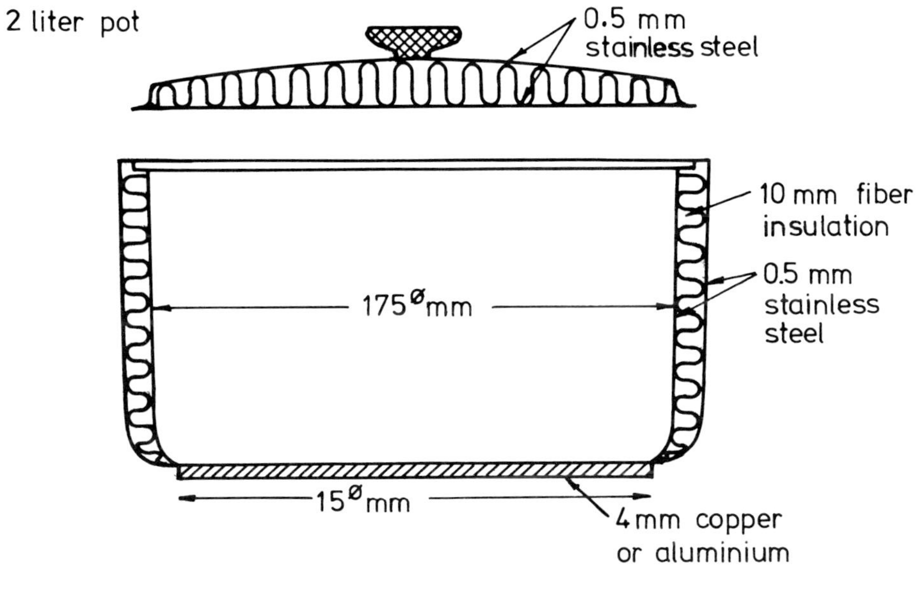

Figure 5. Sketch of a first design of an energy saving 2 liter cooking pot with double walls and a double lid.

Calculation points towards an energy consumption of less than 40 watt for maintaining a temperature around 100°C compared to the normal 200 watt or more. Energy consumption for heating up the pot is, however, only slightly reduced by the new design of the pot. In an average use of the pot including heating up and maintaining a high temperature, the new design points towards an energy saving of 35 to 40% compared to the use of ordinary pots. Preliminary tests seem to confirm those savings.

Eventually we might end up with a pot insulated also in the bottom and with built-in heating elements and control system. In this way we can reduce also the energy consumption for heating up, because we can avoid the heavy bottom and the heavy plate.

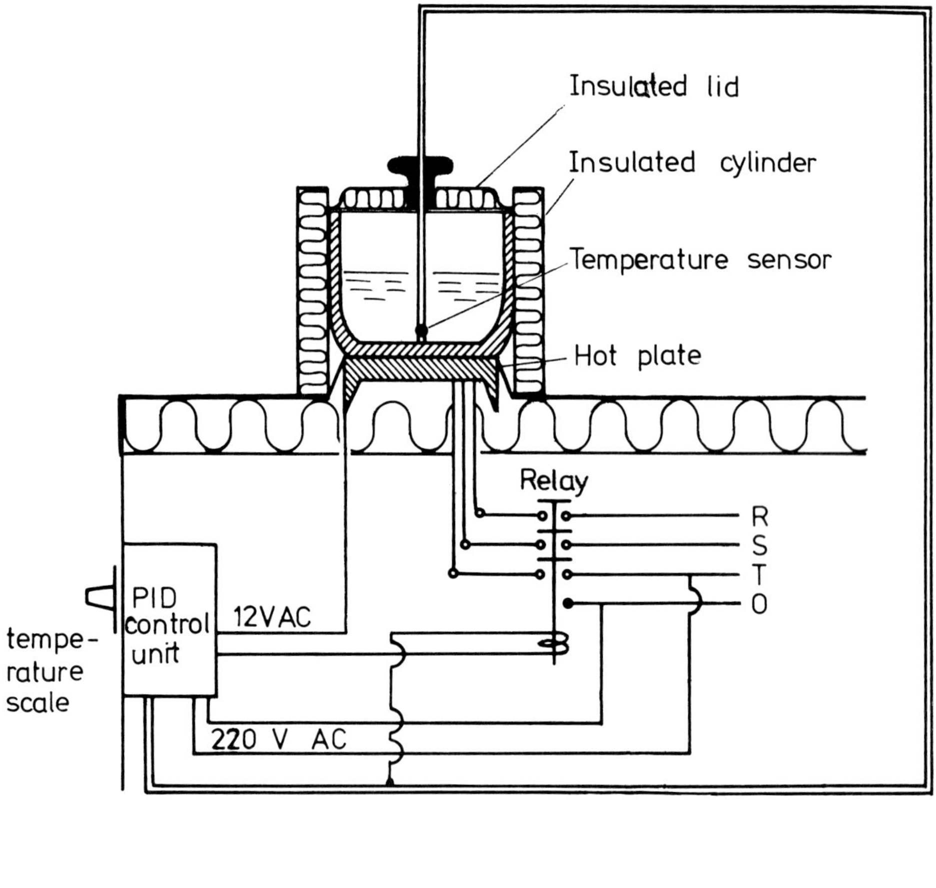

Figure 6. Diagram of a PID controlled hot plate set up for experimental purposes. The pot is insulated with a separate cylinder.

Temperature control is an important element in energy savings in cooking, and this is especially true when the pot is insulated as described above. In this case a surplus of heat supplied to the pot can not dissipate quickly and an overheating or overboiling can often be the result. We plan to install an improved electronic temperature control system. For the first experimental hot plate cooker we will use a PID control system. PID stands for proportional, integration, and differentiation, and is briefly explained as follows. The power input can be regulated <u>proportional</u> to the difference between the temperature we are aiming at, the set value, and the temperature of the sensor. The deviation of the final temperature from the set value, i.e. the off-set value, will then be inversely proportional to the gain of the controller, and in principle the deviation can be infinitely small. However, because of time delay from the heating element via hot plate and pot to the temperature-sensor, the system will show overshoot and oscillations if gain is too high. Often a proportionally controlled system will not be satisfactory for any gain. If a term proportional to the <u>derivative</u> versus time of the temperature is added to the proportional term, the oscillations will be damped, and the gain of the proportional controller can be increased without getting oscillations.

At time-scales longer than the characteristic time-delays of the system, the gain can also be increased with a PID-control system, without risk of oscillations. This is achieved by adding a term proportional to the <u>integral</u> over time of the off-set value of the temperature. For a given system i.e. hotplate, pot and food, it is therefore possible to obtain the desired temperature. If a different system is used, i.e. different pots and different amounts of food, the P, I and D parameters should in general be altered in order to obtain stable control. It will, therefore, probably be necessary to add some sort of adaptive control.

The electrical circuit of the experimental hot plate cooking set up is shown at Fig. 6. The sensor is here inserted into the liquid in the pot from the lid. Later we will probably use a sensor touching the bottom of the pot from below as usual.

We are in the first stage of designing a small efficient oven with a 12 liter inside volume as a supplement to the normal 50 liter oven. For most dishes and baking in an average household (2 to 3 persons) this small size will be sufficient.

An experimental version is designed as shown at Fig. 7. In a small oven like this, the heat distribution is more important than in the larger oven, because the food will be closer to the heat radiating surfaces. We will therefore conduct some experiments with different types of electrical heating element, placed outside and inside the inner casing. The door will have 3 cm of insulation and a small (10x15 cm) double glazed window. The other sides of the oven will be insulated with 6 to 8 cm of insulation.

The heat loss at 200°C, and hence the electricity consumption, is calculated to be less than 150 watt compared to normally around 600 watt for the big ovens. Heating up the small oven is found to require 0.15 kWh compared to 0.6 kWh for the best normal size oven sold today.

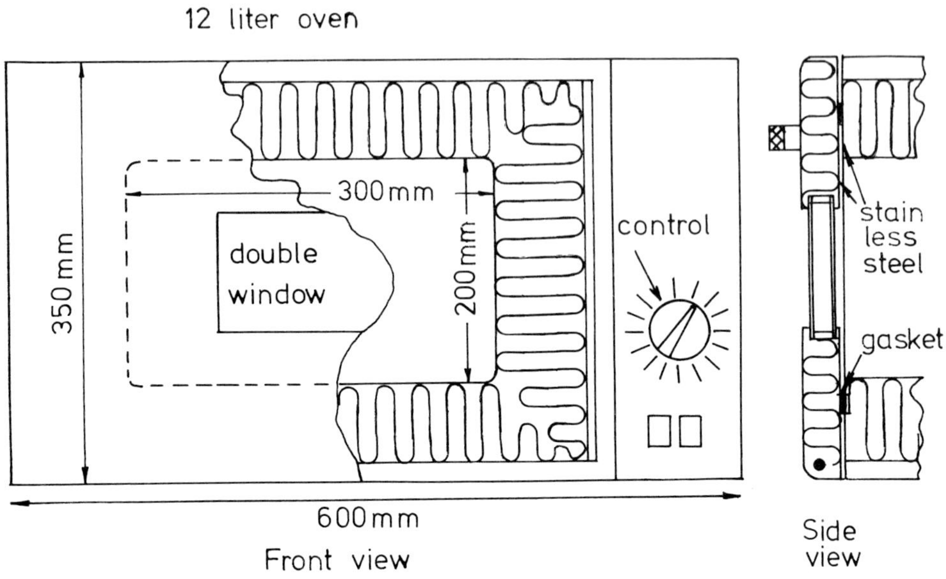

Figure 7. A first design of a 12 liter, well insulated oven.

The final design of the oven is not yet decided. It can be designed as a separate oven to put at the kitchen table or it can be built into a cooker. In order to save space it might even be designed to be plugged into a normal size oven, which will then be available whenever needed.

None of the cooking equipment has been built yet. The designs suggested by us will be discussed over the next couple of months with the manufacturers and other experts in cooking as well as in technology.

No specific calculations of the economy of the energy conservation measures in cooking has been made yet.

CONCLUSION

The project was started September 10. In the three first months we have established the contact with the four cooperating manufacturers Gram (refrigerators), Voss (cookers), Vølund (washing machines), and Danfoss (compressor units and control systems). We have also contacts with the Danish Government Home Economic Council, which is the official institute for testing appliances.

The work of constructing an efficient refrigerator has been started with designing the cabinet for a 209 liter prototype, insulated with 7 cm polyurethane foam compared with normally 3 cm. A compressor unit will be adapted for the low capacity purpose by Danfoss. Concerning the final design of evaporator and condensor as well as control system we aim at a

COP of 1.5 which would lead to an annual electricity consumption of 100 kWh. The economy seems satisfactory with a payback period for consumers of 2 to 3 years and a rate of return of 30 to 40% per year. We will built the refrigerator during the next 6 months.

For cooking we have made first sketch designs of insulated pots and a small 12 liter well insulated oven. An experimental set up has been designed for trying a PID control system for hot plates. The system could be used for an oven as well, or maybe even for a pot with a built-in heating element. No final designs has been made for cooking equipment.

The work on constructing an electricity saving washing machine has not yet started, partly because the most recent model from our cooperator VØlund can accomplish the goal we had.

In the next 3 months we will discuss and probably start building a refrigerator and some cooking pots, and make a final design of an oven.

REFERENCES

1. JØrgen S. NØrgård: "Husholdninger og Energi", DEMO-rapport nr. 4, Polyteknisk Forlag, Lyngby, Denmark, 1979.
2. JØrgen S. NØrgård: "Improved Efficiency in Domestic Electricity Use", Energy Policy, Vol. 7, No. 1, March 1979.
3. JØrgen S. NØrgård: "Electricity Savings through Efficiency of Households Appliances", in Energy Conservation in the Built Environment, proceedings of 2nd International CIB Symposium on Energy Conservation, 1979.
4. JØrgen S. NØrgård and Niels I. Meyer: "The Industrialized Nations must give a Credible Example", in Changing Energy Use Future, Vol. 1, Pergamon Press, 1979.
5. F. Strabo, O. Thun, J.S. NØrgård and S. Prahm: "El-forbrugets udvikling i en landkommune" Nysted Energy Project report No. 5, Nysted Rådhus, DK-4800 Nysted, 1981.
6. JØrgen S. NØrgård and Bente L. Christensen: "Energihusholdning, husholdning, holdning", FDB, Tåstrup, Denmark. To be published, January 1982.
7. JØrgen S. NØrgård: "Low-Electricity Future Household", in IEE conference publication No. 186, London, 1980.
8. Torben Lapp: "Computermodeller af Husholdningsfrysers energiforbrug", Physics Laboratory III, Technical University of Denmark, DK-2800 Lyngby, July 1981.

QUESTION AND DISCUSSION

Morello (Riber) : Are these average refrigerators ? Which kind of improvements reduce the energy losses caused by the opening of the door ? Are the data in Fig. 1 theoretical data ?

Answer : The cases A and B in Fig. 1 represent the average consumption of a 200 liter refrigerator. C is a low energy refrigerator and D is our target.

- With an improved C.O.P. value it takes less electric energy to remove a certain amount of heat which enters the refrigerator when the door has been opened.
- The data in Fig. 1 result from calculations and the testing of different types of refrigerators.

Prof. Boffa (Chairman) : Polymethane is used as the refrigerator insulant. In case of fire toxic isocyanath is released. Should this not be taken into consideration from the safety point of view ? Perhaps another insulant should be used.

Answer : We will in future work take into account problems in case of fire. In the present project however we will continue to use polyurethane. So in principle we don't introduce new hazards but we use of course more insulation material.